physica status solidi (b)

ISSN 0370-1972 * VOL. 126 · NO. 2 · DECEMBER 1984

AKADEMIE-VERLAG · BERLIN

THEORIA CUM PRAXI

Classification Scheme

1. Structure of Crystalline Solids
   1.1 Perfectly Periodic Structure
   1.2 Solid-State Phase Transformations
   1.3 Alloys.-Metallurgy
   1.4 Microstructure (Magnetic Domains See 18; Ferroelectric Domains See 14.4.1)
   1.5 Films
   1.6 Surfaces
2. Non-Crystalline State
3. Crystal Growth
4. Bonding Properties
5. Mössbauer Spectroscopy
6. Lattice Dynamics. Phonons
7. Acoustic Properties
8. Thermal Properties
9. Diffusion
10. Defect Properties (Irradiation Defects See 11)
    10.1 Metals
    10.2 Non-Metals
11. Irradiation Effects (X-Ray Diffraction Investigations See 1 and 10)
12. Mechanical Properties (Plastic Deformations See 10)
    12.1 Metals
    12.2 Non-Metals
13. Electron States
    13.1 Band Structure
    13.2 Fermi Surfaces
    13.3 Surface and Interface States
    13.4 Impurity and Defect States
    13.5 Elementary Excitations (Phonons See 6)
        13.5.1 Excitons
        13.5.2 Plasmons
        13.5.3 Polarons
        13.5.4 Magnons
14. Electrical Properties. Transport Phenomena
    14.1 Metals. Semi-Metals
    14.2 Superconductivity. Superconducting Materials and Devices
    14.3 Semiconductors
        14.3.1 Films
        14.3.2 Surfaces and Interfaces
        14.3.3 Devices. Junctions (Contact Problems See 14.3.4)
        14.3.4 High-Field Phenomena, Space-Charge Effects, Inhomogeneities, Injected Carriers (Electroluminescence See 20.3; Junctions See 14.3.3)
    14.4 Dielectrics
        14.4.1 Ferroelectrics
15. Thermoelectric and Thermomagnetic Properties
16. Photoconductivity. Photovoltaic Effects
17. Emission of Electrons and Ions
    17.1 Field Emission Microscope Investigations
18. Magnetic Properties
    18.1 Paramagnetic Properties
    18.2 Ferromagnetic Properties
        18.2.1 Ferromagnetic Films
    18.3 Ferrimagnetic Properties
    18.4 Antiferromagnetic Properties

(Continued on cover three)

Author Index

physica status solidi (b)

basic research

Volume 126 · Number 2 · Pages 433 to 752, K109 to K178, and A9 to A18

December 1, 1984

PSSB 126(2) 433—752, K109—K178, A9—A18 (1984)

ISSN 0370-1972

AKADEMIE-VERLAG · BERLIN

Subscriptions and orders for single copies should be directed in the GDR:

to the Postzeitungsvertrieb or
to the Akademie-Verlag, DDR-1086 Berlin, Leipziger Str. 3–4;

in the other socialist countries:

to a book-shop for foreign language literature or
to the competent news-distributing agency;

in the FRG and BERLIN (WEST):

to a book-shop or
to the wholesale distributing agency Kunst und Wissen, Erich Bieber OHG, D-7000 Stuttgart 1, Wilhelmstr. 4–6;

in the other Western European countries:

to Kunst und Wissen, Erich Bieber GmbH, CH-8008 Zürich, Dufourstr. 51;

in USA and CANADA:

to Verlag Chemie International Inc., 303 N.W. 12th Avenue, Deerfield Beach, FL 33441, USA;

in other countries:

to the international book- and journal-selling trade,
to Buchexport, Volkseigener Außenhandelsbetrieb der Deutschen Demokratischen Republik, DDR-7010 Leipzig, Postfach 160, or
to the Akademie-Verlag, DDR-1086 Berlin, Leipziger Str. 3–4.

Editorial Note:

"physica status solidi (b)" undertakes that an original paper *accepted for publication* before the 8th of any month will be published within 50 days of this date unless the author requests a postponement. In special cases there may be some delay between *receipt* and *acceptance* of a paper due to the review and, if necessary, revision of the paper.

Schriftleiter und verantwortlich für den Inhalt:
Professor Dr. Dr. h. c. P. Görlich, DDR-1086 Berlin, Leipziger Str. 3–4, Postfach 1233 bzw. DDR-6900 Jena, Schillbachstr. 24.

Verlag:
Akademie-Verlag, DDR-1086 Berlin, Leipziger Str. 3–4; Fernruf: 2236221 und 2236229; Telex-Nr.: 114420; Bank: Staatsbank der DDR, Berlin, Kto.-Nr.: 6836-26-20712.

Chefredakteur:
Dr. H.-J. Hänsch.

Redaktionskollegium:
Prof. Dr. E. Gutsche, Dr. H.-J. Hänsch, Dr. H. Lange, Dr. S. Oberländer.

Anschrift der Redaktion:
DDR-1086 Berlin, Leipziger Str. 3–4; Postfach 1233; Fernruf: 2236279.

Veröffentlicht unter der Lizenznummer 1310 des Presseamtes beim Vorsitzenden des Ministerrates der Deutschen Demokratischen Republik.

Gesamtherstellung:
VEB Druckerei „Thomas Müntzer", DDR-5820 Bad Langensalza.

Erscheinungsweise:
Die Zeitschrift „physica status solidi (b)" erscheint jeweils am 1. eines jeden Monats. Jährlich erscheinen 6 Bände zu je 2 Heften.

Bezugspreis:
Je Band 200,– M zuzüglich Versandspesen (Preis für die DDR: 130,– M).

Bestellnummer dieses Bandes: 1068/126.

Printed in the German Democratic Republic.

AN (EDV) 20831

Contents

Original Papers and Short Notes

Structure

Lattice properties

Defects, atomistic aspects

Magnetism

Extended electronic states and transitions

Localized electronic states and transitions

Errata

Pre-Printed Titles

Systematic List

Subject classification: Corresponding papers begin on the following pages (pages given in italics refer to the principle subject classification):

The Author Index of Volume 126 Begins on Page 753

(It will be delivered together with Volume 127, Number 1.)

Original Papers

phys. stat. sol. (b) **126**, **443** (**1984**)

Subject classification: 2; 9; 21; 21.1

Faculty of Science, Jadavpur University, Calcutta[1])

Partial Structure Factors and Self-Diffusion in Liquid Copper–Lead Alloys

By

S. PALCHAUDHURI, R. N. JOARDER, and R. V. GOPALA RAO

The three partial structure factors, $S_{11}(k)$, $S_{22}(k)$ and $S_{12}(k)$ and also the Bhatia-Thornton structure factors, $S_{NN}(k)$, $S_{CC}(k)$ and $S_{NC}(k)$ are computed for liquid copper–lead alloys at various concentrations and at different temperatures with a square-well potential. The total structure factor obtained from these partials compares satisfactorily with the experimental results. The isothermal compressibilities have been calculated from the partial structures in the long wave limit. The self-diffusion coefficients D_i have also been calculated at those concentrations of Cu for this alloy in the linear trajectory approximation of Helfand. From the self-diffusion coefficients, an estimate of the mutual diffusion coefficients is made to a good approximation.

Mittels eines Potentialkastenmodells werden die drei partiellen Strukturfaktoren $S_{11}(k)$, $S_{22}(k)$ und $S_{12}(k)$ sowie die Bhatia-Thornton-Strukturfaktoren $S_{NN}(k)$, $S_{CC}(k)$ und $S_{NC}(k)$ für flüssige Kupfer–Blei-Legierungen bei verschiedenen Konzentrationen und für verschiedene Temperaturen berechnet. Der aus diesen partiellen Strukturfaktoren erhaltene Gesamtstrukturfaktor stimmt befriedigend mit den experimentellen Ergebnissen überein. Die isotherme Kompressibilität wird aus den partiellen Strukturen im Grenzfall langer Wellen berechnet. Die Koeffizienten der Selbstdiffusion D_i werden für Konzentrationen von Cu berechnet, wo die Legierung sich in der linearen Näherung von Helfand befindet. Aus diesen Selbstdiffusionskoeffizienten werden die gegenseitigen Diffusionskoeffizienten in guter Näherung berechnet.

1. Introduction

Several authors [1] have reported on X-ray and neutron diffraction results of different liquid alloys but the resulting picture is not entirely satisfactory. This is due to the fact that a unique set of partial structure factors cannot be determined from two diffraction measurements. Model calculations in such a case offer a way to find out the three partial structures independently, and a suitable model for a system may well be able to predict to a good accuracy the partial structures. It has been shown [2 to 5] that the structural properties of liquid metals can be explained satisfactorily on the basis of a simple perturbation treatment of the Percus-Yevick (PY) equation with square-well attractive tail. We now like to extend this treatment to the calculation of partial structure factor, $S_{ij}(k)$, of an alloy of liquid copper and lead which does not form any compound as has been suggested by Steeb et al. [6]. Three partial structures, $S_{ij}(k)$, as defined by Ashcroft and Langreth [7] have been computed. We calculated the number–number, concentration–concentration, and number–concentration structure factors, $S_{NN}(k)$, $S_{CC}(k)$, and $S_{NC}(k)$, respectively, as defined by Bhatia and Thornton [8]. The total structure factor $S(k)$ has been computed from the partial structures and compared with the experimental results [6]. In the long-wave limit, the isothermal compressibilities of the system at various concentrations have also been computed.

---

[1]) Calcutta 700032, India.

2. Theory

For binary mixtures, we define the direct correlation function, $C_{ij}(r)$, as

$$\begin{aligned} C_{ij}(r) &= C_{ij}^0(r); & 0 < r \leqq \sigma_{ij}\,, \\ &= -\frac{\varphi_{ij}(r)}{k_B T} = \frac{\varepsilon_{ij}}{k_B T}; & \sigma_{ij} \leqq r \leqq A_{ij}\sigma_{ij}\,, \\ &= 0; & r > A_{ij}\sigma_{ij}\,. \end{aligned} \tag{1}$$

In the above equation $C_{ij}^0(r)$ is the hard-sphere solution of the PY equation for binary liquid mixtures as given by Lebowitz [9]. σ_{ij}, ε_{ij}, and A_{ij} stand for hard-sphere diameter, depth, and breadth, respectively, of the square-well used for the species i. We have taken the values of mixed parameters using the Lorentz-Berthelot mixing rules and the parameters are given by

$$\begin{aligned} \sigma_{12} &= \frac{\sigma_{11} + \sigma_{22}}{2}\,, \\ \varepsilon_{12} &= \sqrt{\varepsilon_{11}\varepsilon_{22}}\,, \\ A_{12} &= \frac{A_{11}\sigma_{11} + A_{22}\sigma_{22}}{\sigma_{11} + \sigma_{22}}\,. \end{aligned} \tag{2}$$

Using Lebowitz's [9] solution of the PY equation for binary mixtures of rigid spheres to represent $C_{ij}(r)$ in the region $0 < r \leqq \sigma_{ij}$, and a square-well potential to represent $C_{ij}(r)$ in the region $\sigma_{ij} \leqq r \leqq A_{ij}\sigma_{ij}$, the Fourier transforms of $C_{ij}(r)$ can be written as

$$\begin{aligned} \varrho_{ii}\hat{C}_{ii}(k) = {} & -\frac{24\eta_{ii}}{(k\sigma_{ii})^6}\{a_i(k\sigma_{ii})^3[\sin k\sigma_{ii} - k\sigma_{ii}\cos k\sigma_{ii}] + \\ & + \beta_i(k\sigma_{ii})^2[2k\sigma_{ii}\sin k\sigma_{ii} - (k^2\sigma_{ii}^2 - 2)\cos k\sigma_{ii} - 2] + \\ & + \gamma_i[(4k^3\sigma_{ii}^3 - 24k\sigma_{ii})\sin k\sigma_{ii} - (k^4\sigma_{ii}^4 - \\ & - 12k^2\sigma_{ii}^2 + 24)\cos k\sigma_{ii} + 24]\} + \frac{24\eta_{ii}\varepsilon_{ii}}{k_B T} \times \\ & \times \left[\frac{-A_{ii}k\sigma_{ii}\cos A_{ii}k\sigma_{ii} + \sin A_{ii}k\sigma_{ii} + k\sigma_{ii}\cos k\sigma_{ii} - \sin k\sigma_{ii}}{(k\sigma_{ii})^3}\right] \end{aligned} \tag{3}$$

and

$$\begin{aligned} \hat{C}_{12}(k) = {} & -\frac{4\pi}{k^6}\{(k^2[a_1 + 2b\sigma_{11} + 4d\sigma_{11}^2(\sigma_{12} + 2\lambda)] - \\ & - 24d\sigma_{12})\,k\sin k\sigma_{12} + (k^2[2b + 12d\sigma_{11}\sigma_{22}] - \\ & - k^4[a_1\sigma_{12} + b\sigma_{11}^2 + d\sigma_{11}^3(\sigma_{12} + 3\lambda)] - 24d) \times \\ & \times \cos k\sigma_{12} + (24d - 2bk^2)\cos\lambda k + 24\lambda dk\sin\lambda k - \\ & - \frac{\varepsilon_{12}}{k_B T}k^3[-A_{12}k\sigma_{12}\cos A_{12}k\sigma_{12} + k\sigma_{12}\cos k\sigma_{12} + \\ & + \sin A_{12}k\sigma_{12} - \sin k\sigma_{12}]\} \end{aligned} \tag{4}$$

with $\eta_{ii} = \pi\varrho_{ii}\sigma_{ii}^3/6$, $\lambda = (\sigma_{22} - \sigma_{11})/2$, and ϱ_{ii} is the number density of the species i in the mixture. The other constants, a_i, β_i, γ_i, etc. are as defined by Lebowitz [9] (see also Ashcroft and Langreth [7]).

The $\hat{C}_{ij}(k)$ are now related to the partial structures, $S_{ij}(k)$, as [7]

$$S_{11}(k) = \left[1 - \varrho_{11}\hat{C}_{11}(k) - \frac{\varrho_{11}\varrho_{22}\hat{C}_{12}^2(k)}{1 - \varrho_{22}\hat{C}_{22}(k)}\right]^{-1}, \tag{5}$$

$$S_{22}(k) = \left[1 - \varrho_{22}\hat{C}_{22}(k) - \frac{\varrho_{11}\varrho_{22}\hat{C}_{12}^2(k)}{1 - \varrho_{11}\hat{C}_{22}(k)}\right]^{-1}, \tag{6}$$

$$S_{12}(k) = (\varrho_{11}\varrho_{22})^{1/2}\,\hat{C}_{12}(k)\,[\{1 - \varrho_{11}\hat{C}_{11}(k)\}\,\{1 - \varrho_{22}\hat{C}_{22}(k)\} - \varrho_{11}\varrho_{22}\hat{C}_{12}^2(k)]^{-1}. \tag{7}$$

The total structure factor, $S(k)$, in terms of these partials is given by

$$S(k) = \sum_{i=1}^{2}\sum_{j=1}^{2} C_i^{1/2}C_j^{1/2}\,\frac{f_i(k)\,f_j(k)}{C_1f_1^2(k) + C_2f_2^2(k)}\,S_{ij}(k), \tag{8}$$

where $f_1(k)$ and $f_2(k)$ are the atomic scattering factors and C_1 and C_2 are the atomic fractions of components 1 and 2, respectively.

In the long-wave limit, (3) and (4) reduce to

$$\varrho_{ii}\hat{C}_{ii}(0) = -24\eta_{ii}\left(\frac{a_i}{3} + \frac{b_i\sigma_{ii}}{4} + \frac{d\sigma_{ii}^3}{6}\right) + \frac{8\eta_{ii}\varepsilon_{ii}(A_{ii}^3 - 1)}{k_BT} \tag{9}$$

and

$$\hat{C}_{12}(0) = \frac{4\pi\varepsilon_{12}\sigma_{12}^3(A_{12}^3 - 1)}{3k_BT} - \frac{4\pi a_1\sigma_{12}^3}{3} - 4\pi\sigma_{11}^3 \times$$

$$\times \left\{\frac{b(\sigma_{11} + 2\sigma_{22})}{12} + \frac{\lambda d\sigma_{11}(3\sigma_{11} + 5\sigma_{22})}{10} + \frac{d\sigma_{11}^2(2\sigma_{11} + 3\sigma_{22})}{30}\right\}. \tag{10}$$

The isothermal compressibility, χ_T, for a binary mixture is related to $\varrho_{ii}\hat{C}_{ii}(0)$ and $\hat{C}_{12}(0)$ by [10]

$$\varrho\chi_Tk_BT = \{1 - C_1\varrho_{11}\hat{C}_{11}(0) - C_2\varrho_{22}\hat{C}_{22}(0) - 2C_1C_2\varrho_{12}\hat{C}_{12}(0)\}^{-1}. \tag{11}$$

Also Bhatia and Thornton [8] have introduced three new structure factors. These authors defined the three correlation functions in terms of fluctuations in (i) the particle density (without regard of the type), $S_{NN}(k)$; (ii) the composition, i.e. the local concentration (in mole fraction), $S_{CC}(k)$, and (iii) the cross-correlation between number and composition, $S_{NC}(k)$. These function are related to $S_{ii}(k)$ and $S_{ij}(k)$ as

$$S_{NN}(k) = C_1S_{11}(k) + C_2S_{22}(k) + 2(C_1C_2)^{1/2}\,S_{12}(k), \tag{12}$$

$$S_{CC}(k) = C_1C_2[C_2S_{11}(k) + C_1S_{22}(k) - 2(C_1C_2)^{1/2}\,S_{12}(k)], \tag{13}$$

$$S_{NC}(k) = C_1C_2\left[S_{11}(k) - S_{22}(k) + \frac{S_{12}(k)\,(C_2 - C_1)}{(C_1C_2)^{1/2}}\right]. \tag{14}$$

3. Results and Discussions

We have computed the structure factors (partials and total) at four different atomic fractions and at different temperatures so as to compare our results with experiment [6]. In the alloy the densities of the components were determined from those of the pure components assuming an ideal law of mixing to hold, e.g.,

$$\varrho_{11} = \frac{C_1\varrho_1^0\varrho_2^0}{C_1(\varrho_2^0 - \varrho_1^0) + \varrho_1^0}, \qquad \varrho_{22} = \frac{(1 - C_1)\,\varrho_1^0\varrho_2^0}{C_1(\varrho_2^0 - \varrho_1^0) + \varrho_1^2},$$

where ϱ_1^0 and ϱ_2^0 are the densities of the pure components under the same conditions.

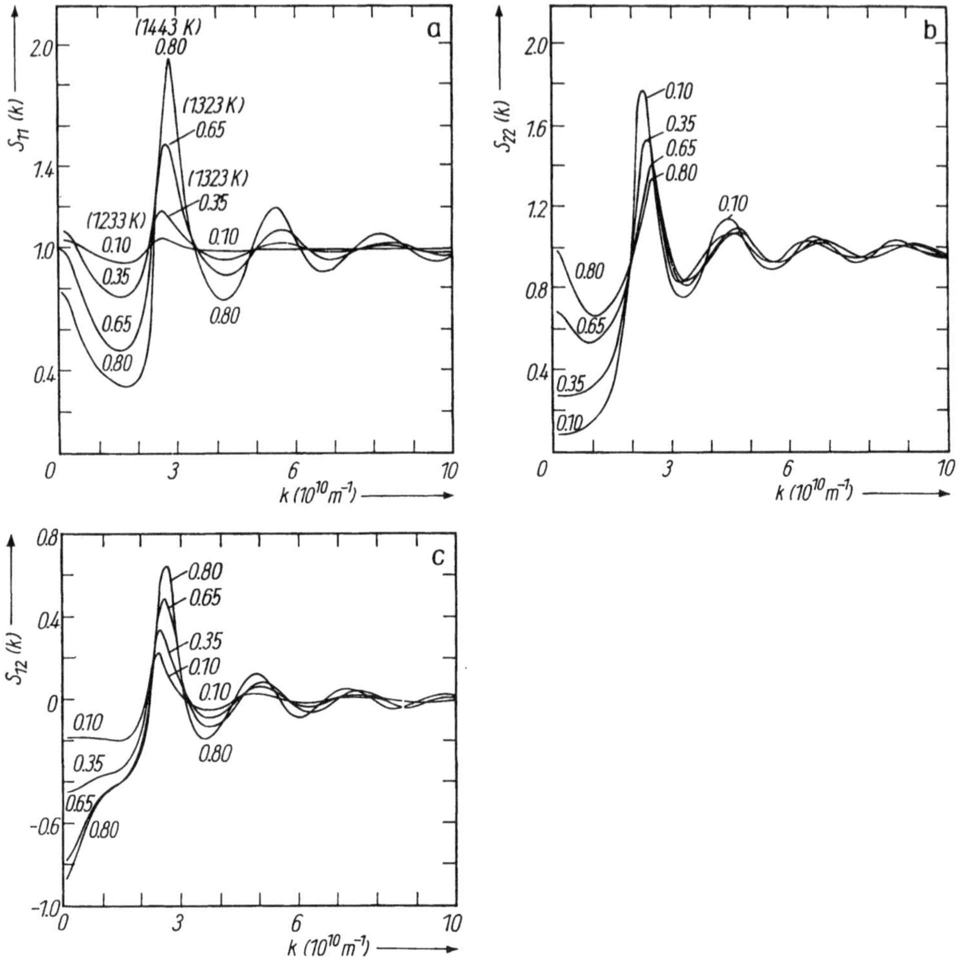

Fig. 1. a) $S_{11}(k)$, b) $S_{22}(k)$, and c) $S_{12}(k)$ vs. k curves for different atomic concentrations of Cu marked at the curves

The potential parameters have been taken from the works of Rao and Murty [11] on pure metals. At higher temperatures we assume that the values of the parameters, except the core diameters, are not changing. The computed partial structure factors $S_{11}(k)$, $S_{22}(k)$, and $S_{12}(k)$ are shown in Fig. 1a to c, respectively, at various concentrations. The concentration is expressed as atomic fraction of copper, C_1.

It is found that in $S_{11}(k)$, as the concentration of Cu increases in the alloy, the first peak height increases gradually and shifts towards the large-k side. In $S_{22}(k)$, the first main peak position is shifting with the increase of Cu concentration, but there is a regular decrease in peak height as the concentration of Cu, C_1 increases from 0.10 to 0.80. These trends in the variation of $S_{11}(k)$ and $S_{22}(k)$ functions are in conformity with the fact that the structure function peak for liquid copper is higher and lies further to the right on the k-axis, as compared to that for liquid Pb. Both $S_{11}(k)$ and $S_{22}(k)$ go to unity in the large-k limit as is expected from (5) and (6). With the increase in C_1 the first peak in $S_{12}(k)$ shifts towards the right and increases in height gradually, while

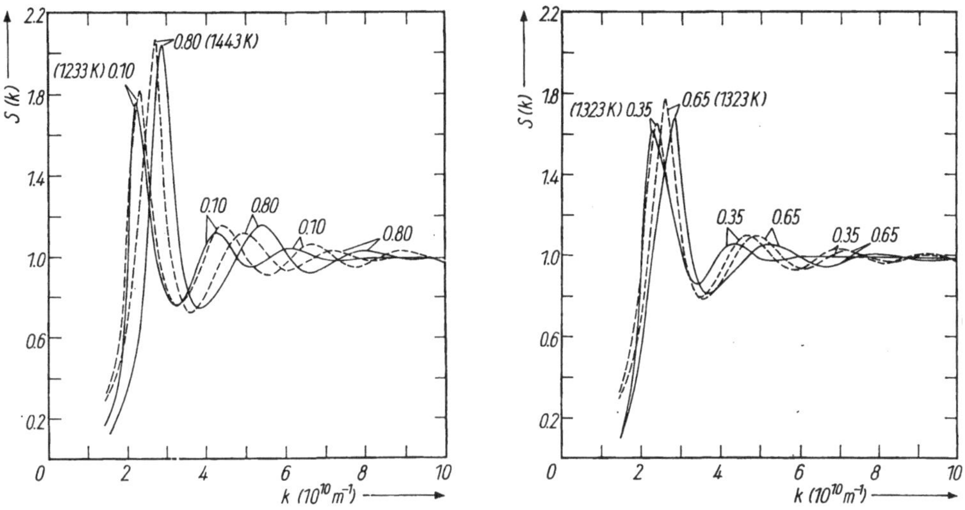

Fig. 2. Total structure factor of (Cu–Pb) alloy at various concentrations of copper. ---- calculated, —— experimental

in the small-k region there is a shoulder as the concentration of Cu increases. The peculiar shoulder characteristics of $S_{12}(k)$ are discussed presently.

Further we have also calculated χ_T, the isothermal compressibility with the help of (9), (10), and (11) and it is found that the compressibility decreases with increase in Cu concentration in the alloy. No comparison could be made as experimental values are not available.

The total structure factors computed from (8) are plotted for different concentrations of copper at different temperatures in Fig. 2. It is gratifying to note that the calculated values are slightly higher than the experimental values even though there is a slight shift in the peak position. It may also be noted that at lower concentrations of Cu the shift is to the right, while at higher concentrations it is to the left. We give in Table 1 the first peak height and its position at various concentrations and temperatures. As can be seen from the table the deviation in peak height is not bigger than 5.47%. It is important to remember that the first peak height primarily determines the transport coefficients [12].

It may be pointed out that the $S_{11}(k)$ and $S_{22}(k)$ curves should be similar to the pure Cu and pure Pb structure functions as is found to be the case. However, in the case of the $S_{12}(k)$ curve there is a marked difference in the low-k region in the sense that a

Table 1

The first main peak height and its position and compressibility

T (K)	concentration of Cu	calc.		exp.		deviation in peak height (%)	position deviation in (10^{10} m^{-1})	χ_T (10^{12}cm^2/dyn)
		k (10^{10} m^{-1})	$S(k)$	k (10^{10} m^{-1})	$S(k)$			
1233	0.10	2.300	1.8155	2.225	1.7473	3.90	0.075	10.3610
1323	0.35	2.400	1.6593	2.300	1.6148	2.75	0.100	8.5418
1323	0.65	2.600	1.7767	2.825	1.6844	5.47	0.225	4.7135
1443	0.80	2.700	2.0653	2.875	2.0426	1.11	0.175	2.0644

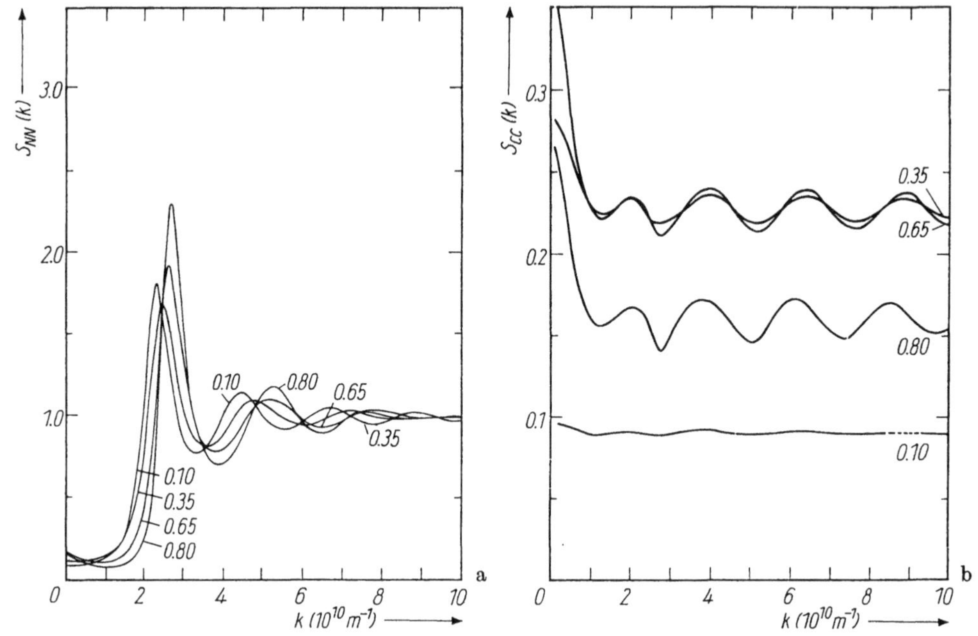

Fig. 3. $S_{NN}(k)$ (a) and $S_{CC}(k)$ (b) vs. k curves for different atomic fractions of copper marked at the curves

shoulder is seen with increasing concentration of Cu. The main peak in the alloy remains at the same position as in pure Pb up to 35% of Cu and there is a broadening in the low-k region of $S(k)$. With 35% concentration of Cu there appears a shoulder in the low-k region and this becomes prominent as the concentration of Cu increases

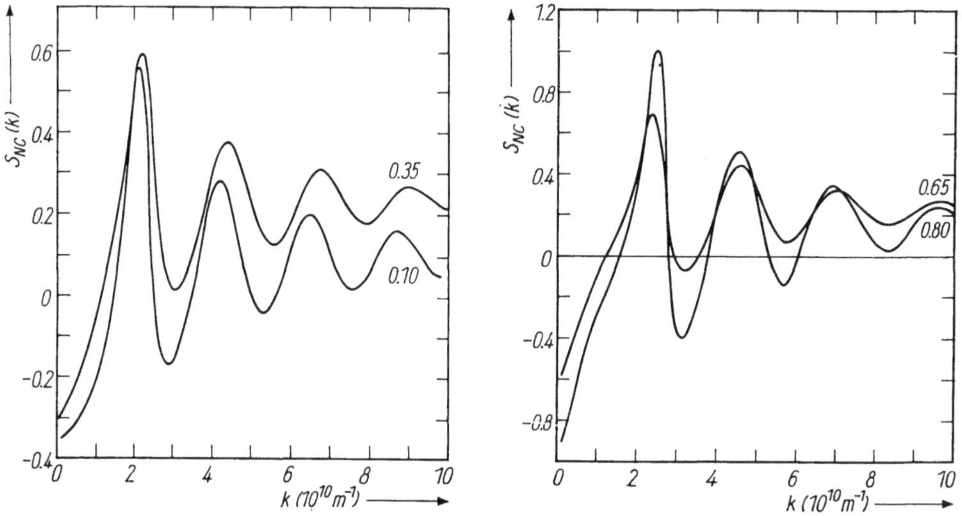

Fig. 4. $S_{NC}(k)$ vs. k curves for different atomic fractions of copper marked at the curves

beyond 35%. Also the first peak shifts to the right which is nearer to that in pure Cu. From this behaviour it may be pointed out that there is a deviation from the uniform statistical distribution in Cu–Pb alloys as has been pointed out by Seeb et al. [6]. There may be some form of micro-segregation beyond 35% Cu concentration. It may also be due to incorporation of s–d hybridization of electrons of Cu and Pb [13].

The Bhatia-Thornton structures, $S_{NN}(k)$, $S_{CC}(k)$, and $S_{NC}(k)$ are calculated for different atomic fractions of Cu. The results have been plotted in Figs. 3 and 4 at a typical set of concentrations for better representation. The variation of $S_{NN}(k)$ with k is similar to that of $S(k)$ for a pure liquid. The peak heights oscillate with increasing concentration of copper. However, a peak of maximum height was obtained with 80% copper concentration. The shift of the first peak position in $S_{NN}(k)$ towards the short wavelength side with the increase in copper concentration is regular for the whole concentration range. The $S_{CC}(k)$ and $S_{NC}(k)$ curves oscillate around a value of C_1C_2 and 0, respectively. It is to be noted specially that in $S_{NC}(k)$ curves there are shoulders at higher concentration of copper. The reason is similar to that which we have given for the $S_{12}(k)$ curves.

4. Self-Diffusion Coefficients

In this section of the paper we present an application of the results obtained in Section 1 to 3. We obtained $g_{ij}(r)$ by a Fourier transformation of the $S_{ij}(k)$ values using the well-known equation

$$g_{ij}(r) - 1 = \frac{1}{2\pi^2 r\varrho} \int [S_{ij}(k) - \delta_{ij}]\, k \sin(kr)\, dk\,. \tag{15}$$

Here δ_{ij} is the well-known Kronecker delta. Using the linear trajectory principle [14] for mixtures due to Davis and Polyvos [15] we obtain the diffusion coefficients

$$D_i = \frac{k_B T}{\xi_i}, \tag{16}$$

where k_B is the Boltzmann constant. ξ_i is the friction coefficient of the species i and is given by

$$\xi_i = \xi_i^H + \xi_i^S + \xi_i^{SH}\,. \tag{17}$$

In (17), ξ_i^H, ξ_i^S, and ξ_i^{SH} are the friction coefficients due to repulsive core interaction, soft interactions, and the cross effects between the hard and soft forces in the potential, respectively, and are given by

$$\xi_i^H = \sum_{j=1}^{2} \frac{8}{3} \sigma_{ij}^2 g_{ij}(\sigma_{ij})\, \varrho_i\, (2\pi\mu_{ij} k_B T)^{1/2}\,, \tag{18}$$

$$\xi_i^S = -\sum_{j=1}^{2} \frac{\varrho_j}{3} \left(\frac{2\pi\mu_{ij}}{k_B T}\right)^{1/2} (2\pi)^{-2} \int_0^\infty k^3\, \hat{V}_{ij}^S(k)\, \hat{G}_{ij}(k)\, dk\,, \tag{19}$$

and

$$\xi_i^{SH} = -\sum_{j=1}^{2} \frac{2\varrho_j}{3} g_{ij}(\sigma_{ij}) \left(\frac{2\mu_{ij}}{\pi k_B T}\right)^{1/2} \int_0^\infty dk (k\sigma_{ij} \cos k\sigma_{ij} - \sin k\sigma_{ij})\, \hat{V}_{ij}^S(k)\,; \tag{20}$$

ϱ_j is the number density of the species j.

$\varrho_j \hat{G}_{ij}(k)$ and $\hat{V}^S_{ij}(k)$ in the above equations are the Fourier transforms of the total correlation function $[g_{ij}(r) - 1]$ and the soft part of the potential, $V^S_{ij}(r)$, respectively. The hard-sphere diameter of species i is σ_i and μ_{ij} is the reduced mass which is related to the atomic masses m_i and m_j as

$$\mu_{ij} = \frac{m_i m_j}{m_i + m_j} . \tag{21}$$

For the particular case of a square-well potential model applied to binary mixtures, the terms $\hat{G}_{ij}(k)$ and $\hat{V}^S(k)$ will be

$$\hat{G}_{ij}(k) = [S_{ij}(k) - \delta_{ij}] (\varphi_i \varphi_j)^{-1/2} , \tag{22}$$

$$\hat{V}^S_{ij}(k) = \frac{4\pi\varepsilon_{ij}}{k^3} (A_{ij}k\sigma_{ij} \cos A_{ij}k\sigma_{ij} - \sin A_{ij}k\sigma_{ij} - k\sigma_{ij} \cos k\sigma_{ij} + \sin k\sigma_{ij}) , \tag{23}$$

where ε_{ij} and A_{ij} represent the depth and breadth, respectively, of the square well used, δ_{ij} is the Kronecker delta function, and $S_{ij}(k)$ are the partial structure factors defined by Ashcroft and Langreth [7].

5. Results and Discussions

The computed results for friction and diffusion coefficients are given in Table 2 and 3, respectively. Unfortunately no experimental data is available for Cu–Pb systems. From Table 2 it is easy to see that hard spheres contribute the major part to the friction constant.

It may be seen from Table 3 that D_{Cu} and D_{Pb} both decrease with increasing concentration of Cu. The diffusion coefficient of pure metal is attained as its mole fraction becomes unity. It is gratifying to note that these values are in satisfactory agreement with previous calculations on pure metals [11]. We have calculated the diffusion coefficients at four concentrations and at three different temperatures for which experimental results are available. In the present case the ratio D_{Cu}/D_{Pb} has been calculated and the ratio varies as the atomic fraction of copper varies from 0.10 to 0.80.

According to the regular solution theory [16, 17] it is expected that this ratio should be a constant because D_{Cu}/D_{Pb} should be independent of composition. Such a constancy in the ratio of diffusion coefficients has been observed by Jacucci and McDonald [18] for rare gas mixtures by computer experiments. In our case the ratio is changing slowly. This is due to the fact that our system is a binary alloy system which contains copper which has an incomplete d-orbital and hence the formation of regular solutions is not expected. Hence this does not yield a constant value.

Table 2
Various friction coefficients for liquid Cu–Pb alloys

concentration (atomic fraction of Cu)	T (K)	ξ_i^H/k_BT (10^5 s/cm^2)		ξ_i^S/k_BT (10^5 s/cm^2)		ξ_i^{SH}/k_BT (10^5 s/cm^2)	
		Cu	Pb	Cu	Pb	Cu	Pb
0.10	1233	0.07501	0.14399	0.01550	0.01326	0.02550	0.02326
0.35	1323	0.07197	0.12860	0.01694	0.01304	0.02719	0.02267
0.65	1323	0.09582	0.16450	0.02631	0.01946	0.04502	0.03603
0.80	1443	0.14011	0.24318	0.03278	0.02539	0.06769	0.05532

Table 3
Self-diffusion and mutual diffusion coefficients for liquid Cu–Pb alloys

concentration (atomic fraction of Cu)	T (K)	D_{Cu} (10^{-5} cm²/s)	D_{Pb} (10^{-5} cm²/s)	D_{Cu-Pb} (10^{-5} cm²/s)	D_{Cu}/D_{Pb}
0.10	1233	8.6189	5.5395	8.3110	1.5558
0.35	1323	8.6121	6.0854	7.7278	1.4152
0.65	1323	5.9824	4.5452	5.0482	1.3162
0.80	1443	4.1564	3.0873	3.3011	1.3463

Several workers [18 to 20] have tried a relation for the mutual diffusion coefficients from the self-diffusion coefficients of the species in the mixture of the type

$$D_{12} = C_2 D_1 + C_1 D_2 + \text{correction term} , \tag{24}$$

where C_1 and C_2 are the atom fractions of the species 1 and 2, respectively, and the correction term accounts for the cross relations between D_1 and D_2. They calculated that this correction term is negligible except for ionic liquids. We have calculated these approximate mutual diffusion coefficients from (24) and these are given in Table 3 (fifth column).

In conclusion it may be pointed out that the potential parameters used in the calculation of partial structure functions and from these the diffusion coefficients were obtained from those of pure metals which in turn were obtained by fitting to the first peak of $S(k)$ of the pure metal. It is very gratifying to note that the diffusion coefficients so calculated give fairly satisfactory values of the pure components in the case of unit mole fraction. In the absence of experimental results the present calculations become all the more important.

Acknowledgements

One of the authors (S.P.) wishes to thank the D. S. T. India for the award of a fellowship during the period of this work. We are extremely thankful to Prof. Dr. S. Steeb, Max-Planck-Institut für Metallforschung, Institut für Werkstoffwissenschaften, Stuttgart, FRG for sending us the data.

References

[1] M. J. Huijben, W. van der Lugt, W. A. M. Reimert, J. Th. M. De Hosson, and C. van Dijk, Physica (Utrecht) B 97, 338 (1979).
[2] R. V. Gopala Rao and A. K. K. Murty, phys. stat. sol. (b) 66, 703 (1974).
[3] R. V. Gopala Rao and A. K. K. Murty, Z. Naturf. **30a**, 383 (1975).
[4] R. V. Gopala Rao and D. Sen, Chem. Phys. Letters **43**, 455 (1976).
[5] R. V. Gopala Rao and B. M. Satpathy, Chem. Phys. Letters **69**, 571 (1980).
[6] P. Lamparter and S. Steeb, Z. Naturf. **35a**, 1178 (1980).
[7] N. W. Ashcroft and D. C. Langreth, Phys. Rev. **159**, 500 (1967).
[8] A. B. Bhatia and D. E. Thornton, Phys. Rev. B **2**, 3004 (1970).
[9] J. L. Lebowitz, Phys. Rev. A **133**, 895 (1964).
[10] J. G. Kirkwood and P. Buff, J. chem. Phys. **19**, 774 (1951).
[11] R. V. Gopala Rao and A. K. Murty, Z. Naturf. **30a**, 619 (1975).
[12] R. V. Gopala Rao and A. K. Murty, Phys. Letters A **51**, 3 (1975).

[13] R. N. JOARDER, S. PALCHAUDHURI, and R. V. GOPALA RAO, Phys. Rev. B **30**, No. 4 (1984).
[14] E. HELFAND, Phys. Fluids **4**, 681 (1961).
[15] H. T. DAVIS and J. A. POLYVOS, J. chem. Phys. **46**, 4043 (1967); J. phys. Chem. **71**, 439 (1967).
[16] R. J. BEARMAN and P. F. JONES, J. chem. Phys. **33**, 1432 (1960).
[17] R. J. BEARMAN, J. phys. Chem. **65**, 1961 (1961).
[18] G. JACUCCI and I. R. MCDONALD, Physica (Utrecht) **A80**, 607 (1975).
[19] I. EBBSJO, P. SCHOFFIELD, K. SKOLD, and I. WALLER, J. Phys. C **7**, 3891 (1974).
[20] M. PARRINELLO, M. P. TOSI, and N. H. MARCH, J. Phys. C **7**, 2277 (1974).

(Received July 17, 1984)

phys. stat. sol. (b) **126**, 453 (1984)

Subject classification: 1.3 and 13; 21.1; 21.1.1

Department of Material Science, Faculty of Engineering, The University of Tokyo[1])

Electronic Theory of Atomic Short-Range Order for Ternary Alloys Using the Pseudopotential Approximation and Its Comparison with Experiments

By

AMDULLA O. O. MEKHRABOV[2]) and M. DOYAMA

An analysis of the electronic theory of atomic short-range ordering for ternary alloys in the pseudopotential approximation is made. The partial ordering energies and short-range order parameters are calculated for ternary Ni–Re–Mo, $Ni_3(Fe_{0.10}, Cr_{0.15})$, and $Ni_3(Fe_{0.20}, Mn_{0.05})$ alloys using the model pseudopotential of Animalu. The comparison of the results between calculated and measured shows a good agreement for all the chosen systems.

Проводится анализ электронной теории атомного ближнего порядка в приближении теории псевдопотенциала для трехкомпонентных сплавов. Расчитаны парциальные энергий упорядочения и параметры ближнего порядка для тройных сплавов переходных металлов Ni–Re–Mo, $Ni_3(Fe_{0,10}\,Cr_{0,15})$ и $Ni_3(Fe_{0,20}\,Mn_{0,05})$ с использованием модельного псевдопотенциала Анималу. Сравнение результатов расчета с экспериментальными данными показывают хорошее согласие теории с экспериментом для всех трех выбранных сплавов.

One of the main problems in solid-state physics is the investigation of the factors determining physical properties of solids. It was experimentally established that the short-range order (SRO), has a considerable effect upon the different physical properties of alloys. The transition of the alloys from the disordered to the ordered state is followed in some cases by the change of the sign of the Hall constant, decrease of the strength, resistance of the shift, electrical resistance, appearance and disappearance of ferromagnetism. Physical properties are also affected when the degree of short- and long-range order changes. Moreover, it has recently become more obvious that the existence of SRO is attributed to the fundamental properties of the atomic, electronic, and spin lattice structures. The data on SRO may therefore be also used for studying the character and the nature of interatomic interactions in the crystals related to them by the structural features of alloys. Investigations of atomic short-range ordering in the alloys are therefore of great practical and theoretical interest. To understand the nature of the behaviour of features with account of ordering for its correct interpretation, the theory of this phenomenon should be developed. No doubt that the nature of SRO may be understood to full extent only in the solid-state quantum theory.

A considerable number of experimental and theoretical works [1, 2] have been devoted to the investigation of atomic short-range order in binary alloys of metals. This is not the case, however, with ternary alloys. At this point, only a few studies

[1]) Bunkyo-ku, Tokyo 113, Japan.

[2]) Permanent address: Department of Physics, The Kirov Azerbaidzhan State University, 23, P. Lumumba St., Baku, USSR.

[3 to 5] are known. The expression connecting the intensity of X-ray diffuse scattering of a polycrystalline solid solution with partial SRO parameters has been obtained in [3]. The attempt to develop the electronic theory of SRO in ternary alloys using the pseudopotential approximation has been made for the first time by Katsnelson et al. in [4]. In the present paper the following equations connecting the partial SRO parameters with ordering energy of pairs of atoms were obtained:

$$\alpha^{AB}(\boldsymbol{k}) = D(AB)\,[1 - \beta U^{AB}(\boldsymbol{k})\,(2C_A C_B + C_C) - \beta U^{AC}(\boldsymbol{k}) \times \times C_C(2C_A - 1) - \beta U^{BC}(\boldsymbol{k})\,C_C(2C_B - 1)]\,, \tag{1}$$

where

$$U^{AB}(\boldsymbol{k}) \equiv V^{AB}(\boldsymbol{k}) - \tfrac{1}{2}\,[V^{AA}(\boldsymbol{k}) + V^{BB}(\boldsymbol{k})] \tag{2}$$

is the ordering energy for a pair of atoms A and B; $\beta = 1/k_B T$; k_B is the Boltzmann constant; T is the absolute temperature; C_A, C_B, and C_C are the concentrations of the A, B, and C-type atoms, respectively; $\boldsymbol{k}$ is the wave vector of an electron from the first Billouin zone; $\alpha^{AB}(\boldsymbol{k})$ is the Cowley SRO parameter [6].

The expressions for the SRO parameters $\alpha^{AC}(\boldsymbol{k})$ and $\alpha^{BC}(\boldsymbol{k})$ can be obtained from (1) by substitution of B by C and A by C, respectively.

The Fourier transform of SRO parameters in the R-space can be written as

$$\alpha^{AB}(R_e) = \sum_{\boldsymbol{k}} \alpha^{AB}(\boldsymbol{k}) \exp(i\boldsymbol{k}\boldsymbol{R}_e)/N\,, \tag{3}$$

where $\boldsymbol{R}_e = \boldsymbol{r}_i - \boldsymbol{r}_j$ is the radius of the coordination sphere l; $\boldsymbol{r}_i$, $\boldsymbol{r}_j$ are the radius-vectors in the i-th and j-th lattice sites, respectively.

In (1) the constant D(AB) would be defined from the following sum rule:

$$\sum_{\boldsymbol{k}} \alpha^{SS'}(\boldsymbol{k})/N = 1\,. \tag{4}$$

Here S and S′ are indices of types of atoms.

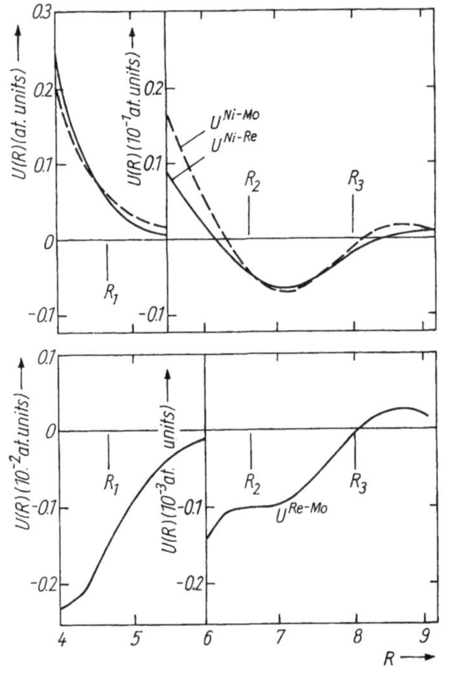

Fig. 1. The dependence of ordering energy of pairs on interatomic distances R (always in at. units) for the Ni–Re, Ni–Mo, and Re–Mo pairs for the Ni–Re–Mo alloy

The Fourier transform of ordering energy included in (2) will be defined by the following expressions:

$$U^{SS'}(\boldsymbol{k}) = \sum_{\boldsymbol{R}} U^{SS'}(R_e) \exp(-i\boldsymbol{k}\boldsymbol{R}_e) , \tag{5}$$

where

$$U^{SS'}(R_e) = \frac{\bar{\Omega}_0}{\pi^2} \int F^{SS'}(q)\, q^2 \frac{\sin(q, R_e)}{qR_e}\, dq , \tag{6}$$

$$F^{SS'}(q) = -\frac{\bar{\Omega}_0}{8\pi} \cdot |W^S(q) - W^{S'}(q)|^2 q^2 \frac{\varepsilon(q) - 1}{\varepsilon^*(q)} + \frac{2\pi}{\bar{\Omega}_0 q^2} |(Z^*)^S - (Z^*)^{S'}|^2 \exp(-q^2/4\eta) , \tag{7}$$

$W^S(q)$ and $W^{S'}(q)$ are the form-factors of unscreened pseudopotentials of S and S′-type atoms; $(Z^*)^S$ and $(Z^*)^{S'}$ are effective valencies of S and S′-type atoms; $\bar{\Omega}_0$ is the average atomic volume; η the Ewald parameter; $\varepsilon(q)$ the static dielectric constant which may be written as in the self-consistent Hartree approximation,

$$\varepsilon(q) = 1 + V(q)\, \chi_0(q) , \tag{8}$$

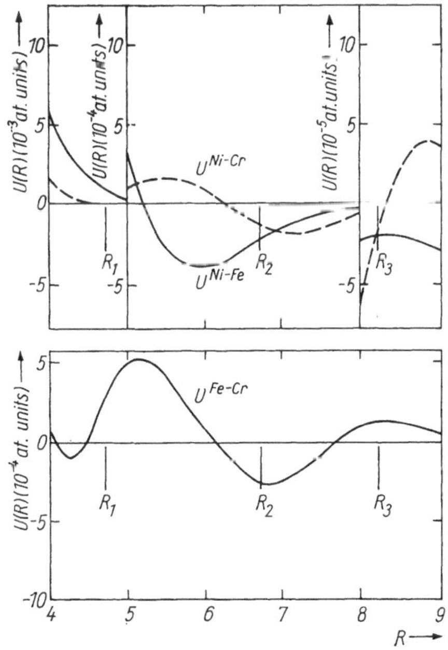

Fig. 2

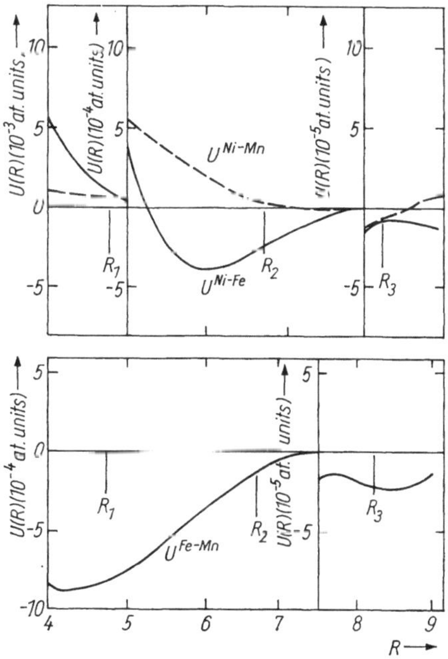

Fig. 3

Fig. 2. The dependence of ordering energy of pairs on interatomic distances R for the Ni–Fe, Ni–Cr, and Fe–Cr pairs for the $Ni_3(Fe_{0.10}Cr_{0.15})$ alloy

Fig. 3. The dependence of ordering energy of pairs on interatomic distances R for Ni–Fe, Ni–Mn, and Fe–Mn pairs for the $Ni_3(Fe_{0.20}Mn_{0.05})$ alloy

Table 1

Ordering energy of pairs and partial short-range order parameters for the Ni–Re–Mo alloy in at. units

pair	$U(R_1)$	$U(R_2)$	$U(R_3)$	$\alpha(R_1)$	$\alpha^{\exp}(R_1)$
Ni–Re	4×10^{-2}	-4.75×10^{-3}	-1.5×10^{-3}	−0.142	−0.150
Ni–Mo	5.5×10^{-2}	-4.7×10^{-3}	-4.0×10^{-4}	0.016	>0
Re–Mo	-1.6×10^{-3}	-1.02×10^{-4}	1.0×10^{-6}	−0.003	<0

where $V(q) = 4\pi\bar{Z}/\Omega_0 q^2$ is the Fourier transform of the electron–electron interaction; $\bar{Z} = C_A Z_A^* + C_B Z_B^* + C_C Z_C^*$; $\chi_0(q)$ is the Lindhard polarization function which is given as in the free-electron approximation,

$$\chi_0(q) = \left(\frac{2}{3} E_F\right)^{-1} \left[\frac{1}{2} + \frac{4k_F^2 - q^2}{8qk_F} \ln \left|\frac{2k_F + q}{2k_F - q}\right|\right]. \tag{9}$$

Here E_F and k_F are the Fermi energy and the Fermi wave number, respectively. $\varepsilon^*(q)$ is the dielectric constant with account of correlation and exchange effects. According to [7], $\varepsilon^*(q)$ may be written as

$$\varepsilon^*(q) = 1 + \frac{V(q)\,\chi_0(q)}{1 - f(q)\,\chi_0(q)}. \tag{10}$$

Here $f(q)$ is the function responsible for the exchange and correlation effects in the interacting electron gas,

$$f(q) = q^2(q^2 + k_F^2 + k_S^2)^{-1}/2\,. \tag{11}$$

In the present paper for the calculation of the partial ordering energies and SRO parameters according to (1) to (11) programmes for computer in Fortran were composed. Direct calculations were made for the ternary alloys Ni–Re–Mo, Ni_3(Fe, Cr), and Ni_3(Fe, Mn). The results are shown in Tables 1 to 3. The sign and the type of SRO were obtained by X-ray diffuse scattering measurements for the alloy Ni–Re–Mo [8] more reliably. The comparison of the calculated data with experiment is possible. Atomic ordering processes in the Ni_3(Fe, Cr) and Ni_3(Fe, Mn) alloys have been studied [9 to 13] qualitatively and there are no experimental data for partial SRO parameters for the alloys.

The unscreened form-factors of pseudopotentials of ions were calculated using the expressions obtained in [14].

The dependence of the partial ordering potentials on the interatomic distances is shown in Fig. 1 for Ni–Re–Mo (3.6 at% Re, 10.4 at% Mo), in Fig. 2 for Ni_3(Fe, Cr) (10 at% Fe, 15 at% Cr), and in Fig. 3 for Ni_3(Fe, Mn) (20 at% Fe, 5 at% Mn). It is

Table 2

Ordering energy of pairs and partial short-range order parameters for the $Ni_3(Fe_{0.10}Cr_{0.15})$ alloy in at. units

pair	$U(R_1)$	$U(R_2)$	$U(R_3)$	$\alpha(R_1)$	$\alpha(R_2)$	$\alpha(R_3)$
Ni–Fe	-1.1×10^{-3}	-2.3×10^{-4}	-1.0×10^{-5}	−0.080	0.044	0.012
Ni–Cr	2.5×10^{-5}	-1.41×10^{-4}	-2.6×10^{-5}	−0.001	0.015	0.004
Fe–Cr	3.2×10^{-4}	-2.35×10^{-4}	1.29×10^{-4}	0.318	0.099	−0.013

Table 3

Ordering energy of pairs and partial short-range order parameters for the $Ni_3(Fe_{0.20}Mn_{0.05})$ alloy in at. units

pair	$U(R_1)$	$U(R_2)$	$U(R_3)$	$\alpha(R_1)$	$\alpha(R_2)$	$\alpha(R_3)$
Ni–Fe	9.38×10^{-4}	-2.49×10^{-4}	-8.8×10^{-6}	−0.105	0.079	0.009
Ni–Mn	6.69×10^{-4}	3.02×10^{-5}	8.7×10^{-6}	−0.111	0.054	0.016
Fe–Mn	-8.26×10^{-4}	-1.19×10^{-4}	-2.31×10^{-5}	−0.111	−0.087	−0.019

seen in the figures that the behaviour of the ordering energies for all the pairs has quasi-oscillatory and sign-changing character. It means that $U(R_i)$ are different for various coordination spheres. The mark at the bottom shows the positions of the first three coordination spheres R_i for the Ni–Re, Ni–Mo, and Re–Mo pairs in Fig. 1 for the Ni–Fe, Ni–Cr, and Fe–Cr pairs in Fig. 2, and for the Ni–Fe, Ni–Mn, and Fe–Mn pairs in Fig. 3.

The values of partial ordering energies and SRO parameters are given in Table 1 for Ni–Re–Mo, in Table 2 for Ni_3(Fe, Cr), and in Table 3 for Ni_3(Fe, Mn) alloys. It is seen from Tables 1 to 3 that the values of ordering energies and their spectrum for different pairs are different not only by the values but also by signs.

Then the calculated values of ordering energies were used for the calculation of partial SRO parameters using expressions (1) to (4). The calculation of $\alpha^{SS'}(R_i)$ was carried out as follows. Since the ternary alloys Ni–Re–Mo, Ni_3(Fe, Cr), and Ni_3(Fe, Mn) have f.c.c. crystal lattice, the Fourier transform of the ordering energy will be defined as

$$\begin{aligned} U_{SS'}(\boldsymbol{k}) &= \sum_{p_l, m_l, n_l} U^{SS'}_{p_l, m_l, n_l}(R_e) \exp(i\boldsymbol{k}\boldsymbol{R}_{p_l, m_l, n_l}) = \\ &= \sum_{p_l, m_l, n_l} U^{SS'}_{p_l, m_l, n_l}(R_e) \cos(2\pi h_1 p_e) \cos(2\pi h_2 m_l) \cos(2\pi h_3 n_e) \\ &= 12U^{SS'}(R_1) \{\tfrac{1}{3} [\cos(\pi h_1) \cos(\pi h_2) + \\ &\quad + \cos(\pi h_2) \cos(\pi h_3) + \cos(\pi h_3) \cos(\pi h_1)]\} + \\ &\quad + 6U^{SS'}(R_2) \{\tfrac{1}{3} [\cos(2\pi h_1) + \cos(2\pi h_2) + \cos(2\pi h_3)]\} + \\ &\quad + 24U^{SS'}(R_3) \{\tfrac{1}{3} [\cos(2\pi h_1) \cos(\pi h_2) \cos(\pi h_3) + \\ &\quad + \cos(2\pi h_2) \cos(\pi h_1) \cos(\pi h_3) + \cos(2\pi h_3) \cos(\pi h_1) \cos(\pi h_2)]\} \\ &= \sum C_e U^{SS'}(R_e) A_e , \end{aligned} \tag{12}$$

where h_1, h_2, h_3 are the continuous coordinates in the reciprocal space; p_e, m_e, n_e are the coordinates of atoms in the l-th coordination sphere; A_l is the sum of expressions $[\cos(2\pi h_1 p_e) \cos(2\pi h_2 m_e) \cos(2\pi h_3 n_e)]$ which is taken for all atoms of the l-th coordination sphere; C_e is the coordination number.

Taking into account the symmetry of the crystal lattice, the calculations of $U^{SS'}(\boldsymbol{k})$ and $\alpha(\boldsymbol{k})$ were carried out in 1/96 of the volume of the crystal unit cell of the reciprocal space. This volume is the volume of a pyramid with the triangle with angles at the points (000), (100), (110) as the basis, and the top of which is the point $(\frac{1}{2}\,\frac{1}{2}\,\frac{1}{2})$. The step of sum is taken 0.1 in the h_i axes or 0.001 over the volume of the unit cell in the reciprocal space.

Let us consider the result of calculation for the Ni–Re–Mo alloy at first. The experimental values of partial SRO parameters are also presented in Table 1. It is seen from Table 1 that the calculated and measured values of SRO parameters coincide very well.

According to [10, 11], concentration disordering of the Ni_3Fe superstructure occurs when Cr up to 6 at% is added to the Ni_3Fe alloy. Further increase of chromium to more than 15 at% causes separation of the solid solution with the appearance of iron-enriched regions and an ordered arrangement of nickel and chromium atoms is also observed. It is seen from Table 2, that in the $Ni_3(Fe_{0.10}Cr_{0.15})$ alloy the concentration of Cr atoms in the nearest neighbor sites of Ni atoms is higher than that of the random alloy $\alpha^{Ni-Cr}(R_1) < 0$, and in the alloy the concentration of Fe atoms in the nearest neighbor sites of Fe atoms is higher than that of the random alloy, i.e. Fe atoms form clusters, $\alpha^{Fe-Cr}(R_1) > 0$.

The result of calculation for the $Ni(Fe_{0.20}Mn_{0.05})$ alloy is shown in Table 3. It is seen from Table 3 that all three SRO parameters at the first coordination sphere in the $Ni_3(Fe_{0.20}Mn_{0.05})$ alloy have a negative sign and almost the same value. It means that in these alloys Mn atoms can substitute Ni and Fe atoms with the same probability and the formation of a three-component superstructure in Ni_3(Fe, Mn) alloys is possible. These results are in good agreement with the result of Young's modulus measurement [12].

Thus, the results of calculation of partial SRO parameters for Ni–Re–Mo, Ni_3(Fe, Cr) and Ni_3(Fe, Mn) alloys shows that the constructed electron theory of SRO for ternary alloys in the pseudopotential approximation allows to determine the partial parameters of SRO with a good accuracy without carrying out complex and laborious experimental investigations.

Acknowledgements

One of the authors (AOOM) wishes to thank Prof. A. A. Katsnelson and Dr. V. M. Silonov from Moscow State University for their interest to this work and for helpful discussions.

References

[1] V. I. Iveronova and A. A. Katsnelson, Short-Range Order in Solid Solution, Izd. Nauka, Moscow 1977.
[2] A. A. Katsnelson, V. M. Silonov, and Farid A. Khwaja, phys. stat. sol. (b) **91**, 11 (1979).
[3] V. I. Iveronova and V. M. Silonov, Proc. 4th All-Union Workshop Ordering of Atoms and Their Influence on the Alloy Properties, Tomsk 1974.
[4] A. A. Katsnelson, A. O. O. Mekhrabov, and V. M. Silonov, Fiz. Metallov i Metallovedenie **52**, 661 (1981).
[5] A. O. O. Mekhrabov, Internal Report VINITI, No. 55, Az-D83 (1983).
[6] J. M. Cowley, J. appl. Phys. **21**, 24 (1950).
[7] J. Hubbard, Proc. Roy. Soc. (London) **A240**, 359 (1957); **A243**, 336 (1957).
[8] A. M. Silonov, A. A. Katsnelson, and V. M. Silonov, Fiz. Metallov i Metallovedenie **51**, 455 (1981).
[9] I. L. Eganyan and Ya. P. Selisskiy, Soviet Phys. – Doklady **12**, No. 3, (1967).
[10] K. G. Binnatov, A. A. Katsnelson, and Yu. L. Rodionov, Fiz. Metallov i Metallovedenie **39**, No. 5 (1975).
[11] V. I. Gomankov and N. I. Nogin, Fiz. Metallov i Metallovedenie **52**, No. 3 (1981).
[12] I. L. Yeganyan and Ya. P. Selisskiy, Fiz. Metallov i Metallovedenie **23**, No. 3 (1967).
[13] V. E. Panin, G. G. Minayeva, V. P. Fadin, and V. V. Prushinskiy, Fiz. Metallov i Metallovedenie **26**, No. 3 (1968).
[14] A. O. E. Animalu, Phys. Rev. B **8**, No. 8 (1973).

(Received July 30, 1984)

phys. stat. sol. (b) **126**, 459 (1984)

Subject classification: 1.2 and 14.4.1; 22.8.1

Institute of Physics, Czechoslovak Academy of Sciences, Prague[1])

On Phase Transitions in $LiNH_4SO_4$

By

V. I. TORGASHEV[2]), V. DVOŘÁK, and F. SMUTNÝ

A Landau-type thermodynamic theory describing the sequence of phase transitions in lithium ammonium sulphate at atmospheric pressure is developed. The form of the phase diagram is derived and compared with that constructed on the basis of published experimental data. The existence of a new monoclinic high pressure phase is proposed. Formulae for permittivity are presented and compared with available experimental data. Soft mode features of Raman and infra-red spectra are briefly discussed.

Развита термодинамическая теория Ландау, описывающая последовательность фазовых переходов в литий-аммоний сульфате при атмосферном давлении. Теоретическая фазовая диаграмма сравнивается с диаграммой, полученной на основе имеющихся экспериментальных данных. Предполагается существование ещё одной моноклинной фазы в области высоких давлений. Приведены формулы для восприимчивостей. Обсуждаются особенности их температурного поведения и дано сравнение с экспериментом. Кратко обсуждается поведение мягких мод в ИК- и КР-спектрах.

1. Introduction

Recently it has been established experimentally [1 to 3] that the following sequence of phases occurs in lithium ammonium sulphate (LAS) with decreasing temperature at atmospheric pressure:

$$\underset{\text{(phase I)}}{\mathrm{Pnma} - \mathrm{D}_{2h}^{16}(v)} \xrightarrow[460\ \mathrm{K}]{T_1} \underset{\text{(phase II)}}{\mathrm{Pn2_1a} \quad \mathrm{C}_{2v}^{9}(v)} \xrightarrow[285\ \mathrm{K}]{T_2} \underset{\text{(phase III)}}{\mathrm{P12_1/a1} - \mathrm{C}_{2h}^{5}(2v)}\,,$$

where v denotes the unit-cell volume of the orthorhombic phase I. Moreover, at hydrostatic pressures above 700 MPa a new phase apparently of $P12_11-C_2^2$ symmetry occurs [4, 5]. All phase transitions in LAS are of first order; the hysteresis at T_2 may amount up to few tens K [6].

In order to describe some properties of LAS several authors [7, 8] have been using the Landau theory starting from the hexagonal prototypic phase $P6_3/mmc-D_{6h}^4$. Indeed, lattice parameters of LAS suggest that phase I might be treated as distorted phase D_{6h}^4 with the pseudohexagonal axis parallel to the x-axis of the orthorhombic coordinate system. In general, there might be several reasons for using such a description: phase I could exhibit a twin structure resulting from a hypothetical hexagonal–orthorhombic phase transition as it is the case of $(NH_4)_2SO_4$ [9] which is isomorphous to LAS. Sometimes all observed phases can be described by means of less of irreducible representations when starting from a hypothetical prototypic phase (e.g. [10]). We have found that no arguments like that apply to LAS and therefore we shall describe properties of LAS within the Landau theory starting from the D_{2h}^{16} symmetry. Since

---

[1]) Na Slovance 2, 180 40 Praha 8, Czechoslovakia.

[2]) On leave of absence from the Institute of Physics, Rostov State University, Rostov-on-Don, USSR.

the space group D_{2h}^{16} is a subgroup of D_{6h}^{4} our thermodynamic potential is essentially that used in [7]; our analysis, however, is more extended.

In Section 2 we present the thermodynamic potential from which a phase diagram is derived and compared with the pressure–temperature diagram of LAS found experimentally. In Section 3 we analyse some macroscopic properties at phase transitions. In Section 4 the origin of the phase C_2^2 with no change of the unit-cell volume is discussed. Some features of infra-red and Raman spectra are briefly discussed in Section 5.

2. Phase Diagram of LAS

Obviously, the symmetry change $D_{2h}^{16} \to C_{2v}^{9}$ can be described by the onset of the polarization component P_y. The static dielectric susceptibility χ_{yy}, however, exhibits above T_1 Curie-Weiss behaviour with extremely small value of Curie-Weiss constant [11]. This experimental evidence suggests that LAS may be classified as a pseudo-proper ferroelectric [12] in which lattice modes carrying P_y are bilinearly coupled to an order parameter ξ of the same symmetry. Small coupling between ξ and P_y results in a small value of the Curie-Weiss constant. In the case of LAS ξ describes the ordering of SO_4 groups which is coupled to displacements of NH_4 and Li ions along the y-axis giving rise to P_y. Thus for the description of the phase transition at T_1 we could choose ξ as the order parameter which is bilinearly coupled to P_y.

The symmetry change $D_{2h}^{16} \to C_{2h}^{5}$ is associated with the doubling of the orthorhombic unit cell along the a-axis. (We note that this doubling is not taken into account in [8].) It can be shown that this symmetry change can be described by a two-component order parameter η_1, η_2 (transforming according to an irreducible representation at the point $\boldsymbol{k} = (1/2)\,\boldsymbol{a}^*$ of the Brillouin zone) provided the spontaneous values satisfy $\eta_1^2 = \eta_2^2$. Since the intersection of the space groups C_{2v}^{9} and C_{2h}^{5} is just C_2^2 we can by means of the order parameters ξ and η_1, η_2 describe all phases observed in LAS. Here we assume that the phase C_2^2 (phase V) which occurs under hydrostatic pressure has doubled period along the a-axis which is not yet confirmed experimentally. From compatibility relations it can be shown that $\xi(\boldsymbol{k}=0)$ and $\eta_i(\boldsymbol{k}=(1/2)\,\boldsymbol{a}^*)$ belong to the same branch of diffusive modes corresponding to a reorientation of SO_4 groups, which becomes doubly degenerate at the Brillouin zone boundary.

The free energy F describing the phase transitions in LAS will be a function of the order parameters ξ, η_i, macroscopic variables P_y, and strain components u_{ik}. Since ξ and P_y are bilinearly coupled, the equilibrium value of ξ is a linear function of P_y and therefore ξ can be eliminated from the free-energy expansion. For simplicity we assume this to be already done and write the expansion of F in terms rather of P_y than of ξ (with a small value of the Curie-Weiss constant). Then the free-energy expansion compatible with the D_{2h}^{16} symmetry and respecting the first-order nature of all phase transitions reads

$$F = a_1(\eta_1^2 + \eta_2^2) + a_2(\eta_1^2 + \eta_2^2)^2 + A_2\eta_1^2\eta_2^2 + a_3(\eta_1^2 + \eta_2^2)^3 + A_3\eta_1^2\eta_2^2(\eta_1^2 + \eta_2^2) + \\ + b_1P_y^2 + b_2P_y^4 + b_3P_y^6 + \gamma_1P_y^2(\eta_1^2 + \eta_2^2) + \dots \,. \tag{1}$$

We have omitted in (1) interactions of the order parameters with u_{ik} since the experimental data necessary to test the formulae (that look rather cumbersome) for elastic constants are scanty. Since all phase transitions in LAS are of first order we put

$$a_2 < 0\,, \qquad b_2 < 0\,, \qquad a_3 > 0\,, \qquad A_3 > 0\,, \qquad b_3 > 0\,. \tag{2}$$

As usually we assume that the coefficients a_1 and b_1 only are temperature dependent such as $a_1 = a_1^0(T - T_a)$, $b_1 = b_1^0(T - T_b)$. Proceeding in the usual way we can now determine from (1) the equilibrium values of the order parameters and the stability conditions in all phases;

in phase I: $P_y = \eta_i = 0$; $a_1 > 0$, $b_1 > 0$, $(i = 1, 2)$;

in phase II: $P_y = (3b_3)^{-1} [(b_2^2 - 3b_1b_3)^{1/2} - b_2]$; $\eta_1 = \eta_2 = 0$, (3)

$$b_1 \leqq \frac{b_2^2}{3b_3}, \qquad a_1 \geqq -\gamma_1(3b_3)^{-1} [-b_2 + (b_2^2 - 3b_1b_3)^{1/2}]; \tag{4a, b}$$

in phase III:

$$P_y = 0; \quad \eta_1^2 = \eta_2^2 \equiv \eta^2 = (6a_3^*)^{-1} \{-a_2^* + [(a_2^*)^2 - 12a_1a_3^*]^{1/2}\}, \tag{5}$$

where

$$a_2^* = 4a_2 + A_2, \qquad a_3^* = 4a_3 + A_3,$$

$$a_1 \leqq \frac{(a_2^*)^2}{12a_3^*}, \tag{6a}$$

$$a_2 + (6a_3 + A_3)\,\eta^2 \geqq 0, \tag{6b}$$

$$a_2^* + 6a_3^*\eta^2 \geqq 0, \tag{6c}$$

$$A_2 + 2A_3\eta^2 \leqq 0, \tag{6d}$$

$$b_1 + 2\gamma_1\eta^2 \geqq 0; \tag{6e}$$

in phase V: $(\eta_1^2 = \eta_2^2 \equiv \eta^2 \neq 0$; $P_y \neq 0)$

$$P_y^2 = -\gamma_1^{-1}(a_1 + a_2^*\eta^2 + 3a_3^*\eta^4), \tag{7a}$$

$$3^3(a_3^*)^2\, b_3\eta^8 + 2 \cdot 3^2 a_2^* a_3^* b_3\eta^6 + 3[6a_1a_3^*b_3 + (a_2^*)^2 b_3 - 2a_3^*b_2\gamma_1]\,\eta^4 + \\ + 2(\gamma_1^3 - a_2^*b_2\gamma_1 + 3a_1a_2^*b_3)\,\eta^2 + (3b_3a_1^2 - 2b_2\gamma_1a_1 + \gamma_1^2b_1) = 0, \tag{7b}$$

$$a_2 + (6a_3 + A_3)\,\eta^2 \geqq 0, \tag{8a}$$

$$A_2 + 2A_3\eta^2 \leqq 0, \tag{8b}$$

$$a_2^* + 6a_3^*\eta^2 \geqq 0, \tag{8c}$$

$$(b_2 + 3b_3P_y^2)\,(a_2^* + 6a_3^*\eta^2) - \gamma_1^2 \geqq 0. \tag{8d}$$

From (6d) and (8b) it follows that A_2 must be negative. (For $A_2 > 0$ two other phases are stable, namely $Pna2_1$–$C_{2v}^9(2v)$ with $\eta_1 \neq 0$, $\eta_2 = P_y = 0$ and $Pb11$–$C_s^2(2v)$ with η_1, $P_y \neq 0$, $\eta_2 = 0$.)

We shall now study the phase diagram corresponding to the free energy (2) with $\eta_1^2 = \eta_2^2 = \eta^2$, i.e.

$$F = 2a_1\eta^2 + a_2^*\eta^4 + 2a_3^*\eta^6 + b_1P_y^2 + b_2P_y^4 + b_3P_y^6 + 2\gamma\eta^2P_y^2. \tag{9}$$

It can be shown that the form of the phase diagram depends substantially on the sign of γ_1 [13]. If $\gamma_1 < 0$ phases III ($\eta_1 = \eta_2 = \eta \neq 0$, $P_y = 0$) and II ($P_y \neq 0$, $\eta_1 = \eta_2 = 0$) can never coexist which is obviously not the case of LAS. Therefore, we have to put $\gamma_1 > 0$. Having assumed that the parameters a_1 and b_1 only are (linearly) temperature and pressure dependent, we plotted our phase diagram in the parameter space on the (a_1, b_1) plane of constant γ_1, $a_2 < 0$, $A_2 < 0$, $b_2 < 0$, $a_3 > 0$, $A_3 > 0$, $b_3 > 0$. Then the phase diagram has the form shown on Fig. 1. Phase I is stable for $a_1 > 0$, $b_1 > 0$. The stability region of phase III is determined by a straight line given by the equality (6a) and by a branch of the parabola (6e). The line of first-order phase transitions between phases I and III is a straight line

$$a_1 = (a_2^*)^2\,(16a_3^*)^{-1}. \tag{10}$$

The stability region of phase II is determined by the conditions (4). The line of first-order phase transitions between phases I and II is a straight line

$$b_1 = b_2^2(4b_3)^{-1} \tag{11}$$

and that one between II and III (the curve B_1B_2 on Fig. 1) is given by

$$b_3^2\{-a_2^* + [(a_2^*)^2 - 12a_1a_3^*]^{1/2}\}^2 \{a_2^* + 2[(a_2^*)^2 - 12a_1a_3^*]^{1/2}\} - -(2a_3^*)^2 [-b_2 + (b_2^2 - 3b_1b_3)^{1/2}]^2 [b_2 + 2(b_2^2 - 3b_1b_3)^{1/2}] = 0\,. \tag{12}$$

The equations of the lines of stability and of first-order transitions for phase V following from (8) are not given here, since we shall not need them in further considerations.

Let us now compare the qualitative features of our phase diagram (Fig. 1) with the phase diagram compiled from available experimental data (Fig. 2). First we have to orientate the temperature (T) and pressure (p) axes in our phase diagram. To do that most simply we assume that the coefficients a_1 and b_1 depend linearly both on T and p. Then we can utilize the experimental fact that the phase transition line III–V is

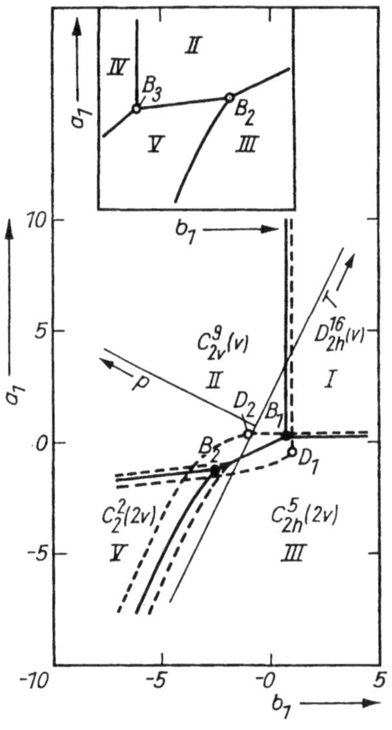

Fig. 1

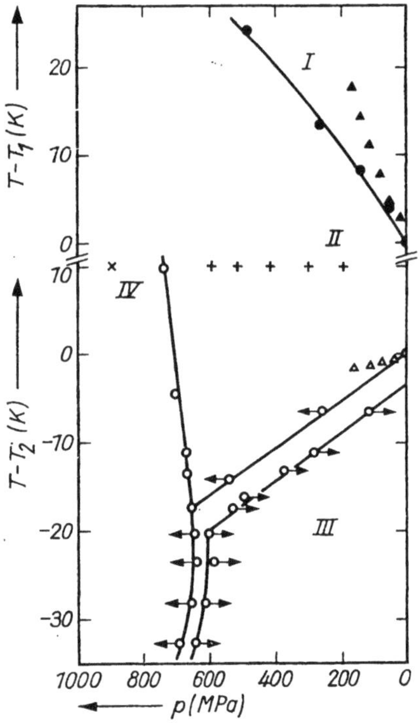

Fig. 2

Fig. 1. The b_1–a_1 phase diagram. Full lines first-order transition lines; dashed lines stability limits. The inset (scale much enlarged) shows the possible form of the b_1–a_1 phase diagram near the B_2 triple point for a model with three order parameters (see Section 4); the symmetry of phase IV is $C_2^2(v)$

Fig. 2. Experimental p–T phase diagram of LAS compiled from [4, 5, 7, 14, 22]. ● and ○ SHG and ^{7}Li NMR data, respectively; ▲ and △ dielectric measurements; × symmetry C_2 and $C_2^2(v)$ determined in [5] and [14], respectively; + behaviour of Raman lines confirms phase II up to 600 MPa [20] almost parallel to the T-axis and hence the T-axis in our phase diagram should be quasiparallel to this phase transition line (see Fig. 1). Furthermore we put the p-axis at such a position that with decreasing temperature at atmospheric pressure we get the observed sequence of phases, i.e. I–II–III. We observe from Fig. 2 that the region of existence of phase II increases with pressure. From extrapolation of the phase transition lines I–II and II–III it follows that phase III could coexist with I only at sufficiently high negative pressure. Both these features are well reproduced by our phase diagram. In agreement with the real phase diagram at high pressures our phase diagram provides the sequence of phases II–V–III with decreasing temperature. The hysteresis observed at the I–II phase transition is smaller than that at II–III and III–V phase transitions. Our choice of free-energy coefficients (see Section 3) makes it possible to reproduce this feature of the phase diagram, too. In general it can be shown that hysteresis regions are asymmetric with respect to the phase transition lines which should be seen on the temperature course of macroscopic quantities near phase transitions.

It should be emphasized that in our model with $\gamma_1 > 0$ the II–V phase transition line can never cross the b_1-axis and therefore in contrast to experimental data (see Fig. 2) it can never be parallel to the T-axis. This means that our II–V phase transition line cannot be identified with the II–IV transition line of Fig. 2. In other words, our phase V ($2v$) cannot be identified with phase IV which at high pressures coexists with phase II. According to NMR data [14] phase IV is of C_2^2 symmetry but with no unit-cell volume doubling, i.e. $C_2^2(v)$. These facts lead us to the conclusion that the phase diagram is not complete and that in fact there should exist two different C_2^2 phases, i.e. with and without unit-cell volume doubling, respectively. Thus in order to describe the phase transition to phase IV–$C_2^2(v)$, still another order parameter should be incorporated in our model and this is postponed to Section 4.

3. Macroscopic Properties at Phase Transitions in LAS

We shall now present the formulae for spontaneous polarization and dielectric susceptibility of LAS which can be derived by a usual thermodynamic procedure.

In phase I: $P_y = 0$;

$$\chi_{yy} = \frac{1}{2b_1} \,.$$

In phase II:

$$\chi_{yy} = 3b_3\{8(b_2^2 - 3b_1b_3)^{1/2}\,[-b_2 + (b_2^2 - 3b_1b_3)]^{1/2}\}^{-1} \,. \tag{13}$$

The spontaneous value of P_y is given by (3).

In phase III:

$$P_y = 0 \,;$$

$$\chi_{yy} = \left\{2b_1 + \frac{2\gamma_1}{3a_3^*}\,[-a_2^* + ((a_2^*)^2 - 12a_1a_3^*)^{1/2}]\right\}^{-1} . \tag{14}$$

The spontaneous value of η is given by (5).

In order to get the temperature course of χ_{yy} from our formulae (13) and (14) we have to choose somehow the coefficients in our model free energy (1). This can be done in the following way: let us take two special points of the phase diagram the coordinates of which can be easily expressed analytically. These points are, for example, the intersections of the stability lines of phases II and III with coordinates

$$D_1\colon \left\{b_1^{(1)} = \frac{b_2^2}{3b_3}\,; \quad a_1^{(1)} = \frac{\gamma_1 b_2}{3b_3}\right\}$$

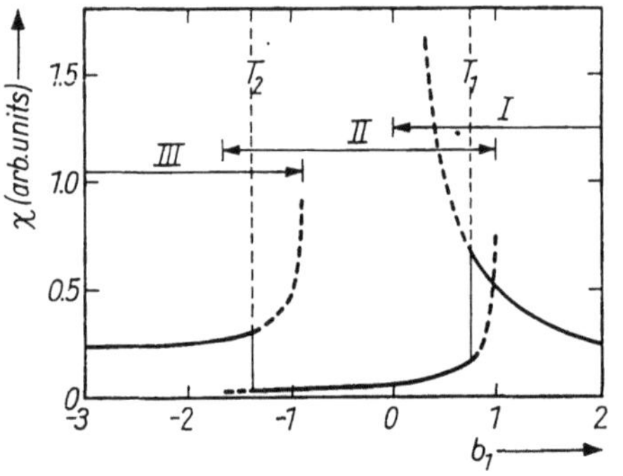

Fig. 3. χ vs. b_1. Arrows indicate stability regions of phases I, II, and III

and

$$D_2: \left\{ b_1^{(2)} = \frac{\gamma_1 a_2^*}{3a_3^*}; \quad a_1^{(2)} = \frac{(a_3^*)^2}{12a_3^*} \right\},$$

respectively (see the points D_1 and D_2 on Fig. 1). Now we put

$$b_1^{(1)} = 1, \quad a_1^{(1)} = -\tfrac{1}{2}, \quad b_1^{(2)} = -\tfrac{1}{4}, \quad a_1^{(2)} = \tfrac{1}{4} \tag{15}$$

which corresponds to a particular scaling of b_1 and a_1-axes. (Note that the scaling selected in this way is different for positive and negative directions of these axes.) The scaling (15) provides four equations for five yet unknown coefficients; we determine them under the arbitrarily chosen condition $b_3 = 1$.

With this choice of coefficients, (13) and (14) take the form

$$(\chi_{yy}^{-1})_{\text{II}} = 8[1 - b_1 + (1 - b_1)^{1/2}], \tag{16}$$

$$(\chi_{yy}^{-1})_{\text{III}} = 2b_1 + 2[1 + (1 - 4a_1)^{1/2}]. \tag{17}$$

The corresponding plot is given on Fig. 3. In phase I near T_1 the susceptibility χ_{yy} exhibits a Curie-Weiss behaviour and at T_1 decreases by a step. Closely above T_2 χ_{yy} is practically temperature independent because the stability line corresponding to the order parameter P_y is distant. According to our model there should be a step-like increase of χ_{yy} at T_2. The theoretical temperature course of χ_{yy} describes well experimental data by Mitsui et al. [11] and Yuzvak et al. [15]. On the other hand, in [16] a Curie-Weiss behaviour has been observed in phase II near T_2 which never will come out with our model. According to [16] $\chi_{yy}(T_2)$ is even higher than $\chi_{yy}(T_1)$; this feature could be possibly explained within our model when setting appropriately the T-axis in the phase diagram essentially due to the fact that negative b_1 may reduce significantly the denominator of (14).

4. The High-Pressure Phase IV

In Section 2 we have concluded that the origin of phase IV–$C_2^2(v)$ is not related to the order parameter η_i. It is easy to show that the symmetry Pnma(v) is reduced to P12$_1$1(v)–$C_2^2(v)$ provided both P_y and a new order parameter u of xz symmetry are frozen in in the lattice simultaneously. Since u is bilinearly coupled to the strain component u_{xz}, the phase transition could be classified as pseudoproper ferroelastic and we can expand the free energy right in u_{xz}. Thus we have to add to the model free energy (1) terms up to the fourth order in u_{xz} at least and the coupling of u_{xz} to P_y and η_i, respectively,

$$c_{55}^0 u_{xz}^2 + B u_{xz}^4 + \delta P_y^2 u_{xz}^2 + \lambda(\eta_1^2 + \eta_2^2)\, u_{xz}^2. \tag{18}$$

Since the phase $C_2^2(v)$ occurs only at high pressures, obviously the system is destabilized by hydrostatic pressure with respect to the strain u_{xz}. We propose the following mechanism of the phase transition induced by p. From the phase diagram (see Fig. 2) it is clear that p acts on the system similarly as an electric field applied along the y-axis, i.e. it extends the region of the polar phase II at the expense of the nonpolar phases I and III, respectively. Therefore, one can expect that in phase II the absolute value of P_y increases under applied hydrostatic pressure p. This effect is clearly seen from (3): b_1 is negative and its absolute value increases with p (see Fig. 1). Now from (18) we find that the effective elastic constant c_{55} becomes in the phase II

$$c_{55} = 2c_{55}^0 + 2\delta P_y^2(p) \tag{19}$$

and may go to zero at sufficiently high p provided the coefficient δ is negative. Thus the phase transition to phase IV which takes place at high hydrostatic pressures is treated as a triggered phase transition [17].

Now if the additional terms (18) are taken into account in F the form of the phase diagram (Fig. 1) changes essentially near the triple point B_2. At least two new phase transition lines occur: (i) A line between phases II and IV which corresponds to that found experimentally (see Fig. 2). It is given by the equation

$$b_1 = -3(c_{55}^0)^2\, b_3\delta^{-2} + 2c_{55}\, |b_2\delta|^{-1} \tag{20}$$

and in agreement with experimental data [4] it is a line of second-order transitions. According to our model this line is parallel to the phase transition line I–II which, however, is not the case found experimentally (see Fig. 2). This discrepancy can be easily removed by taking into account some higher-order terms in F neglected so far.

(ii) A line of first-order phase transition between phases IV and V the equation of which is rather complicated. The existence of these two new lines dramatically changes the nearest region of the triple point B_2. Now two triple points may exist the relative position of which depends not only on the signs of free-energy coefficients but on their magnitudes, too. A possible form of this region of the phase diagram is shown in the inset of Fig. 1. It should be pointed out that the triple point B_3 may have the coordinate $b_1(B_3) > b_1(B_2)$ and in this case the phase transition line II–IV disappears. Since no reliable experimental data near the triple point B_2 are available we shall not further speculate on the form of the phase diagram of LAS which in the case of three order parameters coupled together by nonlinear interactions is a complicated problem on its own.

5. Infra-Red and Raman Spectra

We shall now briefly discuss which soft modes corresponding to order-parameter components P_y, η_1, η_2 could be seen in infra-red and Raman spectra, respectively. The soft mode corresponding to P_y is in phase I of $B_{2u}(y)$ symmetry which is infra-red active. In phase II this mode becomes totally symmetric — $A_1(y; x^2, y^2, z^2)$ — and hence it becomes both infra-red and Raman active. In phase III it is infra-red active only and has the symmetry $A_u(y)$. In this way we can explain the reduction of the Rayleigh wing in the A-spectrum at the II–III phase transition that was observed in Raman spectra of LAS [18] and also the strong low-frequency absorption that occurs in infra-red spectra of LAS [19]. The silent doubly degenerate mode corresponding to η_1, η_2 splits in phase III into Raman active mode of $A_g(x^2, y^2, z^2, xz)$ symmetry and infra-red active mode $B_u(x, z)$. Indeed, a soft mode of A_g symmetry has been observed in the Raman spectra of phase III in the region of 20 cm^{-1} [18]. The quality of infra-red spectra [19] measured so far in phase III do not make any interpretation possible.

6. Conclusion

We have developed a thermodynamic theory which describes the sequence of phase transitions observed in LAS at atmospheric pressure. The coefficients of the free energy can be chosen in such a way that the temperature dependence of permittivity can be well reproduced in the whole temperature range investigated so far. Comparing the theoretical phase diagram with the experimental one we have concluded that besides phase IV another high pressure phase (phase V) should exist in LAS. Phase V is of the same space-group symmetry as phase IV (C_2^2) but its unit-cell volume is doubled. More experimental data are needed for checking all results of our theory, i.e. temperature dependence of elastic constants and high pressure measurements.

Acknowledgement

We thank Dr. J. Holakovský for valuable comments on this paper.

References

[1] K. Itoh, H. Ishikura, and E. Nakamura, Acta cryst. **B37**, 664 (1981).
[2] W. A. Dollase, Acta cryst. **B25**, 2298 (1969).
[3] A. I. Kruglik, M. A. Simonov, and K. S. Aleksandrov, Kristallografiya **23**, 494 (1978).
[4] T. I. Chekmasova, I. S. Kabanov, and V. I. Yuzvak, phys. stat. sol. (a) **44**, K155 (1977).
[5] T. Nakamura, S. Kojima, M. Takashige, and T. Mitsui, Ferroelectrics **25**, 435 (1980).
[6] I. P. Aleksandrova, I. S. Kabanov, S. V. Melnikova, T. I. Chekmasova, and V. I. Yuzvak, Kristallografiya **22**, 321 (1977).
[7] H. Shimizu, A. Oguri, N. Yasuda, and S. Fujimoto, J. Phys. Soc. Japan **45**, 565 (1978); **43**, 349 (1977).
[8] I. P. Aleksandrova, D. Kh. Blat, V. I. Zinenko, and A. I. Kruglik, Kristallografiya **26**, 531 (1981).
[9] A. Sawada, Y. Makita, and Y. Takagi, J. Phys. Soc. Japan **41**, 174 (1976).
[10] V. Dvořák and R. Kind, phys. stat. sol. (b) **107**, K109 (1981).
[11] T. Mitsui, T. Oka, Y. Shiroishi, M. Takashige, K. Iio, and S. Sawada, J. Phys. Soc. Japan **39**, 845 (1975).
[12] V. Dvořák, Ferroelectrics **7**, 1 (1974).
[13] Yu. M. Gufan and E. S. Larin, Fiz. tverd. Tela **22**, 463 (1980).
[14] T. I. Chekmasova and I. P. Aleksandrova, phys. stat. sol. (a) **49**, K185 (1978).
[15] V. I. Yuzvak, L. I. Zherebtsova, and V. B. Shkuryaeva, Kristallografiya **19**, 773 (1974).
[16] U. Syamaprasad and C. P. G. Vallabhan, Phys. Letters **89A**, 37 (1982).
[17] J. Holakovský, phys. stat. sol. (b) **56**, 615 (1973).
[18] H. Poulet and J. P. Mathieu, Solid State Commun. **21**, 421 (1977).
[19] X. Gebraux, J. Mangin, A. Hadni, D. Perrin, and C. D. Tran, Ferroelectrics **40**, 53 (1982).
[20] M. Połomska, Ferroelectrics Letters **44**, 11 (1982).

(Received July 26, 1984)

phys. stat. sol. (b) **126**, 467 (1984)

Subject classification: 8; 10.1; 21

Department of Physics, Panjab University, Chandigarh[1])

Heat of Solution of Hydrogen in Aluminum

By

S. MAHAJAN and S. PRAKASH

The heat of solution of hydrogen in aluminum is investigated accounting for both, the lattice dilation and its relaxation. The lattice contribution is evaluated using a linearly screened local model potential for the ions. The non-linear screening in the density functional formalism along with the exchange and correlation corrections in the jellium model is used to evaluate the proton contribution. The dilation enhances the lattice contribution by about 12% and decreases the protonic contribution by about 5%. The relaxation energy is reduced by about 40% for the dilated lattice. Adding all the contributions, the heat of solution is found to be 20% larger for the dilated lattice than that for the undilated lattice. The calculated heat of solution is found in good agreement with the experimental value. Comparing the results with the calculations of Perrot and Rasolt, the spherical solid model correction is estimated and it is found that it further improves the results.

Die Lösungswärme von Wasserstoff in Aluminium wird sowohl bezüglich der Gitterdilatation als auch ihrer Relaxation untersucht. Der Gitterbeitrag wird unter Benutzung eines linear abgeschirmten lokalen Modellpotentials für die Ionen berechnet. Eine nichtlineare Abschirmung im Dichtefunktionalformalismus zusammen mit Austausch- und Korrelationskorrekturen im Jelliummodell wird zur Berechnung des Protonenbeitrags benutzt. Die Dilatation erhöht den Gitterbeitrag um etwa 12% und verringert den Protonenbeitrag um etwa 5%. Die Relaxationsenergie wird um etwa 40% für das dilatierte Gitter verringert. Durch Addition aller Beiträge wird gefunden, daß die Lösungswärme 20% größer für das dilatierte Gitter im Vergleich zum nichtdilatierten Gitter ist. Die berechnete Lösungswärme ist in guter Übereinstimmung mit den experimentellen Werten. Aus dem Vergleich der Ergebnisse mit den Rechnungen von Perrot und Rasolt, wird die Korrektur für ein Modell sphärischer Festkörper bestimmt, und es wird gefunden, daß es die Ergebnisse weiter verbessert.

1. Introduction

The heat of solution is a basic quantity in the study of metal–hydrogen characteristics. The first theoretical work on hydrogen in Cu was reported by Friedel [1]. The proton with one bound electron stabilized by Slater exchange interaction between bound and parallel-spin conduction electrons was screened in an electron gas by solving the single-particle Schrödinger equation self-consistently. The presence of the ions was accounted for. However, valence–valence exchange interaction and all the correlation corrections were ignored. Popovic et al. [2] developed the full theory of the heat of solution of hydrogen in Al and Mg. The nonlinear self-consistent screening of the proton in jellium is carried out and the host ions are represented by a local pseudopotential. Manninen and Nieminen [3] improved the above investigation replacing the jellium by a spherical solid model proposed by Almbladh and von Barth [4]. In this model the spherical solid potential around the impurity is defined as the spherical average of the difference between the ion pseudopotential and the electrostatic potential of the positive background charge in jellium. This spherical solid potential

[1]) Chandigarh 160014, India.

is added to the effective potential while solving the Schrödinger equation self-consistently for the impurity screening. Kahn et al. [5] further improved these calculations adding non-spherical corrections in the spherical solid model.

In all the above calculations the lattice dilation and relaxation effects are ignored. Recently Perrot and Rasolt [6] incorporated the lattice relaxation effects in the study of the energetics of H in Al, however, the lattice dilation is ignored altogether. Here we present the investigations for the heat of solution of H in Al considering both the lattice dilation and relaxation effects. The jellium model which is quite appropriate [2, 7 to 11] for metals like Al is used for screening calculations. The plan of the paper is as follows: The theory of heat of solution is presented in Section 2. The calculations and results are given in Section 3, and they are discussed in Section 4.

2. Theory

2.1 Heat of solution in the linear response theory

The observed heat of solution of hydrogen in a metal, ΔH_H, is given as

$$\Delta H_\mathrm{H} = 15.86 + \Delta H_\mathrm{el\text{-}p} \quad (\mathrm{eV})\,, \tag{1}$$

where the first term is the sum of the hydrogen molecule dissociation energy and the hydrogen atom ionization energy. The second term is the heat of solution for a free electron and proton in the metal from vacuum. Following Popovic et al. [2] we write $\Delta H_\mathrm{el\text{-}p}$ as the sum of three contributions, i.e.

$$\Delta H_\mathrm{el\text{-}p} = \Delta(N Z_\mathrm{H} V_\mathrm{el}) + \Delta(N V_\mathrm{e}) + \Delta(N V_\mathrm{b})\,, \tag{2}$$

where the first term is the change in volume-dependent energy, the second term the change in electrostatic energy, and the third term the change in the band structure energy due to introduction of a proton and an electron. Here N is the number of atoms in the host matrix and Z_H is the host ionicity.

In the vacancy formation energy and volume calculations, Popovic et al. [12] accounted for both the lattice dilation and lattice relaxations in the linear screening theory. At the first step, the direct and reciprocal space quantities are modified in the general expression for the total energy of the crystal with a vacancy. The total energy is expanded in a power series of lattice dilation and it is minimized with respect to the lattice dilation parameter. This determines the lattice dilation and hence the formation volume. At the second step, the relaxation energy is expressed in terms of the force constants and the relaxation energy is minimized with respect to lattice displacements. The well-documented lattice static method is used to evaluate the relaxation energy. In the present problem the hydrogen is at an interstitial site and is being screened non-linearly. Therefore, the lattice static method is found inapplicable [6]. In view of these difficulties to make the calculations tractable we presume that the lattice dilation is known. Therefore, we modify only the direct space quantities and expand the different contributions in a power series of displacement. The quantities in the reciprocal space are evaluated for the dilated lattice.

If we assume that the lattice dilates uniformly by δ due to the introduction of hydrogen, then the interelectronic distance, atomic volume, and Fermi momentum change as

$$r_\mathrm{s}' \to \frac{r_\mathrm{s}}{1-\delta}\,, \quad \Omega_0' \to \frac{\Omega_0}{1-3\delta}\,, \quad k_\mathrm{F}' \to k_\mathrm{F}(1+\alpha-\delta)\,, \tag{3}$$

where

$$\alpha = \frac{1}{3NZ_H}. \tag{4}$$

Consequently the first term of (2) as defined by Popovic et al. [2] modifies as

$$\Delta(NZ_H V_{el}) = V_{el}^0(k_F) + \frac{k_F}{3} \frac{\partial}{\partial k_F} V_{el}(k_F) - \frac{6\delta\pi Z_H}{\Omega_0} R_m^2 \left(1 - \frac{2D}{3}\right) - $$

$$- NZ_H \delta k_F \frac{\partial}{\partial k_F} V_{el}(k_F), \tag{5}$$

where $V_{el}^0(k_F)$ is the volume-dependent energy per electron for the perfect lattice and R_m and D are the Heine-Abarenkov model potential parameters for the host ions.

The second term of (2) consists of two parts:

(i) the change in electrostatic energy of the perfect crystal lattice $\Delta(NV_e^0)$ due to dilation and

(ii) the electrostatic energy of the proton in the dilated lattice $\Delta(V_e^p)$. To evaluate the first contribution we start with the electrostatic energy of the perfect crystal [2]. The structure factor is expanded in a power series of displacement and the terms up to second order are retained. The evaluation of the second contribution is started from the expression given by Popovic et al. [2]. The second-order perturbation expansion of the structure factor is used in each term. All the terms are added and simplified. The final result obtained is

$$\Delta(NV_e) = \Delta(NV_e^0) + \Delta(V_e^p), \tag{6}$$

where

$$\Delta(NV_e^0) = \lim_{\xi' \to \infty} \frac{1}{N} \sum_{q \neq 0} \sum_i \sum_j \left[i\boldsymbol{q} \cdot \xi_{ij} - \frac{1}{2} (\boldsymbol{q} \cdot \xi_{ij})^2 \right] e^{i\boldsymbol{q} \cdot \boldsymbol{R}_{ij}} F_e(q, \xi'), \tag{7}$$

$$\Delta(V_e^p) = \frac{\alpha_H Z_H^{2/3}}{r_s} - Z_p Z_H \sum_j \frac{\boldsymbol{R}_j \cdot \xi_j}{R_j^3} + Z_p Z_H \sum_j \frac{1}{R_j} \left[\frac{3}{2} \left(\frac{\boldsymbol{R}_j \cdot \xi_j}{R_j^2} \right)^2 - \frac{1}{2} \left(\frac{\xi_j}{R_j} \right)^2 \right], \tag{8}$$

$$F_e(q, \xi') = \frac{2\pi Z_H^2}{\Omega_0 q^2} \exp(-q^2/4\xi'), \tag{9}$$

and

$$\xi_j = \boldsymbol{R}_j' - \boldsymbol{R}_j, \qquad \boldsymbol{R}_{ij} = \boldsymbol{R}_i - \boldsymbol{R}_j, \qquad \xi_{ij} - \xi_i - \xi_j. \tag{10}$$

Here $\boldsymbol{R}_j$ and $\boldsymbol{R}_j'$ are the undisplaced and displaced positions of the host ions, α_H is the Ewald constant which depends upon the position of the proton in the crystal structure. Z_p is the protonic charge and r_s the interelectronic distance. $\boldsymbol{R}_j$ in (8) is measured from the proton position. Wherever $\boldsymbol{q}$ is changed to reciprocal lattice vector, it is calculated for the dilated lattice. In the integration over q where q is continuous from 0 to ∞, the dilation does not alter the results.

Evidently if $\xi_{ij} = 0$, $\Delta(NV_e^0) = 0$. The first term in (7) is not symmetric in $\boldsymbol{R}_i$ and $\boldsymbol{R}_j$, therefore its contribution vanishes when summing over $\boldsymbol{R}_i$ and $\boldsymbol{R}_j$. Only the second term contributes. The first term in (8) is the proton lattice interaction energy for the perfect lattice. The last two terms are the first- and second-order corrections to the electrostatic interaction between proton and lattice due to atomic displacements. These two terms of (8) and $\Delta(NV_e^0)$ given by (7) were absent in the calculations of Popovic et al. [2].

To evaluate the last term of (2) we start with the form factor and structure factor for the host atoms and the impurity and obtain the change in the band structure energy for the constant crystal volume and undisplaced host ions, as done by Popovic et al. [2]. Then the corrections for the displacement of individual ions and the dilation of the crystal are used in the same manner as discussed earlier. It is assumed that the proton and the host ions are screened linearly. It leads to lengthy, but straightforward calculations. The various terms are simplified and collected together. The final result obtained is

$$\begin{aligned}
\Delta(NV_b) = {} & N \sum_{n\neq 0} [|\omega^0(\boldsymbol{k}_n)|^2 g^1(\boldsymbol{k}_n) + 2\omega^0(\boldsymbol{k}_n)\,\omega^1(\boldsymbol{k}_n)\,\{g^0(\boldsymbol{k}_n) + g^1(\boldsymbol{k}_n)\} + \\
& + |\omega^1(\boldsymbol{k}_n)|^2 g^0(\boldsymbol{k}_n)] + \\
& + \frac{1}{N} \sum_{q\neq 0} \sum_i \sum_j \left[i\boldsymbol{q}\cdot\xi_{ij} - \frac{(\boldsymbol{q}\cdot\xi_i)^2}{2} - \frac{(\boldsymbol{q}\cdot\xi_j)^2}{2} \right] e^{i\boldsymbol{q}\cdot\boldsymbol{R}_{ij}} \times \\
& \times [|\omega^0(\boldsymbol{q})|^2 \{g^0(\boldsymbol{q}) + g^1(\boldsymbol{q})\} + 2\omega^0(\boldsymbol{q})\,\omega^1(\boldsymbol{q})\,g^0(\boldsymbol{q})] + \\
& + \frac{1}{N} \sum_{q\neq 0} \sum_i \sum_j (\boldsymbol{q}\cdot\xi_i)(\boldsymbol{q}\cdot\xi_j)\, e^{i\boldsymbol{q}\cdot\boldsymbol{R}_{ij}} |\omega^0(\boldsymbol{q})|^2 g^0(\boldsymbol{q}) + \\
& + \frac{1}{N} \sum_{q\neq 0} \{|\omega_H^0(\boldsymbol{q})|^2 (g^0(\boldsymbol{q}) + g^1(\boldsymbol{q})) + |\omega_H^1(\boldsymbol{q})|^2 g^0(\boldsymbol{q}) + \\
& + 2\omega_H^0(\boldsymbol{q})\,\omega_H^1(\boldsymbol{q})\,\big(g^0(\boldsymbol{q}) + g^1(\boldsymbol{q})\big)\} + \\
& + 2 \sum_{n\neq 0} [\cos(\boldsymbol{k}_n\cdot\boldsymbol{\varrho}_H)\,\{\omega^0(\boldsymbol{k}_n)\,\omega_H^0(\boldsymbol{k}_n)\,\big(g^0(\boldsymbol{k}_n) + g^1(\boldsymbol{k}_n)\big) + \\
& + \big(\omega^0(\boldsymbol{k}_n)\,\omega_H^1(\boldsymbol{k}_n) + \omega^1(\boldsymbol{k}_n)\,\omega_H^0(\boldsymbol{k}_n)\big)\big(g^0(\boldsymbol{k}_n) + g^1(\boldsymbol{k}_n)\big) + \\
& + \omega^1(\boldsymbol{k}_n)\,\omega_H^1(\boldsymbol{k}_n)\,g^0(\boldsymbol{k}_n)\}] + \\
& + \frac{2}{N} \sum_{q\neq 0} \sum_j \left(i\boldsymbol{q}\cdot\xi_j - \frac{(\boldsymbol{q}\cdot\xi_j)^2}{2} \right) e^{i\boldsymbol{q}\cdot(\boldsymbol{R}_j - \boldsymbol{\varrho}_H)} \times \\
& \times [\omega^0(\boldsymbol{q})\,\omega_H^0(\boldsymbol{q})\,\big(g^0(\boldsymbol{q}) + g^1(\boldsymbol{q})\big) + \\
& + \big(\omega^0(\boldsymbol{q})\,\omega_H^1(\boldsymbol{q}) + \omega^1(\boldsymbol{q})\,\omega_H^0(\boldsymbol{q})\big)\,g^0(\boldsymbol{q})] \,, \qquad (11)
\end{aligned}$$

where

$$\omega^1(\boldsymbol{q}) = -3\delta\omega^0(\boldsymbol{q})\,, \qquad \omega_H^1(\boldsymbol{q}) = -3\delta\omega_H^0(\boldsymbol{q})\,,$$

$$g^1(\boldsymbol{q}) = g_d^1(\boldsymbol{q}) + g_e^1(\boldsymbol{q})\,, \qquad g_d^1(\boldsymbol{q}) = 3\delta g^0(\boldsymbol{q}) - \delta k_F \frac{\partial g}{\partial k_F}, \qquad g_e^1(\boldsymbol{q}) = \alpha k_F \frac{\partial g}{\partial k_F},$$

$$\omega^0(q) = -\frac{4\pi Z_H}{\Omega_0 q^2}\left(\frac{D \sin qR_m}{qR_m} + (1 - D)\cos qR_m\right), \qquad \omega_H^0(q) = -\frac{4\pi Z_p}{\Omega_0 q^2},$$

$$g^0(\boldsymbol{q}) = -\frac{1}{2}\,\frac{\Omega_0 q^2}{4\pi}\,\frac{1 - \varepsilon^{-1}(\boldsymbol{q})}{1 - f(\boldsymbol{q})}. \qquad (12)$$

Here $\boldsymbol{k}_n$ is the reciprocal lattice vector and $\boldsymbol{\varrho}_\mathrm{H}$ the proton position coordinate. The quantities $\omega^0(\boldsymbol{q})$ and $\omega^0_\mathrm{H}(\boldsymbol{q})$ are the form factors for host and proton and $g^0(\boldsymbol{q})$ is the screening function for the perfect lattice. $\omega^1(\boldsymbol{q})$, $\omega^1_\mathrm{H}(\boldsymbol{q})$, and $g^1(\boldsymbol{q})$ are the corresponding corrections which arise due to lattice dilation. $\varepsilon(\boldsymbol{q})$ is the modified Hartree dielectric function and $f(\boldsymbol{q})$ accounts for the exchange correlations as suggested by Singwi et al. [13].

Equations (5), (6), and (11) consist of the contribution to the heat of solution from the linearly screened matrix and proton. It is adequate to treat the matrix contribution in the linear response theory, however, the nonlinear screening is to be accounted for the proton contribution. Moreover, some of the terms involve dilation, some involve the displaced position of the host atoms, and others are independent of dilation and displacement. Therefore, to simplify the calculations we rearrange the terms of (2) and rewrite them as

$$\Delta H_\mathrm{el\text{-}p} = [\Delta H^0_\mathrm{d\text{-}i} + \Delta H^\mathrm{p}_\mathrm{d\text{-}i}] + [\Delta H^0_\mathrm{d\text{-}d} + \Delta H^\mathrm{p}_\mathrm{d\text{-}d}] + [\Delta H^0_\mathrm{re} + \Delta H^\mathrm{p}_\mathrm{re}]\,, \tag{13}$$

where

$$\Delta H^0_\mathrm{d\text{-}i} = V^0_\mathrm{el}(k_\mathrm{F}) + \frac{k_\mathrm{F}}{3}\frac{\partial}{\partial k_\mathrm{F}} V_\mathrm{el}(k_\mathrm{F}) + \frac{\alpha_\mathrm{H} Z_\mathrm{H}^{2/3}}{r_\mathrm{s}} + \frac{k_\mathrm{F}}{3Z_\mathrm{H}} \sum_{n\neq 0} |\omega^0(\boldsymbol{k}_n)|^2 \frac{\partial g(\boldsymbol{k}_n)}{\partial k_\mathrm{F}}\,, \tag{14}$$

$$\Delta H^\mathrm{p}_\mathrm{d\text{-}i} = \sum_{n\neq 0} 2\cos(\boldsymbol{k}_\mathrm{n}\cdot\boldsymbol{\varrho}_\mathrm{H})\, \omega^0(\boldsymbol{k}_\mathrm{n})\, \omega^0_\mathrm{H}(\boldsymbol{k}_\mathrm{n})\, g^0(\boldsymbol{k}_\mathrm{n}) + \frac{1}{N}\sum_{q\neq 0} |\omega^0_\mathrm{H}(\boldsymbol{q})|^2\, g^0(\boldsymbol{q})\,, \tag{15}$$

$$\begin{aligned}\Delta H^0_\mathrm{d\text{-}d} = {} & -\frac{6\delta\pi Z_\mathrm{H} R_\mathrm{m}^2}{\Omega_0}\left(1 - \frac{2D}{3}\right) - N Z_\mathrm{H}\delta k_\mathrm{F}\frac{\partial}{\partial k_\mathrm{F}} V_\mathrm{el}(k_\mathrm{F}) - \\ & - 3\delta N \sum_{n\neq 0} |\omega^0(\boldsymbol{k}_\mathrm{n})|^2\, g^0(\boldsymbol{k}_\mathrm{n}) - \delta N k_\mathrm{F} \sum_{n\neq 0} |\omega^0(\boldsymbol{k}_\mathrm{n})|^2 \frac{\partial g(\boldsymbol{k}_\mathrm{n})}{\partial k_\mathrm{F}}\,,\end{aligned} \tag{16}$$

$$\begin{aligned}\Delta H^\mathrm{p}_\mathrm{d\text{-}d} = {} & \sum_{n\neq 0} 2\cos(\boldsymbol{k}_\mathrm{n}\cdot\boldsymbol{\varrho}_\mathrm{H})\, [\omega^0(\boldsymbol{k}_\mathrm{n})\, \omega^0_\mathrm{H}(\boldsymbol{k}_\mathrm{n})\, g^1_\mathrm{d}(\boldsymbol{k}_\mathrm{n}) + \\ & + \big(\omega^0(\boldsymbol{k}_\mathrm{n})\, \omega^1_\mathrm{H}(\boldsymbol{k}) + \omega^1(\boldsymbol{k}_\mathrm{n})\, \omega^0_\mathrm{H}(\boldsymbol{k}_\mathrm{n})\big)\, g^0\,(\boldsymbol{k}_\mathrm{n})] + \\ & + \frac{1}{N}\sum_{q\neq 0} [|\omega^0_\mathrm{H}(\boldsymbol{q})|^2\, g^1_\mathrm{d}(\boldsymbol{q}) + 2\omega^0_\mathrm{H}(\boldsymbol{q})\, \omega^1_\mathrm{H}(\boldsymbol{q})\, g^0\,(\boldsymbol{q})]\,,\end{aligned} \tag{17}$$

$$\begin{aligned}\Delta H^0_\mathrm{re} = {} & -Z_\mathrm{p} Z_\mathrm{H} \sum_j \frac{\boldsymbol{R}_j\cdot\xi_j}{R_j^3} + Z_\mathrm{p} Z_\mathrm{H} \sum_j \frac{1}{R_j}\left[\frac{3}{2}\frac{\boldsymbol{R}_j\cdot\xi_j}{R_j^2} - \frac{1}{2}\left(\frac{\xi_j}{R_j}\right)^2\right] + \\ & + \lim_{\xi'\to\infty} \frac{1}{N}\sum_{q\neq 0}\sum_i\sum_j \left[i\boldsymbol{q}\cdot\xi_{jj} - \frac{(\boldsymbol{q}\cdot\xi_{ij})^2}{2}\right] \mathrm{e}^{i\boldsymbol{q}\cdot\boldsymbol{R}_{ij}}\, F_\mathrm{e}(q, \xi') + \\ & + \frac{1}{N}\sum_{q\neq 0}\sum_i\sum_j \left[i\boldsymbol{q}\cdot\xi_{ij} - \frac{(\boldsymbol{q}\cdot\xi_i)^2}{2} - \frac{(\boldsymbol{q}\cdot\xi_j)^2}{2}\right] \mathrm{e}^{i\boldsymbol{q}\cdot\boldsymbol{R}_{ij}} \times \\ & \times [|\omega^0(\boldsymbol{q})|^2 \big(g^0(\boldsymbol{q}) + g^1(\boldsymbol{q})\big) + 2\omega^0(\boldsymbol{q})\, \omega^1(\boldsymbol{q})\, g^0(\boldsymbol{q})] + \\ & + \frac{1}{N}\sum_{q\neq 0}\sum_i\sum_j (\boldsymbol{q}\cdot\xi_i)\,(\boldsymbol{q}\cdot\xi_j)\, \mathrm{e}^{i\boldsymbol{q}\cdot\boldsymbol{R}_{ij}}\, |\omega^0(\boldsymbol{q})|^2\, g^0(\boldsymbol{q})\,,\end{aligned} \tag{18}$$

$$\begin{aligned}\Delta H^\mathrm{p}_\mathrm{re} = {} & \frac{2}{N}\sum_{q\neq 0}\sum_j \left(i\boldsymbol{q}\cdot\xi_j - \frac{(\boldsymbol{q}\cdot\xi_j)^2}{2}\right) \mathrm{e}^{i\boldsymbol{q}\cdot(\boldsymbol{R}_j - \boldsymbol{\varrho}_\mathrm{H})}\, [\omega^0(\boldsymbol{q})\, \omega^0_\mathrm{H}(\boldsymbol{q})\, \big(g^0(\boldsymbol{q}) + \\ & + g^1_\mathrm{d}(\boldsymbol{q})\big) + \big(\omega^0(\boldsymbol{q})\, \omega^1_\mathrm{H}(\boldsymbol{q}) + \omega^1(\boldsymbol{q})\, \omega^0_\mathrm{H}(\boldsymbol{q})\big)\, g^0(\boldsymbol{q})]\,.\end{aligned} \tag{19}$$

Here the subscripts d-i, d-d, and re represent the dilation-independent, dilation-dependent, and atomic-displacement-dependent contributions. The terms with subscript re consist of both dilation-dependent and dilation-independent terms which will be discussed later. The superscripts (0) and (p) denote the proton-independent and proton-dependent contributions. In the following we (i) account for the nonlinear screening of the proton and calculate the proton-dependent contribution and (ii) evaluate the atomic-displacement-dependent contribution explicitly.

2.2 Nonlinear screening of the proton

We rewrite (15), (17), and (19) separating the terms which represent the charge density of the electron screening cloud around the proton. This gives

$$\Delta H^{\rm p}_{\rm d\text{-}i} = \sum_{q \neq 0} S^0(\boldsymbol{q})\,\omega^0(\boldsymbol{q})\,n^0_{\rm I}(\boldsymbol{q}) + \frac{1}{2N}\sum_{q \neq 0} \exp\,(i\boldsymbol{q}\cdot\boldsymbol{\varrho}_{\rm H})\,\omega^0_{\rm H}(\boldsymbol{q})\,n^0_{\rm I}(\boldsymbol{q})\,, \tag{20}$$

$$\begin{aligned}\Delta H^{\rm p}_{\rm d\text{-}d} = {} & \sum_{q \neq 0} S^0(\boldsymbol{q})\,\omega^0(\boldsymbol{q})\,n^1_{\rm I}(\boldsymbol{q}) - 3\delta \sum_{q \neq 0} S^0(\boldsymbol{q})\,\omega^0(\boldsymbol{q})\,n^0_{\rm I}(\boldsymbol{q}) + \\ & + \frac{1}{2N}\sum_{q \neq 0} \exp\,(i\boldsymbol{q}\cdot\boldsymbol{\varrho}_{\rm H})\,\omega^0_{\rm H}(\boldsymbol{q})\,n^1_{\rm I}(\boldsymbol{q}) - \\ & - 3\delta\,\frac{1}{2N}\sum_{q \neq 0} \exp\,(i\boldsymbol{q}\cdot\boldsymbol{\varrho}_{\rm H})\,\omega^0_{\rm H}(\boldsymbol{q})\,n^0_{\rm I}(\boldsymbol{q})\,,\end{aligned} \tag{21}$$

$$\Delta H^{\rm p}_{\rm re} = \frac{1}{N}\sum_{q \neq 0}\sum_j \left(i\boldsymbol{q}\cdot\boldsymbol{\xi}_j - \frac{(\boldsymbol{q}\cdot\boldsymbol{\xi}_j)^2}{2}\right) e^{i\boldsymbol{q}\cdot\boldsymbol{R}_j}\,\omega^0(\boldsymbol{q})\,[n^0_{\rm I}(\boldsymbol{q})\,(1 - 3\delta) + n^1_{\rm I}(\boldsymbol{q})]\,, \tag{22}$$

where

$$n^0_{\rm I}(\boldsymbol{q}) = 2g^0(\boldsymbol{q})\,\omega^0_{\rm H}(\boldsymbol{q})\exp\,(-i\boldsymbol{q}\cdot\boldsymbol{\varrho}_{\rm H})\,, \tag{23}$$

and

$$n^1_{\rm I}(\boldsymbol{q}) = 2\big(g^1_{\rm d}(\boldsymbol{q})\,\omega^0_{\rm H}(\boldsymbol{q}) + g^0(\boldsymbol{q})\,\omega^1_{\rm H}(\boldsymbol{q})\big)\exp\,(-i\boldsymbol{q}\cdot\boldsymbol{\varrho}_{\rm H})\,. \tag{24}$$

Here $S^0(\boldsymbol{q}) = \delta_{q,k_n}$ is the structure factor for the perfect lattice. The $n^0_{\rm I}(\boldsymbol{q})$ is the displaced electron density around the proton in the undilated lattice and $n^1_{\rm I}(\boldsymbol{q})$ is the correction to it due to lattice dilation.

The first term in (20) is the interaction energy between the electron screening cloud around the proton and the ions in the undilated lattice. The second term is the electron–proton correlation energy. In (21), the first two terms are the corrections to the proton screening cloud–ion interaction and the last two terms are the corrections to electron–proton correlation energy which arise due to lattice dilation. Equation (22) is an additional correction to (20) and it arises due to the displacement of the nearest-neighbour host atoms from their equilibrium position in the perfect lattice. Evidently, if $\xi_i = 0$, the contribution of (22) vanishes. The $\Delta H^{\rm p}_{\rm d\text{-}d}$ and $\Delta H^{\rm p}_{\rm re}$ were absent in the calculations of Popovic et al. [2]. In (20) to (22) $n^0_{\rm I}(\boldsymbol{q})$ and $n^1_{\rm I}(\boldsymbol{q})$ are calculated in the linear screening theory. To account for nonlinear screening of the proton we replace $n^0_{\rm I}(\boldsymbol{q})$ and $n^1_{\rm I}(\boldsymbol{q})$ by the corresponding quantities $n^0(\boldsymbol{q})$ and $n^1(\boldsymbol{q})$ calculated in the nonlinear screening theory and rewrite (20) to (22) as [2]

$$\Delta H^{\rm p}_{\rm d\text{-}i} = \sum_{q \neq 0} S^0(\boldsymbol{q})\,\omega^0(\boldsymbol{q})\,n^0(\boldsymbol{q}) + E_{\rm corr}[n^0(\boldsymbol{q})]\,, \tag{25}$$

$$\Delta H^{\rm p}_{\rm d\text{-}d} = \sum_{q \neq 0} S^0(\boldsymbol{q})\,\omega^0(\boldsymbol{q})\,[n^1(\boldsymbol{q}) - 3\delta n^0(\boldsymbol{q})] + E_{\rm corr}[n^1(\boldsymbol{q})] - 3\delta E_{\rm corr}[n^0(\boldsymbol{q})]\,, \tag{26}$$

$$\Delta H^{\rm p}_{\rm re} = \frac{1}{N}\sum_{q \neq 0}\sum_j \left(i\boldsymbol{q}\cdot\boldsymbol{\xi}_j - \frac{(\boldsymbol{q}\cdot\boldsymbol{\xi}_j)^2}{2}\right) e^{i\boldsymbol{q}\cdot\boldsymbol{R}_j}\,\omega^0(\boldsymbol{q})\,[n^0(\boldsymbol{q})\,(1 - 3\delta) + n^1(\boldsymbol{q})]\,, \quad ($$

where

$$n^0(\boldsymbol{q}) = \int_0^\infty \mathrm{d}r\; 4\pi r^2\, \Delta n^0(r) \frac{\sin qr}{qr}\,, \tag{28}$$

$$n^1(\boldsymbol{q}) = \int_0^\infty \mathrm{d}r\; 4\pi r^2\, \Delta n^1(r) \frac{\sin qr}{qr}\,. \tag{29}$$

Here $\Delta n^0(r)$ is the displaced charge density around the proton calculated self-consistently in nonlinear screening theory for the interelectronic distance of the undilated lattice. $\Delta n^1(r)$ is the correction to $\Delta n^0(r)$ for the change in interelectronic distance due to lattice dilation. The corresponding electron–proton correlation energies $E_{\mathrm{corr}}[n^0(\boldsymbol{q})]$ and $E_{\mathrm{corr}}[n^1(\boldsymbol{q})]$ are estimated using Feynman's theorem [14] and are given as

$$E_{\mathrm{corr}}[n^0(\boldsymbol{q})] = \int_0^1 \frac{\mathrm{d}Z_\mathrm{p}}{Z_\mathrm{p}} \left(-\int \mathrm{d}\boldsymbol{r} \frac{Z_\mathrm{p}}{r}\, \Delta n^0(Z_\mathrm{p}, \boldsymbol{r})\right), \tag{30}$$

and

$$E_{\mathrm{corr}}[n^1(\boldsymbol{q})] = \int_0^1 \frac{\mathrm{d}Z_\mathrm{p}}{Z_\mathrm{p}} \left(-\int \mathrm{d}\boldsymbol{r} \frac{Z_\mathrm{p}}{r}\, \Delta n^1(Z_\mathrm{p}, \boldsymbol{r})\right), \tag{31}$$

where $\Delta n(Z_\mathrm{p}, r)$ is the displaced charge density for the protonic charge Z_p and the integral $[-\int \mathrm{d}\boldsymbol{r}\, (Z_\mathrm{p}/r)\, \Delta n(Z_\mathrm{p}, \boldsymbol{r})]$ is the corresponding electron–proton interaction energy (V_{int}).

2.3 Relaxation energy

The evaluation of the contributions given by (14), (16), (25), and (26) is straightforward. However, the contributions given by (18) and (27) which involve the displacement of each individual atom need to be evaluated for the displacements which minimize ΔH_{re} $(= \Delta H^0_{\mathrm{re}} + \Delta H^\mathrm{p}_{\mathrm{re}})$.

We assume an isotropic displacement of atoms and write

$$\xi_j = \lambda_j \boldsymbol{R}_j\,, \tag{32}$$

where λ_j is the same for all the ions in a nearest-neighbour shell around an impurity. The relaxation energies given by (18) and (27) are simplified using (32). Equations (18) and (27) are added and rearranged as

$$\Delta H_{\mathrm{re}} = \Delta H^{\mathrm{d\text{-}i}}_{\mathrm{re}} + \Delta H^{\mathrm{d\text{-}d}}_{\mathrm{re}}\,, \tag{33}$$

where

$$\begin{aligned}\Delta H^{\mathrm{d\text{-}i}}_{\mathrm{re}} = Z_\mathrm{p} Z_\mathrm{H} \sum_j \frac{1}{R_j} [-\lambda_j + \lambda_j^2] + \\ + \lim_{\xi' \to \infty} \frac{1}{N} \sum_{q \neq 0} \sum_i \sum_j \left[i\boldsymbol{q} \cdot \xi_{ij} - \frac{(\boldsymbol{q} \cdot \xi_{ij})^2}{2}\right] \mathrm{e}^{i\boldsymbol{q}\cdot\boldsymbol{R}_{ij}}\, F(q, \xi') + \\ + \frac{1}{N} \sum_{q \neq 0} \sum_j \left[i\boldsymbol{q} \cdot \xi_j - \frac{(\boldsymbol{q} \cdot \xi_j)^2}{2}\right] \mathrm{e}^{i\boldsymbol{q}\cdot\boldsymbol{R}_j}\, \omega^0(\boldsymbol{q})\, n^0(\boldsymbol{q})\,,\end{aligned} \tag{34}$$

$$\Delta H_{\text{re}}^{\text{d-d}} = -\frac{\delta}{N} \sum_{q \neq 0} \sum_i \sum_j \left[i\boldsymbol{q} \cdot \xi_{ij} - \frac{(\boldsymbol{q} \cdot \xi_i)^2}{2} - \frac{(\boldsymbol{q} \cdot \xi_j)^2}{2} \right] e^{i\boldsymbol{q}\cdot \boldsymbol{R}_{ij}} \times$$

$$\times \left[3\, |\omega^0(\boldsymbol{q})|^2\, g^0(\boldsymbol{q}) + k_{\text{F}}\, |\omega^0(\boldsymbol{q})|^2 \frac{\partial g}{\partial k_{\text{F}}} \right] +$$

$$+ \frac{1}{N} \sum_{q \neq 0} \sum_j \left[i\boldsymbol{q} \cdot \xi_j - \frac{(\boldsymbol{q} \cdot \xi_j)^2}{2} \right] e^{i\boldsymbol{q}\cdot \boldsymbol{R}_j}\, \omega^0(\boldsymbol{q})\, n^1(\boldsymbol{q}) -$$

$$- \frac{3\delta}{N} \sum_{q \neq 0} \sum_j \left[i\boldsymbol{q} \cdot \xi_j - \frac{(\boldsymbol{q} \cdot \xi_j)^2}{2} \right] e^{i\boldsymbol{q}\cdot \boldsymbol{R}_j}\, \omega^0(\boldsymbol{q})\, n^0(\boldsymbol{q})\,, \tag{35}$$

and

$$F(q, \xi') = F_{\text{e}}(q, \xi') + |\omega^0(q)|^2\, g^0(q)\,.$$

Here $\Delta H_{\text{re}}^{\text{d-i}}$ is the relaxation energy at constant crystal volume and $\Delta H_{\text{re}}^{\text{d-d}}$ is the correction to $\Delta H_{\text{re}}^{\text{d-i}}$ due to lattice dilation. In the second term in (35) δ is involved in $n^1(q)$. If the lattice relaxes under constant crystal volume, $\Delta H_{\text{re}}^{\text{d-d}}$ vanishes.

In (34) the first term is the change in electrostatic interaction between the proton and the lattice due to atomic displacements. The second term arises due to the change in band structure and the electrostatic contribution to the binding energy of the metal without the impurity. The last term is the change in band structure energy due to the presence of the impurity. The further changes in the band structure energy due to lattice dilation are given in (35) where the first term is the change in band structure energy of the perfect crystal. The presence of an impurity is accounted for in the second and third terms. The change in the electrostatic energy due to lattice dilation is incorporated by calculating the electrostatic energy for the dilated lattice.

We first simplify $\Delta H_{\text{re}}^{\text{d-i}}$. Using the spherical harmonic expansion of $\exp(i\boldsymbol{q} \cdot \boldsymbol{R}_{ij})$ and the fact that the terms linear in ξ_{ij} will not contribute, the second term of (34) simplifies as

$$\text{II} = -\frac{1}{2} \lim_{\xi' \to \infty} \frac{1}{N} \sum_{q \neq 0} \sum_{ij} q^2 \xi_{ij}^2 \cos^2\theta \sum_{lm} 4\pi i^l\, \text{j}_l(qR_{ij}) \times$$

$$\times\, \text{Y}_{lm}(\theta, \varphi)\, \text{Y}_{lm}(\theta_{ij}, \varphi_{ij})\, F(q, \xi')\,. \tag{36}$$

In writing (36), ξ_{ij} is taken along the Z-axis, (θ, φ) and $(\theta_{ij}, \varphi_{ij})$ are the polar angles for the vectors $\boldsymbol{q}$ and $\boldsymbol{R}_{ij}$, respectively. j_l is the spherical Bessel function of order l and $\text{Y}_{lm}(\theta, \varphi)$ are the spherical harmonics, l and m are the angular and magnetic quantum numbers, respectively.

The sum over q in (36) is replaced by integration and the orthonormality conditions for spherical harmonics are used. This gives

$$\text{II} = -\frac{\Omega_0}{8\pi^2} \lim_{\xi' \to \infty} \sum_i \sum_j \xi_{ij}^2 \int_0^\infty q^4\, \text{d}q \left\{ \frac{2}{3}\, \text{j}_0(qR_{ij}) - \frac{4}{3}\, \text{j}_2(qR_{ij})\, \text{P}_2(\cos\theta_{ij}) \right\} F(q, \xi')\,, \tag{37}$$

where P_2 is the Legendre polynomial of order two. Inclusion of the $q = 0$ term in (37) does not alter (36) as $F(q) = 0$ at $q = 0$. It can readily be shown [6] that for a cubic lattice,

$$S = \sum_i \sum_j \xi_{ij}^2 \, \mathrm{P}_2(\cos \theta_{ij}) \, F(q) =$$

$$= \sum_i \sum_j \lambda_i \lambda_j \left[-\frac{3(\boldsymbol{R}_i \cdot \boldsymbol{R}_{ij}) \, (\boldsymbol{R}_j \cdot \boldsymbol{R}_{ij})}{R_{ij}^2} + (\boldsymbol{R}_i \cdot \boldsymbol{R}_j) \right] F(q) \,, \tag{38}$$

for a given function $F(q)$. Use of (38) and $F(q, \xi')$ in (37) and the simplification of the integral over q in the $\xi' \to \infty$ limit gives

$$\mathrm{II} = -\frac{1}{2} \sum_{ij}{}' \left[\lambda_i^2 R_i^2 - \lambda_i \lambda_j (\boldsymbol{R}_i \cdot \boldsymbol{R}_j)\right] I(R_{ij}) +$$

$$+ \frac{1}{2} \sum_{ij}{}' \lambda_i \lambda_j \left[-\frac{3(\boldsymbol{R}_i \cdot \boldsymbol{R}_{ij}) \, (\boldsymbol{R}_j \cdot \boldsymbol{R}_{ij})}{R_{ij}^2} + (\boldsymbol{R}_i \cdot \boldsymbol{R}_j) \right] J(R_{ij}) \,, \tag{39}$$

where

$$I(R_{ij}) = \frac{\Omega_0}{3\pi^2} \int_0^\infty \mathrm{d}q \, q^4 \mathrm{j}_0(qR_{ij}) \, |\omega^0(q)|^2 \, g^0(q) \,, \tag{40}$$

$$J(R_{ij}) = \frac{\Omega_0}{3\pi^2} \int_0^\infty \mathrm{d}q \, q^4 \mathrm{j}_2(qR_{ij}) \, |\omega^0(q)|^2 \, g^0(q) + \frac{Z_{\mathrm{H}}^2}{R_{ij}^3} . \tag{41}$$

Similarly the third term of (34) is simplified, which yields

$$\mathrm{III} = \sum_j \lambda_j K(R_j) - \tfrac{1}{2} \sum_j \lambda_j^2 L(R_j) \,, \tag{42}$$

where

$$K(R_j) = \frac{\Omega_0}{2\pi^2} \int_0^\infty \left[q^2 \cos qR_j - \frac{q \sin qR_j}{R_j} \right] \omega^0(q) \, n^0(q) \, \mathrm{d}q \,, \tag{43}$$

and

$$L(R_j) = \frac{\Omega_0}{2\pi^2} \int_0^\infty \left[q^3 R_j \sin qR_j + 2q^2 \cos qR_j - \frac{2q \sin qR_j}{R_j} \right] \omega^0(q) \, n^0(q) \, \mathrm{d}q \,. \tag{44}$$

The simplification of (35) requires the explicit expression for δ. It can easily be shown that δ is inversely proportional to N for a point defect [12]. Therefore, we write $\delta = C/N$, where C is a constant which is to be determined with the help of experimental data. Using this value of δ, we estimate the contributions of all the three terms on the right-hand side of (35). In spite of the first and third terms being additive, their contribution was found negligible in comparison to the second-term contribution. Therefore, only the second term of (35) is retained. To simplify the calculations, this term is directly added to the third term of (34) and the integrals are re-evaluated. The results remain the same as given by (42), (43), and (44) except that in (43) and (44) $n_0(q)$ is replaced by $[n_0(q) + n^1(q)]$. With this modification we redefine the integrals given by (43) and (44) as $K'(R_j)$ and $L'(R_j)$, respectively. Adding all the contributions we get the total relaxation energy, ΔH_{re}, as

$$\Delta H_{\rm re} = Z_{\rm p} Z_{\rm H} \sum_j \frac{1}{R_j} [-\lambda_j + \lambda_j^2] - \frac{1}{2} \sum_{ij}' [\lambda_i^2 R_i^2 - \lambda_i \lambda_j (\boldsymbol{R}_i \cdot \boldsymbol{R}_j)] \, I(R_{ij}) + $$
$$+ \frac{1}{2} \sum_{ij}' \lambda_i \lambda_j \left[- \frac{3(\boldsymbol{R}_i \cdot \boldsymbol{R}_{ij})\,(\boldsymbol{R}_j \cdot \boldsymbol{R}_{ij})}{R_{ij}^2} + (\boldsymbol{R}_i \cdot \boldsymbol{R}_j) \right] J(R_{ij}) +$$
$$+ \sum_j \lambda_j K'(R_j) - \frac{1}{2} \sum_j \lambda_j^2 L'(R_j) \,. \tag{45}$$

To minimize $\Delta H_{\rm re}$, (45) is expressed as

$$\Delta H_{\rm re} = \sum_a \lambda_a F_a + \tfrac{1}{2} \sum_a \sum_b \lambda_a \lambda_b \Phi_{ab} \,, \tag{46}$$

where the indices a and b pertain to a particular star of vectors $\boldsymbol{R}_j$ which are stable under the cubic group. The terms F_a and Φ_{ab} involve one-body and two-body interactions, respectively. For sufficiently weak impurity and lattice host potentials (46) reduces to the lattice static method [15]. Differentiating $\Delta H_{\rm re}$ with respect to λ's, we get the set of linear equations

$$\sum_b \Phi_{ab} \lambda_b + F_a = 0 \,, \tag{47}$$

where the dimension of a ($= 1, 2, 3 \ldots$) will depend upon the number of stars taken into account. With the λ's solutions of (47), $\Delta H_{\rm re}$ becomes

$$\Delta H = \tfrac{1}{2} \sum_a \lambda_a F_a \,. \tag{48}$$

The computational procedures for different contributions are described in the following section.

3. Calculations and Results

To estimate the heat of solution defined by (1) all the terms in (13) are calculated separately. The Heine-Abarenkov local model potential parameters $R_{\rm m}$ and D are chosen such that the observed equilibrium lattice parameter and the binding energy per atom for the perfect crystal in second-order perturbation theory are reproduced [12]. The calculated parameters $R_{\rm m}$ and D, the binding energy per atom along with other physical constants for Al are listed in Table 1. The Madelung constant $\alpha_{\rm H}$ is calculated using Ewald's method keeping the proton at an octahedral site in the f.c.c.

Table 1

The lattice parameter (a), binding energy per atom ($V_{\rm T}/N$), and the model potential parameters $R_{\rm m}$ and D for Al.
$r_{\rm s}$ and $k_{\rm F}$ are corresponding interelectronic distance and Fermi momentum

parameters	value (at. units)
a	7.62004
$r_{\rm s}$	2.064
$k_{\rm F}$	0.9295
$V_{\rm T}/N$	−2.082
$R_{\rm m}$	1.38198
D	0.3969

lattice of Al [2, 6, 16]. Using these parameters, the dilation- and proton- independent contribution $\Delta H^0_{d\text{-}i}$ given by (14) is calculated.

The experimental value of dilation for Al–H solid solution is not known, therefore δ is estimated from the experimental values of $\delta\Omega/\Omega_0$ for H in Cu, Ni, and Pd [17, 18]. This value of $\delta\Omega/\Omega_0$ was earlier used to study the electronic structure [19] and the activation energy of H in Al [20]. Substituting this estimated value of δ ($= 0.13/3N$) in (16), $\Delta H^0_{d\text{-}d}$ is calculated explicitly. It is found that $\Delta H^0_{d\text{-}d}$ is additive to $\Delta H^0_{d\text{-}i}$ and it is about 12% of $\Delta H^0_{d\text{-}i}$.

The proton-dependent quantities $\Delta H^p_{d\text{-}i}$ and $\Delta H^p_{d\text{-}d}$ given by (25) and (26) involve explicit calculations for the nonlinear screening of the proton. This has been done [19, 20] by the well-documented Hohenberg-Kohn-Sham [21, 22] density functional formalism. The two-parameter host–impurity trial potential $V_{\text{trial}} = -(Z_p/r)\exp(-\alpha\gamma^\beta)$ is used. The change in electronic charge distribution around the proton, $\Delta n^0(r)$, is calculated for the mean density appropriate to Al metal ($r_s = 2.064$ at. units). The $n^0(q)$ and the correlation energy $E_{\text{corr}}[n^0(q)]$ are calculated with the help of (28) and (30), respectively. The convergence of integrals is assured. To evaluate $n^1(q)$ (29), $\Delta n^1(r)$ is calculated by taking the difference of screening charge densities for $r_s = 2.15$ at. units (dilated) and $r_s = 2.0640$ at. units (undilated). The corresponding electron–proton correlation energy $E_{\text{corr}}[n^1(q)]$ is estimated through (31). The reciprocal lattice vectors for the dilated lattice are generated using $a = 7.93688$ at. units which is calculated from the estimated value of $\delta\Omega/\Omega_0$ for H in Al.

The electron–proton interaction energy defined as $(-\int d\boldsymbol{r}(Z_p/r)\,\Delta n(Z_p, \boldsymbol{r})$ for different proton charges for the dilated and undilated crystal are presented in Table 2. The corresponding electron–proton correlation energies are also mentioned there. The electron–proton interaction increases with the protonic charge. The dilation decreases the electron–proton interaction by about only 1%. The correlation energy $E_{\text{corr}}[n^1(q)]$ which is a correction to $E_{\text{corr}}[n^0(q)]$ due to lattice dilation is found to be 0.0026 at. units which is less than 1% of $E_{\text{corr}}[n^0(q)]$. This concludes that the electron–proton correlation energy remains almost unaltered by lattice dilation.

Using the above values of $n^0(q)$, $n^1(q)$, $E_{\text{corr}}[n^0(q)]$, and $E_{\text{corr}}[n^1(q)]$, the protonic contributions given by (25) and (26) are evaluated. The contribution of (25) is -11.41 eV, while that of (26) is 0.565 eV. Therefore, the dilation reduces the protonic contribution by about 5%.

We next turn towards the calculations of relaxation energy given by (45). First we calculate the dilation-independent relaxation energy given by (34). This is readily

Table 2

The electron–proton interaction energy ($V_{\text{int}} = -\int d\boldsymbol{r}(Z_p/r)\,\Delta n(Z_p, \boldsymbol{r})$) for protonic charges Z_p and the electron–proton correlation energy (E_{corr}) for the interelectronic distance $r_s = 2.064$ at. units (undilated) and $r_s = 2.15$ at. units (dilated) appropriate to Al matrix

Z_p	V_{int}/Z_p (at. units)	
	undilated lattice	dilated lattice
0.25	−0.221	−0.218
0.50	−0.486	−0.483
0.75	−0.805	−0.802
1.00	−1.181	−1.179
E_{corr} (at. units)	−0.521	−0.519

Table 3

Lattice relaxation effects for H in Al. The two series of results refer to H at the octahedral site with and without lattice dilation. In each case two sets of calculations are done, one with three and another with five nearest-neighbour shells

shell		100	111	210	221	300	ΔH_{re} (eV)
without dilation	λ_i	0.0176	−0.0033	0.0039	—	—	−0.044
($\Delta H_{re} = \Delta H_{re}^{d\text{-}i}$)	λ_i	0.0171	−0.0010	0.0032	0.0014	0.0037	−0.043
with dilation	λ_i	0.0140	−0.0027	0.0033	—	—	−0.028
($\Delta H_{re} = \Delta H_{re}^{d\text{-}i} + \Delta H_{re}^{d\text{-}d}$)	λ_i	0.0134	−0.0015	0.0025	0.0004	0.0022	−0.026

done putting $n^1(q) = 0$ in (45). The λ_j defined in (32) are obtained minimizing ΔH_{re} with respect to λ's and solving the set of linear equations given by (47). The proton is kept again at the octahedral site and the nearest-neighbour shell vectors are measured with respect to the proton. The two sets of calculations one with three and the other with five nn shells are carried out and these results for relaxation energies and λ's are presented in Table 3. The results for ΔH_{re} converge. The maximum displacement is for the first nn shell and it is of the order of 2%. Perrot and Rasolt [6] have estimated a 2.4% dilation of the first nn shell in the spherical solid model. This is due to the fact that a part of the crystal potential in the spherical solid model is added to the effective impurity potential as discussed in Section 1, while this is ignored in the jellium model. For the other shells the displacement decreases by an order of magnitude. The trend is similar to that found by Perrot and Rasolt [6]. However, the displacements for these shells (other than the nn shell) in the jellium model are found larger than those in the spherical solid model. The signs of the nn displacements indicate that the second shell is pushed towards the impurity, while other nn shells are pushed away from the impurity. The relaxation energy is −0.043 eV in the five-shell calculations. Perrot and Rasolt [6] found this value −0.060 eV. The difference may be attributed to the difference in the two calculations. The Heine-Abarenkov model potential parameters in the two calculations are also different.

Incorporating $n^1(q)$ explicitly, the relaxation energy for the dilated lattice is calculated adopting the above-discussed procedure. The corresponding results are also presented in Table 3. The systematics of λ's remain unaltered, however, the displacements for the different nn shells have decreased for the dilated lattice. This is an expected result because a part of displacement is already incorporated in the dilation. The relaxation energy is reduced significantly, i.e. about 40% less than that for the undilated lattice.

Perrot and Rasolt [6] pointed out that the zero-point energy ($\approx (\frac{3}{2})\ \hbar\omega$) where ω is the zero-point frequency ,may also be added in the above contributions to get the total heat of solution. In our calculations we found $\hbar\omega = 0.06$ eV which gives $\Delta H^{zp} = (\frac{3}{2})\ \hbar\omega \approx 0.10$ eV. Almost the same contribution is found for the dilated and the undilated lattice.

The different contributions to the heat of solution for dilated and undilated lattices are tabulated in Table 4. An intercomparison shows that the major contribution toward $\Delta H_{el\text{-}p}$ arises from $(\Delta H^0_{d\text{-}i} + \Delta H^0_{d\text{-}d})$ and $(\Delta H^p_{d\text{-}i} + \Delta H^p_{d\text{-}d})$. The contribution of relaxation energy is relatively small. Therefore, a few simplifications which are introduced in the calculations of relaxation energy may not affect our results significantly.

Table 4
The contributions towards the heat of solution ΔH_H (= ΔH_{el-p} + 15.86 + ΔH^{zp} eV) of hydrogen in Al and the experimental value

contributions	undilated lattice (eV)	dilated lattice (eV)
lattice	− 4.0072 ($\Delta H^{o}_{d\text{-}i}$)	− 4.4924 ($\Delta H^{o}_{d\text{-}i} + \Delta H^{o}_{d\text{-}d}$)
protonic	−11.4056 ($\Delta H^{p}_{d\text{-}i}$)	−10.8413 ($\Delta H^{p}_{d\text{-}i} + \Delta H^{p}_{d\text{-}d}$)
relaxation	− 0.043 ($(\Delta H)^{d\text{-}i}_{re}$)	− 0.026 ($\Delta H^{d\text{-}i}_{re} + \Delta H^{d\text{-}d}_{re}$)
ΔH_{el-p}	−15.4558	−15.3597
ΔH^{zp}	0.10	0.10
ΔH_H	0.504	0.601
experimental ΔH_H	= 0.66 eV	

The total heat of solution is obtained by adding all the contributions. It is found 0.504 eV for the undilated lattice and 0.601 eV for the dilated lattice. Therefore, in total the dilation has enhanced the heat of solution by about 20% which is quite significant. The magnitude of relaxation energy is only of the order of 5 to 8%. Therefore, it is emphasized that the dilation correction is more important than the relaxation. The calculated value of ΔH_H agrees fairly well with the experimental value of 0.66 eV [23].

4. Discussion

Perrot and Rasolt [6] used the spherical solid model in the nonlinear screening calculations of the proton. This partially accounts for the presence of the lattice. This part is not accounted for explicitly in our calculations. Therefore, a term-by-term comparison with their calculations is difficult. However, the relaxation of the first nn shell in our calculations for the undilated lattice is in agreement with the calculations of Perrot and Rasolt [6]. Further Perrot and Rasolt [6] found the sum of 15.86 + (lattice + + protonic + relaxation) contribution as 0.56 eV for R_m = 1.23 at. units and D = = 0.398. We repeated our calculations for the same potential parameters and found the above sum 0.49 eV. Therefore, the spherical solid model enhances the heat of solution by about 0.07 eV. If we add this correction to our calculated value 0.601 eV, the results further tend toward the experimental value.

In conclusion, the different contributions to the heat of solution of hydrogen in Al are compared. The lattice and protonic parts are dominating. The corrections due to lattice dilation, the effect of lattice potential on the impurity screening and zero-point vibration are significant. The relaxation energy contribution is minimum. However, addition of all the contributions results in close agreement with the experimental value.

Acknowledgements

The financial support from the Department of Atomic Energy, Bombay, and the University Grants Commission, New Delhi, is gratefully acknowledged.

References

[1] J. FRIEDEL, Phil. Mag. **43**, 153 (1952).
[2] Z. D. POPOVIC, M. J. STOTT, J. P. CARBOTTE, and G. R. PIERCY, Phys. Rev. B **13**, 590 (1976).
[3] M. MANNINEN and R. M. NIEMINEN, J. Phys. F **9**, 1333 (1979).
[4] C. O. ALMBLADH and U. VON BARTH, Phys. Rev. B **13**, 3307 (1976).
[5] L. M. KAHN, F. PERROT, and M. RASOLT, Phys. Rev. B **21**, 5594 (1980).
[6] F. PERROT and M. RASOLT, Phys. Rev. B **23**, 6534 (1981).
[7] C. O. ALMBLADH, U. VON BARTH, Z. D. POPOVIC, and M. J. STOTT, Phys. Rev. B **14**, 2250 (1976).
[8] M. MANNINEN, P. HAUTOJARVI, and R. M. NIEMINEN, Solid State Commun. **23**, 795 (1977).
[9] J.K. NØRSKOV, Solid State Commun. **24**, 691 (1977).
[10] E. ZAREMBA, L. M. SANDER, H. B. SHORE, and J. H. ROSE, J. Phys. F **7**, 1763 (1977).
[11] P. JENA and K. S. SINGWI, Phys. Rev. B **17**, 3518 (1978).
[12] Z. D. POPOVIC, J. P. CARBOTTE, and G. R. PIERCY, J. Phys. F **4**, 351 (1974).
[13] K. S. SINGWI, A. SJOLANDER, M. P. TOSI, and R. H. LAND, Phys. Rev. B **1**, 1044 (1970).
[14] C. H. HODGES and M. J. STOTT, Phys. Rev. B **7**, 73 (1973).
[15] H. KANZAKI, J. Phys. Chem. Solids **2**, 24 (1957).
[16] J. P. BUGEAT and E. LIGEON, Phys. Letters A **71**, 93 (1979).
[17] L. KATZ, M. GUINAN, and R. J. BORG, Phys. Rev. B **4**, 330 (1970).
[18] R. A. ORIANI, Data Collection in: Phase Stability in Metals and Alloys, Ed. P. S. RUDMAN, J. STRINGER, and R. J. JAFFE, Mc Graw-Hill Publ. Co., 1967.
[19] S. MAHAJAN and S. PRAKASH, Nuovo Cimento D **2**, 883 (1983).
[20] S. MAHAJAN, N. SINGH, and S. PRAKASH, J. Phys. F **13**, 1449 (1983).
[21] P. HOHENBERG and W. KOHN, Phys. Rev. **136**, B864 (1964).
[22] W. KOHN and L. J. SHAM, Phys. Rev. **140**, A1133 (1965).
[23] W. EICHENAUER, Z. Metallk. **59**, 613 (1968).

(Received February 21, 1984)

phys. stat. sol. (b) **126**, 481 (1984)

Subject classification: 19

Institute of Chemical Physics, Academy of Sciences of the USSR, Chernogolovka[1])

The Influence of Molecular Motions on Multipulse NMR Spectra in Solids

By

A. K. KHITRIN, V. V. LAIKO, and B. N. PROVOTOROV

A canonical transformation technique generalized for the case of a non-periodical time dependence of the Hamiltonian is used for the analysis of molecular motion influence on multipulse NMR high resolution spectra in solids. The linewidth is shown to be described in some cases by a temperature-dependent effective Hamiltonian.

С помощью обобщения метода канонических преобразований на случай непериодической зависимости гамильтониана от времени рассмотрено влияние молекулярной подвижности на многоимпульсные спектры ЯМР высокого разрешения в твёрдых телах. Показано, что в некоторых случаях ширина линии описывается зависящим от температуры эффективным гамильтонианом.

1. Introduction

The application of spectroscopy to the study of molecular motions is based on both relaxation time and linewidth measurements. NMR spin–lattice relaxation times are determined by the spectral densities of molecular motions [1],

$$T_1^{-1} \sim J^{(1)}(\omega_0) + J^{(2)}(2\omega_0)\,, \tag{1}$$

where ω_0 is the Larmor frequency. There is the same (resonant) contribution to linewidths, too. But in the slow motion region ($\omega_0\tau_c \gg 1$) the linewidth $\Delta\omega$ is contributed mainly by the spectral densities at frequencies of the order of $\Delta\omega$. Since usually $\Delta\omega \ll \omega_0$ it is more convenient to study the slow molecular motions by their influence on $\Delta\omega$. The narrower the NMR linewidth is, the slower motion can be detected. An application of rf pulse sequences can lead to a significant reduction of $\Delta\omega$ [2], so the problem of the influence of molecular motions on multipulse NMR spectra seems attractive.

When the sample is irradiated by a series of rf pulses, the dipole–dipole interaction modulated by the pulses consists of an infinite number of harmonics with frequencies $2\pi n/t_c$ (t_c is the pulse sequence period). To calculate the resonant contribution to the linewidth one has to replace (1) by the sum over all the harmonics. This contribution [3 to 5] vanishes when the correlation time τ_c of molecular motions becomes extremely long ($\tau_c \to \infty$) while the linewidth tends to a finite value. So the region of slow motions requires a more accurate description.

For the analysis of rigid-lattice multipulse NMR spectra it is convenient to use a time-independent effective (or average) Hamiltonian formalism [2, 6 to 8]. The results obtained by such a technique are present as a limiting case in the spin–lattice relaxation theory.

[1]) 142 432 Chernogolovka, USSR.

The influence of slow molecular motions on multipulse spectra can be described by means of an iterational solution of the equation for the density matrix with subsequent extraction of proper terms [9]. But this approach lacks the advantages of the effective Hamiltonian (EH) formalism, and the problem of secular terms [10] cannot be adequately treated in this way. When the lattice Hamiltonian is included into the density matrix equation the periodicity is restored, and it allows one to use the canonical transformation technique [7]. This method has been used in [11] where the result has been expressed in terms of the EH containing the lattice variables. The EH of such a form is somewhat formal and requires an extra analysis to obtain the linewidth.

It is quite natural to perform simultaneous averaging over all kinds of fast motions, i.e. the motions of spins subjected to the influence of both pulse sequence and molecular motions. The result can be expressed in some cases in terms of temperature-dependent EH. The aim of this communication is to develop such an averaging technique by the expansion of the canonical transformation formalism [6, 7] on non-periodic motions, i.e. on the continuous spectrum of oscillations.

2. General Approach

In the interaction representation the density matrix equation is ($\hbar = 1$)

$$\frac{\partial \varrho}{\partial t} = -i[\mathcal{H}_{\mathrm{d}}(t) + F, \varrho]\,, \tag{2}$$

where the dipole–dipole interaction modulated by pulses, $\mathcal{H}_{\mathrm{d}}(t)$, is a periodic function of time (period t_c), F is the lattice Hamiltonian and $\int_0^{t_c} \mathcal{H}_{\mathrm{d}}(t)\,\mathrm{d}t = 0$. The transformation into F interaction representation gives

$$\tilde{\mathcal{H}}_{\mathrm{d}}(t) = \exp(iFt)\,\mathcal{H}_{\mathrm{d}}(t)\exp(-iFt) = \int_{-\infty}^{\infty} \mathrm{d}\omega\,\mathcal{H}_\omega \exp(i\omega t)\,. \tag{3}$$

$\tilde{\mathcal{H}}_{\mathrm{d}}(t)$ contains rapidly and slowly oscillating parts

$$\begin{aligned} \mathcal{H}_{\mathrm{d}}^{\mathrm{f}} &= \fint_{-\infty}^{\infty} \mathrm{d}\omega\mathcal{H}_\omega \exp(i\omega t)\,, \\ \mathcal{H}_{\mathrm{d}}^{\mathrm{s}} &= \int_{-\delta}^{\delta} \mathrm{d}\omega\mathcal{H}_\omega \exp(i\omega t)\,, \end{aligned} \tag{4}$$

where the frequency band $-\delta < \omega < \delta$ is excluded from the first integral in (4) and $\delta \approx ||\mathcal{H}_{\mathrm{d}}||$. The equation for the density matrix takes the form

$$\frac{\partial \tilde{\varrho}}{\partial t} = -i[\mathcal{H}_{\mathrm{d}}^{\mathrm{f}} + \mathcal{H}_{\mathrm{d}}^{\mathrm{s}}, \tilde{\varrho}]\,. \tag{5}$$

To extract the slow part of the density matrix one can perform a canonical transformation which reduces the amplitudes of rapidly oscillating terms,

$$\begin{aligned} \varrho_{\mathrm{s}}^{(1)} &= \exp\{iu_1(t)\}\,\tilde{\varrho}\exp\{-iu_1(t)\}\,, \\ u_1(t) &= \int^{t} \mathcal{H}_{\mathrm{d}}^{\mathrm{f}}(t')\,\mathrm{d}t'\,. \end{aligned} \tag{6}$$

The equation for $\varrho_{\mathrm{s}}^{(1)}$ up to the first order of u_1 is

$$\frac{\partial \varrho_{\mathrm{s}}^{(1)}}{\partial t} = -i\left[\mathcal{H}_{\mathrm{d}}^{\mathrm{s}} + \frac{i}{2}\,[u_1, \mathcal{H}_{\mathrm{d}}^{\mathrm{f}} + \mathcal{H}_{\mathrm{d}}^{\mathrm{s}}]\,, \varrho_{\mathrm{s}}^{(1)}\right] =$$

$$= -i\left[\left[\int_{-\delta}^{\delta} \mathrm{d}\omega\mathcal{H}_\omega \exp(i\omega t) + \frac{i}{2}\left[\fint_{-\infty}^{\infty} \frac{\mathrm{d}\omega}{\omega}\,\mathcal{H}_\omega \exp(i\omega t)\,, \int_{-\infty}^{\infty} \mathrm{d}\omega'\mathcal{H}_{\omega'} \exp(i\omega' t)\right], \varrho_{\mathrm{s}}^{(1)}\right]\right]. \tag{7}$$

3. Lee-Goldburg Experiment

As an example we consider the Lee-Goldburg experiment [12] in which the resonance off-set is used to form in the rotating frame the effective field in the "magic" direction. The Hamiltonian in the interaction representation is

$$\mathcal{H}_{\mathrm{d}}(t) = \mathcal{H}^1 \exp(i\omega_{\mathrm{e}}t) + \mathcal{H}^2 \exp(2i\omega_{\mathrm{e}}t) + \text{c.c.}\,, \tag{8}$$

where

$$[S_{\mathrm{e}}, \mathcal{H}^m] = m\mathcal{H}^m\,, \tag{9}$$

S_{e} is the spin projection on the direction $\boldsymbol{n}_{\mathrm{e}}$ of the effective field $\boldsymbol{\omega}_{\mathrm{e}}$ ($S_{\mathrm{e}} = \boldsymbol{S}\cdot\boldsymbol{n}_{\mathrm{e}}$). The assumption is made that the molecular motions can be described by exponential correlation functions,

$$\mathrm{Sp}\,\{\exp(iFt)\,\mathcal{H}^m \exp(-iFt)\,\mathcal{H}^{-m}\} = \mathrm{Sp}\,\{\mathcal{H}^m\mathcal{H}^{-m}\}\exp\left(-\frac{|t|}{\tau_{\mathrm{c}}}\right). \tag{10}$$

The first term in (7), i.e. $\mathcal{H}_{\mathrm{d}}^{\mathrm{s}}$, stands for the resonant contribution to the linewidth. This contribution can be calculated in a standard way [1]. As an example, the decrement of the longitudinal magnetization damping is

$$\frac{1}{T_{2\mathrm{e}}^{\mathrm{s}}} = \frac{2}{t\,\mathrm{Sp}\,S_{\mathrm{e}}^2}\int_0^t \mathrm{d}t' \int_0^{t'} \mathrm{d}t'' \sum_{m=1,2} m^2 \times$$

$$\times\,\mathrm{Sp}\left\{\int_{-\delta}^{\delta} \mathrm{d}\omega\,\mathcal{H}_{\omega}^m \exp(i\omega t') \int_{-\delta}^{\delta} \mathrm{d}\omega'\,\mathcal{H}_{\omega'}^{-m}\exp(i\omega' t'')\right\} =$$

$$= \frac{2}{\mathrm{Sp}\,S_{\mathrm{e}}^2}\left\{\mathrm{Sp}\,\mathcal{H}^1\mathcal{H}^{-1}\,\frac{\tau_{\mathrm{c}}}{1+\omega_{\mathrm{e}}^2\tau_{\mathrm{c}}^2} + 4\,\mathrm{Sp}\,\mathcal{H}^2\mathcal{H}^{-2}\,\frac{\tau_{\mathrm{c}}}{1+4\omega_{\mathrm{e}}^2\tau_{\mathrm{c}}^2}\right\}. \tag{11}$$

The decrement of the transverse magnetization damping can be obtained in a similar way.

Now we consider the contribution to the linewidth of the second term in (7), which we denote as $\mathcal{H}_{\mathrm{eff}}(t)$. In the case of slow molecular motions ($\tau_{\mathrm{c}} \gg t_{\mathrm{c}} = 2\pi/\omega_{\mathrm{e}}$) only the frequencies near $\pm\,\omega_{\mathrm{e}}$ and $\pm\,2\omega_{\mathrm{e}}$ are important in the $\mathcal{H}_{\omega}$ spectrum. It allows to neglect the $1/\omega$ dependence in the integral. Thus we obtain

$$\mathcal{H}_{\mathrm{eff}}(t) = \frac{1}{\omega_{\mathrm{e}}}\exp(iFt)\,\{[\mathcal{H}^1, \mathcal{H}^{-1}] + \frac{1}{2}[\mathcal{H}^2, \mathcal{H}^{-2}]\}\exp(-iFt)\,. \tag{12}$$

For the rigid lattice ($\tau_{\mathrm{c}} \to \infty$) the dipole–dipole interaction commutes with F and (12) reduces to

$$\mathcal{H}_{\mathrm{eff}}^* = \frac{1}{\omega_{\mathrm{e}}}[\mathcal{H}^1, \mathcal{H}^{-1}] + \frac{1}{2\omega_{\mathrm{e}}}[\mathcal{H}^2, \mathcal{H}^{-2}]\,. \tag{13}$$

$\mathcal{H}_{\mathrm{eff}}^*$ describes the linewidth in the absence of molecular motions.

For rather fast motions ($\tau_{\mathrm{c}} \ll ||\mathcal{H}_{\mathrm{eff}}^*||^{-1}$) $\mathcal{H}_{\mathrm{eff}}(t)$ is to be averaged over the molecular motions. It corresponds to the calculation of the trace over the lattice variables. (The region $\tau_{\mathrm{c}} \ll ||\mathcal{H}_{\mathrm{eff}}^*||^{-1}$ overlaps the slow-motion region ($\tau_{\mathrm{c}} \gg t_{\mathrm{c}}$) discussed above). A similar averaging procedure has been performed in [13] where the EH for a rotating sample has been calculated.

Making use of (7), (8) one can express the result of averaging as

$$\langle \mathcal{H}_{\text{eff}} \rangle = i \sum_{n=1,2} \int_{-\infty}^{\infty} \frac{d\omega}{\omega + n\omega_e} \sum_{i,k,l} \{A_{ik}^n \langle b_{ik}^{\omega} b_{ik}^{-\omega} \rangle + B_{ikl}^n \langle b_{ik}^{\omega} b_{kl}^{-\omega} \rangle \}\,, \tag{14}$$

where b_{ik}^{ω} is the Fourier transform of the dipole–dipole interaction coefficient $b_{ik}(t)$; A_{ik}^n and B_{ikl}^n are the spin operators and the angular bracket indicates averaging over the lattice variables. If we assume that all the correlation functions are simple exponentials, i.e.

$$\langle b_{ik}(t)\, b_{kl}(t + \tau) \rangle = \langle b_{ik} \rangle \langle b_{kl} \rangle + (\langle b_{ik} b_{kl} \rangle - \langle b_{ik} \rangle \langle b_{kl} \rangle) \exp\left(-\frac{|\tau|}{\tau_c}\right), \tag{15}$$

where τ_c is the correlation time of the molecular motions, (14) can be expressed as

$$\begin{aligned} \langle \mathcal{H}_{\text{eff}} \rangle = {} & \frac{\omega_e \tau_c^2}{1 + \omega_e^2 \tau_c^2} (\langle [\mathcal{H}^1, \mathcal{H}^{-1}] \rangle - [\langle \mathcal{H}^1 \rangle, \langle \mathcal{H}^{-1} \rangle]) + \\ & + \frac{2\omega_e \tau_c^2}{1 + 4\omega_e^2 \tau_c^2} (\langle [\mathcal{H}^2, \mathcal{H}^{-2}] \rangle - [\langle \mathcal{H}^2 \rangle, \langle \mathcal{H}^{-2} \rangle]) + \\ & + \frac{1}{\omega_e} [\langle \mathcal{H}^1 \rangle, \langle \mathcal{H}^{-1} \rangle] + \frac{1}{2\omega_e} [\langle \mathcal{H}^2 \rangle, \langle \mathcal{H}^{-2} \rangle]\,. \end{aligned} \tag{16}$$

(We note here that for anisotropic motions $\mathcal{H}^m$ in (11) should be replaced by $\mathcal{H}^m - \langle \mathcal{H}^m \rangle$ as well.) So the evolution of the EH with the motions becoming faster can be schematically represented as $\mathcal{H}_{\text{eff}}^* \to \langle \mathcal{H}_{\text{eff}}^* \rangle \to \mathcal{H}_{\text{eff}}^*(\langle \mathcal{H}_d \rangle)$, where the second transition is described by (16).

4. Application to Multipulse Sequences

As the simplest example of multipulse sequence which brings to the residual interactions of the same order of magnitude we take the multipulse spin-locking sequence $90_y^0 - (\tau - \varphi_x - \tau)^n$ when the resonance off-set is used to form an effective field ω_e [6] in the "magic" direction. In this case

$$\begin{aligned} \mathcal{H}_d(t) = {} & \mathcal{H}_d^1 \sum_{n=-\infty}^{\infty} c_n \exp\left\{-i\left(\frac{n\pi}{\tau} - \omega_e\right)t\right\} + \\ & + \mathcal{H}_d^2 \sum_{n=-\infty}^{\infty} b_n \exp\left\{-i\left(\frac{n\pi}{\tau} - 2\omega_e\right)t\right\} + \text{c.c.}\,; \end{aligned} \tag{17}$$

$$c_n = \frac{(-1)^n \sin \omega_e \tau}{n\pi + \omega_e \tau}\,, \qquad b_n = \frac{(-1)^n \sin 2\omega_e \tau}{n\pi + 2\omega_e \tau}\,. \tag{18}$$

To obtain an expression similar to (11) it is necessary to summarize over the harmonics. The result is

$$\begin{aligned} \frac{1}{T_{2e}^s} = \frac{2}{9} M_2 \tau_c \Bigg\{ 2 - \alpha & \frac{\sin^2(\omega_e \tau) \sinh(2/\alpha)}{\cosh(2/\alpha) - \cos(2\omega_e \tau)} - \\ & - \alpha \frac{\sin^2(2\omega_e \tau) \sinh(2/\alpha)}{\cosh(2/\alpha) - \cos(4\omega_e \tau)} \Bigg\}\,; \qquad \alpha = \frac{\tau_c}{\tau}\,. \end{aligned} \tag{19}$$

The dependence $T_{2e}^s(\tau_c)$ is close to that given by (11) since in the sum over harmonics the terms are proportional to $1/n^2$ and the first few terms are dominant. For the EH as well, an equation similar to (16) can be obtained.

The approach discussed above can be used for the analysis of any other line-narrowing pulse sequence. As an example, we consider the WHH-4 sequence [2]. To obtain the EH in this case it is necessary to perform the canonical transformation (6) up to terms of $||\mathcal{H}_d||^2 t_c^2$ order (t_c is the period of pulse sequence). The equation for the slow part of the density matrix has then the form of (7),

$$\frac{\partial \varrho_s}{\partial t} = - i[\mathcal{H}_d^s + \mathcal{H}_{eff}(t) , \varrho_s] . \tag{20}$$

The $\mathcal{H}_d^s$ contribution to the relaxation rate can be calculated by use of the Bloembergen, Purcell, and Pound theory [1] generalized for the case of many harmonics [3], or by the equivalent iterational solution [4, 5],

$$\frac{1}{T_{2e}^s} = \frac{2}{3} M_2 \tau_e \left\{1 - \frac{5 \cosh (\tau/\tau_c) \sinh^2 (\tau/\tau_c)}{(\tau/\tau_c) \sinh (3\tau/\tau_c)}\right\} \approx$$

$$\approx \frac{2}{3} M_2 \frac{\tau_c}{1 + (\pi\tau_c/2\tau)^2} , \tag{21}$$

where τ is the interval between the nearest pulses ($\tau_c = 6\tau$). This contribution dominates at $\tau_c \approx t_c$. For slow molecular motions the EH contribution should be taken into account.

When $\tau_c \gg t_c$, the EH can be expressed as (see (12))

$$\mathcal{H}_{eff}(t) = \exp (iFt) \, \mathcal{H}_{eff}^* \exp (-iFt) , \tag{22}$$

where

$$\mathcal{H}_{eff}^* = \frac{\delta}{\sqrt{3}} S_e + \frac{t_c^2}{648} [\mathcal{H}_d^x - \mathcal{H}_d^y, [\mathcal{H}_d^x, \mathcal{H}_d^y]] , \tag{23}$$

S_e is the spin projection on the effective field, i.e. on [111] direction, δ the resonance off-set, $\mathcal{H}_d^p$ the secular part of $\mathcal{H}_d$ with respect to the p-axis. $\mathcal{H}_{eff}^*$ contains as well the terms describing different imperfections (pulse duration, errors in angles, phases, etc.) which we omitted here.

First we consider the case $\delta = 0$. There are two contributions to the linewidth. One of them is given by (21) and another one is described by the EH. If $\tau_c \gg ||\mathcal{H}_{eff}^*||^{-1}$, the Hamiltonian (23) gives the second contribution to the linewidth. When $\tau_c < < ||\mathcal{H}_{eff}^*||^{-1}$ averaging of EH occurs and it becomes equal to $\langle \mathcal{H}_{eff}^* \rangle$. At $\tau_c \ll t_c$ the dipole–dipole interactions averaged over the molecular motions are to be substituted into (23) $(\langle \mathcal{H}_{eff} \rangle = \mathcal{H}_{eff}^*(\langle \mathcal{H}_d \rangle))$. A simple consideration leads to the conclusion that for rather slow motions ($\tau_c \lessgtr ||\mathcal{H}_{eff}^*||^{-1}$) the linewidth can be less than for the rigid lattice [9]. This effect is however screened and the reason is as follows. The rigid-lattice lineshape in such experiments is close to a Lorentzian [2], and $||\mathcal{H}_{eff}^*||$ essentially exceeds the linewidth. It follows that at $\tau_c \approx ||\mathcal{H}_{eff}^*||^{-1}$ (when line narrowing is expected) the direct relaxation given by (21) dominates and the extra diminishing of the linewidth is not observable.

For $\delta \neq 0$ one can use (21) to (23) to find the longitudinal and transverse magnetization damping rates. An extra minimum in the dependence of the longitudinal relaxation rate on τ_c which is expected here at $\delta\tau_c \approx 1$ [9, 11] is also difficult to observe because of the reason discussed above.

References

[1] A. Abragam, Principles of Nuclear Magnetism, Clarendon Press, Oxford/London 1961.
[2] U. Haeberlen and J. S. Waugh, Phys. Rev. **175**, 453 (1968).
[3] U. Haeberlen and J. S. Waugh, Phys. Rev. **185**, 420 (1969).
[4] W. Gründer, H. Schmiedel, and D. Freude, Ann. Phys. 27, 409 (1971).
[5] W.-K. Rhim, D. P. Burum, and D. D. Elleman, J. chem. Phys. **68**, 692 (1978).
[6] Yu. Ivanov, B. N. Provotorov, and E. B. Feldman, Zh. eksper. teor. Fiz. **75**, 1847 (1978).
[7] B. N. Provotorov and E. B. Feldman, Zh. eksper. teor. Fiz. 79, 2206 (1980).
[8] L. L. Buishvili and M. G. Menabde, Zh. eksper. teor. Fiz. 77, 2435 (1979).
[9] V. E. Zobov and A. V. Ponomarenko, in: Radiospektroskopiya tverdogo tela, Krasnoyarsk 1979 (p. 70).
[10] N. N. Bogolyubov and Yu. A. Mitropolskii, Asimptoticheskie metody v teorii nelineinykh kolebanii, Moscow 1963.
[11] L. N. Erofeev, A. K. Khitrin, B. N. Provotorov, and V. P. Tarasov, Phys. Letters A 87, 443 (1982).
[12] M. Lee and W. I. Goldburg, Phys. Rev. A **140**, 1261 (1965).
[13] V. V. Laiko and A. K. Khitrin, Khim. Fiz. **1**, 123 (1983).

(Received June 21, 1984)

phys. stat. sol. (b) 126, 487 (1984)

Subject classification: 6

Institute of Semiconductor Physics,
Academy of Sciences of the USSR, Siberian Branch, Novosibirsk[1])

The Photoexcited Flow of Nonequilibrium Phonons along a Crystal Surface

By

M. D. BLOKH and M. V. ENTIN

It is shown, that for a light beam of oblique incidence to the crystal surface, a nonequilibrium phonon flux along the surface can exist. The transverse acoustic phonons of the lower branch of spectrum whose free path length is large at low temperatures give the main contribution to the flow. Such phonons arise either immediately at infrared absorption or as the result of the decay of optical and longitudinal acoustic phonons. The phonon energy flux is calculated and shown to be of the same order of magnitude as the incident light flux. The phonon–electron drag current in the metal film bordering the unexposed surface of the dielectric is considered.

Показано, что при наклонном падении светового пучка на поверхность кристалла возможен поток фононов вдоль поверхности. Основной вклад в поток дают поперечные акустические фононы нижней ветви спектра, которые при низких температурах имеют большую длину свободного пробега. Такие фононы появляются как непосредственно при инфракрасном поглощении света, так и при распадах оптических и продольных акустических фононов. Найдена величина потока энергии фононов и показано, что она имеет порядок величины падающего светового потока. Вычислен ток фонон-электронного увлечения в металлической пленке, граничащей с неосвещенной поверхностью диэлектрика.

1. Introduction

Recently the interest in methods of photoexcitation of nonequilibrium phonons in solids has been increasing [1 to 3]. The large free path length of transverse acoustic (TA) phonons at low temperatures is the reason for this interest because it permits to register the time and space distribution of the phonons. In optical excitation of long lifetime phonons so far phonon flows spreading isotropically from the source into the depth of the sample were investigated. The significance of polarization and direction of the light falling on the crystal surface has not been paid attention to before. In this paper we show that if the light beam falls on the surface under an angle and illuminaties it homogeneously, a phonon flow along the surface can exist. The direction of the flow is combined of the unit normal to the surface $\boldsymbol{N}$ and the polarization vector of the radiation $\boldsymbol{E}$ (or wave vector $\boldsymbol{k}$ for unpolarized light).[2])

Let us explain the physics of the phenomenon under consideration. Let acoustic phonons arise in the sample volume as a result of light absorption. The phonons move mainly along the electric intensity vector $\boldsymbol{E}$ of the radiation because the probability of the phonon photoexcitation with wave vector $\boldsymbol{q}$ contains a contribution proportional to $|\boldsymbol{qE}|^2$. This asymmetry of phonon excitation probability is even about $\boldsymbol{q}$ and therefore it by itself cannot be the reason for phonon number and energy fluxes. The latter appear by taking into account the sample boundaries which scatter

---

[1]) Prospekt Akademika Lavrenteva 13, 630090 Novosibirsk, USSR.

[2]) For electrons the phenomenon like this is called surface photogalvanic effect [4].

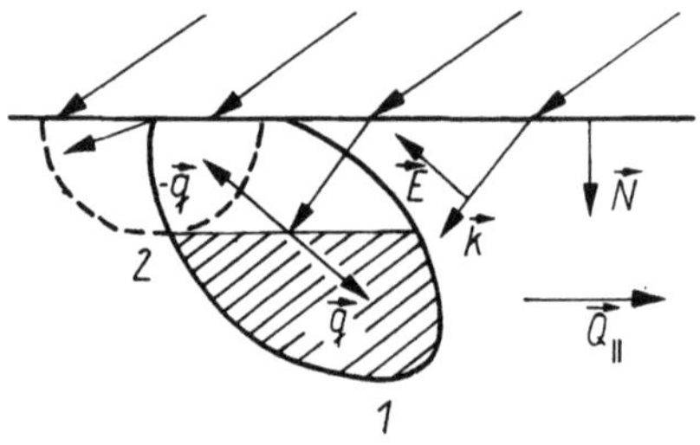

Fig. 1. The ellipse represents a direction diagram of photoexcited TA phonons. The shaded part 1 concerns phonons moving away from the surface. The direction diagram after collision with the surface is marked by 2. The difference in diagrams 1 and 2 leads to the phonon flow along the surface to the right

phonons. The light falling at an angle on the crystal surface, the vector $\boldsymbol{E}$ makes an angle (unequal to zero and $\pi/2$) with $\boldsymbol{N}$ and the momentum of photoexcited phonons moving from the boundary after scattering is directed mainly along the surface (Fig. 1). These are the phonons which create particle and energy fluxes along the crystal surface. Opposite surfaces obviously cause fluxes in opposite directions. Nevertheless, a resulting flux exists if the sample surfaces are different (e.g., in mirror coefficient) or the light is absorbed in the near-surface region.

The value of the effect is obviously proportional to the phonon free path length l. The phonons of lower TA branches at low temperature have the greatest l and as a result the energy flux along the surface is determined by these phonons. Transverse acoustic phonons can arise either immediately under the influence of light or as a result of longitudinal acoustic (LA) phonon decay according to the schemes

$$\begin{array}{ll} \text{photon} \to \text{TA} + \text{TA}\,, & \\ \text{photon} \to \text{TA} + \text{LA}\,, & \end{array} \quad \text{(I)} \qquad \begin{array}{ll} \text{LA} \to \text{TA} + \text{TA}\,, & \\ \text{LA} \to \text{TA} + \text{LA}\,. & \end{array} \quad \text{(II)}$$

An optical phonon can be taken as an intermediate stage in processes (I).

The decay process of high-frequency LA phonons arising at light absorption is characterized by the time $10^2\omega_D^{-1}$ (ω_D is the Debye frequency) [5] and is the most rapid relaxation process in the phonon system. In this process the anisotropy of the phonon distribution function is partly preserved. The formed TA phonons cannot decay any more but can be scattered by isotropic impurities and the crystal surface. Their distribution function $n_t(\boldsymbol{q})$ can be found from the kinetic equation containing TA phonons generated by the light and the decaying LA phonons with appropriate boundary conditions.

We calculate the phonon energy flux

$$\boldsymbol{Q} = \int \frac{\partial \omega_q}{\partial \boldsymbol{q}} \hbar\omega_q n_t(\boldsymbol{q}) \frac{d^3q}{(2\pi)^3}$$

in pure crystals assuming homogeneous and stationary illumination, low crystal temperature, and using the isotropic medium approximation with one degenerate TA branch. We suppose that the phonon photoexcitation level is low, then combinational and induced decay processes can be neglected.

2. Kinetic Equation and Phonon Generation Rate

As TA phonons decay rapidly, spatially inhomogeneous and relaxation processes can be neglected in the kinetic equation for the longitudinal acoustic phonon distribution function n_l. Taking into consideration decays only we write the kinetic equation for LA phonons as follows:

$$\sum_{q', q''} |M_{lt}(q, q', q'')|^2 [n_l(q') \delta_{q', q+q''} \delta(\omega_l' - \omega_l - \omega_t'') -$$
$$- n_l(q) \delta_{q', q-q''} \delta(\omega_l - \omega_l' - \omega_t'')] -$$
$$- \sum_{q', q''} |M_{tt}(q, q', q'')|^2 n_l(q) \delta_{q', q-q''} \delta(\omega_l - \omega_t' - \omega_t'') + g_l(q) = 0 , \quad (1)$$

where M_{lt}, M_{tt} are the matrix elements of decay LA → LA + TA and LA → TA + + TA, respectively, averaged over TA phonon polarizations, g_l is the rate of phonon generation by light.

The kinetic equation for the TA phonon distribution function $n_t(q)$ can be written as

$$\frac{s_t q_z}{q} \frac{\partial n_t}{\partial z} + \frac{n_t - \langle n_t \rangle}{\tau} = g_t + \dot{n}_t , \quad (2)$$

where s_t is the transverse sound velocity, the z-axis is normal to the surfaces of the sample being a plane-parallel plate of thickness d ($0 \leqq z \leqq d$), τ is a phonon momentum relaxation time defined by impurity (in particular, isotropic) scattering, g_t the TA phonon photogeneration rate, $\dot{n}_t$ the TA phonon "internal" generation rate caused by LA phonon decays,

$$\dot{n}_t(q) = \sum_{q', q''} |M_{lt}(q, q', q'')|^2 n_l(q') \delta_{q', q+q''} \delta(\omega_l' - \omega_l'' - \omega_t) +$$
$$+ |M_{tt}(q, q', q'')|^2 n_l(q'') \delta_{q'', q+q'} \delta(\omega_l'' - \omega_t' - \omega_t) . \quad (3)$$

The term with $\langle n_t \rangle$ in (2) does not make a contribution to the phonon flux since it is spherically symmetric, so this term can be neglected without loss of generality.

The boundary conditions for n_t at the crystal surface must account for the possibility of phonons moving into the border medium and their scattering by surface roughnesses. Generally speaking, the number of phonons before and after interaction with the surface are unequal. We describe phenomenologically all the above processes by means of mirror coefficients $P_{1,2}$ for illuminated and nonilluminated surfaces, respectively,

$$n_t(q_{||}, q_z > 0)|_{z=0} = P_1 n_t(q_{||}, - q_z)|_{z=0} ,$$
$$n_t(q_{||}, q_z < 0)|_{z=d} = P_2 n_t(q_{||}, - q_z)|_{z=d} . \quad (4)$$

Here and further on, the mark || indicates the projection of a vector on the sample plane.

We consider the phonon generation in case of infrared absorption for one- and two-phonon excitation. In the first case the light gives rise to an optical phonon, decaying rapidly to acoustic phonons. In both cases the final expressions for $g_{l,t}$ are distinguished only by a multiplier. Let us obtain the generation rates $g_{l,t}$ for the process photon → TA + LA in centrosymmetrical crystals which will only be considered. The Hamiltonian of the light interaction with lattice vibrations in the isotropic medium approximation can be written as follows:

$$H = A \sum_{q, j, j'} [(q e_{qj}) (e_{qj'} e) - (q e_{qj'}) (e_{qj} e)] (c_{qj'} + c^+_{-qj'}) (c_{-qj} + c^+_{qj}) , \quad (5)$$

e_{qj} being unit vectors of phonon polarization, e the unit vector of light polarization, j the vibration branch number, c_{qj} phonon operators, the constant A will be further expressed through the radiation intensity. It is not difficult to be convinced that the Hamiltonian of this type satisfies the operations of time and space inversion.

After averaging over TA phonon polarizations we obtain for the generation probabilities

$$g_l(\boldsymbol{q}) = g_t(-\boldsymbol{q}) = \frac{\pi \Lambda^2}{\hbar^2 s_t(\xi + 1)} [q^2 + (\boldsymbol{q}\boldsymbol{e}) (\boldsymbol{q}\boldsymbol{e}^*)] \,\delta(q - q_0) \,, \tag{6}$$

where $\xi = s_l/s_t$ is the ratio of longitudinal to transverse sound velocity, $q_0 = \omega_0/s_t(\xi + 1)$, ω_0 the light frequency. The directions of phonon movement and light polarization are correlated according to (6). Obviously the correlation remains with the crystal symmetry taken into consideration.

Assuming that the light absorption is caused only by phonon excitation one can write $\hbar\omega_0 \int g_l(\boldsymbol{q})\, \mathrm{d}^3q/(2\pi)^3 = \alpha I$, where α is the absorption coefficient, I the light intensity. It follows that $\Lambda^2 = 3\pi\hbar\alpha I/2q_0^5$. As $I = I_0\, \mathrm{e}^{-\alpha z}$, $g_{l,\,t}$ and $\dot{n}_t$ are also exponentially dependent on z according to (6), (1), and (3).

Solving (1) with boundary conditions (3) gives the following expression for the phonon energy flux along the crystal surface (averaged over sample thickness):

$$\boldsymbol{Q}_{\|} = \frac{\hbar s_t^2}{d} \int \frac{\mathrm{d}^3q}{(2\pi)^3}\, \boldsymbol{q}_{\|} \tau \theta(q_z) (\bar{g}_t + \dot{\bar{n}}_t) \left\{ \frac{l(1-k)}{1 - \alpha^2 l^2} + \frac{l(1-k')}{2(1 - P_1 P_2 k'^2)} \times \right.$$
$$\left. \times \left[\frac{P_1(k - k') + P_1 P_2 k k' - 1}{1 - \alpha l} - \frac{P_2(1 - k k') - k + P_1 P_2 k'}{1 + \alpha l} \right] \right\}, \tag{7}$$

$l = (|q_z|/q)\, s_t \tau$, $k = \mathrm{e}^{-\alpha d}$, $k' = \mathrm{e}^{-d/l}$, $\theta(q_z)$ being the unit step function, a primed quantity means its magnitude at $z = 0$. From (7) the energy flux is determined by the anisotropic parts of $g_t(\boldsymbol{q})$ and $\dot{n}_t(\boldsymbol{q})$.

3. Phonon Energy Flux and Phonon–Electron Drag Current in a Metallic Film

Before substitution into (3) for the "internal" generation $\dot{n}_t$ the LA phonon distribution function must be obtained from (1). Let us solve it in the form of a series expansion in spherical harmonics $n_l(\boldsymbol{q}) = \sum\limits_{L,M} n_{LM}(q)\, Y_{LM}(\Omega)$, where Ω is the spherical angle between $\boldsymbol{q}$ and $\boldsymbol{E}$. From (6) it is seen that for the calculation of $Q_{\|}$ one must obtain only the harmonic $n_{2M}(q)$ which defines the anisotropic part of $n_l(\boldsymbol{q})$.

In (1) and (3) squared matrix elements can be present in the form $|M_{lt}(\boldsymbol{q}, \boldsymbol{q}', \boldsymbol{q}'')|^2 = A_{lt} q q' q''$, $|M_{tt}(\boldsymbol{q}, \boldsymbol{q}', \boldsymbol{q}'')|^2 = A_{tt} q q' q''$. The magnitudes A_{lt} and A_{tt} can be connected with the LA phonon lifetime relative to the decay into LA + TA and TA + TA, respectively,

$$\tau_{lt}^{-1} = A_{lt} q^5 \frac{8\xi^2(5\xi^2 - 5\xi + 2)}{60\pi^2(\xi + 1)^5 s_t}, \qquad \tau_{tt}^{-1} = A_{tt} q^5 \frac{15\xi^4 - 10\xi^2 + 3}{960\pi^2 s_t}.$$

For the second harmonic of the LA phonon distribution function one can write the following equation:

$$A_l n_{2M}(q) - A_{lt} \frac{\xi^2}{16} \sqrt{\frac{5}{\pi}} \int\limits_1^{\frac{\xi+1}{\xi-1}} (\varkappa - 1)^2 \{3[\varkappa^2 + 1 - \xi^2(\varkappa - 1)^2]^2 - 4\varkappa^2\}\, n_{2M}(\varkappa q)\, \mathrm{d}\varkappa =$$
$$= \frac{(2\pi)^2}{q^5}\, g_{2M}(q) \,, \tag{8}$$

where $A_l = (\tau_{lt}^{-1} + \tau_{tt}^{-1})\, q^{-5}$, g_{2M} a spherical harmonic of the phonon photoexcitation probability (6). Equation (8) can be solved by means of Fourier transformation relative to a variable $t = \ln q$. As a result of such a solution we are sure that the integral term of (8) makes a relatively small contribution to the total energy flux therefore it will be neglected. Then n_{2M} is simply expressed by g_{2M} and (3), (6), (7) give the final formula for the energy flux (averaged over sample thickness). In the general case the formula is very cumbersome, so let us consider simplifying possibilities.

For the one-phonon absorption of light with the appearance of a resonance optical phonon the absorption coefficient is large, therefore one can regard $\alpha^{-1} \ll d, l$. If the sample is so thin that $d \ll l$ (film), the expression for energy flux along the plate surface becomes

$$Q_{\parallel} = K \frac{(1 - P_1)(1 + P_2)}{1 - P_1 P_2} I_0 \operatorname{Re}[\boldsymbol{e}_{\parallel}(\boldsymbol{N}\boldsymbol{e}^*)] , \tag{9}$$

where

$$K = \frac{1}{4(\xi + 1)} \left[1 + \frac{6\xi(4\xi + 1)}{(3\xi - 1)(\xi + 1)} \frac{\tau_1}{\tau_{1t}} + \right.$$
$$\left. + \frac{\xi^2(0.24\xi^6 - 0.54\xi^4 + 5.7\xi^2 + 4.44)}{5\xi^2 - 1} \frac{\tau_1}{\tau_{tt}}\right],$$

where τ_1 is the LA phonon decay time. The magnitude of $Q_{\parallel}$ is seen from this formula to be of the same order as the density of the light flux falling on the sample surface. In the opposite case $l \ll d$ we obtain

$$Q_{\parallel} = K(1 - P_1) I_0 l d^{-1} \operatorname{Re}[\boldsymbol{e}_{\parallel}(\boldsymbol{N}\boldsymbol{e}^*)] . \tag{10}$$

In two-phonon absorption the absorption coefficient is small so we consider that $\alpha^{-1} \gg d, l$.

For the film we have

$$Q_{\parallel} = K \frac{P_1 - P_2}{1 - P_1 P_2} I_0 \alpha d \operatorname{Re}[\boldsymbol{e}_{\parallel}(\boldsymbol{N}\boldsymbol{e}^*)] . \tag{11}$$

Note that for nonpolarized light in (9) to (11) the light polarization vector has to be replaced by the photon unit wave vector.

We calculate the temperature gradient caused by the phonon flow in a thermally isolated plate. Assume that the deviation from equilibrium is small (the energy of nonequilibrium phonons is smaller than that of equilibrium ones). Then, because of the interaction of photoexcited phonons with equilibrium ones, a temperature gradient of the equilibrium phonons is established. The longitudinal thermal conductivity of the plate is also connected with equilibrium phonons and for $d \ll l$ it is determined by the formula

$$\varkappa_{\parallel} = \frac{(1 + P_1)(1 + P_2) \hbar d}{4(1 - P_1 P_2)} \sum_j s_j^2 \int \frac{d^3q}{(2\pi)^3} \frac{q_{\parallel}^2}{q_z} \frac{\partial n_{0j}}{\partial T} , \tag{12}$$

where n_{0j} is the equilibrium distribution function of phonon. Cutting the integration (12) at small angles by the parameter d/l we obtain

$$\varkappa_{\parallel} = \frac{(1 + P_1)(1 + P_2) d}{16(1 - P_1 P_2)} \ln \frac{l}{d} \sum_j s_j C_j , \tag{13}$$

where $C_j = 2\pi^2 (T/s_j)^3/15\hbar^3$ is the specific heat capacity of the jth branch of the phonon spectrum. From (9) and (13) the temperature gradient can be calculated by $\nabla T = Q_{\parallel}/\varkappa_{\parallel}$. For the estimation of the effect we use the following typical parameters: $s_t = 3 \times 10^5$ cm/s, $\xi = 1.7$, $T = 2$ K, $d = 100$ μm, $l/d \approx 10$, and obtain $\nabla T/I_0 \approx$ ≈ 1 (cm · K/W).

As a result of the interaction with the light and the surface the phonons receive a directed momentum along the sample surface. If the phonon path length is so large that $l \gg d$ and the dielectric plate borders on a conducting film, the phonon momentum due to phonon–electron drag can be passed to the electrons resulting in a current in the film. Assume that electrons receive the whole momentum of the phonons passing into the metal film. In this case the current in the film is easily calculated. Let Π_2 ($\Pi_2 + P_2 \leqq 1$) mark the coefficient of phonon transmission into the metal film joined with the nonilluminated surface of the dielectric. The component of the force along the surface per unit film surface is

$$\boldsymbol{F}_{\parallel} = \Pi_2 \hbar s_t \int \frac{d^3q}{(2\pi)^3} \frac{q_z \boldsymbol{q}_{\parallel}}{q} \theta(q_z)\, n_t(\boldsymbol{q}, z = d)\,. \tag{14}$$

Substituting the solution of (2) into (14) we obtain

$$F_{\parallel} = \frac{3K\Pi_2(1 - P_1)\, I_0}{4(1 - P_1P_2)\, s_t} \operatorname{Re}\,[\boldsymbol{e}_{\parallel}(\boldsymbol{N}\boldsymbol{e}^*)]\,. \tag{15}$$

Pay attention to the fact that the photoexcited phonons act upon the electrons by a force which inversely proportional to the sound velocity s_t. But for the case of an immediate action of the light upon the electrons the radiation pressure force is inversely proportional to the light velocity. So the preliminary transformation of light into phonons permits the radiation pressure effect to be increased by a factor of c/s_t.

The current density is connected with the force (15) by the formula $j_{\parallel} = \sigma F_{\parallel}/en_e a$, σ is the conductivity, n_e the electron concentration, a the metal film thickness. Supposing the electron free path length to be $l_e \approx 10^{-4}$ cm, their velocity $v_F \approx 10^8$ cm/s, $a \approx 10^{-4}$ cm, we obtain $j \approx 1$ A/cm². For a typical metal the arising electric field force is $E \approx 1$ μV/cm.

4. Concluding Remarks

Strictly speaking, our theory is valid for isotopically pure crystals. Isotopic impurities which occur in most crystals influence the phonon flow propagation for two reasons.

Firstly the isotopic phonon scattering leads to branch conversion TA → LA. In general the conversion creates a large incoming term in the kinetic equation (1) for LA phonons. However, we should take into account only those LA phonons which remember the directed momentum arising in the photoexcitation as other phonons do not make a contribution to the resulting energy flux along the surface. As shown in [5] the impurity scattering TA → TA frequency is much higher than the conversion scattering TA → LA frequency. So transverse phonons manage to be scattered by isotopic impurities many times and forget completely the initial direction of the momentum before the conversion TA → LA had taken place. For the same reasons the conversion LA → TA can be neglected in (2) for TA phonons and boundary conditions on the surface.

Secondly, the free path length of the first-generation phonon excited immediately by the light decreases considerably due to the isotopic scattering. Therefore, the main contribution to the flux may be due to the phonons of the next generation arising as a result of the decay. Such phonons have the greater path length and keep partially the memory about the directed momentum of the first-generation phonons. Quantitative estimations of the results are considerably variable depending on crystal type and experimental conditions determining the decay scheme.

Acknowledgements

We are sincerely grateful to Dr. Y. B. Levinson for discussing the paper and critical remarks and to Dr. I. A. Ghilinskii for his attention to the work.

References

[1] R. G. Ulbrich, V. Narayanamurti, and M. A. Chin, Phys. Rev. Letters **45**, 1432 (1980).
[2] P. Hu, V. Narayanamurti, and M. A. Chin, Phys. Rev. Letters **46**, 192 (1981).
[3] D. B. McWhan, P. Hu, M. A. Chin, and V. Narayanamurti, Phys. Rev. B **26**, 4774 (1982).
[4] L. I. Magarill and M. V. Entin, Zh. eksper. teor. Fiz. **81**, 1001 (1981).
[5] N. M. Queseinov and Y. B. Levinson, Solid State Commun. **45**, 371 (1983).

(Received July 23, 1984)

phys. stat. sol. (b) **126**, 495 (1984)

Subject classification: 4; 12.2; 22.5.2

Department of Physics, Shree Varshney College, Aligarh[1])

The Role of the Zero-Point Vibration towards Crystal Properties of Rubidium Halide Crystals

By

R. P. SINGH, J. P. S. RANA, A. DAUD, and G. B. VARSHNEY

The contribution of zero point vibration towards the cohesive energy, molecular force constant, reststrahlen frequency, third order elastic constants, bulk modulus and its pressure dependence are evaluated for rubidium halide crystals. The Hafemeister-Flygare (HF) type short-range overlap repulsive potential effective up to the next nearest neighbour is used. The zero point energy is estimated by the well known Narayan and Ramaseshan relation and the van der Waals (vdW) coefficients are calculated by the Slater-Kirkwood variational (SKV) method. The results show a close agreement with their experimental data.

Der Beitrag der Nullpunktsschwingungen zur Kohäsionsenergie, der molekularen Kraftkonstante, der Reststrahlenfrequenz, der elastischen Konstanten dritter Ordnung, dem Elastizitätsmodul und seiner Druckabhängigkeit werden für Rubidiumhalogenidkristalle berechnet. Dazu wird das kurzreichweitige abstoßende Hafemeister-Flygare(HF)-Überlappungspotential bis zu den übernächsten Nachbarn benutzt. Die Nullpunktenergie wird aus der wohlbekannten Narayan-Ramaseshan-Beziehung und die van der Waals-Koeffizienten (vdW) mit der Slater-Kirkwood-Variationsmethode (SKV) berechnet. Die Ergebnisse zeigen eine gute Übereinstimmung mit experimentellen Werten.

1. Introduction

In the recent past, several authors [1 to 4] have used various potentials for investigating the static and thermodynamical properties of some ionic crystals. In our view, a potential containing all the components, namely zero-point vibration, long-range Coulomb, van der Waals three-body, and overlap repulsive interactions is expected to yield better results than those obtained when one or more components are dropped. In earlier studies [1 to 4] one or the other component of the potential has been dropped. The results obtained from them have significant deviation from experimental data. To eliminate this unsatisfactory situation we have incorporated all the components in our model potential. The van der Waals coefficients due to dipole–dipole and dipole–quadrupole interactions have been estimated from the Slater-Kirkwood variational (SKV) [6] method. The overlap repulsive potential of Hafemeister-Flygare (HF) type effective up to next nearest neighbours (NNN) has been considered. The present study yields results in excellent agreement with the experimental ones justifying the inclusion of all the components in the model potential. Theory and technique of evaluation of model parameters have already been given in our earlier paper [7].

[1]) Aligarh 202001, India.

2. Results and Discussions

The input data listed in Table 1 have been used to evaluate the van der Waals coefficients whose values are presented in Table 3. These values of van der Waals

Table 1

Value of the input data for rubidium halide crystals

property	RbF	ref.	RbCl	ref.	RbBr	ref.	RbI	ref.
r_0	2.7985	[12]	3.2456	[11]	3.3993	[11]	3.6275	[11]
C_{11}	6.55	[13]	4.499	[14]	3.863	[15]	3.21	[15]
C_{12}	1.35	[13]	0.767	[14]	0.474	[15]	0.360	[15]
C_{44}	0.965	[13]	0.497	[14]	0.4085	[15]	0.292	[15]
α_+	1.022	[16]	1.136	[16]	1.321	[16]	1.433	[17]
α_-	0.759	[16]	2.97	[16]	4.5	[16]	5.31	[17]
r_+	1.645	[18]	1.645	[18]	1.645	[18]	1.645	[18]
r_-	1.133	[18]	1.648	[18]	1.809	[18]	2.048	[18]
θ_{D_0}	—		168.8	[19]	136.4	[19]	107.9	[19]
V_0	—		1.59	[20]	1.49	[20]	—	
ε_∞	1.94	[20]	2.20	[20]	2.36	[20]	2.61	[20]
ε_0	5.99	[20]	4.53	[20]	4.51	[20]	4.55	[20]
ν_{L_0}	8.544	*)	5.284	[21]	3.804	[21]	3.1513	[21]

*) Calculated from the Lyddane-Sachs-Teller (LST) relation.
Polarizabilities are in 10^{-1}nm^3, ionic radii in 10^{-1}nm, elastic constants in 10^{10}N/m^2, and frequencies (ν_{L_0}) in 10^{12}Hz,

Table 2

Model parameters for rubidium halide crystals

property	RbF	RbCl	RbBr	RbI
b (10^{-19} J)	0.235	0.224	0.274	0.228
ϱ (10^{-1} nm)	0.315	0.324	0.358	0.371
$f(r)$	−0.005	−0.003	−0.001	−0.002
$f'(r)$	0.016	0.009	0.003	0.006
$f''(r)$	−0.050	−0.029	−0.009	−0.016
$f'''(r)$	0.158	0.089	0.026	0.044

Table 3

Value of van der Waals coefficients c_{ij} (10^{-79} J m^6) and d_{ij} (10^{-99} J m^8)

property	RbF	RbCl	RbBr	RbI
c_{ij}	37.443	132.097	215.310	284.512
c_{ii}	61.631	72.259	90.578	102.409
c_{jj}	23.529	265.498	563.291	843.150
d_{ij}	15.002	67.390	118.770	149.653
d_{ii}	19.835	24.522	33.143	39.037
d_{jj}	10.946	167.598	384.752	535.729
C	323.877	1176.318	2010.682	2730.582
D	104.514	491.013	871.100	1149.652

Table 4
Cohesive energy (in 10^3 J/mol) of rubidium halide crystals

property	RbF	RbCl	RbBr	RbI
Φ_{Z_0}	—	3.164	2.556	2.022
Φ_C	−868.901	−749.205	−715.330	−670.330
Φ_T	25.779	13.658	5.108	9.008
Φ_V	−40.884	−63.107	−81.741	−74.599
Φ_R	92.692	96.828	120.064	103.000
Φ tot.	−791.314	−698.662	−669.343	−630.899
$\Phi_{\text{exp.}}$	−777.907 [22]	−684.960 [22]	−658.165 [22]	−621.740 [22]
Φ_{other}	−831.498 [3]	−726.828 [3]	−698.358 [3]	−671.563 [3]

coefficients have been used to calculate the short-range parameters (b, ϱ) presented in Table 2. The parameter $r_0 f_0'$ and its derivatives appearing in the expressions of third-order elastic (TOE) constants are obtained from the relation $f(r) = f_0 \exp(-r/\varrho)$ [8]. The cohesive energy and its components are shown in Table 4. The zero-point

Table 5a
Static and thermodynamical properties of rubidium halide crystals

property	RbF	RbCl	RbBr	RbI
bulk modulus	0.308	0.201	0.160	0.131
B_T (10^{11} N/m²)	0.271*)	0.158*)	0.134*)	0.106*)
pressure derivatives	4.97	5.24	5.04	5.20
of bulk modulus	5.09*)	5.62*)	5.59*)	5.60*)
(dB_T/dp)	5.22**)	4.53**)	4.54**)	4.54**)
molecular force constant f (10^1 N/m) without Lorentz field	4.66	4.22	4.26	3.46
molecular force constant f (10^1 N/m) with Lorentz field	2.28	2.24	2.04	1.87

*) Values are from experiment [9, 10].
**) Values are from [1].

Table 5b
Values of reststrahlen frequency (10^{12} Hz) with Lorentz field effects

solid	calc.	exp.	ref.	er or (%)
RbF	4.725	4.862	[21]	2.817
RbCl	3.648	3.820*)		4.502
RbBr	2.743	2.752*)		0.327
RbI	2.366	2.387	[21]	0.879

*) Calculated from the Lyddane-Sachs-Teller (LST) relation.

Table 6

Third-Order Elastic (TOE) constants (in 10^{11} N/m^2) of rubidium halide crystals

property	RbF	RbCl	RbBr	RbI
C_{111}	−7.102	−7.564	−6.528	−5.098
C_{112}	−0.460	−0.288	−0.194	−0.105
C_{166}	−0.421	−0.369	−0.345	−0.185
C_{123}	0.188	0.482	0.101	0.077
C_{144}	0.142	0.417	0.096	0.069
C_{456}	0.119	0.385	0.093	0.065

energy calculated from $\Phi_{z_0} = 9k\theta_{D_0}/4$, for all rubidium halides are also given in Table 4. The other six force constant parameters are calculated as usual.

On comparing our values of cohesive energy of rubidium halides and those of other investigators [3] with the experimental values, it is found that the values of other investigators show a marked deviation while ours are very close to them. The values of B_T and its pressure derivative are evaluated by the usual method and presented in Table 5a with relevant references [1, 9, 10]. The values of B_T and its pressure dependence are in good agreement with their experimental data and show a marked improvement over those obtained by other authors. The values of third-order elastic (TOE) constants are presented in Table 6 and analysis of these results is presented in Table 7 which reveals that the Cauchy violation in higher-order elastic constants is very well accounted for rubidium halides with NaCl structure. They follow the same systematic trend, as is observed in case of sodium halides. It is also observed that C_{111}, C_{112}, C_{166} are in general negative whereas the values of C_{123}, C_{144}, C_{456} are positive. As no experimental data is available for (TOE) constants, our results in this regard are of academic interest only. The values of reststrahlen frequencies for all the rubidium halides are presented in Table 5b. The Lorentz field effect on this property has been found to improve the results from 26% to 44%.

Table 7

Cauchy discrepancies of third-order elastic (TOE) constants for rubidium halides

Cauchy discrepancy	RbF	RbCl	RbBr	RbI
$[(C_{112} - C_{166}) : C_{112}]/C_{166}$	0.203	0.768	2.245	4.146
$[(C_{123} - C_{456}) : C_{123}]/C_{456}$	3.076	0.527	0.862	2.425
$[(C_{144} - C_{456}) : C_{144}]/C_{456}$	1.357	0.203	0.304	0.903

Acknowledgements

The authors are thankful to Prof. R. K. Singh and Dr. S. P. Sanyal, Department of Physics, Jabalpur University, Jabalpur, for encouragement and useful suggestions. One of us (GBV) is grateful to CSIR, New Delhi for award of a SRF.

References

[1] V. C. Jain and J. Shanker, Indian J. pure and appl. Phys. **17**, 713 (1978).
[2] R. K. Singh and Shailendra Tiwari, J. Phys. Soc. Japan **49**, 1046 (1980).
[3] R. P. Singh and J. Shanker, phys. stat. sol. (b) **93**, 373 (1979).
[4] S. P. Sanyal and R. K. Singh, phys. stat. sol. (b) **105**, 193 (1981).
[5] M. J. L. Sangster, V. Schroder, and R. M. Atwood, J. Phys. C **11**, 1523 (1978).
[6] J. C. Slater and J. G. Kirkwood, Phys. Rev. **37**, 682 (1931).
[7] R. P. Singh, J. P. S. Rana, and G. B. Varshney, phys. stat. sol. (b) **121**, 493 (1984).
[8] W. Cochran, CRC Crit. Rev. Solid State Sci. **2**, 1 (1971).
[9] C. S. Smith and L. S. Cain, J. Phys. Chem. Solids **36**, 205 (1975).
[10] R. W. Roberts and C. S. Smith, J. Phys. Chem. Solids **31**, 2397 (1970); 619 (1970).
[11] M. Born and K. Huang, Dynamical Theory of Crystal Lattice, Clarendon Press, Oxford 1954.
[12] C. Kittel, Introduction of Solid State Physics, Wiley 1956 (p. 80).
[13] S. Haussühl, Z. Phys. **159**, 223 (1960).
[14] B. J. Marshali, D. O. Pederson, and G. G. Dosris, J. Phys. Chem. Solids **28**, 1061 (1967).
[15] J. T. Lewis, A. Lehoczpy, and C. V. Briscoe, Phys. Rev. **161**, 877 (1967).
[16] L. Pauling, Proc. Roy. Soc. A **114**, 181 (1927).
[17] J. R. Tessmann, A. K. Khan, and W. Shockley, Phys. Rev. **92**, 890 (1953).
[18] P. A. Sysio, Acta cryst. B **15**, 2374 (1969).
[19] B. N. N. Achar and G. R. Barsch, phys. stat. sol. (a) **6**, 247 (1971).
[20] R. P. Lowndes and D. H. Martin, Proc. Roy. Soc. London A **308**, 473 (1969); A **316**, 351 (1970).
[21] G. Raunio and S. Rolandson, J. Phys. C **3**, 1013 (1970); J. Phys. C **4**, 958 (1971); phys. stat. sol. **40**, 749 (1970).
[22] M. F. C. Ladd, J. chem. Phys. **75**, 1 (1931).

(Received July 13, 1984)

phys. stat. sol. (b) **126**, 501 (1984)

Subject classification: 14.3; 6; 20.1; 22.9

Chelyabinsk State University[1])

Charge Transfer and Electron–Molecular Vibration Coupling in Tetramerized Quasi-1D Semiconductors

By

V. M. Yartsev

Electron–molecular vibration coupling in tetramerized organic ion-radical semiconductors is considered. Eigenvalues and eigenfunctions of the molecular tetramer with two electrons are calculated in terms of extended Hubbard model parameters. The energy inequivalence of the monomer sites is introduced explicitly. The complex frequency dependent conductivity of the tetramerized compounds is found to produce a number of sharp maxima due to indirect excitation of the totally symmetric (a_g) internal molecular modes of vibrations. These peaks are shown to have a fine structure if the charge density is not shared equally by the constituent molecules. The calculated reflectance is compared with the measured one for $TEA(TCNQ)_2$ single crystals.

Рассмотрено взаимодействие электронов с внутримолекулярными колебаниями в тетрамеризующихся органических ион-радикальных полупроводниках. Найдены собственные значения и собственные функции двух электронов в тетрамере молекул в рамках обобщенной модели Хаббарда с явным учетом различия энергий электрона помещенного на неэквивалентные узлы. Показано, что частотная зависимость комплексной электропроводности имеет ряд резких максимумов, соответствующих косвенному возбуждению полносимметричных внутримолекулярных колебаний. Неравномерное распределение электронной плотности в тетрамере приводит к расщеплению этих пиков на дублеты. Проведено сравнение результатов с экспериментальной зависимостью коэффицента отражения от частоты для монокристаллов $TEA(TCNQ)_2$.

1. Introduction

Molecular organic solids based on acceptors like tetracyanoquinodimethane (TCNQ) crystallize in face-to-face arrays of planar TCNQ molecules, with significant π-overlap along the stack indicated by close contacts of 0.3 to 0.35 nm. Possessing a variety of cations at their disposal, chemists obtained compounds with either *regular* stacks of TCNQ, containing a single charge transfer (CT) interaction t or *alternating* (dimerized, trimerized, tetramerized) stacks with several t's (see review [1]).

The theory of the optical properties of dimerized [2, 3] and trimerized [4] quasi-1d crystals was constructed to explain the role of electron–molecular vibration (EMV) coupling in these solids.[2]) Tetramerized compounds [6 to 13] exhibit optical properties [14 to 20] with both common features: (i) strong dependence on the polarization of external radiation, (ii) sharp lines corresponding to the totally symmetric (a_g) modes of internal TCNQ vibrations, (iii) broad CT peak (peaks?), and peculiarities: i) the pronounced fine structure (doublets) of molecular vibration lines, ii) CT peak is shifted to lower energy (≈ 2500 cm^{-1}).

---

[1]) Molodogvardeitsev 70b, Chelyabinsk, USSR.

[2]) Recently, this cluster-type approach was used [5] to calculate magnetic and thermodynamic quantities also for dimers and trimers.

The optical spectrum of TEA $(TCNQ)_2$ [3]) was interpreted [21] with the help of the phase phonon theory by Rice [22], but we have shown recently [23] that this theory is more appropriate for the organic compounds with *regular* stacks of TCNQ molecules, like MEM $(TCNQ)_2$ [3]) at $T > 355$ K or TTF-TCNQ [24]. The detailed cluster-type theory was absent due to lack of knowledge of exact wave functions and eigenvalues of tetramerized systems.

In this paper the method developed earlier for trimers [4] is applied to the compounds in which the ion-radical molecules are present as *quasi-isolated tetramers* ... ABBAABBA ..., where transfer integrals between neighbouring molecules are assumed to satisfy the relations $|t_{AA}| \ll |t_{BB}| \equiv t' \lessapprox |t_{AB}| \equiv t$. The insignificance of interaction between tetramers is proved by experiments [25] on the influence of irradiation on the IR spectrum and follows from the structural data [8].

Section 2 describes the EMV coupling in a tetramer and introduces the internal tetrameric modes of vibration. The eigenvalues and the eigenfunctions of the system "two electrons on four sites" are calculated in Section 3. In Section 4 we derive the frequency-dependent conductivity which can be used to explain the fine structure of internal molecular excitations. Finally, the obtained relations are discussed and compared with reflectance from TEA $(TCNQ)_2$ single crystals.

2. EMV Coupling in a Tetramer

We assume that the optical properties of a tetramerized 1D organic compound may be understood if an assembly of non-interacting tetramers in an externally applied electric field $\boldsymbol{E}$ is described by the Hamiltonian (see, e.g., [2])

$$H = H_e + H_v + \sum_{\alpha, i} g_\alpha n_i Q_{\alpha i} - \boldsymbol{E}\boldsymbol{p} . \tag{1}$$

Here

$$H_e = \Delta(n_1 + n_4) - \Delta(n_2 + n_3) + \frac{U}{2} \sum_{i,\sigma} n_{i,\sigma} n_{i,-\sigma} + V(n_1 n_2 + n_2 n_3 + n_3 n_4) + \\ + t' \sum_\sigma (c^+_{2\sigma} c_{3\sigma} + c^+_{3\sigma} c_{2\sigma}) + t \sum_\sigma (c^+_{1\sigma} c_{2\sigma} + c^+_{2\sigma} c_{1\sigma} + c^+_{3\sigma} c_{4\sigma} + c^+_{4\sigma} c_{3\sigma}) \tag{2}$$

and

$$H_v = \sum_{i,\alpha} \tfrac{1}{4} (P^2_{\alpha i} + Q^2_{\alpha i}) \, \omega_\alpha \tag{3}$$

describe, respectively, the radical electrons and the symmetric molecular vibrations of each monomer in the absence of vibronic coupling. EMV coupling is described by the third term in (1). We use the extended variant of the Hubbard model [26] in order to take into account the Coulomb repulsion, V, between the electrons on neighbouring sites. Also, the energy inequivalence Δ of the monomer sites in the tetramer is explicitly introduced in (2). The nature of this kind of inequivalence is discussed in detail for a dimer [3], where asymmetric positions of the cations lead to the shift of charge density in the dimer, the latter resulting in the distortion of molecules as the internal bond lengths readjust to the new local charge densities due to EMV coupling. This process of increased stability is counterbalanced by the elastic, restoring forces associated with the internal molecular distortion. The inhomogeneity of electron density distribution can be verified by examining the internal bond lengths in molecules, which indicate [27] the charge density value. In a tetramer the catalytic role in the formation of the distortion gap may be played by V- and t'-interactions. The dif-

[3]) TEA triethylammonium, MEM methyl-ethyl-morfolinium.

ferences in internal bond lengths of molecules A and B in tetramers are observed in all tetramerized compounds [6 to 13]. Other notations in (1) to (3) are the same as in [4].

It follows from (1) that the equations of motion for the modes are

$$\ddot{Q}_{\alpha i} + \omega_\alpha^2 Q_{\alpha i} = -2 g_\alpha n_i \omega_\alpha \,. \tag{4}$$

We introduce the tetrameric normal mode coordinates

$$S_\alpha = \frac{1}{2}\,(Q_{\alpha 1} + Q_{\alpha 2} + Q_{\alpha 3} + Q_{\alpha 4})\,, \qquad s_\alpha = \frac{1}{2}\,(Q_{\alpha 1} - Q_{\alpha 2} - Q_{\alpha 3} - Q_{\alpha 4})\,,$$
$$R_\alpha = \frac{1}{\sqrt{2}}\,(Q_{\alpha 1} - Q_{\alpha 4})\,, \qquad r_\alpha = \frac{1}{\sqrt{2}}\,(Q_{\alpha 2} - Q_{\alpha 3})\,. \tag{5}$$

According to (4) the equations of motion for these modes are

$$\ddot{S}_\alpha + \omega_\alpha^2 S_\alpha = -g_\alpha \omega_\alpha (n_1 + n_2 + n_3 + n_4)\,, \tag{6}$$
$$\ddot{s}_\alpha + \omega_\alpha^2 s_\alpha = -g_\alpha \omega_\alpha (n_1 - n_2 - n_3 + n_4)\,, \tag{7}$$
$$\ddot{R}_\alpha + \omega_\alpha^2 R_\alpha = -\sqrt{2}\, g_\alpha \omega_\alpha (n_1 - n_4)\,, \tag{8}$$
$$\ddot{r}_\alpha + \omega_\alpha^2 r_\alpha = -\sqrt{2}\, g_\alpha \omega_\alpha (n_2 - n_3)\,. \tag{9}$$

Equation (6) describes in-phase oscillations of a_g modes of the four monomers and their frequencies are just ω_α. The modes s_α correspond to in-phase oscillations of molecules A of a tetramer out-of-phase with the vibrations of molecules B. In this case the electron density moves symmetrically and no resulting electric dipole moment arises, s_α cannot be excited by the external electric field.

Equations (8) and (9) show that finite values of R_α and r_α produce electric dipole moments, respectively,

$$\boldsymbol{p}_1 = (2\boldsymbol{a} + \boldsymbol{a}')\, e\, \frac{n_1 - n_4}{2} \tag{10}$$

and

$$\boldsymbol{p}_2 = \boldsymbol{a}'\, e\, \frac{n_2 - n_3}{2}\,, \tag{11}$$

where the vector $\boldsymbol{a}(\boldsymbol{a}')$ is specified by the mode of the intermolecular overlap between molecules AB (BB) and its magnitude will be of the order of the intermolecular separation. Consequently, an externally applied electromagnetic field can excite antisymmetric tetrameric modes R_α and r_α.

Now we can rewrite the Hamiltonian (1) in the form

$$H = H_0 + H_{\text{int}}\,, \tag{12}$$

where

$$H_{\text{int}} = \left(\sum_\alpha \frac{g_\alpha}{\sqrt{2}}\, R_\alpha - \frac{2\boldsymbol{a} + \boldsymbol{a}'}{2}\, e\boldsymbol{E}\right)(n_1 - n_4) + \left(\sum_\alpha \frac{g_\alpha}{\sqrt{2}}\, r_\alpha - \frac{\boldsymbol{a}'}{2}\, e\boldsymbol{E}\right)(n_2 - n_3) \tag{13}$$

describes the interaction of the applied field $\boldsymbol{E}$ with the electrons and subsequent excitation of R_α and r_α. The Hamiltonian H_0 describes noninteracting tetrameric modes of vibration and electronic excitations which will be explicitly treated in the following section.

The Hamiltonian (13) in case of weak EMV coupling [2] may be treated [28] as a perturbation in the linear response theory. The Fourier components of ground state expectation values of $(n_1 - n_4)$ and $(n_2 - n_3)$ are

$$\begin{aligned}(n_1 - n_4)_\omega &= \chi_{11}(\omega)\left[\boldsymbol{E}_\omega \frac{2\boldsymbol{a} + \boldsymbol{a}'}{2} e - \sum_\alpha \frac{g_\alpha}{\sqrt{2}} R_{\alpha\omega}\right] + \chi_{12}(\omega)\left[\boldsymbol{E}_\omega \frac{\boldsymbol{a}'}{2} e - \sum_\alpha \frac{g_\alpha}{\sqrt{2}} r_{\alpha\omega}\right],\\ (n_2 - n_3)_\omega &= \chi_{21}(\omega)\left[\boldsymbol{E}_\omega \frac{2\boldsymbol{a} + \boldsymbol{a}'}{2} e - \sum_\alpha \frac{g_\alpha}{\sqrt{2}} R_{\alpha\omega}\right] + \chi_{22}(\omega)\left[\boldsymbol{E}_\omega \frac{\boldsymbol{a}'}{2} e - \sum_\alpha \frac{g_\alpha}{\sqrt{2}} r_{\alpha\omega}\right],\end{aligned} \tag{14}$$

where

$$\begin{aligned}\chi_{11}(\omega) &= \sum_\beta \frac{|\langle\beta|\, n_1 - n_4\,|1\rangle|^2\, 2\omega_{\beta 1}}{\omega_{\beta 1}^2 - \omega^2 - i\omega\gamma_e},\\ \chi_{22}(\omega) &= \sum_\beta \frac{|\langle\beta|\, n_2 - n_3\,|1\rangle|^2\, 2\omega_{\beta 1}}{\omega_{\beta 1}^2 - \omega^2 - i\omega\gamma_e},\\ \chi_{12}(\omega) &= \sum_\beta \frac{\langle 1|\, n_1 - n_4\,|\beta\rangle\, \langle\beta|\, n_2 - n_3\,|1\rangle\, 2\omega_{\beta 1}}{\omega_{\beta 1}^2 - \omega^2 - i\omega\gamma_e} = \chi_{21}(\omega)\end{aligned} \tag{15}$$

are the reduced CT electronic polarizabilities in which $\omega_{\beta 1} = E_\beta - E_1$, E_β and $|\beta\rangle$ are the exact eigenstates and eigenvalues of H_e (2), $\beta = 1$ labels the ground state. With the use of (8) to (11) and the relation $-i\omega(\boldsymbol{p}_{1\omega} + \boldsymbol{p}_{2\omega}) = \boldsymbol{E}_\omega \sigma(\omega)$, the frequency-dependent conductivity may be found as

$$\sigma(\omega) = -i\omega N_t e^2 \frac{\chi_{11}A^2 + 2\chi_{12}AB + \chi_{22}B^2 + D(\chi_{12}^2 - \chi_{11}\chi_{22})\,(A^2 + B^2)}{1 - (\chi_{11} + \chi_{22})\,D + (\chi_{11}\chi_{22} - \chi_{12}^2)D^2}, \tag{16}$$

where

$$D(\omega) = \sum_\alpha \frac{g_\alpha^2 \omega_\alpha}{\omega_\alpha^2 - \omega^2 - i\omega\gamma_\alpha}. \tag{17}$$

Here N_t is the number of tetramers per unit volume, γ_α and γ_e denote the natural width of the originally uncoupled vibrational modes and electronic excitations, $A = (2a + a')/2$, $B = a'/2$.

3. Eigenstates and Eigenvalues of Two Radical Electrons per Tetramer

In order to calculate matrix elements in (15) one should find eigenstates of electronic Hamiltonian H_e (2). We choose them as linear combinations

$$|\beta\rangle = \sum_k a_k^\beta \Psi_k \tag{18}$$

of basis states Ψ_k, the latter being all possible distributions of two radical electrons on four sites, cf. Table 1. The coefficients a_k^β can be obtained as a solution of the system of 28 linear equations generated from (2) and (18). This problem is simplified by employing spin projection and spatial symmetries [4]: we introduce the corresponding unitary, hermitian operators

$$\hat{I}_\sigma = \prod_{i=1}^{4} (c_{i\uparrow}^+ c_{i\downarrow} + c_{i\downarrow}^+ c_{i\uparrow} + c_{i\uparrow} c_{i\downarrow} c_{i\downarrow}^+ c_{i\uparrow}^+ - c_{i\uparrow}^+ c_{i\downarrow}^+ c_{i\downarrow} c_{i\uparrow}) \tag{19}$$

and

$$\hat{I}_s = \prod_{\sigma=\uparrow,\downarrow} \prod_{i=1}^{2} (c_{i\sigma}^+ c_{l\sigma} + c_{l\sigma}^+ c_{i\sigma} + c_{i\sigma} c_{l\sigma} c_{l\sigma}^+ c_{i\sigma}^+ - c_{i\sigma}^+ c_{l\sigma}^+ c_{l\sigma} c_{i\sigma}); \qquad l = 5 - i\,. \tag{20}$$

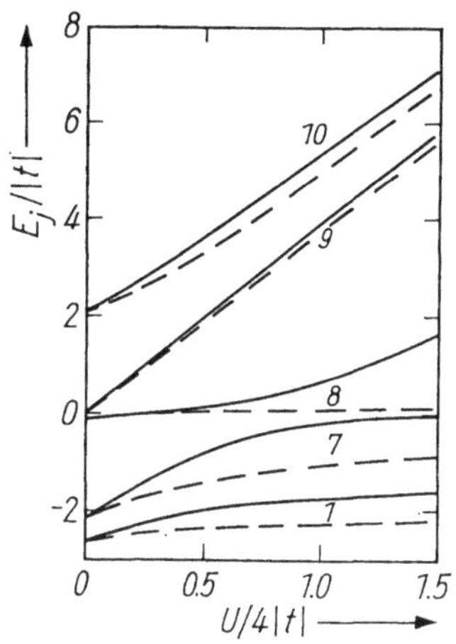

Fig. 1. Excited state energies (7 to 10) which are accessible by optical transitions from the ground state (1) for the two-electron system in a tetramer: $t'/t = 0.4$, $\Delta/|t| = 0.25$, $V = 0$ (dashed lines); $0.4U$ (solid lines)

Operators (19) and (20) commute with the Hamiltonian (2), hence all eigenvectors of the Hamiltonian (2) can be distributed in the subspaces corresponding to the combinations of the eigenvalues of $\hat{I}_\sigma$ and $\hat{I}_s$: $I_\sigma = 1$, $I_s = 1$; $I_\sigma = 1$, $I_s = -1$; $I_\sigma = -1$, $I_s = 1$; $I_\sigma = -1$, $I_s = -1$. The ground state belongs to the ($I_\sigma = 1$, $I_s = 1$) case and is shown in Fig. 1 as a function of parameters U, V, t, t'. The ground state wave function is

$$|1\rangle = a_1^1(\Psi_1 + \Psi_4) + a_2^1(\Psi_2 + \Psi_3) + a_5^1(\Psi_5 + \Psi_{10} + \Psi_{11} + \Psi_{16}) + \\ + a_6^1(\Psi_6 + \Psi_9 + \Psi_{12} + \Psi_{17}) + a_7^1(\Psi_7 + \Psi_{13}) + a_8^1(\Psi_8 + \Psi_{14}) . \quad (21)$$

In Table 2 we present the magnitudes of coefficients a_k^1 and the charge density distribution in a tetramer for several parameter values. The mean charges on mole-

Table 1

Possible distributions of two electrons on monomers (i) in the basis states Ψ_k

k	i 1	2	3	4	k	i 1	2	3	4
1	↑↓	0	0	0	15	0	↓	0	↑
2	0	↑↓	0	0	16	0	0	↓	↑
3	0	0	↑↓	0	17	↑	↑	0	0
4	0	0	0	↑↓	18	↑	0	↑	0
5	↑	↓	0	0	19	↑	0	0	↑
6	↑	0	↓	0	20	0	↑	↑	0
7	↑	0	0	↓	21	0	↑	0	↑
8	0	↑	↓	0	22	0	0	↑	↑
9	0	↑	0	↓	23	↓	↓	0	0
10	0	0	↑	↓	24	↓	0	↓	0
11	↓	↑	0	0	25	↓	0	0	↓
12	↓	0	↑	0	26	0	↓	↓	0
13	↓	0	0	↑	27	0	↓	0	↓
14	0	↓	↑	0	28	0	0	↓	↓

Table 2

Coefficients a_k^1 in the series of the ground state wave functions on the basis states Ψ_k and mean charges for the two-electron system in a tetramer

$\frac{V}{U}$	$\frac{\Delta}{\|t\|}$	$\frac{t'}{t}$	$\frac{U}{4\|t\|}$	$\frac{E_1}{\|t\|}$	a_1^1	a_2^1	a_5^1	a_6^1	a_7^1	a_8^1	q_A	q_B
0	0	0.8	1	−2.43	0.059	0.148	0.190	0.319	0.263	0.361	0.420	0.580
0	0.25	0.8	1	−2.56	0.051	0.173	0.182	0.301	0.197	0.427	0.329	0.671
0.4	0.25	0.2	1	−1.67	0.012	0.035	0.038	0.383	0.352	0.281	0.544	0.456
0.4	0.25	0.8	1	−1.96	0.041	0.138	0.132	0.362	0.294	0.309	0.474	0.526
0.4	0.25	0.8	1.5	−1.74	0.024	0.082	0.100	0.384	0.342	0.247	0.550	0.450

cules A and B are calculated from the relations (see Table 1)

$$q_A = 2[(a_1^1)^2 + (a_5^1)^2 + (a_6^1)^2 + (a_7^1)^2]\,,$$
$$q_B = 2[(a_2^1)^2 + (a_5^1)^2 + (a_6^1)^2 + (a_8^1)^2]\,. \tag{22}$$

Applying operators $(n_1 - n_4)$ and $(n_2 - n_3)$ to the ground state function gives

$$(n_1 - n_4)\,|1\rangle = 2a_1^1(\Psi_1 - \Psi_4) + a_5^1(\Psi_5 - \Psi_{10} + \Psi_{11} - \Psi_{16}) + {} + a_6^1(\Psi_6 - \Psi_9 + \Psi_{12} - \Psi_{15})\,, \tag{23}$$

$$(n_2 - n_3)\,|1\rangle = 2a_2^1(\Psi_2 - \Psi_3) + a_5^1(\Psi_5 - \Psi_{10} + \Psi_{11} - \Psi_{16}) - {} - a_6^1(\Psi_6 - \Psi_9 + \Psi_{12} - \Psi_{15})\,. \tag{24}$$

These functions are orthogonal to all excited eigenstates except the ones from the subspace $I_\sigma = 1$, $I_s = -1$. Corresponding eigenvalues are given by the roots of the following equation:

$$4(U - E)\,Et^2 - [(U - E)^2 - 4\Delta^2]\,[(V - E)\,E + t'^2] = 0\,. \tag{25}$$

Consequently, optical transitions from the ground state of the two-electron system in a tetramer are allowed only to the four states shown in Fig. 1. The corresponding matrix elements have the general form ($\beta = 7$ to 10)

$$\langle\beta|\, n_1 - n_4\,|1\rangle = 4(a_1^1a_1^\beta + a_5^1a_5^\beta + a_6^1a_6^\beta)\,, \tag{26}$$

$$\langle\beta|\, n_2 - n_3\,|1\rangle = 4(a_2^1a_2^\beta + a_5^1a_5^\beta - a_6^1a_6^\beta)\,. \tag{27}$$

The coefficients a_k^β are given in the Appendix.

4. Fine Structure

Reflectance spectra of TEA $(TCNQ)_2$ at 100 K [16] and other tetramerized compounds [19, 20] reveal that the peaks associated with resonant excitation of totally symmetric a_g modes are doublets rather than singlets. The lower the temperature, the more pronounced is the fine structure [19]. We proposed in [29] an explanation of this phenomenon based upon the dependence of a_g-mode frequencies on the charge density. There are indications that this dependence is linear [30, 31]. As follows from Table 2 the charge density is different at TCNQ molecules A and B, and we assume that their a_g modes have *different* frequencies $\omega_{\alpha A}$ and $\omega_{\alpha B}$, respectively. EMV coupling constants g_α may be different, too. The equations of motion (8) and (9) for the modes

R_α and r_α are modified,

$$\begin{aligned} \ddot{R}_\alpha + \omega_{\alpha A}^2 R_\alpha &= -\sqrt{2}\, g_{\alpha A}\omega_{\alpha A}(n_1 - n_4), \\ \ddot{r}_\alpha + \omega_{\alpha B}^2 r_\alpha &= -\sqrt{2}\, g_{\alpha B}\omega_{\alpha B}(n_2 - n_3), \end{aligned} \tag{28}$$

and (14) takes the form

$$\begin{aligned} (n_1 - n_4)_\omega\, [1 - D_A\chi_{11}] &= \chi_{11} e E_\omega A + \chi_{12}[eE_\omega B + D_B(n_2 - n_3)_\omega], \\ (n_2 - n_3)_\omega\, [1 - D_B\chi_{22}] &= \chi_{22} e E_\omega B + \chi_{12}[eE_\omega A + D_A(n_1 - n_4)_\omega], \end{aligned} \tag{29}$$

where

$$D_{A,B} = \sum_\alpha \frac{g_{\alpha A,B}^2\, \omega_{\alpha A,B}}{\omega_{\alpha A,B}^2 - \omega^2 - i\omega\gamma_\alpha}. \tag{30}$$

We can find $(n_1 - n_4)_\omega$ and $(n_2 - n_3)_\omega$ from (29) and with the help of (10), (11) obtain the conductivity of tetramerized compounds,

$$\sigma(\omega) = -i\omega N_t e^2 \frac{\chi_{11}A^2 + 2\chi_{12}AB + \chi_{22}B^2 + (\chi_{12}^2 - \chi_{11}\chi_{22})\,(A^2 D_B + B^2 D_A)}{1 - \chi_{11}D_A - \chi_{22}D_B + (\chi_{11}\chi_{22} - \chi_{12}^2)\, D_A D_B}, \tag{31}$$

where χ_{ij} are given by (15) as before. The frequencies $\omega_{\alpha A}$ and $\omega_{\alpha B}$ may be estimated from the bond length analysis [27] and the linear dependence of frequency on charge density [30, 31]. Note that local fields may shift the frequencies up to 10 cm^{-1} [30] in different TCNQ compounds.

5. Discussion

To obtain a physical insight into the origin of electronic transitions which are described by (16) or (31), we turn off the EMV coupling (putting all g_α equal to zero) and consider a tetramer as two dimers interacting with transfer integral t'. The ground state wave function (21) may be viewed as a superposition of the cases "two electrons on one dimer, none on the other" and "one electron on each dimer". The relative probability is given by the ratio of coefficients a_1^1, a_2^1, a_5^1 to a_6^1, a_7^1, a_8^1 (see Table 2). Excitations to the higher energy states (E_9, E_{10}) occur in dimers with two electrons, the final state containing two electrons on one molecule, B and A, respectively. For typical parameter values maxima of Re σ produced by these excitations are at least an order of magni-

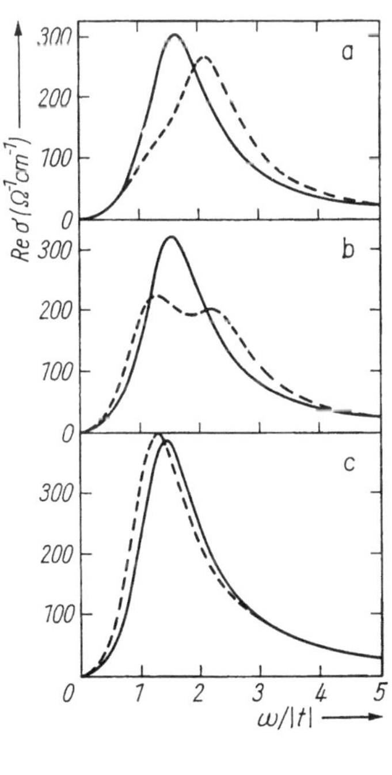

Fig. 2. Conductivity of tetramerized compounds in absence of EMV coupling ($g_\alpha = 0$). Hamiltonian (2) parameters are: $\Delta/|t| = 0.25$; $U = 4|t|$, $V = 0$ (dashed lines), $0.4U$ (solid lines). a) $t'/t = 0.2$, b) 0.4, c) 0.8

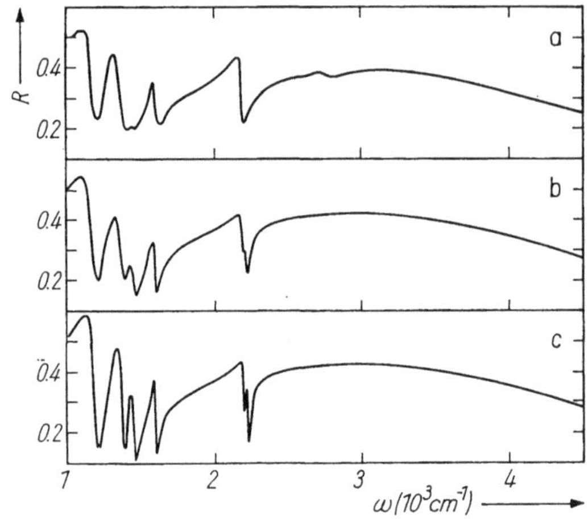

Fig. 3. a) Experimental data for reflectance R in TEA $(TCNQ)_2$ at 300 K [16]; b, c) calculated reflectance (see text)

tude smaller than those for transitions from the ground state E_1 to E_7 and E_8. In the atomic limit ($U \gg 4|t|$) the excitations $E_1 \to E_7$ and $E_1 \to E_8$ correspond to CT excitations between BB and AB molecules, respectively.

In Fig. 2 we show the change of the conductivity spectrum of the tetramerized compounds due to an increment of "interdimer" interaction, t'/t. The role of Coulomb interaction of electrons on neighbouring sites, V, is also examined in Fig. 2, where dashed lines correspond to $V = 0$ and solid lines to $V = 0.4\ U$, as estimated by Hubbard [26]. The absolute values of $\mathrm{Re}\,\sigma$ were calculated with the use of TEA $(TCNQ)_2$ structural data [7]: $N_t = 0.72 \times 10^{21}\ \mathrm{cm}^{-3}$, $A = 0.584$ nm, $B = 0.198$ nm.

Now we undertake a more detailed comparison of our results with experimental data on reflectance of TEA $(TCNQ)_2$ single crystals [16] shown in Fig. 3a. The reflectance may be calculated from the dielectric function which is related to the complex conductivity in a standard way,

$$\varepsilon(\omega) = \varepsilon_\infty + \frac{4\pi i}{\omega}\,\sigma(\omega)\,, \tag{32}$$

where a constant dielectric function ε_∞ has been introduced to account for the effect of the remaining high-frequency electronic transitions of the molecular constituents of the crystals. For TEA $(TCNQ)_2$ we take $\varepsilon_\infty = 2$ [18]. The matrix elements (26), (27) and transition energies, $\omega_{\beta 1}$, depend on the Hubbard model parameters U, V, t, t', Δ. We chose $|t|$ equal to 1580 cm^{-1} (0.2 eV) as determined for the similar compound MEM $(TCNQ)_2$ [3]. The transfer integral was shown to be proportional to the overlap between atomic π-orbitals of neighbouring TCNQ molecules [32]. The overlap between molecules BB and AB was calculated for the structural data [7] and their ratio was found to be 0.8. The energy inequivalence $2\Delta = 0.1$ eV of molecules A and B was estimated as the deformation energy resulting in differences of internal bond lengths (elastic force constants, see [7]). A value of 1 to 1.4 eV [33 to 36] is commonly accepted for the Coulomb interaction of two electrons deposited on one TCNQ molecule and V/U is taken equal to 0.4 [26, 36]. These values were used to calculate eigenvalues of the two-electron system in a tetramer and, consequently, transition energies $\omega_{\beta 1}$ $|t| = 1.44$, 3.53, 7.56, 8.84 for $\beta = 7$ to 10, respectively.

Then, using a_k^β given in Table 2 and in the Appendix, we obtain the electronic polarizabilities (15). The optimum values for the parameter g_α and γ_α are listed in Table 3. Also included are experimental values of a_g vibrational modes of $TCNQ^\circ$ [37] and $TCNQ^-$ [38], employed in (30) as $\omega_{\alpha A}$ and $\omega_{\alpha B}$, respectively.

Table 3
Vibronic band parameters (in cm^{-1}) in TEA $(TCNQ)_2$ at 300 K

$\omega_{\alpha A}$ [37]	$\omega_{\alpha B}$ [38]	g_α	γ_α
2229	2206	210	20
1602	1615	250	20
1454	1391	500	50
1207	1196	340	50

The reflectance calculated from (31), (32) is shown in Fig. 3b. In Fig. 3c we plot $R(\omega)$ for the same parameters, but all γ_α are taken smaller by a factor of two. The fine structure becomes more apparent in qualitative agreement with experimental data on TEA $(TCNQ)_2$ at lower temperatures [16] and reflectance from MTPP $(TCNQ)_2$ and MTPA $(TCNQ)_2$ single crystals [19, 20].

Acknowledgement

The author is grateful to V. L. Bonch-Bruevich for helpful discussions.

Appendix

The eigenfunctions, which belong to the $I_\sigma = 1$, $I_s = -1$ subspace have the form

$$|\beta\rangle = a_1^\beta(\Psi_1 - \Psi_4) + a_2^\beta(\Psi_2 - \Psi_3) + a_5^\beta(\Psi_5 - \Psi_{10} + \Psi_{11} - \Psi_{16}) + + a_6^\beta(\Psi_6 - \Psi_9 + \Psi_{12} - \Psi_{15}) ,$$

where

$$a_1^\beta = \frac{2|t|}{U + 2\Delta - E_\beta}\, a_5^\beta , \qquad a_2^\beta = \frac{2|t|}{U - 2\Delta - E_\beta}\, a_5^\beta , \qquad a_6^\beta = \frac{t'}{E_\beta}\, a_5^\beta ,$$

$$a_5^\beta = \left\{4 + 8t^2[(U + 2\Delta - E_\beta)^{-2} + (U - 2\Delta - E_\beta)^{-2}] + 4\left(\frac{t'}{E_\beta}\right)^2\right\}^{-1/2} ;$$

$\beta = 7, 8, 9, 10$.

References

[1] I. F. Shchegolev, phys. stat. sol. (a) **12**, 9 (1972).
[2] M. J. Rice, Solid State Commun. **31**, 93 (1979).
[3] M. J. Rice, V. M. Yartsev, and C. S. Jacobsen, Phys. Rev. B **21**, 3437 (1980).
[4] V. M. Yartsev, phys. stat. sol. (b) **112**, 279 (1982).
[5] W. N. Mei and Y. C. Lee, Phys. Rev. B **24**, 1111 (1981).
[6] J. Jaud, D. Chasseau, J. Gaultier, and C. Hauw, C.R. Acad. Sci. (France) **278 C**, 769 (1974).
[7] H. Kobayashi, Y. Ohashi, F. Marumo, and Y. Saito, Acta cryst. **B26**, 459 (1970).
[8] A. T. McPhail, G. M. Semeniuk, and D. B. Chesnut, J. Chem. Soc. (A) **13**, 2174 (1971).
[9] P. Delhaes, G. Keryer, S. Flandrois, and J. P. Manceau, phys. stat. sol. (b) **80**, 125 (1977).
[10] T. Konno and Y. Saito, Acta cryst. **B29**, 2815 (1973).
[11] S. Huizinga, J. Kommandeur, G. A. Sawatzky, B. T. Thole, K. Kopinga, W. J. M. de Jonge, and J. Roos, Phys. Rev. B **19**, 4723 (1979).
[12] G. J. Ashwell, S. C. Wallwork, S. R. Baker, and P. I. C. Bertier, Acta cryst. **B33**, 1174 (1975).
[13] D. Chasseau, J. Gaultier, C. Hauw, and J. Jaud, C.R. Acad. Sci. (France) **276 C**, 751 (1973).
[14] R. M. Vlasova, A. I. Gutman, L. D. Rotenshtein, and N. F. Kartenko, phys. stat. sol. (b) **47**, 435 (1971).

[15] M. D. KAPLUNOV, T. P. PANOVA, and YU. G. BORODKO, phys. stat. sol. (a) **13**, K67 (1972).
[16] M. V. BELOUSOV, A. M. VAINRUB, and R. M. VLASOVA, Fiz. tverd. Tela **18**, 2637 (1976).
[17] A. M. VAINRUB, R. M. VLASOVA, and V. N. SEMKIN, Zh. eksper. teor. Fiz., Pisma **20**, 7 (1977).
[18] A. BRAU, P. BRÜESCH, J. P. FARGES, W. HINZ, and D. KUZE, phys. stat. sol. (b) **62**, 615 (1974).
[19] R. ŠWIETLIK and A. GRAJA, J. Phys. (France) **44**, 617 (1983).
[20] J. PETZELT, K. KRÁL, N. RYŠAVÁ, L. DOBIÁŠOVA, J. KROUPA, J. OSWALD, A. SZYSZKOWSKI, and A. GRAJA, Solid State Commun. **32**, 1315 (1979).
[21] M. J. RICE, L. PIETRONERO, and P. BRUESCH, Solid State Commun. **21**, 757 (1977).
[22] M. J. RICE, Phys. Rev. Letters **31**, 36 (1976).
[23] V. M. YARTSEV and C. S. JACOBSEN, Phys. Rev. B **24**, 6167 (1981).
[24] S. ETEMAD, Phys. Rev. B **24**, 4959 (1981).
[25] K. KAMARÁS, K. HOLCZER, and A. JÁNOSSY, phys. stat. sol. (b) **102**, 467 (1980).
[26] J. HUBBARD, Phys. Rev. B **17**, 494 (1978).
[27] S. FLANDROIS and D. CHASSEAU, Acta cryst. B **33**, 2744 (1977).
[28] L. D. LANDAU and E. M. LIFSHITS, Statisticheskaya Fizika, Izd. Nauka, Moscow 1976 (§ 125) (in Russian).
[29] V. M. YARTSEV and M. J. RICE, phys. stat. sol. (b) **100**, K97 (1980).
[30] S. MATSUZAKI, R. KUWATA, and K. TOYODA, Solid State Commun. **33**, 403 (1980).
[31] M. TOKUMOTO, N. KOSHIZUKA, H. ANZAI, and T. ISHIGURO, J. Phys. Soc. Japan **51**, 332 (1982).
[32] A. J. BERLINSKY, J. F. CAROLAN, and L. WEILER, Solid State Commun. **15**, 795 (1974).
[33] D. B. TANNER, J. S. MILLER, M. J. RICE, and J. J. RITSKO, Phys. Rev. B **21**, 5835 (1980).
[34] J. B. TORRANCE, Ann. New York Acad. Sci. **313**, 210 (1978).
[35] K. YAKUSHI, T. KUSAKA, and H. KURODA, Chem. Phys. Letters **68**, 139 (1979).
[36] S. MAZUMDAR and Z. G. SOOS, Phys. Rev. B **23**, 2810 (1981).
[37] A. GIRLANDO and C. PECILE, Spectrochim. Acta **A29**, 1859 (1973).
[38] R. BOZIO, I. ZANON, A. GIRLANDO, and C. PECILE, J. Chem. Soc., Faraday Trans. II **74**, 235 (1978).

(Received September 27, 1984)

phys. stat. sol. (b) **126**, 511 (1984)

Subject classification: 4 and 13; 12.2; 22.1; 22.1.1; 22.1.2

AT & T Bell Laboratories, Murray Hill[1])

Cohesive Properties of Ge, Si, and Diamond Calculated with Minimum Basis Sets

By

B. HOLLAND[2]), H. S. GREENSIDE[3]), and M. SCHLÜTER

Ab initio calculations are presented of the cohesive properties of Ge, Si, and diamond. Electron–electron interactions are treated within the local density functional approximation and electron–ion interactions are described by norm-conserving pseudopotentials. The wave functions are expanded in local Gaussian orbitals. Acceptable results for equilibrium lattice constant, bulk modulus, and cohesive energy are obtained with few (eight to twelve) Gaussian functions per atom. The results, which are comparable to converged plane wave or local orbital calculations suggest the feasibility of studies of cohesive properties of systems with large numbers of atoms.

Ab-initio-Berechnungen der Kohäsionseigenschaften von Ge, Si und Diamant werden durchgeführt. Die Elektron–Elektron-Wechselwirkung wird mit der Näherung des lokalen Dichtefunktionals behandelt und die Elektron–Ion-Wechselwirkung mit normerhaltenden Pseudopotentialen beschrieben. Die Wellenfunktionen werden nach lokalen Gaußorbitalen entwickelt. Annehmbare Ergebnisse für die Gleichgewichtsgitterkonstante, den Elastizitätsmodul und die Kohäsionsenergie werden mit wenigen (acht bis zwölf) Gaußfunktinonen pro Atom erhalten. Die Ergebnisse, die vergleichbar sind mit Rechnungen auf der Grundlage konvergierender ebener Wellen oder lokaler Orbitale, zeigen die Durchführbarkeit von Berechnungen der Kohäsionseigenschaften von Systemen mit großer Atomzahl.

The study of cohesive properties of elemental solids and simple compounds has advanced considerable in recent years [1]. The successes are based on the use of the local density functional (LDF) theory [2] for the description of exchange-correlation effects. Within this approximation an effective one-electron wave equation has to be solved self consistently. This, however, represents a sizable numerical task, even for solids with few atoms per unit cell. A large number of techniques has been developed for this purpose [3 to 13]. A major difficulty is the correct description of the strong electron–ion interaction in extended systems. The concept of norm-conserving pseudopotentials [14] has been introduced to eliminate chemically inactive core electrons and thus to provide calculational advantages at no loss of accuracy.

To solve LDF-pseudopotential Hamiltonians plane wave basis sets have traditionally been employed [4]. This representation has the advantage of being complete, chemically "unbiased", and computationally simple. However, disadvantages arise when large numbers of plane waves are required for the description of highly localized orbitals. Conversely, a number of studies have used a localized Gaussian orbital basis to solve the wave equation [5, 6, 8, 11 to 13]. We use here this method and directly follow the approach developed and described in [5, 11] to investigate the question about the minimum Gaussian set necessary to obtain acceptable cohesive properties.

---

[1]) Murray Hill, New Jersey 07974, USA.

[2]) Present address: University of Warwick, Coventry, Great Britain.

[3]) Present address: Princeton Plasma Physics Laboratory, Princeton University, Princeton, New Jersey 48544, USA.

Table 1
Convergence behavior of the ground state energy E_{tot} for a typical Si calculation (20 Gaussian orbitals per atom)

iteration No.	ΔE (eV/atom)	E_{tot} (eV/atom)
0	3.262	−107.614
1	2.865	−107.632
2	0.367	−107.689
3	0.102	−107.691

The quantity ΔE is given in the text.

The electron gas data of Ceperley and Alder [15] as parametrized by Perdew and Zunger [16] are used in the LDA description of exchange and correlation. Norm-conserving pseudopotentials, including scalar-relativistic effects as calculated and tabulated in [17] are employed. Since these potentials themselves are fitted to Gaussians, all Hamiltonian matrix elements can be evaluated analytically in real space. The total energy, although it can be evaluated in real space, is for simplicity evaluated in Fourier space as described in [5, 11]. In contrast to the procedure described in [11] we do not terminate the calculations after the zeroth iteration but iterate until full self-consistency ($\Delta E_{tot} \lessapprox 10^{-3}$ eV) is reached (usually after three to four iterations). A measure for self-consistency in the spirit of the discussion of [11] is given by the term

$$\Delta E = \int \mathrm{d}r\, \varrho(r)_{out}\, \{V_H[\varrho(r)_{out}] - V_H[\varrho(r)_{in}] + V_{xc}[\varrho(r)_{out}] - V_{xc}[\varrho(r)_{in}]\}\,,$$

where ϱ_{in} and ϱ_{out} are input and output charge densities. In Table 1 we show how this term decreases as a function of iterations using a standard feedback scheme. Also shown in Table 1 is the total energy E_{tot} which being the variational quantity converges rapidly. In the present case the iterations were initiated with the superposition of empirical atomic potentials rather than charge densities. The use of the latter would presumably decrease the number of necessary iterations by one.

The choice of an optimum Gaussian basis is a rather complex problem. A straightforward solution is, of course, to minimize the *crystal* energy as a function of the Gaussian decay lengths by iterating to self-consistency at each step. This is a rather time-consuming process which, to our experience, can be circumvented. We start by inspecting the numerically obtained wave functions of the *free* atoms of the constituents. Since the long-range tail of the atomic wave function will be drastically altered inside the solid, only the short- and intermediate range shape of the wave function (i.e. $r \lessapprox d_0/2$) is of interest here and should be well approximated by Gaussians. Thus, rather than explicitly minimizing the crystal energy we minimize the free atom energy in a Gaussian basis set, allowing no long-range variational freedom. These calculations are speedy and yield the desired short- and intermediate range Gaussian decay lengths for use in the crystalline calculations. It has been our experience that the minimum necessary number of Gaussians and their decay lengths obtained in that fashion also nearly minimize the crystalline total energy. Similar experiences have been reported by Feibelman [18] for molecular calculations. The minimum cut-off in angular momentum space (s, p, d, etc.) cannot, of course, be obtained from atom calculations and has to be determined by explicit crystalline convergence tests.

Table 2
Calculated cohesive properties for diamond, Si, and Ge

	a_0 (10^{-10} m)	B (Mbar)	E_{coh} (eV)
C (exp.)	3.57	4.42	7.37
8 (sp) set 1			
$\alpha = 0.47, 2.45$	−5%	20%	−16%
20 (spd) set 2			
$\alpha = 0.47, 2.45$	−3%	15%	−1%
12 (sp) set 3			
$\alpha = 0.32, 0.88, 3.13$	<1%	8%	6%
Si (exp.)	5.43	0.99	4.63
8 (spd) set 1			
$\alpha = 0.2, 0.6$	<1%	10%	−17%
20 (spd) set 2			
$\alpha = 0.2, 0.6$	1%	−2%	10%
Ge (exp.)	5.65	0.77	3.85
8 (spd) set 1			
$\alpha = 0.2, 0.6$	−3%	−1%	−17%
20 (spd) set 2			
$\alpha = 0.2, 0.6$	−2%	−8%	7%

In Table 2 we present several results for a_0, the equilibrium lattice constant B, the bulk modulus, and E_{coh}, the cohesive energy. The quantities have been only approximately evaluated from a parabola fit to three chosen lattice constant values. This introduces no sizable error into a_0 and E_{coh}, but it is known to affect B on a 10% scale.

For Ge and Si we find that the "standard" basis set [5] of two decay lengths with ten (s, p, d and r^2) functions each, gives very good results, rather close to converged plane wave and LAPW calculations.[4]) The cohesive energy is obtained from a comparison to numerically integrated free atoms in spin-polarized s^2p^2 configurations. The atomic spin-polarization energy is −0.56 eV for Ge and −0.61 eV for Si. Note that the solid is overbound by some 7 to 10%, i.e. about 0.5 eV/atom which is typically found for LDA calculations [1]. If we restrict the 20 orbitals to an eight-orbital set, i.e. s with $\alpha = 0.2, 0.6$ and $p_{x,y,z}$, $d_{xy,yz,zx}$ with $\alpha = 0.2$, lattice constant and bulk modulus do not drastically change. Because of the decreased variational freedom, however, E_{coh} decreased by about 1 eV resulting in underbinding of the solid. Further restrictions of the basis set result in more drastic deterioration of the cohesive properties. We therefore conclude that about eight functions per atom represent the minimum basis for calculating reasonably accurate structural properties of Si and Ge. As will be discussed below, this basis also gives quite reasonable band structures. The results are thus rather promising, since they seem to allow us to study systems of up to 50 to 100 atoms with generally available computers.

For diamond the wave functions are more localized, in particular the p-states since they feel no core orthogonality repulsion. This necessitates the inclusion of an extra short decay length Gaussian. On the other hand, as also found in [11] the inclusion of d-states is not necessary. The three decay lengths of $\alpha = 0.32, 0.88$, and 3.13, which

[4]) See [5] for diamond and Si. The plane wave results for Ge are unpublished and also obtained from Bachelet [19].

Table 3

Calculated band structures (in eV) of diamond, Si, and Ge using local orbital basis sets (see Table 2 for explanations)

	diamond				Si			Ge		
	PW's	LO set 1	LO set 2	LO set 3	PW's	LO set 1	LO set 2	PW's	LO set 1	LO set 2
Γ_1	−21.45	−20.74	−21.62	−21.64	−11.93	−11.98	−12.01	−12.70	−12.38	−12.79
$\Gamma_{25'}$	0	0	0	0	0	0	0	0	0	0
$\Gamma_{15'}$	5.40	4.98	5.35	5.39	2.53	2.57	2.49	2.57	2.53	2.50
$\Gamma_{2'}$	13.38	13.57	13.35	13.34	3.29	3.43	3.15	0.09	0.61	− 0.11
X_1	−12.65	−13.24	−12.93	−12.94	− 7.78	− 7.89	− 7.89	− 8.84	− 8.61	− 8.97
X_4	− 6.22	− 6.85	− 6.38	− 6.54	− 2.88	− 3.01	− 2.85	− 3.05	− 3.03	− 3.03
X_1	4.63	9.49	5.17	5.22	0.61	1.32	0.80	0.70	1.80	1.01
X_4	16.73	16.60	16.84	16.49	9.97	10.11	10.00	9.46	9.47	9.49
$L_{2'}$	−15.57	−15.76	−15.85	−15.78	− 9.52	− 9.68	− 9.68	−10.62	−10.35	−10.73
L_1	−13.35	−13.56	−13.41	−13.63	− 7.00	− 7.06	− 7.02	− 7.56	− 7.35	− 7.62
$L_{3'}$	− 2.81	− 3.23	− 2.87	− 2.97	− 1.20	− 1.27	− 1.21	− 1.38	− 1.37	− 1.40
L_3	8.31	10.25	8.35	8.89	3.31	4.58	3.35	3.74	4.66	3.77
L_1	8.92	9.41	8.81	9.35	1.48	1.94	1.52	0.19	0.88	0.24

The results are compared to plane wave calculations for diamond, Si, and Ge (cf. Footnote[4])).

we found by minimizing the free atom energy are rather close to the values quoted in [11] of $\alpha = 0.25$, 0.94, and 3.5 which were obtained by a constrained search of the minimum crystalline energy. The cohesive energy for diamond contains a zero-point motion correction of 0.18 eV/atom for the solid and a -1.13 eV spin-polarization energy for the free atom which is obtained with the Ceperley-Alder form for exchange and correlation. This value differs somewhat from the -1.40 eV quoted in [11]. An accidental cancellation of differences in numbers, however, leads to an identical cohesive energy of 7.84 eV/atom. The results for E_{coh} quoted here in Table 2 differ from earlier calculations [3] using the same method and similar potentials and decay lengths due to a numerical error in the evaluation of the absolute value of E_{coh} in [5].

In Table 3 we compare the band structure results of our calculations to those obtained from converged plane wave calculations. While the larger basis set results compare favorably with the plane wave results, deviations increase for the smaller basis sets. The eight-orbital set for diamond yields rather sizable differences for the X_1 conduction band minimum. Similar discrepancies are also found in an all-electron full-core calculation by Zunger and Freeman [8].

As widely discussed in the literature, LDA energy gaps are too small compared to experiment. This can be seen here from the "converged" results in Table 3. The effect is particularly striking for Ge ($E_g \approx 0.9$ eV) where LDA yields about zero gap, provided scalar-relativistic effects are included. These effects shift the Γ_2' conduction band edge (s-like) down by about0.5 eV, an effect which has often been overlooked in the past, but which has recently been stressed in an independent calculation (cf. Footnote [4])).

In summary, we have investigated the possibility of reducing the size of local orbital basis sets used in calculating cohesive properties of semiconductors. We have found that with about eight to twelve functions per atom reasonable quantitative results can be obtained. This opens the possibility to study complex systems such as dislocations, grain boundaries, and other extended defects with rather larger numbers of atoms involved.

References

[1] M. Schlüter and L. J. Sham, Phys. today, **35**, 36 (1982).
[2] P. Hohenberg and W. Kohn, Phys. Rev. **136**, B 864 (1964).
W. Kohn and L. J. Sham, Phys. Rev. **140** A, 1133 (1965).
[3] V. L. Moruzzi, J. F. Janak, and A. R. Williams, Calculated Electronic Properties of Metals, Pergamon Press, 1978.
[4] J. Ihm and M. L. Cohen, Phys. Rev. B **21**, 1527 (1980).
[5] G. B. Bachelet, H. S. Greenside, G. A. Baraff, and M. Schlüter, Phys. Rev. B **24**, 4745 (1981).
[6] B. Harmon, W. Weber, and D. R. Hamann, Phys. Rev. B **25**, 1109 (1982).
[7] P. Glötzel, B. Segall, and O. K. Andersen, Solid State Commun. **36**, 403 (1980).
[8] A. Zunger and A. J. Freeman, Phys. Rev. B **15**, 5049 (1977).
[9] V. Mednick and L. Kleinman, Phys. Rev. B **22**, 5768 (1980).
[10] B. Delley, A. J. Freeman, and D. E. Ellis, Phys. Rev. Letters **50**, 488 (1983).
[11] J. R. Chelikowsky and S. G. Louie, Phys. Rev. B **29**, 3470 (1984).
[12] R. C. Chaney, T. K. Tung, C. C. Lin, and E. E. Lafon, J. chem. Phys. **52**, 361 (1970).
[13] P. J. Feibelman, D. R. Hamann, and J. A. Appelbaum, Phys. Rev. B **20**, 1433 (1979).
[14] D. R. Hamann, M. Schlüter, and C. Chiang, Phys. Rev. Letters **43**, 1494 (1980).
[15] D. M. Ceperley and B. J. Alder, Phys. Rev. Letters **45**, 566 (1980).
[16] J. Perdew and A. Zunger, Phys. Rev. B **23**, 5048 (1981).
[17] G. B. Bachelet, D. R. Hamann, and M. Schlüter, Phys. Rev. B **26**, 4199 (1982).
[18] P. J. Feibelman, private communication.
[19] G. B. Bachelet, private communication.

(Received September 25, 1984)

phys. stat. sol. (b) 126, 517 (1984)

Subject classification: 11; 10.1; 21.1; 21.6

M. I. Kalinin Leningrad Polytechnical Institute[1])

Athermal Vacancy Cluster Formation in F.C.C. Metals by High Energy PKA

By

A. I. MELKER and S. N. ROMANOV

Formulas describing the total cross sections for athermal formation of tri- and tetravacancies by a projectile atom are deduced using the impulse approximation. For Cu and Au, the formation probability of various configurations of these vacancy clusters created by a projectile atom with an energy $E = 10$ to 100 keV is calculated. It is shown that in Cu the 120°- and 180°-clusters (linear vacancy chains) are the most probable configurations, whereas in Au the most of trivacancies are the 120°- and 60°-ones and among tetravacancies prevail the 180°-ones.

В импульсном приближении выведены формулы полных поперечных сечений атермического образования три- и тетравакансий, созданных налетающим атомом. Для Cu и Au вычислена вероятность образования различных конфигураций этих вакансионных кластеров при энергии налетающего атома $E = 10$ до 100 keV. Показано, что в Cu наиболее вероятными конфигурациями являются 120°- и 180°-градусные три- и тетравакансии (линейные цепочки вакансий), тогда как в Au большинство тривакансий составляют 120°- и 60°-градусные, а среди тетравакансий преобладают 180°-градусные.

1. Introduction

It is well known [1, 2] that swelling is connected with an alternative formation of volume and plane vacancy clusters (vacancy voids and vacancy dislocation loops, respectively). The existing swelling theories use as the basis one of the following void formation models:

1) homogeneous formation,
2) heterogeneous formation via impurity atoms (usually gas atoms),
3) void formation within collision cascades.

For pure metals there is only the choice between the first and the third models. However, it has been recently shown [3] that the collision cascade void formation model agrees better with the experimental swelling results than the homogeneous model. The kinetic theories do not take into consideration the formation mechanism of vacancy voids and vacancy dislocation loops and postulate that the vacancy cluster growth is due to mobile defect trapping. Considering the vacancy cluster formation we see that a trivacancy is a branching [1] because a tetravacancy can already be either a volume or plane defect. Nevertheless, for nearly all metals it is still unknown which three- and tetravacancy configurations are more preferable energetically and which configurations may be formed in the collision cascades at the dynamic stage of nucleation.

The simplest vacancy cluster is a divacancy. The probability of its formation has been calculated in [4, 5] for a number of f.c.c. and b.c.c. metals. This cluster was generated by a primary knock-on atom (PKA) of the energy range 0.5 to 100 keV.

[1]) Ul. Polytekhnicheskaya 29, 195251 Leningrad, USSR.

In the present paper the same method [5] was used to calculate the formation probabilities of the three- and tetravacancy configurations in f.c.c. metals.

2. Impact Displacement Area

The high-energy PKA collisions may be divided into two classes. Let b be a collision diameter and D_{hkl} the inter-atomic distance in $\langle hkl \rangle$ direction. Consider the moment of the nearest approach. If the distance between PKA and a crystal lattice atom is $r_{\min} \approx b \ll D_{hkl}$, the PKA scatters at high angles. If $b < r_{\min} < \varrho_d \sim D_{hkl}$, where ϱ_d is a displacement impact parameter corresponding to a displacement impulse [5], the PKA scatters at small angles. In the first case the collision cascade branches and a single vacancy is formed in the branching point from which two new high-energy cascade branches grow. In the second case single vacancies and their clusters are created along the PKA trajectory and the impulses of knock-on atoms are normal to this trajectory [5]. This picture is in good agreement with the computer simulation results on the structure of high-energy collision cascades obtained by the dynamic method (for example, see Fig. 8 from [6] where a 100 keV collision cascade in α-iron is shown).

We are concerned with the process of the vacancy formation along a high-energy branch. The projectiles have various impact parameters. In impulse approximation [5] the total cross section of single vacancy formation is equal to the impact displacement area

$$\sigma(V_1) = \pi \varrho_d^2 . \tag{1}$$

An $\langle hkl \rangle$ divacancy is formed if the projectile comes through the shadow created by the impact displacement area of the first atom on the analogous area of the second one (see Fig. 1 from [5]). The shadow area depends on the θ-angle between the projectile impulse drawn through the first impact displacement area centre and the $\langle hkl \rangle$ direction. Upon integrating the shadow area over the θ-angle range and multiplying the result by the number of different orientations of a divacancy in a crystal, we can obtain the total cross section of divacancy formation,

$$\sigma(V_2, hkl) .$$

There are four different trivacancy configurations in a f.c.c. lattice (Fig. 1). These configurations can be characterized by the angle at the top of an isosceles triangle. The possible values of χ are 180°, 120°, 90°, and 60°. A 180°-trivacancy is put along a $\langle 110 \rangle$ direction, if $\chi = 120°$ or 60° the trivacancy is placed over a $\{111\}$ plane, and in the case of $\chi = 90°$ the trivacancy plane is $\{100\}$.

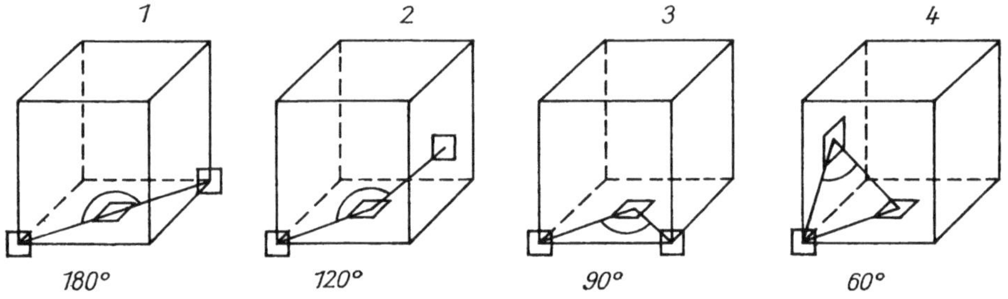

Fig. 1. Trivacancy configurations in a f.c.c. lattice

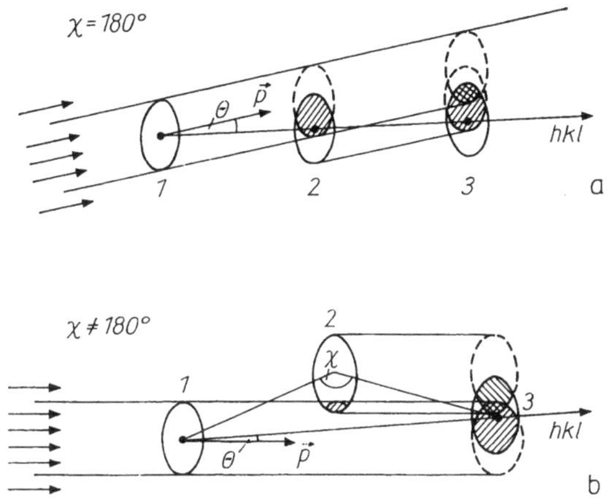

Fig. 2. Geometry of trivacancy formation, a) line configuration, b) plane configuration

Consider the probability for trivacancy formation. The impact displacement areas of the first and second atoms produce the shadow on the displacement area of the third one (Fig. 2a). Similarly as in astronomy we call these shadows semishadows. If a projectile is coming through the overlapping area of two semishadows, it is obvious that a trivacancy is formed. Upon averaging the square of the shadow area over the θ-angle range and multiplying the obtained result by the number of the different lattice orientations of a chosen trivacancy configuration, the total cross section of this configuration formation, $\sigma(V_3, \chi)$, can be found in the way similar to [5]. It is not difficult to be certain that the shadow area for a 180°-trivacancy is the same as for a divacancy with the doubled interatomic distance between single vacancies (Fig. 2b). Hence the total cross section of a 180°-trivacancy may be computed directly from the formulas in [5] for a divacancy after replacing D_{hkl} by $2D_{hkl}$.

In general, the characteristic angle χ is not equal to 180° (Fig. 2b), the projectile impulse is not put over the trivacancy plane, and the real shadow form distinguishes from the one shown in Fig. 2b. The straight calculation of the total cross section for trivacancy formation leads to large computational difficulties. Therefore, it is better to proceed in a different way.

3. Effective Cross Section of Cluster Formation

According to [5] the effective total cross section of divacancy formation can be written in the form

$$\sigma(V_2, hkl) = w_{12}(hkl)\,\sigma(V_1)\,, \tag{2}$$

where $w_{12}(hkl)$ is the probability of the transition from a monovacancy to a divacancy along $\langle hkl\rangle$. For each divacancy put into connection the top of the displacement cone [5] coincides with the first vacancy, the symmetry axis is directed along $\langle hkl\rangle$, and the top angle is equal to $2\theta_d$. If the parallel particle beam passes through the impact displacement area of the first vacancy, $\theta_d = \arcsin 2\varrho_d/D_{hkl}$ is the maximum angle at which the beam can still create the second vacancy. We draw a unit sphere around the first vacancy. Then the displacement cone cuts a displacement solid angle Ω_d out of the sphere and we can write down the transition probability in the form

$$w_{12}(hkl) = Z_2(hkl)\,\frac{\Omega_d}{4\pi}\,\langle w\rangle\,. \tag{3}$$

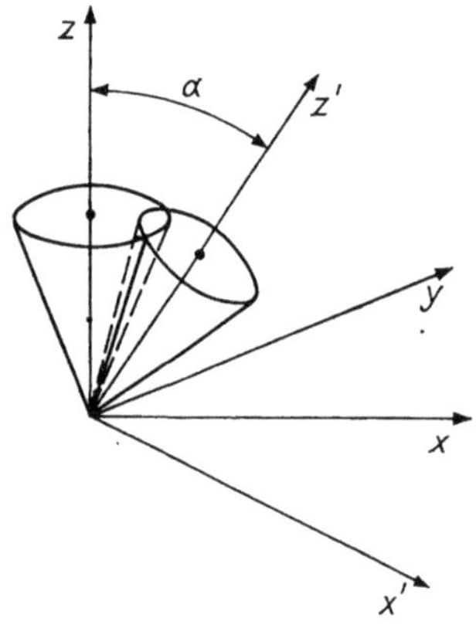

Fig. 3. Displacement cones intersection in the case of trivacancy formation

Here $Z_2(hkl)$ is the number of different orientations of a divacancy in a crystal, $\Omega_d/4\pi$ the intersection probability for the particle beam and a displacement cone; $\langle w \rangle$ the average probability for a beam particle to hit a displacement cone provided that the beam and the cone intersect.

We place in a crystal lattice site the tops of two displacement cones whose symmetry axes coincide with the divacancies forming a trivacancy (Fig. 3). If the two neighbouring displacement cones overlap, it means that a projectile can create an athermal trivacancy. As in the case of a divacancy the effective cross section of a trivacancy formation can be written in the form

$$\sigma(V_3, \chi) = Z_3 \frac{\Omega_d(12)}{4\pi} \langle w_1 \rangle \langle w_2 \rangle \sigma(V_1) . \tag{4}$$

Here Z_3 is the orientation number of trivacancy, $\Omega_d(12)$ the solid angle of overlapping for two displacement cones, $\langle w_i \rangle$ the average probability for the i-th particle of the beam to hit the i-th displacement cone.

The transition probabilities and the displacement solid angles of divacancies were discussed in [5], therefore, by using formula (3) it is not difficult to calculate $\langle w_i \rangle$. The orientation number can be obtained by the following method. Let us take one of the two vacancies forming a divacancy as the initial one and find the number of ways Z_{23} by means of which the third vacancy can join the divacancy. Then $Z_3 = Z_2(hkl)\, Z_{23}$. For example, if the initial vacancy is along $\langle 110 \rangle$, then the number of ways of joining the third vacancy equals to 1, 4, 2, and 4 for $\chi = 180°, 120°, 90°$, and $60°$, respectively. Therefore, the problem of finding the effective total cross section of trivacancy formation is reduced to calculating the solid angle of overlapping, $\Omega_d(12)$.

4. Solid Angle of Overlapping

The intersection lines of the unit sphere and displacement cones are circles. These circles are placed on the plane the distance of which from the coordinate origin is equal to $p_i = \cos \theta_i$. Here $2\theta_i$ is the plane angle at the top of the i-th displacement cone. Let $\boldsymbol{n}_i$ be directed along the i-th displacement cone symmetry axis, $\boldsymbol{n}_1$ coinciding with the z-axis, and the angle between $\boldsymbol{n}_2$ and z-axis be equal to $\alpha = (\pi - \chi)/2$ (Fig. 3). If the y-axis is common, then $\boldsymbol{n}_1 = (0, 0, 1)$ and $\boldsymbol{n}_2 = (\sin \alpha, 0, \cos \alpha)$. Let the second unit vector $\boldsymbol{n}_2$ be known. Then the equation of the plane on which the second circle is placed can be written as follows:

$$x \sin \alpha + z \cos \alpha = p_2 . \tag{5}$$

On solving the system consisting of (5) and the unit sphere equation,

$$x^2 + y^2 + z^2 = 1 , \tag{6}$$

we get that

$$a_2 y^2 + z^2 + c_{21} z + c_{20} = 0 , \tag{7}$$

where

$$a_2 = \sin^2 \alpha , \qquad c_{21} = 2p_2 \cos \alpha , \qquad c_{20} = p_2^2 - \sin^2 \alpha .$$

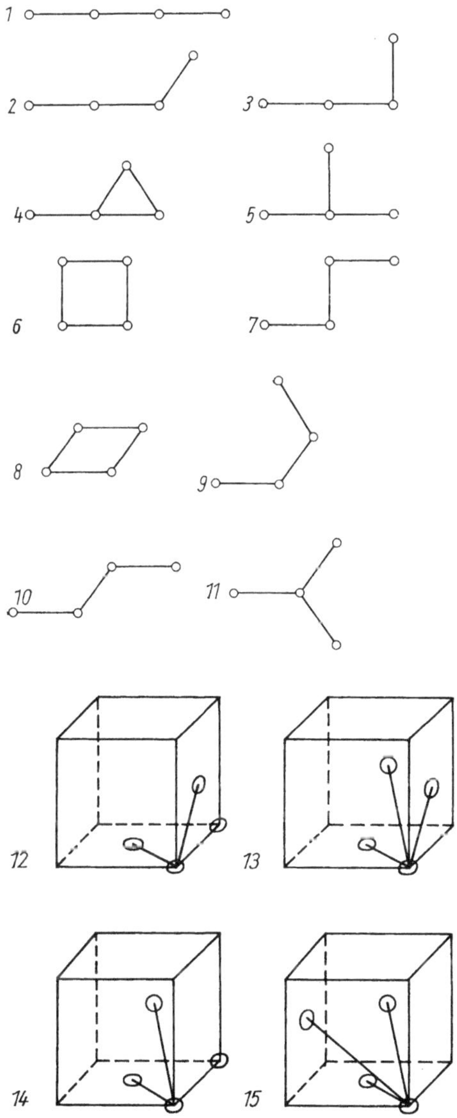

Fig. 4. Tetravacancy configurations in a f.c.c. lattice

Introducing spherical coordinates $y = \sin\theta \sin\varphi$, $z = \cos\theta$, where θ, φ are the respective angles, (7) can be rewritten in the following way:

$$\varphi_2(\theta) = \arcsin\left(-C_2(\theta)/A_2(\theta)\right)^{1/2}, \tag{8}$$

where

$$A_2(\theta) = a_2 \sin^2\theta\,, \qquad C_2(\theta) = \cos^2\theta + c_{21}\cos\theta + c_{20}\,. \tag{9}$$

Hence, the solid angle of overlapping for two displacement cones equals to

$$\Omega_d(12) = 2\int_{\alpha-\theta_2}^{\theta_1} \sin\theta\, d\theta \int_0^{\varphi_2(\theta)} d\varphi =$$

$$= 2\int_{t_1}^{t_2} \arcsin\left(-C_2(t)/A_2(t)\right)^{1/2} dt\,, \tag{10}$$

where $t_1 = \cos\theta_1$; $t_2 = \cos(\alpha - \theta_2)$.

Now we consider tetravacancy configurations in a f.c.c. crystal lattice. These configurations may be linear, plane, or volume ones. As in the case of a linear trivacancy, the formation of a linear tetravacancy is identical to that of a divacancy with a trebled interatomic distance (Fig. 4.1). Therefore, the formulas for divacancy formation [5] may be used to calculate this process.

There is also a subset of plane tetravacancy configurations among all the configuration sets (Fig. 4.2 to 4.5). For this subset the effective cross sections of different tetravacancy formation can be computed in the same way as for trivacancies. The other plane tetravacancies have three displacement cones (Figs. 4.6 to 4.11). Their symmetry axes are placed on the same plane. In this case the effective total cross section is determined by the solid angle of overlapping of the two extreme cones. This is mathematically equivalent to trivacancy formation.

The position of the third displacement cone for volume tetravacancy configurations (Fig. 4.12 to 4.15) depends on two rotations around the y- and z-axes (β and γ angles, respectively (Fig. 5)), therefore $\boldsymbol{n}_3 = (\sin\beta\cos\gamma, \sin\beta\sin\gamma, \cos\beta)$. The equation of the third plane containing the third circle is

$$x\sin\beta\cos\gamma + y\sin\beta\sin\gamma + z\cos\beta = p_3\,. \tag{11}$$

Similarly to the case of the second circle we find that the following equation takes place for the third circle:

$$A_3(\theta)\sin^2\varphi - 2B_3(\theta)\sin\varphi + C_3(\theta) = 0\,, \tag{12}$$

where

$$A_3(\theta) = a_3\sin^2\theta\,,$$

$$2B_3(\theta) = (b_{30} + b_{31}\cos\theta)\sin\theta\,,$$

$$C_3(\theta) = c_{32}\cos^2\theta + c_{31}\cos\theta + c_{30}\,,$$

$$a_3 = \sin^2\beta\,, \qquad b_{30} = 2p_3\sin\beta\sin\gamma\,, \qquad b_{31} = \sin 2\beta\sin\gamma\,,$$

$$c_{32} = \sin^2\beta\cos^2\gamma + \cos^2\beta\,, \qquad c_{31} = -2p_3\cos\beta\,,$$

$$c_{30} = p_3^2 - \sin^2\beta\cos^2\gamma\,.$$

Thus, we have

$$\varphi(\theta) = \arcsin\frac{B_3(\theta) + \left(B_3^2(\theta) - A_3(\theta)\,C_3(\theta)\right)^{1/2}}{A_3(\theta)} \tag{13}$$

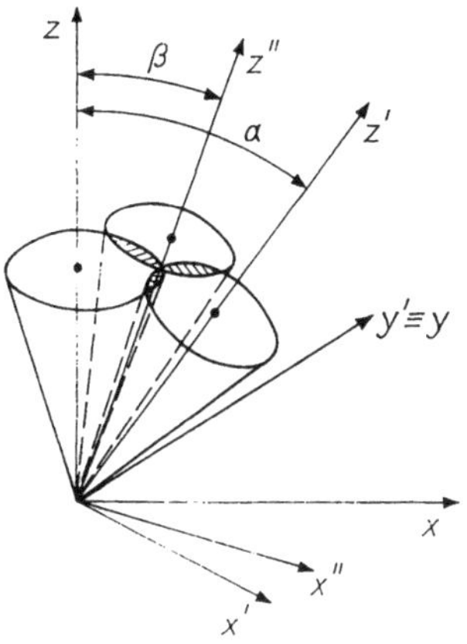

Fig. 5. Displacement cones intersection in the case of tetravacancy formation

and the solid angle of the three cones overlapping is equal to

$$\Omega_{\mathrm{d}}(123) = \int\limits_{\alpha-\theta_2}^{\theta_{\mathrm{m}}} \sin\theta\, \mathrm{d}\theta \int\limits_{\varphi_3(\theta)}^{\varphi_2(\theta)} \mathrm{d}\varphi =$$

$$= \int\limits_{t_{\mathrm{m}}}^{t_2} \left[\arcsin\left(-C_2(t)/A_2(t)\right)^{1/2} - \arcsin \frac{B_3(t) + \left(B_3(t) - A_3(t)\, C_3(t)\right)^{1/2}}{A_3(t)}\right] \mathrm{d}t\,. \tag{14}$$

Here $t_{\mathrm{m}} = \cos\theta_{\mathrm{m}}$ is the intersection point of the second and third circles, corresponding to the minimum of angle θ. From (8) and (12) one can obtain the equation for t_{m}

$$\left(C_3 - \frac{C_2 A_3}{A_2}\right)^2 + \frac{4 B_3^2 C_2}{A} = 0\,. \tag{15}$$

This equation can be rewritten as follows:

$$At^4 + Bt^3 + Ct^2 + Dt + E = 0\,, \tag{16}$$

where

$$A = a_2(c_{32} - a_3/a_2)^2 + b_{31}^2\,,$$

$$B = 2a_2(c_{32} - a_3/a_2)\,(c_{31} - c_{21}a_3/a_2) + b_{31}(b_{31}\,c_{21} + 2b_{30})\,,$$

$$C = a_2(c_{31} - c_{21}a_3/a_2)^2 + 2(c_{32} - a_3/a_2)\,(c_{30} - c_{20}a_3/a_2) + \\ + b_{30}^2 + 2b_{30}b_{31}c_{12} + b_{31}^2c_{20}\,,$$

$$D = 2a_2(c_{31} - c_{21}a_3/a_2)\,(c_{30} - c_{20}a_3/a_2) + b_{30}^2c_{21} + 2b_{30}b_{31}c_{20}\,,$$

$$E = a_2(c_{30} - c_{20}a_3/a_2)^2 + b_{30}^2c_{20}\,.$$

5. Results and Discussion

The formation probabilities for athermal vacancy clusters in copper and gold are given in Table 1. One may see the following features. In copper the most popular configurations are linear vacancy chains (120°- and 180°-clusters). Their formation probability

Table 1

configuration number	E (keV)	Cu		Au	
		σ (10^{-20} m^2)	w	σ (10^{-20} m^2)	w
trivacancy					
1	10	0.309	0.107	1.204	0.188
	30	0.176	0.081	0.845	0.157
	50	0.131	0.070	0.709	0.144
	75	0.102	0.061	0.613	0.134
	100	0.084	0.056	0.551	0.128
2	10	0.427	0.153	2.567	0.402
	30	0.205	0.097	1.588	0.296
	50	0.139	0.076	1.258	0.256
	75	0.100	0.062	1.041	0.227
	100	0.078	0.053	0.907	0.209
3	10	0.048	0.017	0.253	0.040
	30	0.019	0.007	0.178	0.033
	50	0.012	0.006	0.144	0.029
	75	0.007	0.004	0.120	0.026
	100	0.004	0.003	0.105	0.024
4	10	0.111	0.040	1.849	0.289
	30	0.036	0.017	1.540	0.286
	50	0.017	0.009	0.451	0.092
	75	0.007	0.004	0.340	0.074
	100	0.003	0.002	0.281	0.065
tetravacancy					
1	10	0.136	0.047	0.525	0.082
	30	0.078	0.036	0.370	0.069
	50	0.058	0.031	0.311	0.063
	75	0.045	0.027	0.269	0.059
	100	0.037	0.025	0.242	0.056
2	10			0.349	0.055
	30	0.065	0.031	0.221	0.041
	50	0.038	0.021	0.175	0.036
	75	0.025	0.015	0.143	0.031
	100	0.017	0.012	0.123	0.029
3	10	0.029	0.011	0.191	0.030
	30	0.010	0.005	0.123	0.023
	50	0.005	0.003	0.111	0.022
	75	0.003	0.002	0.087	0.019
	100	0.001	0.001	0.072	0.017
4	10	0.057	0.020	0.428	0.067
	30	0.021	0.010	0.251	0.047
	50	0.011	0.006	0.195	0.040
	75	0.005	0.003	0.157	0.034
	100	0.003	0.002	0.134	0.031
12	10	—	—	0.010	0.0016
	30	—	—	0.064	0.0011
	50	—	—	0.035	0.0007
	75	—	—	0.014	0.0003
	100	—	—	0.0007	0.0002

idminishes with increasing energy of a projectile, and at high energy they are practically the only athermal clusters. However, there is a somewhat different picture for gold. At low energy, mono- and divacancies are not generated at all, the most of trivacancies are 120°- and 60°-ones and 180°-configurations prevail over other tetravacancies. This difference is connected with the form of the repulsive potential. For gold, the pre-exponential constant of the Born-Mayer potential is approximately ten times that of copper [7]. As a result, the displacement impact parameter in gold is again half that in copper [5], therefore, the displacement cones in gold are overlapping to a greater extent.

References

[1] Yu. V. Konobeev, in: Modelirovanie na EVM defektov v kristallakh, FTI Akad. Nauk SSSR, Leningrad 1979 (p. 40).
[2] I. A. Akhiezer and L. N. Davydov, Metallofizika **3**, No. 1, 3, No. 3, 3 (1981).
[3] S. I. Golubov and Yu. V. Konobeev, Zh. tekh. Fiz., Pisma 7, 1420 (1981).
[4] S. N. Romanov and A. I. Melker, Zh. tekh. Fiz. **51**, 1252 (1981).
[5] A. I. Melker and S. N. Romanov, phys. stat. sol. (b) **122**, K17 (1984).
[6] V. V. Kirsanov, Defekty v kristallakh i ikh modelirovanie na EVM, Izd. Nauka, 1980 (p. 134).
[7] M. W. Thompson, Defects and Radiation Damage in Metals, Cambridge University Press, 1969.

(Received April 30, 1984)

phys. stat. sol. (b) 126, 527 (1984)

Subject classification: 10

Department of Mechanical Engineering and Engineering Materials Group, University of Maryland, College Park[1])

The Role of Surface Dislocations in the Continuum Theory of Lattice Defects

By

M. J. MARCINKOWSKI

The close relationship between the concept of surface dislocations and the continuum theory of lattice defects is clearly demonstrated. In fact, it is shown that without the incorporation of surface dislocations into the classical continuum theories, it is not possible to obtain a complete and self-consistent understanding of elasto–plastic bending, grain boundary structure, and the behavior of deformed finite bodies in general.

Es wird deutlich die enge Beziehung zwischen der Vorstellung von Oberflächenversetzungen und der Kontinuumstheorie von Gitterdefekten dargestellt. Tatsächlich wird gezeigt, daß ohne die Einbeziehung von Oberflächenversetzungen in die klassische Kontinuumstheorie es nicht möglich ist, ein vollständiges und selbstkonsistentes Verständnis der elasto–plastischen Verbiegung, der Korngrenzenstruktur und des Verhaltens von deformierten Körpern im allgemeinen zu erhalten.

1. Introduction

It has been shown repeatedly [1 to 5] that the concept of surface dislocations allows any problem dealing with the mechanical behavior of matter to be solved with a high degree of mathematical rigor and with deep physical insight. For the specific case of grain boundaries, this approach was found to be indispensable in clarifying the nature of such boundaries [6, 7]. It would therefore appear quite logical at this point to inquire into the reasons as to why the concept of surface dislocations does not appear in the beautifully abstract continuum theories of lattice defects such as developed by Kröner [8, 9]. The purpose of this study is to show that not only are surface dislocations compatible with such continuum theories, but that they, in fact, give them a completeness which has been hitherto lacking. For the sake of clarity and simplicity, the theory will be limited to the linear approximation.

2. Background

When a body is deformed in a purely elastic manner, the elastic strains e_{km} must obey the following compatibility condition [10, 11]:

$$4(\partial_{nl} e_{km})_{[nm][lk]} = 0 \,, \tag{1}$$

where the square brackets indicate that the antisymmetric part is to be taken with respect to the enclosed subscripts. Equation (1) insures that neither gaps nor material overlap is present within the body as a result of the deformation. If, on the other hand, both elastic and plastic strain coexist, as in the case of a dislocated body, then the compatibility condition becomes [11]

$$4(\partial_{nl} e^{\mathrm{T}}_{km})_{[nm][lk]} = 0 \,, \tag{2}$$

[1]) College Park, Maryland 20742, USA.

where e^{T}_{km} is the total strain, and is given by

$$e^{\mathrm{T}}_{km} = e_{km} + e^{\mathrm{P}}_{km} \tag{3}$$

where e^{P}_{km} denotes the plastic strain. Combining (2) and (3) yields

$$4(\partial_{nl} e_{km})_{[nm][lk]} = -G_{nmlk}\,, \tag{4}$$

where G_{nmlk} is the incompatibility tensor defined by

$$G_{nmlk} \equiv 4(\partial_{nl} e^{\mathrm{P}}_{km})_{[nm][lk]}\,. \tag{5}$$

Equation (4) states that neither the elastic, nor the plastic distortions, taken by themselves, are compatible, but that the incompatibility occasioned by the plastic strain is accommodated by a complementary elastic strain. It follows from (5) that for a compatible pure plastic distortion, $G_{nmlk} = 0$.

It can be shown [8, 11] that the incompatibility tensor gives rise to a dislocation density α_{nk} which can be expressed as follows:

$$\alpha_{nk} \equiv \varepsilon_{nml}(\partial_m e^{\mathrm{P}}_{lk} + \varepsilon_{qlk} \varkappa^{\mathrm{P}}_{mq})\,, \tag{6}$$

where ε_{nml} is the permutation tensor, while $\varkappa^{\mathrm{P}}_{mn}$ is the plastic bend-twist given by

$$\varkappa^{\mathrm{P}}_{mn} = \partial_m \omega^{\mathrm{P}}_n\,, \tag{7}$$

where ω^{P}_n is the plastic rotation which may be written as

$$\omega^{\mathrm{P}}_n = \tfrac{1}{2}\,\varepsilon_{nlk} \beta^{\mathrm{P}}_{lk}\,, \tag{8}$$

where β^{P}_{lk} is the plastic distortion given by

$$\beta^{\mathrm{P}}_{lk} = \partial_l u^{\mathrm{P}}_k\,, \tag{9}$$

where u^{P}_k is the plastic displacement. The plastic strain in (6) can also be written in terms of the displacements as follows:

$$e^{\mathrm{P}}_{lk} = \partial_{(l} u^{\mathrm{P}}_{k)}\,, \tag{10}$$

where the parentheses indicate that the symmetric part is to be taken with respect to the enclosed subscripts. Because of the complimentary nature of the elastic and plastic strains embodied in (4), α_{nk} can be written in terms of the elastic strain and bend-twist as follows [11]:

$$\alpha_{nk} = -\varepsilon_{nml}(\partial_m e_{lk} + \varepsilon_{qlk} \varkappa_{mq}) \tag{11}$$

rather than in terms of the corresponding plastic quantities given by (6). In addition, the elastic counterparts of (7) to (10) are given by

$$\varkappa_{mn} = \partial_m \omega_n\,, \tag{12}$$

$$\omega_n = \tfrac{1}{2}\,\varepsilon_{nlk} \beta_{lk}\,, \tag{13}$$

$$\beta_{lk} = \partial_l u_k\,, \tag{14}$$

$$e_{lk} = \partial_{(l} u_{k)}\,, \tag{15}$$

respectively.

3. Elastic Bending

Following de Wit [11], we shall first interpret the equations of the previous section in terms of simple elastic bending. This case has previously been treated to a limited extent by means of surface dislocations [12, 13]. In particular, consider the two-dimensional elastically bent body depicted in Fig. 1a which may be described by the following displacements:

$$u_1 = A x_1 x_2\,, \qquad u_2 = -\tfrac{1}{2} A x_1^2\,. \tag{16}$$

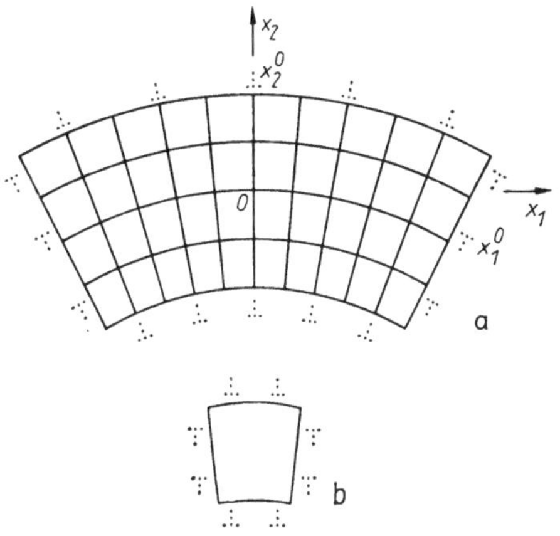

Fig. 1. a) Elastic bending in terms of surface dislocations. b) Infinitesimal element associated with a)

For simplicity, Poisson's ratio ν has been assumed equal to zero. Equation (16) yields for the quantities given by (12) to (15),

$$\begin{aligned}
\beta_{11} &= Ax_2 , & \beta_{12} &= -Ax_1 , \\
\beta_{21} &= Ax_1 , & \beta_{22} &= 0 , \\
e_{11} &= Ax_2 , & e_{12} &= 0 , \\
e_{21} &= 0 , & e_{22} &= 0 , \\
\omega_3 &= -Ax_1 , & \varkappa_{13} &= -A .
\end{aligned} \tag{17}$$

For the present case, (11) reduces to

$$\alpha_{31} = -\partial_1 e_{21} + \partial_2 e_{11} + \varkappa_{13} \tag{18a}$$

or in view of (17)

$$\alpha_{31} = 0 + A - A = 0 . \tag{18b}$$

Intuitively, one would expect a zero dislocation density to be associated with an elastically deformed body. On the other hand, what is the meaning to be attached to the individual non-vanishing terms in (18b)? This is best seen by referring to the infinitesimal body in Fig. 1b which may be viewed as cut out from the larger body of Fig. 1a. The horizontal faces of the infinitesimal body have associated with them surface dislocations (drawn dotted) of density $+A$ corresponding to the $\partial_2 e_{11}$ term of (18a), whereas the vertical faces possess a dislocation density $-A$ corresponding to the $\varkappa_{13}$ term of (18a). Since the dislocations are continuously distributed, and since α_{31} corresponds to the dislocation density at a point, it vanishes in accordance with (18b). Strictly speaking, we must use discontinuous functions to express the dislocation densities associated with Fig. 1b, which in turn leads to a reversal of signs in (18b). The reason for this will be made clear in the following paragraphs.

It follows from the above discussion that equations (17) apply to an infinite body. For a finite body of thickness $2x_2^0$ and length $2x_1^0$, they must be modified as follows:

$$\begin{aligned}
\beta_{11} &= Ax_2 \mathrm{H}(-x_2 + x_2^0) , & \beta_{12} &= -Ax_1 \mathrm{H}(-x_1 + x_1^0) , \\
\beta_{21} &= Ax_1 \mathrm{H}(-x_1 + x_1^0) , & \beta_{22} &= 0 , \\
e_{11} &= Ax_2 \mathrm{H}(-x_2 + x_2^0) , & e_{12} &= 0 , \\
e_{21} &= 0 , & e_{22} &= 0 , \\
\omega_3 &= -Ax_1 \mathrm{H}(-x_1 + x_1^0) , & \varkappa_{13} &= Ax_1 \delta(x_1 - x_1^0) ,
\end{aligned} \tag{19}$$

where $H(x_k - x_k^0)$ are Heaviside functions defined by [4]

$$H(-x_k + x_k^0) = \begin{cases} 0 & \text{if} \quad x_k - x_k^0 > 0 \\ 1 & \text{if} \quad x_k - x_k^0 < 0 , \end{cases} \tag{20}$$

while

$$\partial_k H(-x_k + x_k^0) = -\delta(x_k - x_k^0) , \tag{21}$$

where $\delta(x_k - x_k^0)$ is the Dirac delta function defined such that $\delta(x_k - x_k^0) = 0$ everywhere except for $x_k = x_k^0$. In view of (19), (18a) becomes

$$\alpha_{31} = 0 - Ax_2\,\delta(x_2 - x_2^0) + Ax_1\,\delta(x_1 - x_1^0) . \tag{22}$$

Because of the following property of the Dirac delta function [14]

$$\int_{-\infty}^{+\infty} f(x_k)\,\delta(x_k - x_k^0)\,\mathrm{d}x_k = f(x_k^0) \tag{23}$$

(22) can be integrated to yield an expression for the surface dislocation density given by

$$\alpha_{31}^{S} = 0 - Ax_2^0 + Ax_1^0 . \tag{24}$$

A further integration of α_{31}^{S} along the surface of the upper right quadrant of Fig. 1a yields

$$b_1 = 0 - Ax_2^0x_1^0 + Ax_1^0x_2^0 = 0 , \tag{25}$$

which is simply the powerful law of the conservation of the Burgers vector $\boldsymbol{b}_1$ [4, 5]. Equation (25) could have also been arrived at more directly by means of the following surface integral [8]:

$$b_k = \int_S \alpha_{nk}\,\mathrm{d}S_n . \tag{26}$$

It might at first seem that the surface dislocations depicted in Fig. 1a are to a certain degree fictitious since there appears to be no extra half planes associated with them. This however is not true since the extra half planes may be visualized as being contained in the surroundings of the finite body and terminate thereon. In this respect, the elastic deformation associated with a finite body may be viewed as the response to the plastic distortions occasioned by the surface dislocations so as to satisfy the compatibility condition given by (2). Further credence to this view is given by the fact that the surface dislocation density associated with Fig. 1a can be given in terms of plastic distortions as embodied by (6). This is done by writing

$$e_{11}^{P} = Ax_2^0 H(x_2 - x_2^0) , \qquad \omega_3^{P} = -Ax_1^0 H(x_1 - x_1^0) , \tag{27}$$

where

$$H(x_k - x_k^0) = \begin{cases} 0 & \text{if} \quad x_k - x_k^0 < 0 \\ 1 & \text{if} \quad x_k - x_k^0 > 0 , \end{cases} \tag{28}$$

which when substituted into (6) and integrated yields a result identical to that of (25). The sign convention adopted for the Burgers vector is that of SF/RH [11, 15]. In general, a second set of secondary surface dislocations with Burgers vectors normal to those shown in Fig. 1a would be required if Poisson's ratio is taken different from zero [4, 5, 13]. However, these secondary dislocations contribute only minimally to the surface tractions.

4. Elasto–Plastic Bending

The classical picture of elasto–plastic bending, and one which was heralded as a triumph of dislocation theory, is depicted in Fig. 2a [11, 15, 16]. Unfortunately, it is

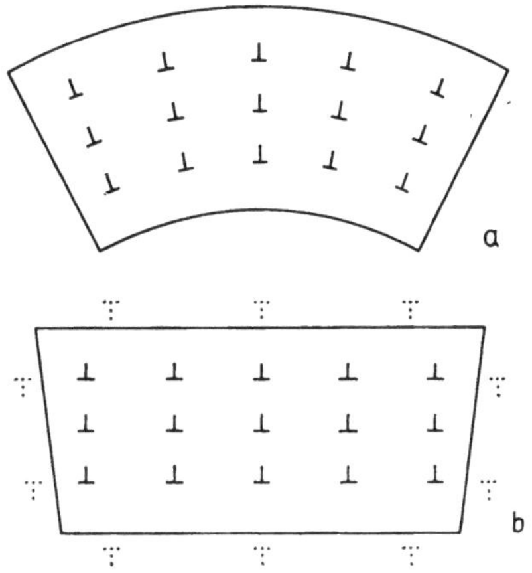

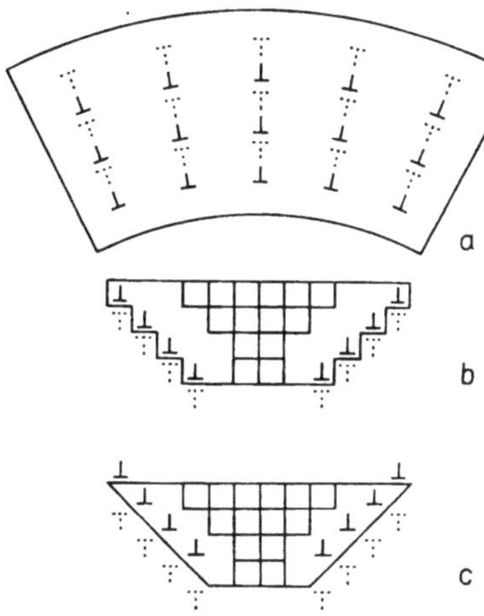

Fig. 2 Fig. 3

Fig. 2. a) Classical picture of elasto-plastic bending. b) Non-bent counterpart of a)

Fig. 3. a) Surface dislocation model of elasto-plastic bending. Element which has undergone b) plastic, c) elasto-plastic deformation

at best incomplete and at worst deceptive. This is clear from the fact that the following powerful law of conservation of Burgers vectors is violated [4, 5]:

$$\sum b_{\mathrm{L}} + \sum b_{\mathrm{S}} = 0\,, \tag{29}$$

where the subscripts L and S refer to lattice and surface dislocations respectively. Clearly, $b_{\mathrm{S}} = 0$ in Fig. 2a, while the b_{L} are all of the same sign. If the bending in Fig. 2a were to be eliminated, while still maintaining the same lattice dislocation density, then Fig. 2b would obtain. A set of surface dislocations (shown dotted) would clearly be required to satisfy the stress-free boundary conditions on the surface of the body [4]. Equation (29) would thus be satisfied. The question now remaining is: where are the surface dislocations associated with an elasto-plastically bent body? The answer is provided by Fig. 3a, where it can be seen that each lattice dislocation has associated with it a corresponding surface dislocation, i.e., a dipole. The reasoning here is similar to that used in the solution of the grain boundary problem [6, 7].

In order to see how the configuration in Fig. 3a arises, let us first consider the following plastic distortions and their accompanying strains:

$$\begin{aligned}
\beta_{11}^{\mathrm{P}} &= Ax_2\,, & \beta_{12}^{\mathrm{P}} &= -Ax_1\,,\\
\beta_{21}^{\mathrm{P}} &= Ax_1\,, & \beta_{22}^{\mathrm{P}} &= 0\,,\\
e_{11}^{\mathrm{P}} &= Ax_2\,, & e_{12}^{\mathrm{P}} &= 0\,,\\
e_{21}^{\mathrm{P}} &= 0\,, & e_{22}^{\mathrm{P}} &= 0\,,\\
\omega_3^{\mathrm{P}} &= -Ax_1\,, & \varkappa_{13}^{\mathrm{P}} &= -A\,.
\end{aligned} \tag{30}$$

These distortions may be viewed in terms of the infinitesimal volume element shown in Fig. 3b. From (6), we may write the corresponding dislocation density in terms of the plastic strains as follows:

$$\alpha_{31} = \partial_1 e_{21}^{\mathrm{P}} - \partial_2 e_{11}^{\mathrm{P}} - \varkappa_{13}^{\mathrm{P}}\,, \tag{31a}$$

which in terms of equations (30) yields

$$\alpha_{31} = 0 - A + A = 0 . \tag{31b}$$

The term $-A$ in the above equation arises from the lattice dislocations in Fig. 3b which give rise to the lateral distortions, whereas the $+A$ term arises from the surface dislocations which give rise to the rotations of the vertical faces. Both sets of dislocations may be viewed as combining with each other so as to generate the ledge or stepped surfaces, as well as giving rise to a corresponding elimination of the elastic distortions, in accordance with the following relation proceeding toward the right [4, 6, 7]:

$$b_L + b_S \rightleftarrows L , \tag{32}$$

where L denotes the strength of the ledge.

In order to make Fig. 3b consistent with Fig. 3a, the ledges associated with the infinitesimal element must be eliminated as depicted in Fig. 3c. This effect is equivalent to allowing the surface and lattice dislocations to uncouple in accordance with the reaction of (32) towards the left. This in turn generates elastic strains. Fig. 3c thus correctly represents an infinitesimal volume element of Fig. 3a with the corresponding set of distortions and strains given by

$$\begin{gathered}
\beta_{11}^{P} = Ax_2 , \quad \beta_{12}^{P} = 0 , \quad \beta_{21}^{P} = 0 , \quad \beta_{22}^{P} = 0 , \\
e_{11}^{P} = Ax_2 , \quad e_{12}^{P} = 0 , \quad e_{21}^{P} = 0 , \quad e_{22}^{P} = 0 , \\
\omega_3^{P} = 0 , \quad \varkappa_{13}^{P} = 0 , \\
\beta_{11} = 0 , \quad \beta_{12} = -Ax_1 , \quad \beta_{21} = Ax_1 , \quad \beta_{22} = 0 , \\
e_{11} = 0 , \quad e_{12} = 0 , \quad e_{21} = 0 , \quad e_{22} = 0 , \\
\omega_3 = -Ax_1 , \quad \varkappa_{13} = -A .
\end{gathered} \tag{33}$$

When the above relations are applied to (31a) we obtain

$$\alpha_{31} = 0 - A + 0 = -A , \tag{34}$$

which represents the lattice dislocation density of Fig. 3c denoted by the solid symbols. As in the case of pure elastic bending, the elastic distortions occasioned by the surface dislocations in Fig. 3c may be viewed in terms of an array of half planes which surround the element and which terminate on it. This allows us to write

$$\omega_3^{P} = -Ax_1 , \quad \varkappa_{13} = -A , \tag{35}$$

which when substituted into (31a), gives a surface dislocation density of

$$\alpha_{31}^{S} = 0 - 0 + A = A . \tag{36}$$

Combining (36) and (34) yields

$$\alpha_{31} + \alpha_{31}^{S} = 0 \tag{37}$$

in agreement with Fig. 3c and the law of the conservation of Burgers vectors. It should be pointed out here that we are dealing with a continuum so that the surface and lattice dislocations of Fig. 3a are of infinitesimal strength and separation as well as continuously distributed, which causes the stress fields associated with the dipoles to virtually vanish. Such is not the case within a real crystal where the lattice dislocations are finite in number and quantized. Since the surface dislocations have the same total strength as the lattice dislocations, but are not quantized, they are spread over a distance on the order of the dipole separation. It is undoubtedly for this reason that only the lattice dislocations have been able to be viewed directly by various experimental techniques such as etch pits [15].

5. Plastic Bending

A pure plastic bending is illustrated in Fig. 4 [11, 17] with the same total displacements given by (16) for the elastic bending. An infinitesimal volume element associated with this particular distortion is illustrated in Fig. 3b. The relevant distortions and related quantities have already been given by (30) and (31). It is clear that no lattice dislocations remain in the body shown in Fig. 4; instead, they have left the crystal and combined with their corresponding surface dislocations to form ledges, in accordance with the reaction towards the right given by (32). These ledges manifest themselves as a sequence of steps on the surface of the body in Fig. 4.

6. Grain Boundaries

When the lattice dislocations associated with the bent body in Fig. 3a are vertically aligned at its center, a symmetric tilt-type grain boundary is created. To see how this comes about, let us first consider the body in Fig. 5a which has been subjected to a simple plastic shear. As in the case of Fig. 3c, the ledges are next removed, in accordance with (32) taken toward the left, to create pairs of dipoles as depicted in Fig. 5b. The resulting distortions may be described as follows:

$$\beta^{\mathrm{P}}_{11} = 0\,, \qquad \beta^{\mathrm{P}}_{12} = 0\,, \qquad \beta^{\mathrm{P}}_{21} = A\mathrm{H}(-x_1)\,, \qquad \beta^{\mathrm{P}}_{22} = 0\,,$$

$$e^{\mathrm{P}}_{11} = 0\,, \qquad e^{\mathrm{P}}_{12} = \tfrac{1}{2}\,A\mathrm{H}(-x_1)\,, \qquad e^{\mathrm{P}}_{21} = \tfrac{1}{2}\,A\mathrm{H}(-x_1)\,, \qquad e^{\mathrm{P}}_{22} = 0\,,$$

$$\omega^{\mathrm{P}}_{3} = -\tfrac{1}{2}\,A\mathrm{H}(-x_1)\,, \qquad \varkappa^{\mathrm{P}}_{13} = \tfrac{1}{2}\,A\delta(x_1)\,,$$

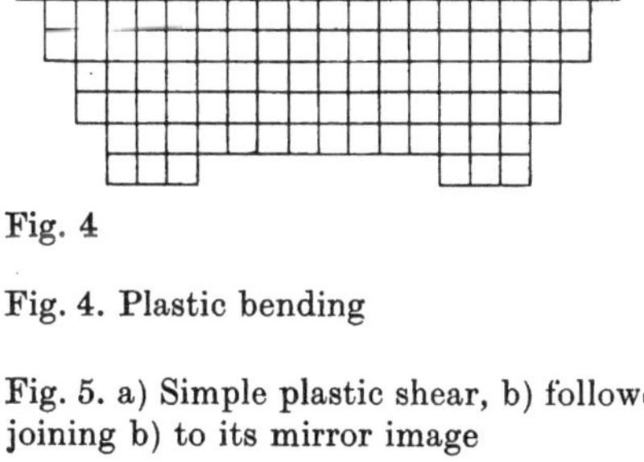

Fig. 4

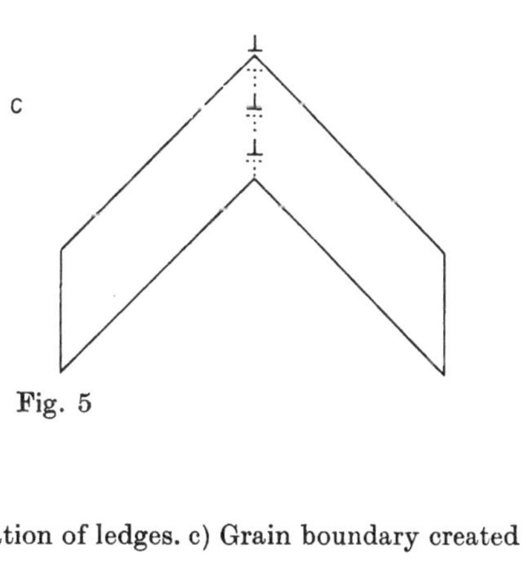

Fig. 5

Fig. 4. Plastic bending

Fig. 5. a) Simple plastic shear, b) followed by elimination of ledges. c) Grain boundary created by joining b) to its mirror image

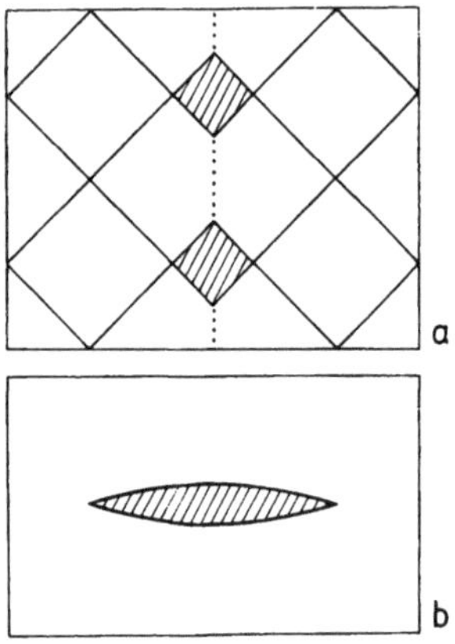

Fig. 6. a) Partially coalesced grain boundary. b) Body containing a crack

$$\beta_{11} = 0\,, \qquad \beta_{12} = -A\mathrm{H}(-x_1)\,, \qquad \beta_{21} = A\mathrm{H}(-x_1)\,, \qquad \beta_{22} = 0\,,$$
$$e_{11} = 0\,, \qquad e_{12} = 0\,, \qquad e_{21} = 0\,, \qquad e_{22} = 0\,,$$
$$\omega_3 = -A\mathrm{H}(-x_1)\,, \qquad \varkappa_{13} = A\delta(x_1)\,, \tag{38}$$

where

$$\mathrm{H}(-x_1) = \begin{cases} 0 & \text{if} \quad x_1 > 0 \\ 1 & \text{if} \quad x_1 < 0\,, \end{cases} \tag{39}$$

and where x_1 is measured with respect to the coordinate system drawn in Fig. 5a. Equations (38), together with (31a), yield the following lattice dislocation density:

$$\alpha_{31} = -\tfrac{1}{2}A\delta(x_1) - 0 - \tfrac{1}{2}A\delta(x_1) = -A\delta(x_1)\,, \tag{40}$$

whereas (18a) gives the following surface dislocation density:

$$\alpha_{31}^{\mathrm{S}} = 0 + 0 + A\delta(x_1) = A\delta(x_1)\,. \tag{41}$$

The body in Fig. 5b can be joined to its mirror image along the dislocated interfaces to form the symmetric tilt boundary shown in Fig. 5c.

The grain boundary depicted in Fig. 5c is fully coalesced [4, 6, 7]; that is, no gaps are contained within it. It is only under this special condition that equations (38) apply. In reality, the grain boundary will not be fully coalesced, as illustrated (dotted line) in Fig. 6a. Voids (shown shaded) will therefore appear in the boundaries, and their size will depend upon the surface energy of the voids. Similar voids or cracks appear during the fracture process when a body is subjected to an external stress such as shown by the shaded region of Fig. 6b. In principle, both of these internal surfaces can be represented in terms of the continuum theory of lattice defects, where the surface boundary conditions are satisfied by an appropriate set of surface dislocations, similar to those used for the finite body in Fig. 1a. In these latter cases however, the surface energy, or tension, plays an important role in the theory [4, 5].

7. Summary and Conclusions

It has been shown that the concept of surface dislocations is fully compatible with the continuum theory of lattice defects. In fact, it is not possible to arrive at a complete understanding of elasto-plastic bending or the structure of grain boundaries without the incorporation of surface dislocations into the theory. Perhaps one of the main reasons why this incorporation was not made earlier by other investigators is due to the fact that the continuum theory of lattice defects has been for the most part developed in terms of infinite bodies. As soon as the bodies are made finite, the surface dislocations enter into the theory in a quite natural fashion.

Acknowledgements

The research time and manuscript preparation facilities for this study were made available by the Mechanical Engineering Department of the University of Maryland. The author would also like to express his appreciation to Prof. E. Kröner, Institut für Theoretische und Angewandte Physik, University of Stuttgart, for his suggestion that a link be established between the concept of surface dislocations and the classical continuum theory of defects.

References

[1] M. J. Marcinkowski, phys. stat. sol. (b) **93**, 703 (1979).
[2] M. J. Marcinkowski, phys. stat. sol. (a) **60**, 109 (1980).
[3] M. J. Marcinkowski, phys. stat. sol. (a) **78**, 401 (1983).
[4] M. J. Marcinkowski, Unified Theory of the Mechanical Behavior of Matter, John Wiley & Sons, New York 1979.
[5] K. Jagannadham and M. J. Marcinkowski, Unified Theory of Fracture, Trans. Tech. Publications, Aedermannsdorf (Switzerland) 1983.
[6] M. J. Marcinkowski, phys. stat. sol. (a) **73**, 409 (1982).
[7] M. J. Marcinkowski, J. Mater. Sci. **18**, 827 (1983).
[8] E. Kröner, in: Mechanik der Deformierbaren Medien, Ed. A. Sommerfeld, 5th ed., Chap. 9, Akademische Verlagsgesellschaft, Leipzig 1964.
[9] E. Kröner, Kontinuumstheorie der Versetzungen und Eigenspannungen, Springer-Verlag, Berlin/Göttingen/Heidelberg 1958.
[10] Y. C. Fung, Foundations of Solid Mechanics, Prentice-Hall, Inc., Englewood Cliffs (N.J.) 1965.
[11] R. de Wit, Internat. J. Eng. Sci. **19**, 1475 (1981).
[12] M. J. Marcinkowski, Meccanica **13**, 151 (1978).
[13] K. Jagannadham and M. J. Marcinkowski, phys. stat. sol. (a) **54**, 715 (1979).
[14] I. M. Gelfand and G. E. Shilov, Generalized Functions, Vol. 1, Properties and Operations, Academic Press, New York 1964.
[15] J. P. Hirth and J. Lothe, Theory of Dislocations, McGraw-Hill Publ. Co., New York 1968.
[16] A. H. Cottrell, Dislocations and Plastic Flow in Crystals, Clarendon Press, Oxford 1953.
[17] F. R. N. Nabarro, Theory of Crystal Dislocations, Clarendon Press, Oxford 1967.

(Received August 28, 1984)

phys. stat. sol. (b) **126**, 537 (1984)

Subject classification: 18.2 and 18.4

A. F. Ioffe Physio-Technical Institute, Academy of Sciences of the USSR, Leningrad[1])

The Parisi-Sompolinsky Solution for an Infinite-Ranged Spin Glass Model with m-Component Spins

By

A. V. GOLTSEV

The Parisi-Sompolinsky solution is considered for the m-component Sherrington-Kirkpatrick model in the general case $H > 0$ and $J_0 > 0$. Self-consistent equations are derived which determine the free energy, the magnetization, and the functions $q(x)$ and $\Delta(x)$. The stability problem is discussed.

Für den allgemeinen Fall $H > 0$ und $J_0 > 0$ wird die Parisi-Sompolinsky-Lösung für das m-komponentige Sherrington-Kirkpatrick-Modell untersucht. Es werden selbstkonsistente Gleichungen abgeleitet, die die freie Energie, die Magnetisierung und die Funktionen $q(x)$ und $\Delta(x)$ bestimmen. Das Stabilitätsproblem wird diskutiert.

1. Introduction

The correct solution of the Sherrington-Kirkpatrick model for Ising spin glasses [1] may be found by using the replica method [2] and the dynamic approach [3]. In the framework of the Parisi solution [2] the order parameter is the function $q(x)$, $x \in (0, 1)$. Sompolinsky [3] introduced two order parameter functions $q(x)$ and $\Delta(x)$ and gave a physical interpretation of these functions. His solution is invariant under a general scale transformation [4 to 6]. In particular, the Parisi solution corresponds to the new scale: $X(x) = -\dot{\Delta}(x)/\dot{q}(x)$. The Sompolinsky solution may be obtained also by use of the replica method [7].

In the present paper we study the Parisi-Sompolinsky solution for the m-component SK model in the general case $H > 0$ and $J_0 > 0$. In Section 2 we derive self-consistent equations which determine the free energy (F), the magnetization ($\boldsymbol{M}$), and the order parameter functions $q(x)$ and $\Delta(x)$. In Section 3 we discuss some questions of the stability problem and prove the existence of massless modes on each time scale.

2. Self-Consistent Equations for the Parisi-Sompolinsky Solution

We consider a systems of N interacting m-component spins $\boldsymbol{S}$ with Hamiltonian

$$\mathcal{H} = -\frac{1}{2} \sum_{ij} J_{ij} \boldsymbol{S}_i \boldsymbol{S}_j - \sum_i \boldsymbol{S}_i \boldsymbol{H} \,, \tag{1}$$

where the J_{ij} are independent random variables distributed normally with a common variance J^2/N and mean value J_0/N, the cartesian component S_μ satisfies $\sum_{\mu=1}^m S_\mu^2 = m$. For notation simplicity, we take the convention $J = 1$. Making use of the

[1]) 194021 Leningrad, USSR.

replica method, the averaged free energy per spin can be written as [1]

$$\beta\mathcal{F} = -\lim_{n\to 0}\frac{1}{n}\max\left\{-\frac{1}{4}\beta^2\sum_{\alpha,\beta}\sum_{\mu,\nu}(y_{\mu\nu}^{\alpha\beta})^2 - -\frac{1}{2}\beta J_0\sum_\alpha (M_\mu^\alpha)^2 + \ln \operatorname*{tr}_{\{S_\mu^\alpha\}}\exp\left[\frac{1}{2}\beta^2\sum_{\alpha,\beta}\sum_{\mu,\nu}y_{\mu\nu}^{\alpha\beta}S_\mu^\alpha S_\nu^\beta + + \beta J_0\sum_{\alpha,\mu}M_\mu^\alpha S_\mu^\alpha + \beta H\sum_\alpha S_1^\alpha\right]\right\}, \tag{2}$$

where the magnetic field is applied in direction 1, the upper indices (α, β) are replica indices and run from 1 to n, the lower indices (μ, ν) are indices of cartesian components and run from 1 to m. In the extremum of (2)

$$\begin{aligned} y_{\mu\nu}^{\alpha\alpha} &= \langle S_\mu^\alpha S_\nu^\alpha\rangle = y_\mu\delta_{\mu,\nu}\,,\\ y_{\mu\nu}^{\alpha\beta} &= \langle S_\mu^\alpha S_\nu^\beta\rangle = q_\mu^{\alpha\beta}\delta_{\mu,\nu} \qquad (\alpha \neq \beta)\,,\\ M_\mu^\alpha &= \langle S_\mu^\alpha\rangle = M_\mu\,, \end{aligned} \tag{3}$$

where $y_{\mu\geqq 2} \equiv y_2$, $q_{\mu\geqq 2}^{\alpha\beta} = q_2^{\alpha\beta}$. Considering the case $H = J_0 = 0$, de Almeida et al. [8] obtained the instability of the replica symmetric solution $(q^{\alpha\beta} = q)$ for all $T < T_c = 1$. In the general case $m > 1$, $H > 0$, and $J_0 > 0$ the critical temperature for replica symmetry breaking is determined by the equation [9]

$$T^2 = \langle (S_\mu^\alpha)^2 (S_\mu^\beta)^2\rangle \qquad (\mu \geqq 2)\,.$$

At T above T_c a solution with non-zero longitudinal order parameter $q_1^{\alpha\beta} = q_1$ and zero transverse order parameter $q_2^{\alpha\beta}$ is stable. For T below T_c the replica symmetry is broken and the transverse order parameter $q_2^{\alpha\beta}$ is non-zero [10 to 12].

Let us consider the Parisi-Sompolinsky solution for the m-component spin glasses. Using the Duplantier method [13], we obtain

$$\begin{aligned}\beta\mathcal{F} = &-\tfrac{1}{4}\beta^2\,\{2my_2 - y_1^2 - (m-1)\,y_2^2 - 2mq_2(1) + q_1^2(1) + \\ &+ (m-1)\,q_2^2(1) + 2\int_0^1 dx(q_1(x)\,\dot{\Delta}_1(x) + (m-1)\,q_2(x)\,\dot{\Delta}_2(x))\} - \\ &- [f(0, \beta\boldsymbol{H} + \beta J_0\boldsymbol{M} + \beta\boldsymbol{z})]_z\,,\end{aligned} \tag{4}$$

$$M_\mu = [M_\mu(0, \beta\boldsymbol{H} + \beta J_0\boldsymbol{M} + \beta\boldsymbol{z})]_z\,, \tag{5}$$

$$q_\mu(x) = [Q_\mu(x; 0, \beta\boldsymbol{H} + \beta J_0\boldsymbol{M} + \beta\boldsymbol{z})]_z\,, \tag{6}$$

where

$$\dot{\Delta}(x) \equiv \frac{d\Delta(x)}{dx}\,,$$

$$[f(\boldsymbol{z})]_z \equiv \int_{-\infty}^{\infty}\prod_{\mu=1}^{m}\frac{dz_\mu}{(2\pi q_\mu(0))^{1/2}}\exp\left[\frac{-\sum_\mu z_\mu^2}{2q_\mu(0)}\right]f(\boldsymbol{z})\,.$$

The function $f(y, \boldsymbol{h})$, $\boldsymbol{h} = (h_1, \dots h_m)$, defined on the interval $y \in (0, 1)$, satisfies the equation

$$\dot{f}(y, \boldsymbol{h}) = -\tfrac{1}{2}\beta^2\sum_\nu(\dot{q}_\nu(y)\,\partial_\nu^2 f(y, \boldsymbol{h}) - \dot{\Delta}_\nu(y)\,(\partial_\nu f(y, \boldsymbol{h}))^2)\,, \tag{7}$$

where $\dot{f} = \partial f/\partial y$, $\partial_\nu f = \partial f/\partial h_\nu$, with the boundary condition

$$f(1, \boldsymbol{h}) = \ln g(\boldsymbol{h}) \,, \tag{8}$$

$$g(\boldsymbol{h}) = \operatorname*{tr}_{\{S_\mu\}} \exp\left[\boldsymbol{hS} + \tfrac{1}{2}\beta^2(y_1 - y_2 - q_1(1) + q_2(1))\, S_1^2\right] . \tag{9}$$

The function $\boldsymbol{M}(y, \boldsymbol{h})$, $y \in (0, 1)$, obeys the equations

$$\dot{M}_\mu = \hat{L} M_\mu \,, \tag{10}$$

$$M_\mu(1, \boldsymbol{h}) = G_\mu(\boldsymbol{h}) \equiv \partial_\mu \ln g(\boldsymbol{h}) \,, \tag{11}$$

where the differential operator $\hat{L}$ is

$$\hat{L} = -\tfrac{1}{2}\beta^2 \sum_\nu \left(\dot{q}_\nu(y)\, \partial_\nu^2 - 2\dot{\Delta}_\nu(y)\, M_\nu(y, \boldsymbol{h})\, \partial_\tau\right) . \tag{12}$$

The function $Q_\mu(x; y, \boldsymbol{h})$, $y \in (0, x)$, satisfies (10) with the boundary condition

$$Q_\mu(x; x, \boldsymbol{h}) = M_\mu^2(x, \boldsymbol{h}) \,. \tag{13}$$

It is easy to show that $M_\mu(y, \boldsymbol{h}) = \partial_\mu f(y, \boldsymbol{h})$. Equations (5) (6), (10) to (13) have been derived independently by de Almeida and Lage [14] and Goltsev [15, 16] for the Ising case $m = 1$ and have been generalized for the case $m > 1$ in our previous paper [17].

The function $\Delta_\mu(x)$ determines the local susceptibility $\chi_\mu(x)$ measured at the frequency ω_x [3],

$$\begin{aligned} \chi_\mu(x) &= \beta(1 - q_\mu(1) + \Delta_\mu(x)) \\ &= [\chi_\mu(x; 0, \beta\boldsymbol{H} + \beta J_0 \boldsymbol{M} + \beta\boldsymbol{z})]_{\boldsymbol{z}} \,, \end{aligned}$$

where the function $\chi_\mu(x; y, \boldsymbol{h})$ obeys (10) with the boundary condition

$$\chi_\mu(x; x, \boldsymbol{h}) = \beta\, \partial_\mu M_\mu(x, \boldsymbol{h}) \,.$$

The value y_μ is equal to

$$y_\mu = [Y_\mu(0, \beta\boldsymbol{H} + \beta J_0 \boldsymbol{M} + \beta\boldsymbol{z})]_{\boldsymbol{z}} \,,$$

where the function $Y_\mu(y, \boldsymbol{h})$ satisfies (10) with the boundary condition

$$Y_\mu(1, \boldsymbol{h}) = g^{-1}(\boldsymbol{h})\, \partial_\mu^2 g(\boldsymbol{h}) \,.$$

Now we consider the physical meaning of the field $\boldsymbol{h}$ and the function $\boldsymbol{M}(x, \boldsymbol{h})$. The differential equation (10) may be replaced by a functional integral equation [16],

$$\begin{aligned} M_\mu(x, \boldsymbol{h}) = \int_{-\infty}^{\infty} \prod_{x<y<1} \prod_{\nu=1}^{m} \frac{Dz_\nu(y)}{(2\pi\dot{q}_\nu(y))^{1/2}} \exp\left[-\frac{1}{2} \sum_{\nu=1}^{m} \int_x^1 \mathrm{d}y\, \frac{z_\nu^2(y)}{\dot{q}_\nu(y)}\right] \times \\ \times G_\mu(\boldsymbol{H}(1, x, \{\boldsymbol{z}\})) \,, \end{aligned} \tag{14}$$

where the effective field $H_\mu(y, x, \{\boldsymbol{z}\})$, $y < x$, satisfies a self-consistent equation

$$\begin{aligned} H_\mu(y, x, \{\boldsymbol{z}\}) = \beta \int_x^y \mathrm{d}t\, z_\mu(t) - \beta^2 \int_x^y \mathrm{d}t\, \dot{\Delta}_\mu(t) \times \\ \times M_\mu(t, \boldsymbol{H}(t, x, \{\boldsymbol{z}\})) + h_\mu \,. \end{aligned} \tag{15}$$

Let h be equal to the effective field which remains frozen on the time scale τ_x,

$$\begin{aligned} h_\mu = \beta \int_0^x \mathrm{d}t\, z_\mu(t) - \beta^2 \int_0^x \mathrm{d}t\, \dot{\Delta}_\mu(t)\, M_\mu(t, \boldsymbol{H}(t, x, \{\boldsymbol{z}\})) + \\ + \beta H_\mu + \beta J_0 M_\mu + \beta z_\mu \equiv \tilde{H}_\mu(x, 0, \{\boldsymbol{z}\}) \,. \end{aligned} \tag{16}$$

$M_\mu(x, \tilde{\boldsymbol{H}}(x, 0, \{\boldsymbol{z}\})) \equiv M_\mu(x, \{\boldsymbol{z}\})$ is equal to the local magnetization which remains frozen on the time scale τ_x. The local magnetization $M_\mu(x, \{\boldsymbol{z}\})$ was introduced by Sompolinsky [3] for the Ising case.

For the function $Q_\mu(x; y, \boldsymbol{h})$ we have

$$Q_\mu(x; y, \boldsymbol{h}) = \int\limits_{-\infty}^{\infty} \prod_{y<t<x} \prod_{\nu=1}^{m} \frac{Dz_\nu(t)}{(2\pi \dot{q}_\nu(t))^{1/2}} \times$$

$$\times \exp\left[-\frac{1}{2}\sum_{\nu=1}^{m} \int\limits_{y}^{x} \mathrm{d}t \frac{z_\nu^2(t)}{\dot{q}_\nu(t)}\right] M_\mu^2 (x, \boldsymbol{H}(x, y, \{\boldsymbol{z}\})) . \tag{17}$$

Therefore, the function $q_\mu(x)$ is

$$q_\mu(x) = \int\limits_{-\infty}^{\infty} \left(\prod_{0<t<x} \prod_{\nu=1}^{m} \frac{Dz_\nu(t)}{(2\pi \dot{q}(t))^{1/2}}\right) \left(\prod_{\nu=1}^{m} \frac{\mathrm{d}z_\nu}{(2\pi q_\nu(0))^{1/2}}\right) \times$$

$$\times \exp\left[-\frac{1}{2}\sum_{\nu} \int\limits_{0}^{x} \mathrm{d}t \frac{z_\nu^2(t)}{\dot{q}_\nu(t)} - \frac{1}{2}\sum_{\nu} \frac{z_\nu^2}{q_\nu(0)}\right] M_\mu^2(x, \tilde{\boldsymbol{H}}(x, 0, \{\boldsymbol{z}\})) \equiv$$

$$\equiv [M_\mu^2(x, \{\boldsymbol{z}\})]_{\{\boldsymbol{z}\}} . \tag{18}$$

Equation (18) may be used to obtain some exact relations. Differentiating the left- and right-hand sides of (18) with respect to x, we obtain

$$\dot{q}_\mu(x) = \beta^2 \sum_{\nu} \dot{q}_\nu(x) \, [(\partial_\nu M_\mu(x, \{\boldsymbol{z}\}))^2]_{\{\boldsymbol{z}\}} , \tag{19}$$

where

$$\partial_\nu M_\mu(x, \{\boldsymbol{z}\}) \equiv \partial_\nu M_\mu(x, \boldsymbol{h})_{\boldsymbol{h}=\tilde{\boldsymbol{H}}(x, 0, \{\boldsymbol{z}\})} .$$

For the Ising case (19) has the form

$$1 = \beta^2 \left[\left(\frac{\partial}{\partial h} M(x, \{\boldsymbol{z}\})\right)^2\right]_{\{\boldsymbol{z}\}}$$

and means the marginal dynamic stability [18] and the existence of massless modes on each time scale. The existence of massless modes in the Heisenberg spin glass is proved in the next section.

For the case $H = 0$ and $J_0 < 1$ (19) yields that the static (zero-field) susceptibility $\chi(0)$ is equal to unity at all $T < T_c = 1$.

Let us differentiate (19) with respect to x,

$$\begin{aligned} T^2 \ddot{q}_\mu(x) = {} & \sum_{\nu=1}^{m} \ddot{q}_\nu(x) \, [\partial_\nu M_\mu(x, \{\boldsymbol{z}\}))^2]_{\{\boldsymbol{z}\}} + \\ & + \beta^2 \sum_{\nu=1}^{m} \sum_{\gamma=1}^{m} \dot{q}_\nu(x) \, \dot{q}_\gamma(x) \, [\partial_\nu \, \partial_\gamma M_\mu(x, \{\boldsymbol{z}\}))^2]_{\{\boldsymbol{z}\}} + \\ & + 2\beta^2 \sum_{\nu=1}^{m} \sum_{\gamma=1}^{m} \dot{q}_\nu(x) \, \dot{\Delta}_\gamma(x) \, [(\partial_\nu M_\mu(x, \{\boldsymbol{z}\})) \times \\ & \times (\partial_\gamma M_\mu(x, \{\boldsymbol{z}\})) \, (\partial_\nu M_\gamma(x, \{\boldsymbol{z}\}))]_{\{\boldsymbol{z}\}} . \end{aligned} \tag{20}$$

For $H = 0$ and $J_0 < 1$ (20) has the simple form

$$\frac{\dot{\Delta}(x)}{\dot{p}(x)} = -\frac{1}{2} \frac{\sum\limits_{\nu,\gamma} [(\partial_\nu \partial_\gamma M_\mu(x, \{\boldsymbol{z}\}))^2]_{\{\boldsymbol{z}\}}}{\sum\limits_{\nu,\gamma} [(\partial_\nu M_\mu(x, \{\boldsymbol{z}\})) \, (\partial_\gamma M_\mu(x, \{\boldsymbol{z}\})) \, (\partial_\nu M_\gamma(x, \{\boldsymbol{z}\}))]_{\{\boldsymbol{z}\}}} . \tag{21}$$

Unfortunately, the physical meaning of (21) is not clear at present.

3. The Stability Problem

The stability analysis of the Parisi-Sompolinsky solution is a difficult problem even for the Ising spin glass model [5, 19 to 21]. De Dominicis and Kondor [5, 20] performed the full stability analysis of the Ising spin glass at T close to T_c. They found three families of eigenfunctions which give rise to nonnegative eigenvalues. Now we consider some questions of the stability problems for the Heisenberg spin glass model at $H = J_0 = 0$.

The stabiliy matrix has the form

$$M_{ij\ kl}^{\alpha\beta\ \gamma\nu} = T^2\delta_{\alpha,\gamma}\delta_{\beta,\nu}\delta_{i,k}\delta_{j,l} + \tfrac{1}{2}\left(\langle S_i^\alpha S_j^\beta\rangle \langle S_k^\gamma S_l^\nu\rangle - \langle S_i^\alpha S_j^\beta S_k^\gamma S_l^\nu\rangle\right). \tag{22}$$

An eigenvector $R_{ij}^{\alpha\beta}$ of the matrix (22) may be presented in the form

$$R_{ij} = \begin{cases} P_{ij}^{\alpha\beta} & \text{if} \quad i \neq j\,, \\ f_i^{\alpha\beta} & \text{if} \quad i = j\,, \quad \alpha \neq \beta\,, \\ R_j^\alpha & \text{if} \quad i = j\,, \quad \alpha = \beta\,. \end{cases} \tag{23}$$

We suppose $R_{ij}^{\alpha\beta} = R_{ji}^{\beta\alpha}$. The eigenvalue equation

$$\sum_{\gamma,\nu} \sum_{k,l} M_{ij\ kl}^{\alpha\beta\ \gamma\nu} R_{kl}^{\gamma\nu} = \lambda R_{ij}^{\alpha\beta}$$

then becomes

$$(\lambda - T^2)\, P_{ij}^{\alpha\beta} + \sum_{\gamma,\nu} \langle S_i^\alpha S_j^\beta S_i^\gamma S_j^\nu\rangle\, P_{ij}^{\gamma\nu} = 0\,, \tag{24}$$

$$2(\lambda - T^2)\, f_i^{\alpha\beta} + \sum_{j=1}^{m} \sum_{\gamma,\nu} \left(\langle S_i^\alpha S_i^\beta S_j^\gamma S_j^\nu\rangle - q^{\alpha\beta} q^{\gamma\nu}\right) f_j^{\gamma\nu} + $$
$$+ \sum_{j=1}^{m} \sum_{\gamma} \left(\langle S_i^\alpha S_i^\beta (S_j^\gamma)^2\rangle - q^{\alpha\beta}\right) R_j^\gamma = 0\,, \tag{25a}$$

$$2(\lambda - T^2)\, R_i^\alpha + \sum_{j=1}^{m} \sum_{\gamma} \left(\langle (S_i^\alpha)^2 (S_j^\gamma)^2\rangle - 1\right) R_j^\gamma + $$
$$+ \sum_{j=1}^{m} \sum_{\gamma\neq\nu} \left(\langle (S_\alpha^i)^2 (S_j^\gamma S_j^\nu)\rangle - q^{\gamma\nu}\right) f_j^{\gamma\nu} = 0\,. \tag{25b}$$

As follows from (24) and (25), fluctuations which lie in a subspace of vectors $P_{ij}^{\alpha\beta}$ do not interact with fluctuations which lie in the subspace of vectors $(f_i^{\alpha\beta}, R_i^\alpha)$. The spectrum of the stability matrix contains eigenvalues determined by (24) and (25).

At first we study (25). Eigenvectors of the first class are independent of the cartesian component index, $\tilde{R}_i^{\alpha\beta} = \tilde{R}^{\alpha\beta} = (f^{\alpha\beta}, R^\alpha)$. In this case (25) takes the form

$$2(\lambda - T^2)\, f^{\alpha\beta} + \sum_{j=1}^{m} \sum_{\gamma\neq\nu} \left(\langle S_i^\alpha S_i^\beta S_j^\gamma S_j^\nu\rangle - q^{\alpha\beta} q^{\gamma\nu}\right) f^{\gamma\nu} = 0\,, \tag{26}$$
$$(\lambda - T^2)\, R^\alpha = 0\,.$$

Consequently, the eigenvector $R^{\alpha\beta} = (0, R^\alpha)$ gives rise to the eigenvalue $\lambda = T^2$. There are also three eigenvalue families given by (26). For eigenfunctions of the first family [20] we obtain exact equation

$$\int_x^1 \mathrm{d}y f(y)\, K(y, x) + \int_0^x \mathrm{d}y f(y)\, K(x, y) = \lambda f(x)\,, \tag{27}$$

where at $y > x$

$$K(y, x) = \sum_{j=1}^{m} \left[(\partial_i Q_j(y; x, \{z\}))\, M_i(x, \{z\})\right]_{\{z\}}\,.$$

For T close to $T_c = 1$ we have $K(y, x) \approx q(x)$ and (27) takes the same form which De Dominicis and Kondor studied for $m = 1$ [5, 20]. An analagous situation occurs also for eigenfunctions of the second family. The spectrum of eigenvalues of the third family may be exactly obtained by use of non-symmetric vectors

$$\tilde{f}^{\alpha\beta}_{k,p,r} = \begin{cases} \neq 0, & \alpha \cap \beta = k\,, \\ 0, & \alpha \cap \beta \neq k, \end{cases} \tag{28a}$$

$$f^{\alpha\beta}_{k,p,r} = \tilde{f}^{\alpha'\beta'}_{k,p,r} \quad \text{if} \quad \alpha \cap \alpha' > p, \qquad \beta \cap \beta' > r\,, \tag{28b}$$

$$\sum_{\alpha(\alpha\cap\alpha_0 \geqq p)} \tilde{f}^{\alpha\beta}_{k,p,r} = \sum_{\beta(\beta\cap\beta_0 \geqq r)} \tilde{f}^{\alpha\beta}_{k,p,r} = 0 \quad \text{for all} \quad \alpha_0, \beta_0\,, \tag{28c}$$

where $\alpha \cap \beta = k$ means that $q^{\alpha\beta} = q(k)$; in the continuous limit $n \to 0$ the indices k, p, and r satisfy the inequality $0 \leqq k \leqq p, r \leqq 1$. Then the symmetric vectors $\tilde{f}^{\alpha\beta}_{k,p,r} = \tilde{f}^{\alpha\beta}_{k,p,r} + \tilde{f}^{\beta\alpha}_{k,p,r}$ give rise to eigenvalues

$$\lambda(k, p, r) = T^2 - \sum_{j=1}^{m} < \Big(\sum_{\alpha'(\alpha'\cap\alpha > p)} S_i^\alpha (S_j^{\alpha'} - S_j^{\alpha''}) \Big) \times$$
$$\times \Big(\sum_{\beta'(\beta'\cap\beta > r)} S_i^\beta (S_j^{\beta'} - S_j^{\beta''}) \Big) \rangle\,, \tag{29}$$

where $\alpha \cap \beta = \alpha' \cap \beta' = k, \alpha \cap \alpha'' = p, \beta \cap \beta'' = r$. In the continuous limit $n \to 0$ (29) becomes

$$\lambda(k, p, r) = T^2 - T^2 \sum_{j=1}^{m} [\chi_{ij}(p; k, \{z\})\, \chi_{ij}(r; k, \{z\})]_{\{z\}}\,, \tag{30}$$

where $\chi_{ij}(p; k, \{z\}) = \chi_{ij}(p; k, \tilde{\boldsymbol{H}}(k, 0, \{z\}))$ and the function $\chi_{ij}(x; y, \boldsymbol{h})$ satisfies (10) with the boundary condition

$$\chi_{ij}(x; x, \boldsymbol{h}) = \beta\, \partial_i M_j(x, \boldsymbol{h}) = \beta\, \partial_j M_i(x, \boldsymbol{h})\,.$$

At $k = p = r = x$

$$\lambda(x, x, x) = T^2 - \sum_j [(\partial_j M_i(x, \{z\}))^2]_{\{z\}}\,.$$

Equation (19), yields that $\lambda(x, x, x) = 0$. Consequently there are massless modes on each time scale τ_x.

Now we consider fluctuations which break the symmetry between the cartesian directions. Let us introduce vector $R_i^{\alpha\beta} = (f_i^{\alpha\beta}, R_i^\alpha)$ which obeys the equation

$$\sum_{i=1}^{m} \tilde{R}_i^{\alpha\beta} = 0\,. \tag{31}$$

For example, if $\tilde{R}_i^{\alpha\beta} = \tilde{R}^{\alpha\beta}$ for $i \neq i_0$ and $\tilde{R}_{i_0}^{\alpha\beta} = -(m-1)\, \tilde{R}^{\alpha\beta}$ then the vector $\tilde{R}_i^{\alpha\beta}$ satisfies (31). Eigenvalues of the second class are determined by the equations

$$2(\lambda - T^2)\, f^{\alpha\beta} + \sum_{\gamma \neq \nu} \langle S_i^\alpha S_i^\beta (S_i^\gamma S_i^\nu - S_j^\gamma S_j^\nu) \rangle\, f^{\gamma\nu} +$$
$$+ \sum_\gamma \langle S_i^\alpha S_i^\beta ((S_i^\gamma)^2 - (S_j^\gamma)^2) \rangle\, R^\gamma = 0\,,$$
$$2(\lambda - T^2)\, R^\alpha + \sum_\gamma \langle (S_i^\alpha)^2 ((S_i^\gamma)^2 - (S_j^\gamma)^2) \rangle\, R^\gamma +$$
$$+ \sum_{\gamma \neq \nu} \langle (S_i^\alpha)^2 (S_i^\gamma S_i^\nu - S_j^\gamma S_j^\nu) \rangle\, f^{\gamma\nu} = 0\,, \tag{32}$$

where $i \neq j$; $i, j = 1, \dots m$. Equations (32) follow from (25). It is interesting to note that at $T > T_c$ the replica symmetry is unbroken and an eigenvector $\tilde{R}^{\alpha\beta} = (0, R)$

gives rise to an eigenvalue

$$\lambda = T^2 - \frac{m}{m+2}. \tag{33}$$

A possible instability at $T < T^* = (m/(m+2))^{1/2} < 1$ was discussed by Toulouse and Gabay [22]. However, for $T < T_c = 1$ the replica symmetry is broken. As a consequence, the vector $(0, R^\alpha = R)$ is no eigenvector. The spectrum of eigenvalues given by (32) contains also three families. For the first family at T close to T_c we have the eigenvector $\tilde{R}^{\alpha\beta} = (f^{\alpha\beta}_{(1)}, R^\alpha = R)$ and the eigenvalue equation has the form

$$2q(x) \int_x^1 dy\, f(y) + 2 \int_0^x dy\, q(y)\, f(y) - \\ - 4aq(x) \int_0^1 dy\, q(y)\, f(y) = \lambda f(x)\,, \tag{34}$$

where

$$a = \frac{m^2}{[(m+2)^2 (\lambda + 2t - 2/(m+2))]};$$

$t = 1 - T/T_c \ll 1$. For $m \gg 1$ and $T < T^*$ we find that (34) has only non-negative eigenvalues. Consequently, at $T < T_c$ the interaction of fluctuations $f^{\alpha\beta}$ and $R^\alpha = R$ eliminates the instability at $T = T^*$.

For eigenvectors of the second family $R^{\alpha\beta} = (f^{\alpha\beta}_{(2)}, R^\alpha)$ at $T \approx T_c$ the eigenvalue equation has the form

$$2 \int_0^x dy\, q(y)\, f(y, z) + 2q(x) \int_x^1 dy\, f(y, z) - 2q(x) \int_z^1 dy\, y \frac{\partial}{\partial y} f(z, y) - \\ - 4aq(x) \left\{ \int_0^1 dy\, q(y)\, f(y, z) + q(z) \int_z^1 dy\, y \frac{\partial}{\partial y} f(z, y) \right\} - \lambda f(x, z) \tag{35}$$

for $z < x$, and for $x < z$

$$0 = -(\lambda + xq(x) + \int_x^z dy\, q(y))\, f(x, z) + \\ + q(x) \int_x^1 dy (f(y, x) + f(y, z)) + \int_0^x dy\, q(y)\, (f(y, x) + f(y, z)) - \\ - \int_z^1 dy\, y \frac{\partial}{\partial y} [q(z)\, f(x, y) + q(x)\, f(z, y)] + xq(x)\, f(x, x) + \\ + \int_x^z dy\, q(y)\, f(x, y) - 2aq(x) \left\{ \int_0^1 dy\, q(y)\, (f(y, x) + f(y, z)) - \\ - q(x) \int_x^1 dy\, y \frac{\partial}{\partial y} f(x, y) - q(z) \int_z^1 dy\, y \frac{\partial}{\partial y} f(z, y) \right\}. \tag{36}$$

Eigenvalues of the third family ($R^{\alpha\beta}_{k,p,r} = (f^{\alpha\beta}_{k,p,r}, 0)$) give rise to the eigenvalues

$$\beta^2\lambda_{(2)}(k, p, r) = 1 - [\chi_{ii}(p; k, \{z\})\, \chi_{ii}(r; k, \{z\}) - \\ - \chi_{ij}(p; k, \{z\})\, \chi_{ij}(r; k, \{z\})]_{\{z\}}\,, \tag{37}$$

where $i \neq j$. The eigenvalues $\lambda_{(2)}(k, p, r)$ are non-negative and equal to zero for $k = p = r = 0$.

The third class of eigenvalues is determined by (24) and contains also three families. At $T \approx T_c$ for the first family ($P^{\alpha\beta}_{ij} = f^{\alpha\beta}_{(1)}$, $P^{\alpha\alpha} = P$) the eigenvalue equation has the form (34) with $a = (\lambda + 2t)^{-1}$. For the second family $P^{\alpha\beta}_{ij} = (f^{\alpha\beta}_{(2)}, P^{\alpha\alpha})$ the eigenvalue equation has the form (35) and (36) with $a = (\lambda + 2t)^{-1}$. The eigenvalues of the third family are given by

$$\beta^2\lambda_{(3)}(k, p, r) = 1 - [\chi_{ii}(p; k, \{z\})\, \chi_{jj}(r; k, \{z\}) + \\ + \chi_{ij}(p; k, \{z\})\, \chi_{ij}(r; k, \{z\})]_{\{z\}}\,, \tag{38}$$

where $i \neq j$.

For the case $H \neq 0$ ($\boldsymbol{H} = (H, 0, \dots 0)$) we prove only the existence of massless modes on each time scale. Let us consider the eigenvector $R^{\alpha\beta}_{ij} = (0, f^{\alpha\beta}_i(k, p, r), 0)$ where $f^{\alpha\beta}_i = f^{\alpha\beta}_2$ if $i \neq 1$, which gives rise to the eigenvalue

$$\lambda(k, p, r) = \tfrac{1}{2}\,(2T^2 - B - c_{11}) - [\tfrac{1}{4}\,(B - c_{11})^2 + Ac_{21}]^{1/2}\,, \tag{39}$$

where

$$c_{ij}(k, p, r) = [\chi_{ij}(p; k, \{z\})\, \chi_{ij}(r; k, \{z\})]_{\{z\}}\,,$$

$$A = \sum_{i=2}^{m} c_{i1}\,, \qquad B = \sum_{i=2}^{m} c_{i2}\,.$$

Equation (19) gives

$$T^2\dot{q}_1(x) = \dot{q}_1(x)\,\tilde{c}_{11} + \dot{q}_2(x)\,\tilde{A}\,, \\ T^2\dot{q}_2(x) = \dot{q}_1(x)\,\tilde{c}_{21} + \dot{q}_2(x)\,\tilde{B}\,, \tag{40}$$

where $\tilde{c}_{ij} = c_{ij}(x, x, x)$. Substituting (40) into (39) at $k = p = r = x$ we obtain $\lambda(x, x, x) = 0$.

In addition we present some results at T close to T_c and $H = J_0 = 0$. Equation (19) yields that the function $q(x)$ obeys the equation

$$t = xq(x) - \frac{3}{m+2}\, q^2(x) + \int_x^1 \mathrm{d}x\, q(x) \tag{41}$$

up to terms of order $O(t^3)$. Therefore

$$q(x) = \begin{cases} t + \dfrac{3}{m+2}\, t^2 + O(t^3), & x_1 \leqq x < 1\,, \\ \dfrac{m+2}{6}\, x + O(tx, x^2), & 0 < x < x_1\,; \end{cases}$$

$$x_1 = \frac{6}{m+2}\, t + O(t^2)\,,$$

in agreement with [23]. Using (30), (37), and (38) we obtain

$$\lambda(k, p, r) = \frac{3}{(m+2)}\,(q^2(p) + q^2(r) - 2q^2(k))\,,$$

$$\lambda_{(2)}(k, p, r) \approx \lambda_{(3)}(k, p, r) = \frac{3}{(m+2)}\left(q^2(p) + q^2(r) - \frac{2(m+6)}{3(m+2)}\, q^2(k)\right). \tag{42}$$

For the Ising spin glass ($m = 1$), the eigenvalue $\lambda(k, p, r)$ has been obtained by Kondor and De Dominicis [5].

4. Conclusion

In the present paper we derive self-consistent differential equations (4) to (11) which determine the free energy, the magnetization, and the function $q(x)$ and $\Delta(x)$. These equations may be replaced by functional integral equations (14) to (18) which allow to obtain the condition of marginal dynamic stability for the Heisenberg spin glasses.

In our stability analysis we find that there are three classes of eigenvalues of the stability matrix. Each class contains three eigenvalues families. Exact expressions for eigenvalues of the third families are obtained and the existence of massless modes on each time scale is proved. Eigenvalue equations are derived for the first and second families of these classes. For $m \gg 1$ we find no instability at $T < T^* = (m/(m+2))^{1/2}$. Consequently the phase diagram of the Ising m-component SK-model is qualitatively the same as for the Ising spin glass.

References

[1] D. Sherrington and S. Kirkpatrick, Phys. Rev. Letters **35**, 1972 (1975).
[2] G. Parisi, Phys. Rev. Letters **43**, 1754 (1979); J. Phys. A **13**, 1887 (1980).
[3] H. Sompolinsky, Phys. Rev. Letters **47**, 935 (1981).
[4] H.-J. Sommers, J. Phys. A **16**, 447 (1983).
[5] I. Kondor and C. de Dominicis, J. Phys. A **16**, L73 (1983).
[6] C. de Dominicis, M. Gabay, and B. Duplantier, J. Phys. A **15**, L47 (1982).
[7] C. de Dominicis, M. Gabay, and H. Orland, J. Physique, Letters **42**, L-523 (1981).
[8] J. R. L. de Almeida, R. C. Jones, J. M. Kosterlitz, and D. J. Thouless, J. Phys. C **11**, L875 (1978).
[9] G. Toulouse, M. Gabay, T. C. Lubensky, and J. Vannimenus, J. Physique, Letters **43**, L-109 (1982).
[10] M. Gabay and G. Toulouse, Phys. Rev. Letters **47**, 201 (1981).
[11] M. A. Moore and A. J. Bray, J. Phys. C **15**, L301 (1982).
[12] D. M. Cragg, D. Sherrington, and M. Gabay, Phys. Rev. Letters **49**, 158 (1982).
[13] B. Duplantier, J. Phys. A **14**, 283 (1981).
[14] J. R. L. de Almeida and E. J. S. Lage, J. Phys. C **16**, 939 (1983).
[15] A. V. Goltsev, J. Phys. A **16**, L339 (1983).
[16] A. V. Goltsev, J. Phys. A **17**, 237 (1984).
[17] A. V. Goltsev, J. Physique, Lettres **44**, L-697 (1983).
[18] H. Sompolinsky and A. Zippelius, Phys. Rev. B **25**, 6860 (1982).
[19] D. J. Thouless, J. R. L. de Almeida, and J. M. Kosterlitz, J. Phys. C **13**, 3271 (1980).
[20] C. de Dominicis and I. Kondor, Phys. Rev. B **27**, 606 (1983).
[21] A. V. Goltsev, J. Phys. A **16**, 1337 (1983).
[22] G. Toulouse and M. Gabay, J. Physique, Lettres **42**, L-103 (1981).
[23] M. Gabay, T. Garel, and C. de Dominicis, J. Phys. C **15**, 7165 (1982).

(Received July 26, 1984)

phys. stat. sol. (b) 126, 547 (1984)

Subject classification: 13.3 and 19

Institute of Physics, Czechoslovak Academy of Sciences, Prague[1])

Angular Dependence of Spin-Wave Resonance in Thin Films with Asymmetrical Boundary Conditions

Analytical Treatment

By

M. JIRSA and V. KAMBERSKÝ

The boundary problem is solved for an isotropic homogeneous infinite insulator film with spins pinned asymmetrically by a uniaxial surface anisotropy. An arbitrarily oriented external magnetic field is assumed. Formally a simple condition for the values of the resonant wave-vector $\boldsymbol{k}_1$ is obtained. The analytical expressions for the standing spin-wave resonance (SSWR) mode linewidths and intensities are derived and discussed. The second part of the paper is devoted to the SSWR in thin metallic films with thicknesses $2d \ll \delta$ (the skin depth). The influence of conductivity on the mode positions, linewidths, and intensities is investigated analytically with the help of the spin-wave approximation.

Das Randwertproblem einer isotropen homogenen unbegrenzten Isolatorschicht mit Spins, die asymmetrisch durch eine einachsige Oberflächenanisotropie festgehalten sind, wird gelöst. Ein willkürlich orientiertes äußeres Magnetfeld wird vorausgesetzt. Eine formal einfache Bedingung für den Wellenvektor $\boldsymbol{k}_1$ der Resonanzwelle wird erhalten. Analytische Ausdrücke für die Linienbreite und Intensität der stehenden Spinwellenresonanz (SSWR) werden abgeleitet und diskutiert. Der zweite Teil des Artikels ist der SSWR in dünnen Metallschichten mit der Dicke $2d \ll \delta$ (Skintiefe) gewidmet. Der Einfluß der elektrischen Leitfähigkeit auf die Position, Linienbreite und Intensität der SSWR wird analytisch mit Hilfe der Spinwellenapproximation untersucht.

1. Introduction

In the previous paper [1] the characteristics of the standing spin-wave resonance (SSWR)—positions, linewidths, and intensities of the resonance modes—were determined analytically with the help of the spin-wave approximation for a film with symmetric boundary conditions in an arbitrarily oriented external magnetic field.

However, the case of spins symmetrically pinned on the film surfaces is rarely met in praxis. Most films have different boundary conditions on the free surface and the film–substrate interface [2 to 8]. The positions and intensities of the SSWR modes excited in such a film magnetized along its normal were derived by Ondris and Gärtner [9] and more precisely in [10].

The angular dependence of the SSWR spectrum may be investigated in two steps: first irrespective of the angular dependence of the effective surface pinning parameters and secondly taking this dependence into account.

The first step was made in [11]. The vibrations of a finite linear chain were used as a model for standing spin waves in a ferromagnetic film. The allowed modes were discussed for various combinations of the effective pinning parameters.

It occurs that the angular (or other) dependence of the effective surface pinning mechanism may introduce new effects [12] which cannot be predicted from the scheme with the "rigid" set of boundary conditions used in [11].

[1]) Na Slovance 2, 18040 Prague Liben, Czechoslovakia.

The experiments made on yttrium–iron-garnet (YIG) films prepared in different ways [2 to 7] were interpreted in terms of the surface inhomogeneity (SI) model introduced by Puszkarski for the case of an unidirectional surface anisotropy [2, 13]. Later on the concept was modified using the tensorial form of surface anisotropy [4, 6, 14]. Recently the polynomial form was introduced [15, 16], however, only the first (isotropic) and the third (uniaxial) terms occur to be effective in the interpretation of the experiments on YIG films [15, 17]. In this case the polynomial notation is equivalent to the tensorial one.

The presence of the uniaxial part in the surface anisotropy has been corroborated by most experiments. However, the corresponding term has been mostly used in connection with the circular precession approximation which oversimplifies the angular dependence of the effective uniaxial surface anisotropy term and, consequently, may lead to underestimation of its part in the total effective pinning of the surface spins.

Spalek and Maksymowicz [18] investigated the model of a film with an uniaxial anisotropy and a surface region of different magnetization. The generalized pinning parameters formally satisfying the Rado-Weertman boundary condition [19] were introduced taking into account the ellipticity of spin precession. Unfortunately, the elliptical precession used in the formulation of the problem was omitted in the subsequent treatment which resulted into the same situation as in the above-mentioned papers.

The aim of our paper is to investigate the angular dependence of the (asymmetric) uniaxial surface anisotropy and its influence on the SSWR spectrum. The treatment may help to correct interpretation of the measurements even in the more realistic case of a film possessing another pinning mechanism besides the uniaxial surface anisotropy, e.g. a surface layer of different properties (such a layer can be due to noncollinearity of the surface spins [20], a diffusion zone in the substrate–film interface [4, 5, 15, 21], etc.).

The explicit (matrix) form of the condition giving the allowed values of the resonant spin-wave wave vector k_1 for an arbitrary type of an asymmetric spin-pinning including the effect of elliptical precession is given in [22].

The solution of the boundary problem for a perfectly homogeneous insulator film with asymmetrically pinned spins due to the uniaxial surface anisotropy, presented in the first part of the present paper, results in a simpler form of the condition for k_1-values than in [22]. The analytical formula for the mode intensity is also given. In the second part the case of a metallic film is discussed and the analytical formulae for the mode linewidth (including the exchange-conductivity broadening) and the intensity are derived in the spin-wave approximation. Arbitrary values of the surface anisotropy constants and an arbitrary orientation of the external magnetic field are allowed.

2. Theory

Let us first summarize the assumptions and the relations valid both for the insulating and metallic films.

A homogeneously magnetized infinite film of thickness $2d \ll \delta$ is assumed to possess uniaxial surface anisotropy with an easy or hard axis perpendicular to the film plane. The values of the surface anisotropy constant K_s on both surfaces differ. Both the symmetric and antisymmetric driving fields (parallel to the x-axis) are taken into account.

The orientation of the film with respect to the coordinate system, the external magnetic field, and the magnetization is shown in Fig. 1 of [1]. The Gilbert equation is used to describe the motion of the magnetic moment,

$$\frac{dM}{dt} = -\gamma M \times \left(H + \frac{2A}{M_0^2} \nabla^2 M - \frac{\alpha}{\gamma M_0} \frac{\partial M}{dt}\right), \tag{1}$$

where H is the internal magnetic field, M and M_0 the magnetic polarization vectors in the instantaneous and equilibrium positions, respectively, α the Gilbert damping constant, γ the gyromagnetic ratio, and A the exchange stiffness constant.

The microwave magnetic field and polarization are assumed to depend on time as $h, m \sim e^{i\omega t}$, where ω is the angular microwave frequency.

Decomposition of (1) into components and linearization gives

$$\frac{i\omega}{\gamma} m_x = -\hat{Q} m_\varphi + M_0 \cos \varphi h_y , \tag{2}$$

$$-\frac{i\omega}{\gamma} m_\varphi = -\hat{P} m_x + M_0 h_x , \tag{3}$$

where $m_\varphi = \sqrt{m_x^2 + m_y^2}$.

The operators $\hat{P}$ and $\hat{Q}$ are

$$\hat{P} = H_0 \cos(\varphi - \eta) - \frac{M_0}{\mu_0} \cos^2 \varphi - \frac{2A}{M_0} \frac{\partial^2}{\partial z^2} + \frac{i\alpha\omega}{\gamma}, \tag{4}$$

$$\hat{Q} = \hat{P} + \frac{M_0}{\mu_0} \sin^2 \varphi .$$

The Maxwell equations may be rearranged into the form

$$m_x = -\mu_0 \left(1 + \frac{i\delta^2}{2} \frac{\partial^2}{\partial z^2}\right) h_x , \tag{5}$$

$$m_\varphi \cos \varphi = -\mu_0 \left(1 + \frac{i\delta^2}{2} \frac{\partial^2}{\partial z^2}\right) h_y , \tag{6}$$

where δ is the classical skin depth, $\delta = \sqrt{2/\mu_0 \omega \sigma}$, σ is the electric conductivity, and μ_0 the permeability of vacuum. The boundary conditions for the m-components are [23]

$$\frac{\partial m_x}{\partial z} + \frac{K_s}{A} \cos^2 \varphi \, m_x = 0 , \tag{7}$$

$$\frac{\partial m_\varphi}{\partial z} + \frac{K_s}{A} \cos 2\varphi m_\varphi = 0 ,$$

where K_s is the (uniaxial) surface anisotropy constant.

2.1 Insulator films

In this case $\sigma \to \infty$ and $\delta \to \infty$. The condition for a non-trivial solution of (2), (3) (secular equation) is

$$\left(\frac{\omega}{\gamma}\right)^2 - PQ = 0 , \tag{8}$$

where P and Q are the characteristic values of the operators $\hat{P}$ and $\hat{Q}$, respectively.

Equation (8) has two spin-wave roots $k_{1,2}$, where k_2 is always purely imaginary. The resonance condition is given by the real part of (8).

The values of the wave vector $\boldsymbol{k}_1$ (determining the mode position through the term $2Ak_1^2/M_0$) may be obtained by solving the boundary problem. This means to solve (7) with m_x and m_φ expressed in terms of contributions corresponding to two waves propagating in the film interior,

$$\begin{aligned} m_x &= m_{x1} \cos (k_1 z + \psi_1) + m_{x2} \cos (k_2 z + \psi_2) , \\ m_\varphi &= m_{x1} \Omega_1 \cos (k_1 z + \psi_1) + m_{x2} \Omega_2 \cos (k_2 z + \psi_2) . \end{aligned} \tag{9}$$

The ellipticity factors $\Omega_{1,2}$ are

$$\Omega_{1,2} = \frac{\Omega}{\sin^2 \varphi \pm \sqrt{\sin^4 \varphi + \Omega^2}} , \tag{10}$$

where Ω is the reduced frequency, $\Omega = 2\omega\mu_0/\gamma M_0$, and the phase shifts $\psi_{1,2}$ represent the asymmetry of the solution. As a result we obtain the following condition for the reduced wave vectors $y_1 (y_1 = k_1 d)$:

$$(y_1 \operatorname{tg} y_1 - \beta_{\text{ef}}^{\text{S}}) (y_1 \operatorname{cotg} y_1 + \beta_{\text{ef}}^{\text{A}}) - \beta_{\text{A}}^2 F_{\text{int}} = 0 , \tag{11}$$

where the constants $\beta_{\text{ef}}^{\text{S,A}}$ are

$$\beta_{\text{ef}}^{\text{S,A}} = \frac{\beta_{\text{S}}^2 - \beta_{\text{A}}^2}{\beta_{\text{S}}} \frac{\Omega_{\text{I}} \mp \dfrac{\beta_{\text{S}}}{Y_2^{\text{S,A}}} \Omega_{\text{II}} \cos 2\varphi \cos^2 \varphi}{\Omega_{\text{II}} \mp \dfrac{\beta_{\text{S}}^2 - \beta_{\text{A}}^2}{\beta_{\text{S}} Y_2^{\text{S,A}}} \Omega_{\text{III}}} . \tag{12}$$

Here $\Omega_{\text{I}} = \Omega_1 \cos 2\varphi - \Omega_2 \cos^2 \varphi$, $\Omega_{\text{II}} = \Omega_1 - \Omega_2$; $\Omega_{\text{III}} = \Omega_1 \cos^2 \varphi - \Omega_2 \cos 2\varphi$, and

$$\beta_{\text{S,A}} = (d/2A) [K_s(+d) \pm K_s(-d)] .$$

The interaction function F_{int} is

$$F_{\text{int}} = \frac{Y_1^{\text{A}} + \dfrac{\beta_{\text{S}}^2 - \beta_{\text{A}}^2}{Y_2^{\text{S}}} \cos 2\varphi \cos^2 \varphi}{\beta_{\text{S}} - \dfrac{\beta_{\text{S}}^2 - \beta_{\text{A}}^2}{Y_2^{\text{S}}} \dfrac{\Omega_{\text{III}}}{\Omega_{\text{II}}}} \; \frac{Y_1^{\text{S}} + \dfrac{\beta_{\text{S}}^2 - \beta_{\text{A}}^2}{Y_2^{\text{A}}} \cos 2\varphi \cos^2 \varphi}{\beta_{\text{S}} + \dfrac{\beta_{\text{S}}^2 - \beta_{\text{A}}^2}{Y_2^{\text{A}}} \dfrac{\Omega_{\text{III}}}{\Omega_{\text{II}}}} , \tag{13}$$

where $Y_i^{\text{S}} = y_i \operatorname{tg} y_i$ and $Y_i^{\text{A}} = y_i \operatorname{cotg} y_i$, $i = 1, 2$.

The phase shifts $\psi_{1,2}$ are obtained from the following relations:

$$\operatorname{tg} \psi_1 = - \frac{\beta_{\text{A}} \Omega_{\text{II}} [Y_1^{\text{S}} Y_2^{\text{A}} + (\beta_{\text{S}}^2 - \beta_{\text{A}}^2) \cos 2\varphi \cos^2 \varphi] \operatorname{cotg} y_1}{\beta_{\text{S}} \Omega_{\text{II}} [Y_1^{\text{A}} Y_2^{\text{A}} + (\beta_{\text{S}}^2 - \beta_{\text{A}}^2) \cos 2\varphi \cos^2 \varphi] + (\beta_{\text{S}}^2 - \beta_{\text{A}}^2) (Y_1^{\text{A}} \Omega_{\text{III}} + Y_2^{\text{A}} \Omega_{\text{I}})} , \tag{14}$$

$$\operatorname{tg} \psi_2 = - \frac{\beta_{\text{A}} \Omega_{\text{II}} [Y_1^{\text{A}} Y_2^{\text{S}} + (\beta_{\text{S}}^2 - \beta_{\text{A}}^2) \cos 2\varphi \cos^2 \varphi)] \operatorname{cotg} y_2}{\beta_{\text{S}} \Omega_{\text{II}} [Y_1^{\text{A}} Y_2^{\text{A}} + (\beta_{\text{S}}^2 - \beta_{\text{A}}^2) \cos 2\varphi \cos^2 \varphi] + (\beta_{\text{S}}^2 - \beta_{\text{A}}^2) (Y_1^{\text{A}} \Omega_{\text{III}} + Y_2^{\text{A}} \Omega_{\text{I}})} . \tag{15}$$

As the wave vectors y_1 and y_2 are interrelated, due to the secular equation (8),

$$y_2^2 = y_1^2 - \frac{M_0^2 d^2}{2\mu_0 A} \sqrt{\sin^4 \varphi + \Omega^2} , \tag{16}$$

the values of y_1 satisfying the boundary conditions may be obtained by solving the transcendental equation (11) together with (16).

It is seen that for $\beta_A = 0$ (equally pinned spins) the interaction between the roots of different symmetry does not exist and (11) separates into two simple relations for the symmetrical (S) and the antisymmetrical (A) modes. The phase shifts $\psi_{1,2}$ are zero in this case (see (14), (15)) and, consequently, the modes have pure S or A character; $\beta_{ef}^{S,A}$ are then identical with the reduced effective surface anisotropy constants derived in [1].

If $\beta_A \neq 0$ the solutions of (11) have a mixed character and the phase shifts $\psi_{1,2}$ are nonzero. In the case $\beta_A = \pm\beta_S$ (spins fully free on one surface) (11) becomes identically fulfilled and cannot therefore be the condition for the determination of the values of y_1. The phase shifts $\psi_{1,2}$ then satisfy the relations $\psi_i = y_i - p_i\pi$ for the S modes and $\psi_i = y_i + \pi/2 - p_i\pi$ for the A modes, $i = 1, 2$ and $p_{1,2}$ are integers. The condition for y_1 is then

$$y_1 \operatorname{tg} 2y_1 = 2\beta_S \frac{\Omega_I \mp \frac{\beta_S}{Y_2^{S,A}} \Omega_{II} \cos 2\varphi \cos^2 \varphi}{\Omega_{II} \mp \frac{\beta_S}{Y_2^{S,A}} \Omega_{III}}. \tag{17}$$

If $\beta_S \to \infty$ (spins fully free on one surface and fully pinned on the other), $y_1 = (n + 1/2)\pi/2$ as may be expected.

In the following the modes are classified as S (quasi-symmetrical) or A (quasi-antisymmetrical), according to the value of y_1; ψ_1 is assumed to lie between 0 and $\pi/2$.

Similarly as in the case of symmetrically pinned spins [1], the amplitude of the wave k_2 is much smaller than that of the wave k_1. In the determination of the mode intensity the wave k_2 can be therefore neglected. We have then

$$\begin{aligned} m_x &= m_{x1} \cos(k_1 z + \psi_1); \quad m_\varphi = m_{x1}\Omega_1 \cos(k_1 z + \psi_1) \quad \text{for the S modes,} \\ m_x &= m_{x1} \sin(k_1 z + \psi_1); \quad m_\varphi = m_{x1}\Omega_1 \sin(k_1 z + \psi_1) \quad \text{for the A modes.} \end{aligned} \tag{9'}$$

Equating the absorbed and the dissipated power (both integrated over the film thickness) one obtains the value of m_{x1}. This is equivalent to the projection of (2) and (3) onto the characteristic functions $f(z) = \cos(k_1 z + \psi_1)$ for the S modes and $\sin(k_1 z + \psi_1)$ for the A modes. The projection gives

$$m_{x1} = -\frac{h_d M_0 Q}{(\omega/\gamma)^2 - PQ} \frac{J^{S,A}}{I^{S,A}}, \tag{18}$$

where

$$I^{S,A} = \frac{1}{2d} \int_{-d}^{+d} f^2(z) \, dz = \frac{1}{2}\left(1 \pm \frac{\sin 2y_1}{2y_1} \cos 2\psi_1\right), \tag{19}$$

$$J^{S,A} = \frac{1}{2d} \int_{-d}^{+d} f(z) \, dz = g^{S,A} \frac{\sin y_1}{y_1}, \tag{20a}$$

for the symmetric driving field, $g^S = \cos\psi_1$ and $g^A = \sin\psi_1$,

$$J^{S,A} = \frac{1}{2d} \int_{-d}^{+d} \frac{z}{d} f(z) \, dz = \frac{\mp g^{A,S}}{y_1}\left(\frac{\sin y_1}{y_1} - \cos y_1\right) \tag{20b}$$

for the antisymmetric driving field. The indexes S and A indicate the correspondence to the S or A modes, respectively. The power absorbed per unit area of the film is

$$R = \tfrac{1}{2} \int_{-d}^{+d} h_x \dot{m}_x \, dz = i h_d \omega \, d m_{x1} J^{S,A}. \tag{21}$$

The mode linewidth ΔH may be evaluated by equating the real and imaginary parts of the denominator in (18) for $H_0 = H_{0n} \pm \Delta H/2$, where H_{0n} is the resonance field of the n-th mode (see [24]). As the resonance line in thin films is Lorentzian, the inflection linewidth ΔH_{inf} is the $1/\sqrt{3}$ multiple of ΔH [1, 24],

$$\Delta H_{\text{inf}} = \frac{2}{\sqrt{3}} \frac{\alpha\omega}{\gamma} K_0 , \tag{22}$$

where K_0 is the geometrical factor,

$$K_0 = \left[\cos(\varphi - \eta) - \frac{\sin(\varphi - \eta) \sin 2\varphi}{\sin^2 \varphi + \sqrt{\sin^4 \varphi + \Omega^2} - \dfrac{4A\mu_0}{M_0^2} k_1^2} \left(2 - \frac{\sin^2 \varphi}{\sqrt{\sin^4 \varphi + \Omega^2}}\right)\right]^{-1} . \tag{23}$$

Substituting (22) into (21) one obtains

$$R = \frac{2d}{\sqrt{3}} h_d^2 \frac{M_0 K_0}{\Delta H_{\text{inf}}} \frac{\sin^2 \varphi + \sqrt{\sin^4 \varphi + \Omega^2}}{2\sqrt{\sin^4 \varphi + \Omega^2}} \left(\frac{J^2}{I}\right)^{\text{S,A}} = R_0 \left(\frac{J^2}{I}\right)^{\text{S,A}} . \tag{21'}$$

The integrals $I^{\text{S,A}}$, $J^{\text{S,A}}$ may be lucidly evaluated in terms of the effective surface anisotropy constant β_e defined as

$$\beta_e = \begin{cases} y_1 \operatorname{tg} y_1 & \text{for the S modes ,} \\ -y_1 \operatorname{cotg} y_1 & \text{for the A modes .} \end{cases} \tag{24}$$

For $\beta_S \neq \pm\beta_A$ and with respect to (11) β_e is

$$\begin{aligned} \beta_e &= \beta_{\text{ef}}^S + \beta_A^2 \frac{F_{\text{int}}}{\beta_{\text{ef}}^A + Y_1^A} && \text{for the S modes ,} \\ &= \beta_{\text{ef}}^A + \beta_A^2 \frac{F_{\text{int}}}{\beta_{\text{ef}}^S - Y_1^S} && \text{for the A modes ,} \end{aligned} \tag{24'}$$

for $\beta_S = \pm\beta_A$ (spins fully free on one surface) one obtains for β_e (see (17))

$$\beta_e \doteq \frac{\beta_S}{2} \frac{\Omega_{\text{I}} + \dfrac{\beta_S}{|y_2|} \Omega_{\text{II}} \cos 2\varphi \cos^2 \varphi}{\Omega_{\text{II}} + \dfrac{\beta_S}{|y_2|} \Omega_{\text{III}}} \left(1 - F(y_1)\right) , \tag{24''}$$

where $F(y_1) = \operatorname{tg}^2 y_1$ for the S modes and $F(y_1) = \operatorname{cotg}^2 y_1$ for the A modes.

In contrast to the case of equally pinned spins, the modes of both symmetries are excited both by the symmetric and antisymmetric driving field,

$$\left(\frac{J^2}{I}\right)^{\text{S}} = 2 \cos^2 \psi_1 \left(\frac{\beta_e}{y_1}\right)^2 \frac{1}{y_1^2 + \beta_e(\beta_e + \cos 2\psi_1)} , \tag{25a}$$

$$\left(\frac{J^2}{I}\right)^{\text{A}} = 2 \sin^2 \psi_1 \frac{1}{y_1^2 + \beta_e(\beta_e + \cos 2\psi_1)} \tag{25b}$$

for the symmetric driving field and

$$\left(\frac{J^2}{I}\right)^{\mathrm{S}} = 2 \sin^2 \psi_1 \frac{\left(1 - \frac{\beta_e}{y_1^2}\right)^2}{y_1^2 + \beta_e(\beta_e + \cos 2\psi_1)}, \tag{25c}$$

$$\left(\frac{J^2}{I}\right)^{\mathrm{A}} = 2 \cos^2 \psi_1 \frac{\left(\frac{1 + \beta_e}{y_1}\right)^2}{y_1^2 + \beta_e(\beta_e + \cos 2\psi_1)} \tag{25d}$$

for the antisymmetric driving field.

For a more transparent outline two special cases will be treated. In both cases the condition $\beta_S \ll y_1$ is assumed to apply. The inequality restricts the calculations to the range of the higher-order modes.

First, the additional condition $\beta_A \ll \beta_S$ is assumed. Then $\sin^2 \psi_1 \doteq (\beta_A/y_1)^2 \, (\Omega_I/\Omega_{II})^2$, $\cos^2 \psi_1 \doteq 1$, $\beta_e \doteq \beta_{ef}^{S,A}$ and the absorbed power can be written as

$$R^S = 2R_0 \frac{\beta_e^2}{y_1^2[y_1^2 + \beta_e(\beta_e + 1)]}, \tag{26a}$$

$$R^A = 2R_0 \left(\frac{\Omega_I}{\Omega_{II}}\right)^2 \frac{\beta_A^2}{y_1^2[y_1^2 + \beta_e(\beta_e + 1)]} \tag{26b}$$

for the symmetric driving field and

$$R^S = 2R_0 \left(\frac{\Omega_I}{\Omega_{II}}\right)^2 \frac{\beta_A^2}{y_1^2[y_1^2 + \beta_e(\beta_e + 1)]}, \tag{26c}$$

$$R^A = 2R_0 \frac{(1 + \beta_e)^2}{y_1^2[y_1^2 + \beta_e(\beta_e + 1)]} \tag{26d}$$

for the antisymmetric driving field.

In this special case

a) the absorptions corresponding to the modes excited by driving fields of opposite symmetry (26b, c) have the same values and are both proportional to β_A^2; consequently they can be used for the determination of the asymmetry of spin-pinning; b) the absorptions of the modes excited by the driving field of the same symmetry (26a, d) have the same dependence on β_e as in the case of the symmetrical spin-pinning (see [1]), only the "asymmetrical" expression for β_e (12) is used instead of the "symmetrical" one [1].

In the second example it is assumed that $\beta_A = \pm\beta_S$ (spins are fully free on one surface). Then $\psi_i = y_i - p_i\pi$ for the S modes and $\psi_i = y_i + \pi/2 - p_i\pi$ for the A modes, $i = 1, 2$. The absorbed power is

$$R^S \doteq 2R_0\beta_e^2 y_1^{-4}, \tag{27a}$$

$$R^A \doteq 2R_0\beta_e^2 y_1^{-4} \tag{27b}$$

for the symmetric driving field and

$$R^S \doteq 2R_0\beta_e^2 y_1^{-4}, \tag{27c}$$

$$R^A \doteq 2R_0(1 + \beta_e) y_1^{-4} \tag{27d}$$

for the asymmetric driving field, where β_e is given by (24"). It is seen that in contrast to the A modes the intensity of the S modes is practically not influenced by the symmetry of the driving field. The spectra of both the S and the A modes have the same envelope in the symmetric driving field.

2.2 Metallic films

As was shown in [1], the conductivity does not significantly influence the boundary conditions for the hf magnetization in thin films ($2d \ll \delta$). The values of the wave vector k_{eff} defined by the relation

$$H_{\text{oh}} - H_{0n} = \frac{2A}{M_0} k_{\text{eff}}^2 , \tag{28}$$

where H_{oh} and H_{0n} are the resonance fields of the homogeneous mode in an insulator film and the n-th mode in a metallic film, respectively, is practically identical with the corresponding "insulator" values of k_1. This is true for any SSWR mode including the first one and for any value of symmetrical spin-pinning.

In the film with asymmetrically pinned spins the situation is similar. We use, therefore, the "insulator" relations (11) or (17) for the determination of y_{eff} in a metallic film, $y_{\text{eff}} = k_{\text{eff}} d$. Also the "insulator" notation will be used for metallic films, i.e., k_1 instead of k_{eff}, etc. Expressions (14) and (15) will serve for the determination of ψ_1 and ψ_2, respectively.

In metallic films, in addition to the spin waves, two electromagnetic waves are excited. All the waves must satisfy the Gilbert equation ((2) and (3)) and the Maxwell equations (5) and (6).

The electromagnetic waves have wave vectors $k_3 \approx k_4 \approx \delta^{-1}$ in films with $2d \ll \delta$, so that $|k_3| \approx |k_4| \ll |k_1|$. Their characteristic functions are, therefore, nearly constant over the film thickness. The contribution of the electromagnetic waves to the hf magnetic field in the film can be then replaced by a constant term enabling to satisfy the boundary condition

$$h_x = h_e + h_{x1}\big(f(z) - f(d)\big) , \tag{29a}$$

$$h_y = h_{y1}\big(f(z) - f(d)\big) , \tag{29b}$$

where $h_e = h_d$ for the symmetric driving field and $h_e = h_d z/d$ for the antisymmetric driving field.

It is obvious that the nearly constant terms do not practically influence the absorption. The same is true for the nonresonant wave k_2. The components of $\boldsymbol{m}$ can be then formally expressed by (9′).

The above-mentioned approximation is called the spin-wave approximation.

Proceeding as before and bearing in mind that the magnitudes h_{x1}, h_{y1} must be imaginary to fulfil the boundary condition and contribute therefore to power dissipation, an expression for the absorbed power can be derived, which is formally the same as (21′). For the inflection linewidth ΔH_{inf} one obtains

$$\Delta H_{\text{inf}} = \frac{2}{\sqrt{3}} \left[\frac{\alpha\omega}{\gamma} + \frac{2M_0}{\mu_0} \left(\frac{d}{\delta}\right)^2 \frac{1}{y_1^2} \big(1 - G(y_1)\big) \frac{Q_0 + P_0 \cos^2\varphi}{Q_0 + P_0} \right] K_0 , \tag{30}$$

where $G(y_1) = f(d)\, J^{\text{S,A}}/I^{\text{S,A}}$, I and J are given by (19) and (20), respectively, P_0 and Q_0 are the real parts of P and Q, respectively, $P_0 = \frac{1}{2}\left(\sqrt{\sin^4\varphi + \Omega^2} - \sin^2\varphi\right)$, $Q_0 = \frac{1}{2}\left(\sqrt{\sin^4\varphi + \Omega^2} + \sin^2\varphi\right)$.

The first term in the square brackets in (30) corresponds to the Gilbert damping, the other is the exchange-conductivity broadening. Whereas the phenomenological damping term is influenced by spin-pinning only through the geometrical factor K_0 (see (23)) the dependence of the exchange-conductivity broadening on the boundary conditions is more complex (for given angle η both k_1 (or y_1) and φ depend on the surface anisotropy constants).

3. Summary

From the treatment presented in this paper the following conclusions can be made:

I. For the insulator films: a) The linewidth is angle-dependent through the geometrical factor K_0 (see also [24]). b) As K_0 depends on spin-pinning (k_1 and φ are functions of the surface anisotropy constants) the linewidth is influenced by boundary conditions. c) The mode intensity (absorption peak height) is angle-dependent, but this dependence is not influenced by that of the linewidth since K_0 cancels out. d) The mode intensity strongly depends on the effective surface anisotropy constant. The higher-order S modes vanish at the critical angle where $\beta_e = 0$; the critical angle for the A modes (excited by the antisymmetric driving field) is that at which $\beta_e = -1$. It is obvious that the values of the critical angles are strongly influenced by the asymmetry of spin-pinning.

II. For metallic films ($2d \ll \delta$): a) Conductivity influences the SSWR spectra only through the exchange-conductivity broadening of the resonance line. The integral intensity (area of the absorption curve) is the same as in the insulator film. b) The angular and the boundary-condition dependences of the Gilbert and exchange-conductivity parts of the linewidth differ, similarly as in the case of the symmetrically pinned spins [1].

It should be noted that the analytical formulae presented here can be applied successfully only to modes which do not overlap. Overlapping of adjacent modes is very frequent in films with asymmetrically pinned spins as the modes of both symmetries are excited and the resonance spectrum is therefore twice denser.

The analytical expressions derived in this paper are not simple as the situation which is described is not simple. The main contributions of these formulae may be seen in the possibility to trace the influences of different parameters on the SSWR spectrum, and in simpler fitting of the experimental data in comparison with exact solutions [1].

References

[1] M. Jirsa, phys. stat. sol. (b) **125**, 187 (1984).
[2] J. T. Yu, R. A. Turk, and P. E. Wigen, AIP Conf. Proc. **10**, 144 (1972).
[3] R. D. Henry, P. J. Besser, D. M. Heinz, and J. E. Mee, IEEE Trans. Magnetics **9**, 535 (1973).
[4] R. A. Turk, J. T. Yu, and P. E. Wigen, AIP Conf. Proc. **18**, 1403 (1973).
[5] S. Bhagat, H. Lessoff, C. Vittoria, and C. Guenzer, phys. stat. sol. (a) **20**, 731 (1973).
[6] J. T. Yu, R. A. Turk, and P. E. Wigen, Solid State Commun. **14**, 283 (1974).
[7] J. T. Yu, R. A. Turk, and P. E. Wigen, Phys. Rev. B **11**, 420 (1975).
[8] L. J. Maksymowicz and D. Sendorek, J. Magnetism and magnetic Mater. **37**, 177 (1983).
[9] M. Ondris and H. Gaertner, Z. angew. Phys. **24**, 200 (1969).
[10] M. Jirsa, Thesis, Prague 1981, unpublished.
[11] H. Puszkarski, phys. stat. sol. (b) **50**, 87 (1972).
[12] M. Jirsa, phys. stat. sol. (b) **124**, 609 (1984).
[13] H. Puszkarski, Acta phys. Polon. **A38**, 217, 899 (1970).
[14] P. E. Wigen and H. Puszkarski, Solid State Commun. **18**, 363 (1976).
[15] P. E. Wigen, T. S. Stakelon, H. Puszkarski, and P. Yen, AIP Conf. Proc. **29**, 670 (1976).
[16] P. Yen, T. S. Stakelon, and P. E. Wigen, Phys. Rev. B **19**, 4575 (1979).
[17] H. R. Summers and P. E. Wigen, J. appl. Phys. **50**, 2015 (1979).
[18] J. Spalek and A. Z. Maksymowicz, Solid State Commun. **15**, 559 (1974).
[19] G. T. Rado and J. R. Weertman, J. Phys. Chem. Solids **11**, 315 (1959).
[20] T. S. Stakelon, J. appl. Phys. **49**, 1592 (1978).
[21] O. G. Ramer and C. H. Wilts, phys. stat. sol. (b) **73**, 443 (1976).
[22] J. R. Banavar and F. Keffer, Phys. Rev. B **20**, 1221 (1979).
[23] R. F. Soohoo, Phys. Rev. **131**, 594 (1963).
[24] M. Okochi and H. Nosé, J. Phys. Soc. Japan **27**, 312 (1969).

(Received August 16, 1984)

phys. stat. sol. (b) **126**, 557 (1984)

Subject classification: 13.1 and 19; 21.1

Institute of Metal Physics, Academy of Sciences of the USSR, Ural Scientific Centre, Sverdlovsk[1])

Electronic Structure of V_3Si and V_3Ge

Comparison of NMR Data with Band-Structure Calculations

By

A. V. SKRIPOV and A. P. STEPANOV

The parameters of electronic structure near the Fermi energy obtained from ^{51}V NMR measurements in V_3Si and V_3Ge are compared with the results of band-structure calculations. It is found that the band-structure calculations of Klein et al. combined with the model of electron lifetime broadening can satisfactorily describe the experimental temperature dependence of spin–lattice relaxation rate in V_3Si and V_3Ge. For both the compounds the experimental values of the electronic density of states at the Fermi level, $N(E_F)$, are about twice the theoretical ones. The d-orbital content of $N(E_F)$ in V_3Si derived from the experiment is consistent with the predictions of Mattheiss and Weber.

Параметры электронной структуры V_3Si и V_3Ge вблизи энергии Ферми, найденные из измерений ЯМР ^{51}V, сравниваются с результатами зонных расчетов. Показано, что зонные расчеты Клейна и др. в сочетании с моделью релаксационного уширения пиков плотности электронных состояний позволяют удовлетворительно описать наблюдаемую температурную зависимость скорости спин–решеточной релаксации в V_3Si и V_3Ge. Для обоих соединений экспериментальные значения плотности состояний на уровне Ферми приблизительно в два раза превышают теоретические. Экспериментальные данные по d-орбитальным вкладам в $N(E_F)$ для V_3Si согласуются с предсказаниями Маттхейса и Вебера.

1. Introduction

The anomalous physical properties of some intermetallic compounds with the A-15 structure are widely believed to be due to the fine structure in the electronic density of states $N(E)$ near the Fermi energy E_F. The existence of such fine structure is the key feature of a number of theoretical models [1 to 3] proposed to explain the properties of these compounds. The fine structure in $N(E)$ has been also obtained in first-principles band-structure calculations [4 to 10] for some A-15 compounds. However, at present it is not quite clear how well these calculations describe the details of $N(E)$ near E_F. The aim of the present work is to show how the nuclear magnetic resonance (NMR) measurements in A-15 compounds can be used to test some results of band-structure calculations. The measurements of the nuclear spin–lattice relaxation time T_1 are especially suitable for this purpose since, in contrast to such $N(E_F)$-dependent quantities as magnetic susceptibility and specific heat, T_1 has no $N(E_F)$-independent contributions and is also sensitive to the orbital content of $N(E)$ near E_F. For comparison of experimental data with theoretical band-structure results we have chosen the isoelectronic compounds V_3Si and V_3Ge with significantly differing physical properties. The ^{51}V NMR properties of high-quality samples of these compounds including a number of V_3Si single crystals have been studied in our previous works

[1]) Ul. S. Kovalevskaya 18, 620219 Sverdlovsk, USSR.

[11 to 13]. In the present work we give a detailed description of the analysis of NMR data enabling us to determine some microscopic properties of A-15 compounds. We also compare the predictions of band-structure calculations with our experimental results on $N(E_F)$, the orbital content of $N(E)$ near E_F, and the temperature dependence of NMR properties governed by the shape of $N(E)$ near E_F.

2. Analysis of NMR Data

Combining the experimental data on the temperature dependence of the ^{51}V spin-lattice relaxation rate $R = (T_1T)^{-1}$, the isotropic (K^{iso}) and anisotropic (K^a) components of the Knight shift, and the magnetic susceptibility χ, one can separate different contributions to these quantities and estimate the density and the orbital content of electron states at the Fermi level. The analysis is based on the following equations derived in the tight-binding approximation [14, 11]:

$$\chi(T) = \chi_s + \chi_{orb} + \chi_d(T)\,, \tag{1}$$

$$\chi_d(T) = 2\mu_B^2 \frac{N_d(T)}{1 - JN_d(T)}, \tag{2}$$

$$K^{iso}(T) = K_s + K_{orb}^{iso} + K_d(T)$$

$$= \frac{1}{\mu_B} [H_s\chi_s + H_{orb}\chi_{orb} + H_d\chi_d(T)]\,, \tag{3}$$

$$K^a(T) = K_{orb}^a + K_{dip}^a(T)\,, \tag{4}$$

$$K_{dip}^a(T) = \frac{H_{orb}\chi_d(T)}{7\mu_B}\left(F_\sigma + \frac{1}{2}F_\pi - F_{\delta_1} - F_{\delta_2}\right), \tag{5}$$

$$\begin{aligned} R(T) = 2h\gamma^2k_BN_d^2(T)\,\{&H_d^2(F_\sigma^2 + \tfrac{1}{2}F_\pi^2 + F_{\delta_1}^2 + F_{\delta_2}^2) + \\ &+ H_{orb}^2[F_\pi(3F_\sigma + F_{\delta_1} + F_{\delta_2}) + \\ &+ \sin^2\theta\,(4F_{\delta_1}F_{\delta_2} + \tfrac{1}{4}F_\pi^2 - \tfrac{3}{2}F_\sigma F_\pi - \tfrac{1}{2}F_{\delta_1}F_\pi - \tfrac{1}{2}F_{\delta_2}F_\pi)]\}\,. \end{aligned} \tag{6}$$

The subscripts s, orb, d, and dip denote the s-spin, d-orbital, d-spin, and dipole contributions, respectively, and H_i are the corresponding hyperfine fields at nuclear sites; γ is the gyromagnetic ratio of a nucleus, θ the angle between the local symmetry axis z (V atom chains for our V_3X compounds) and the magnetic field $\boldsymbol{H}$, and J the effective electron–electron interaction parameter. The susceptibilities in (3) are taken per atom. $N_d(T)$ is the effective density of d-electron states at the Fermi level,

$$N_d^p(T) = -\int N_d^p(E)\,\frac{\partial f}{\partial E}\,dE\,; \qquad p = 1, 2\,, \tag{7}$$

where f is the Fermi-Dirac function. F_α are the fractional contributions of d-electron states of orbital symmetry $\sigma(3z^2 - r^2)$, $\pi(xz, yz)$, $\delta_1(x^2 - y^2)$, and $\delta_2(xy)$ to $N_d(T)$,

$$F_\sigma + F_\pi + F_{\delta_1} + F_{\delta_2} = 1\,. \tag{8}$$

We have neglected the s-contribution to the spin–lattice relaxation rate. According to our estimates, for the typical ratios $N_s(E_F)/N_d(E_F) \leqq 0.05$ in A-15 compounds this contribution is almost entirely cancelled by the negative s–d contribution [14]. We have also neglected the effect of electron–electron interaction on R since the dominant contribution to the relaxation rate in transition metals is usually due to the orbital hyperfine interaction which is not affected by exchange.

The temperature dependence of the measured quantities is determined by the functions $N_d(T)$ and $N_d^2(T)$. There are several sources of the temperature dependence of

$N_d^p(T)$, all of them being effective if $N_d(E)$ strongly depends on energy near the Fermi level. The most evident one originates from the temperature dependence of the distribution function f. However, in order to account for the observed temperature dependences of NMR properties and χ in A-15 compounds with this mechanism alone, a fine structure in $N_d(E)$ on a scale of ≈ 50 K is required, whereas the band calculations [4 to 10] give the $N_d(E)$ structure on a scale of 200 to 1000 K. It has been suggested [15, 8] that the broadening of $N(E)$ peaks due to electron-lifetime effects may provide an essential contribution to the temperature dependences of physical properties. In this case, $N_d(E)$ in (7) should be replaced by

$$N_d(E, T) = \int N_d(E')\, S\big(E, E', \Gamma(T)\big)\, dE' , \tag{9}$$

where $S\big(E, E', \Gamma(T)\big)$ is a broadening function with the characteristic width $\Gamma = \hbar/\tau$, and τ is the electron relaxation time. $\tau(T)$ can be estimated from

$$\varrho^{-1} = \frac{\Omega_p^2 \tau}{4\pi} , \tag{10}$$

using experimental data on resistivity $\varrho(T)$ and the Drude plasma frequency $\Omega_p(E)$ obtained from the band-structure calculations [7]. It has been found [15] that for A-15 compounds Γ is considerably higher than $k_B T$ for all temperatures above T_c. Hence, the thermal broadening of f can be neglected, and $\partial f/\partial E$ in (7) can be replaced by a δ-function. This approximation will be used throughout the present paper. It implies, in particular, that $N_d^2(T) = [N_d(T)]^2$. We do not take into account the effects of phonons on the thermal occupation of crystalline eigenstates. Such effects in systems with $N(E)$ structure on a scale of phonon frequencies and with strong electron–phonon interaction have been considered by Pickett [16]. They result in an additional broadening of f. However, the parameter-free calculation of χ including these effects has failed to reproduce the experimental temperature dependence of χ in Nb_3Sn [17].

In order to estimate the microscopic parameters entering (1) to (6) we used the following procedure. The first step is to determine $\chi_d(T)$ and $N_d(T)$ from the experimental $\chi(T)$ and $R(T)$ data in a "self-consistent" way, using (1), (2), and (6). The temperature-independent term $\chi_s + \chi_{orb}$ in (1) is treated as a variable parameter to make the ratio $R(T)/N_d^u(T)$ constant in the temperature range studied. For J we have adopted the estimate of Clogston [18], $J \approx 0.14$ eV. This is close to the value of 0.17 eV obtained from specific heat and χ measurements in single-crystalline V_3Si [19]. In principle, J may be treated as the second variable parameter. We assume that hyperfine fields and admixture coefficients F_α are temperature independent. This may be justified by the fact that the experimental K^{iso} versus χ and K^a versus χ plots (with temperature as implicit parameter) in V_3Si are linear [11] (cf. (3) and (5)).

The next step is to determine hyperfine magnetic fields and different contributions to K^{iso}. H_d is obtained directly from the experimental $\chi(T)$ and $K^{iso}(T)$ data using the slope of the K^{iso} versus χ plot and (3). χ_s is estimated in the free-electron model, and H_s is taken the same as in V metal [20]. Then χ_{orb}, K_s, K_{orb}^{iso}, and H_{orb} are obtained from (1) and (3).

The final step is to determine the admixture coefficients F_α. K_{orb}^a is obtained from the intercept of the K^a versus χ_d plot using (4) and (5). The slope of this plot yields the linear combination $F_\sigma + \frac{1}{2} F_\pi - F_{\delta_1} - F_{\delta_2}$, according to (5). The value of this linear combination is used together with (6) and (8) to find F_σ, F_π, and $F_\delta = F_{\delta_1} + F_{\delta_2}$. Since F_{δ_1} and F_{δ_2} enter (5), (6), and (8) in the same way we cannot make any distinction between them. For V_3Si the values of F_σ, F_π, and F_δ are found to be insensitive to the $F_{\delta_1}/F_{\delta_2}$ ratio [11], since F_δ is small compared to F_σ and F_π. As usually, we neglect the difference between H_{orb} values in the expressions for the Knight shift and for the spin–lattice relaxation rate [14]. $\theta = \arccos (1/\sqrt{3})$ in (6), according to our experimental arrangement ($\boldsymbol{H} \parallel [111]$) in the course of relaxation measurements on single-crystalline V_3Si.

The procedure used in earlier attempts to determine F_α in some A-15 compounds from NMR measurements [21, 22] is based on the electric-field-gradient and R data at a single temperature (77 K). Hence, this procedure lacks the "self-consistency" properties of our treatment. Some other aspects of the data treatment [21, 22] have been criticized in [11 to 13].

3. Results and Discussion

The shape of the $N(E)$ function near the Fermi energy can be tested using experimental temperature dependences of NMR properties and magnetic susceptibility. Combining the band-structure calculations with the lifetime-broadening model [15] and $\varrho(T)$ measurements one can find the temperature dependences of $N_d(T)$ and $N_d^2(T)$ from (7), (9), and (10), and compare them with those obtained from the experimental $\chi(T)$, $K(T)$, or $R(T)$ data. Since, in contrast to χ and K, the spin–lattice relaxation rate contains only contributions directly determined by the effective density of electron states at the Fermi level, the $R(T)$ data are particularly suitable for such a comparison. We have used the results of [23] where $N(E_F)$ and $\Omega_p(E_F)$ as functions of Γ are derived from the self-consistent band-structure calculations [7] with a Lorentzian broadening function. The $\varrho(T)$ data are then used to relate Γ with temperature for a given sample and to obtain the temperature dependence of $N_d^2(T) \sim R(T)$. Fig. 1 and 2 show the $R(T)/R(300\text{ K})$ curves calculated in such a way for the three samples: single-crystalline V_3Si with $T_c = 16.9$ K, $\varrho(17\text{ K}) = 3.8\ \mu\Omega$ cm, $\varrho(300\text{ K})/\varrho(17\text{ K}) = 25$, the same sample irradiated by fast neutrons ($T_c = 16.3$ K, $\varrho(17\text{ K}) = 10.5\ \mu\Omega$ cm, $\varrho(300\text{ K})/\varrho(17\text{ K}) = 9.5$), and polycrystalline V_3Ge ($T_c = 6.3$ K, $\varrho(7\text{ K}) = 7\ \mu\Omega$ cm, $\varrho(300\text{ K})/\varrho(7\text{ K}) = 12.3$). The $R(T)$ data for these samples have been partially reported in our previous papers [12, 13] where the experimental details can be found. The experimental points for the unirradiated V_3Si between 17 and 23 K and for V_3Ge between 6.5 and 9 K are not included in Fig. 1 and 2 since in these temperature regions $R(T)$ is believed to be strongly affected by

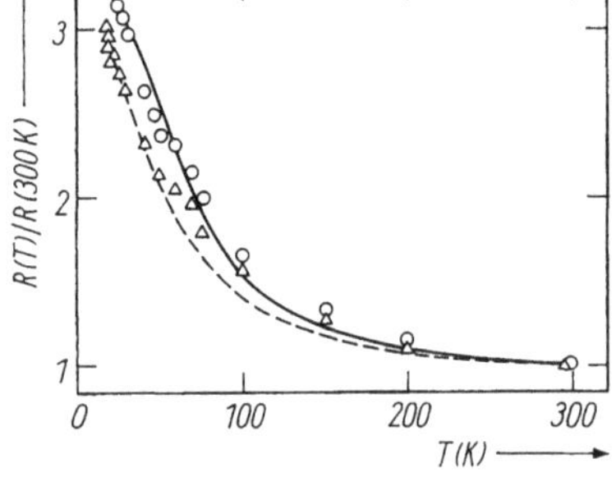

Fig. 1

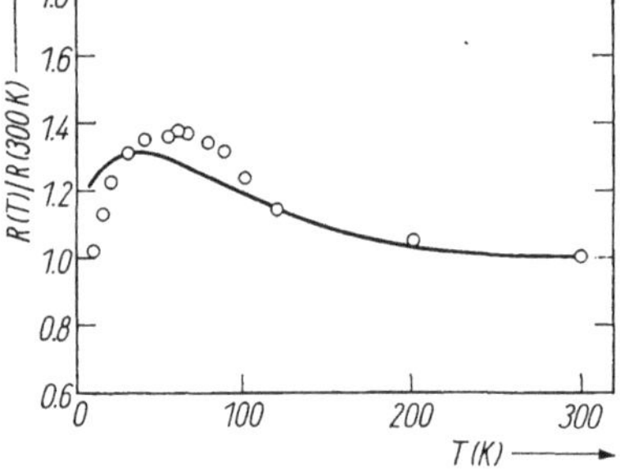

Fig, 2

Fig. 1. Temperature dependence of the relative spin–lattice relaxation rate $R(T)/R$ (300 K) in unirradiated V_3Si (solid line calculations, ○ experiment) and neutron-irradiated V_3Si (dashed line calculations, △ experiment)

Fig. 2. Temperature dependence of the relative spin–lattice relaxation rate $R(T)/R$ (300 K) in V_3Ge. Solid line calculations, points experiment

Table 1

Hyperfine magnetic fields and different contributions to χ, K^{iso}, and K^a in V_3Si and V_3Ge

sample	H_d	H_{orb}	χ_s	χ_{orb}	χ_d (25 K)	K_s	K_{orb}^{iso}	K_d (25 K)	K_{orb}^a	K_{dip}^a (25 K)
	(10^5 Oe)		(10^{-4} cm^3/mol)			(%)			(%)	
V_3Si	−0.68	1.75	0.2	5.8	9.6	0.07	0.60	−0.185	−0.055	0.115
V_3Ge	−0.60	1.90	0.2	5.4	4.9	0.07	0.62	−0.095	—	—

the structural transition and superconducting fluctuations, respectively. As can be seen from Fig. 1 and 2 the agreement between the calculated and experimental $R(T)/R(300\text{ K})$ data is quite reasonable for both V_3Si and V_3Ge. In particular, the band-structure results give the correct values for the relative changes in R between 300 and 25 K in unirradiated and neutron-irradiated V_3Si and also predict a maximum in $R(T)$ for V_3Ge at low temperatures. It should be noted that the maximum is predicted only for clean V_3Ge with $\varrho_0 \lesssim 15\ \mu\Omega$ cm. This may explain the fact that the $\chi(T)$ maximum [24] in V_3Ge was not observed in early experiments [25].

The numerical results of the analysis of experimental data for V_3Si and V_3Ge are presented in Tables 1 to 3. The results for the neutron-irradiated V_3Si are not included since the applicability of the data treatment described in Section 2 to disordered systems is problematic [12]. It can be seen from Table 1 that both H_d and H_{orb} for V_3Ge are close to the corresponding values for V_3Si. This is consistent with the fact that hyperfine fields at V nuclei are nearly the same for different vanadium-based A-15 compounds [26]. The anisotropic component of the Knight shift in V_3Ge is too small to be reliably determined from NMR measurements on polycrystalline samples. Consequently, we could not use (5) to determine the admixture coefficients F_α for V_3Ge.

In Table 2 the theoretical $N(E_F)$ values (converted to units of states/eV atom spin) are compared with our experimental data.[2]) The values of $N_d(25\text{ K})$ and $N_d(10\text{ K})$ are taken as the representative experimental densities of states at the Fermi level in V_3Si and V_3Ge, respectively. It can be seen that for both compounds the experimental values are about twice the theoretical ones. On the other hand, there is a reasonable agreement between our experimental data and $N(E_F)$ values determined from specific-

Table 2

Theoretical and experimental values of the density of electron states at the Fermi level (in units of states/eV atom spin) for V_3Si and V_3Ge

sample	band-structure calculations					HC + M[a]) [27]	this work
	[4]	[5]	[7]	[9]	[10]		
V_3Si	0.75	0.80	0.92	0.92	1.10	1.76	2.14
V_3Ge	0.74	0.77	0.53	—	—	0.91	1.25

[a]) Heat capacity data combined with McMillan's equation.

[2]) In our previous works we have used units of states/eV × atom V spin, differing from the present ones by a factor of $\frac{4}{3}$.

heat measurements using McMillan's equation. This situation seems to be common for a number of vanadium-based A-15 compounds [27]: though both the experimental and theoretical $N(E_F)$ values are fairly consistent within themselves, the theoretically determined values are too low compared with the experimental ones. One of the possible origins of this discrepancy is the lack of an adequate account of electron–electron interaction in theory and/or experiment.

The orbital content of d-electron states in V_3Si near E_F was calculated only in two papers [4, 9] where the tight-binding approximation was used to fit the augmented-plane-wave energy-band results. In Table 3 the results of these papers (normalized to the d-electron density of states at E_F) are compared with our data on F_α[3]). In the earlier work of Mattheiss [4] the δ_1-orbital was predicted to be the dominant component of $N(E_F)$. As has been noted in [11, 12] this is totally inconsistent with the experimental NMR data, particularly with the sign of the temperature dependence of K^a. In the later work of Mattheiss and Weber [9] it has been shown that the addition of V atom s- and p-orbitals to the tight-binding basis significantly changes the results near E_F, the σ-orbital becoming the dominant component of $N(E_F)$. This conclusion is in qualitative agreement with our results on F_α. Thus, there is strong evidence, both from theory and experiment, that $N(E_F)$ in V_3Si is dominated by $\sigma(3z^2 - r^2)$ orbitals. The Fermi-electron-density maxima originating from the overlap of these orbitals lie on V atom chains.

Table 3

Theoretical and experimental values of the fractional admixture coefficients at the Fermi level for V_3Si

<table>
<tr><th>source</th><th>F_σ
$(3z^2 - r^2)$</th><th>F_π
(xz, yz)</th><th>F_{δ_1}
$(x^2 - y^2)$</th><th>F_{δ_2}
(xy)</th></tr>
<tr><td>band calculations [4]</td><td>0.13</td><td>0.28</td><td>0.45</td><td>0.14</td></tr>
<tr><td>band calculations [9]</td><td>0.44</td><td>0.26</td><td>0.19</td><td>0.11</td></tr>
<tr><td>experiment (this work)</td><td>0.67</td><td>0.31</td><td colspan="2">0.02</td></tr>
</table>

4. Conclusions

The results of the present work demonstrate that NMR measurements in A-15 compounds can provide a critical test of the calculated band structure near E_F. The temperature dependences of spin–lattice relaxation rate obtained from the band-structure calculations [7] are in reasonable agreement with the experimental ones for V_3Ge and V_3Si with different residual resistivities. In particular, the band-structure calculations successfully predict the low-temperature maximum of the spin–lattice relaxation rate in V_3Ge. It may be concluded that these calculations yield the realistic $N(E)$ profiles near E_F. The agreement between the theoretical [9] and experimental data on the orbital content of $N_d(E_F)$ in V_3Si may also be regarded as reasonable taking into account the simplifying assumptions involved in the analysis of NMR data. On the other hand, the experimental values of $N(E_F)$ for V_3Si and V_3Ge are about twice the calculated ones. This discrepancy indicates that some important problems remain to be solved in order to get a consistent picture of the electronic structure of V_3Si and V_3Ge.

---

[3]) Our present numerical results on F_α differ from those reported in [11 to 13] since in these works the factor $\frac{1}{2}$ at F_π in (5) was missing.

Acknowledgements

The authors wish to thank A. D. Shevchenko and A. V. Pavlyuchenko for providing the samples used in this study and for resistivity measurements.

References

[1] J. Labbe and J. Friedel, J. Phys. Radium **27**, 153 (1966).
[2] R. W. Cohen, G. D. Cody, and J. J. Halloran, Phys. Rev. Letters **19**, 840 (1967).
[3] L. P. Gorkov and O. N. Dorokhov, J. low-Temp. Phys. **22**, 1 (1976).
[4] L. F. Mattheiss, Phys. Rev. B **12**, 2161 (1975).
[5] T. Jarlborg and G. Arbman, J. Phys. F **7**, 1635 (1977).
[6] A. T. van Kessel, H. W. Myron, and F. M. Mueller, Phys. Rev. Letters **41**, 181 (1978).
[7] B. M. Klein, L. L. Boyer, D. A. Papaconstantopoulos, and L. F. Mattheiss, Phys. Rev. B **18**, 6411 (1978).
[8] W. E. Pickett, K. M. Ho, and M. L. Cohen, Phys. Rev. B **19**, 1734 (1979).
[9] L. F. Mattheiss and W. Weber, Phys. Rev. B **25**, 2248 (1982).
[10] T. Jarlborg, A. A. Manuel, and M. Peter, Phys. Rev. B **27**, 4186 (1983).
[11] A. V. Skripov, A. P. Stepanov, V. A. Marchenko, V. M. Pan and A. D. Shevchenko, Zh. eksper. teor. Fiz. **77**, 2313 (1979); Soviet Phys. – J. exper. theor. Phys. **50**, 1113 (1979).
[12] A. V. Skripov, A. P. Stepanov, V. M. Pan, A. D. Shevchenko, and I. G. Michailov, phys. stat. sol. (b) **102**, 671 (1980).
[13] A. V. Skripov and A. P. Stepanov, Fiz. tverd. Tela **23**, 966 (1981).
[14] A. Narath, in: Hyperfine Interactions, Ed. A. J. Freeman and R. B. Frankel, Academic Press, 1967.
[15] L. R. Testardi and L. F. Mattheiss, Phys. Rev. Letters **41**, 1612 (1978).
[16] W. E. Pickett, Phys. Rev. B **26**, 1186 (1982).
[17] W. E. Pickett, Physica (Utrecht) **107B**, 703 (1982).
[18] A. M. Clogston, Phys. Rev. **136**, A8 (1964).
[19] A. Guha, M. P. Sarachik, F. W. Smith, and L. R. Testardi, Phys. Rev. B **18**, 9 (1978).
[20] Y. Yafet and V. Jaccarino, Phys. Rev. **135**, A1630 (1964).
[21] F. Y. Fradin and D. Zamir, Phys. Rev. B **7**, 4861 (1973).
[22] F. Y. Fradin and J. D. Williamson, Phys. Rev. B **10**, 2803 (1974).
[23] C. M. Soukoulis and D. A. Papaconstantopoulos, Phys. Rev. B **26**, 3673 (1982).
[24] A. V. Pavlyuchenko, Fiz. tverd. Tela **19**, 1431 (1977).
[25] H. J. Williams and R. C. Sherwood, Bull. Amer. Phys. Soc. **5**, 430 (1960).
[26] L. A. G. M. Wulffers, Thesis, Leiden 1977.
[27] H. B. Radousky, T. Jarlborg, G. S. Knapp, and A. J. Freeman, Phys. Rev. B **26**, 1208 (1982).

(Received December 22, 1983)

phys. stat. sol. (b) 126, 565 (1984)

Subject classification: 11; 13; 22.1.2

Institute of Nuclear Physics, Tomsk Polytechnical Institute[1])

Angular Distributions of Fast Electrons at Planar Channeling in Silicon Crystals

By

V. V. KAPLIN, S. B. NURMAGAMBETOV, V. I. GRIDNEV, E. I. ROZUM, S. PAK, S. R. UGLOV, and S. A. VOROBIEV

Detailed experimental and theoretical investigations for the orientation dependence of the angular distribution of 5.6 MeV electrons transmitted through a silicon crystal in (110) and (111) planar directions are performed. The realization of atomic-type and molecular-type quantum states of channeled electrons as well as the effect of preferential population of the definite states are observed

Проведено детальное экспериментальное и теоретическое исследование ориентационной зависимости угловых распределений 5,6 MeV электронов прошедших через кристалл кремния в (110) и (111) плоскостных направлениях. Наблюдались реализация атомного и молекулярного типа квантовых состояний каналирования электронов, а также эффект преимущественного заселения определенных квантовых состояний.

1. Introduction

A detailed investigation of the channeling effects of relativistic electrons supplements the well-known theory of the electron wave diffraction in crystals with new physical ideas [1]. The theory of fast-electron channeling based on the conception of an averaged potential of atomic rows or planes [2] operates with ideas about the existence of quantum states of transverse bound motion of particles transmitted through a crystal. In this case the complete system "channeled electron in a crystal target" is very similar to a two-dimensional quasi-atom moving through the crystal with relativistic velocity and having a discrete spectrum of transverse energies. Considering that in this case an electron is fairly well localized on a separate atomic string or plane there is reason to discuss definite prospects of the creation of new devices for generation of channeling radiation in a wide spectral range and for control and formation of electron beams on the basis of a crystal target. The quantum character of the transverse motion basically differentiates the flux of channeled particles which have definite connection with the crystal lattice, from a particle flux in a random target.

In the first case the character of the electron interaction with crystal atoms and impurities will be defined by the preferential population of the quantum states of the channeled electron. This phenomenon can be used effectively for governing the electron flux creating atomic excitations or defects in crystal lattice, as well as for the location of impurities in the crystal. In the crystal structure of diamond type the (111) atomic planes are arranged by pairs, thus to describe the electron channeling one ought to use the presentation of the molecular quantum states rather than the atomic one. For the first time molecular-type states of channeled electrons were expected in [3], in which their contribution to the formation either the angular distri-

---

[1]) P. O. Box 25, 634050 Tomsk, USSR.

bution of particles behind the crystal or the orientation dependence of Rutherford scattering of electrons has been discussed. Recently, when investigating hard γ-radiation from channeled 900 MeV electrons [4] and X-ray radiation from 4 MeV electrons [5] in $\langle 110 \rangle$ diamond crystals the important role of molecular-type states in the channeling process of electrons was shown.

In the present work a detailed experimental and theoretical investigation was performed for the orientation dependence of the angular distributions of electrons with total energy of $E_0 = 5.6$ MeV having been transmitted through a silicon crystal of about 2 μm thickness in (110) and (111) directions. The obtained results demonstrate the realization of the preferential population of definite quantum states of channeled electrons, which determine the shape of the angular distribution of transmitted electrons at different crystal orientations. The realization of bound states of atomic type and of molecular type near the atomic planes was observed and discussed in the framework of the quantum theory of electron channeling.

2. Method of Theoretical Calculation

For a detailed consideration of the formation of the angular distribution profiles we have calculated the band structure of the spectrum of transverse energies for channeled electrons. We have also calculated the wave functions, the distribution densities, and the orientation dependences of the quantum state excitations of electron transverse energy. The calculation was made by numerical integration of the Schrödinger equation with the periodic potential of the system of (110) atomic planes in Molière approximation with accounting for two neighbour planes. For (111) atomic planes the periodic potential was taken in the form

$$U_p(x) = \sum_m V_p(x - md_p) , \tag{1}$$

where

$$V_p(x) = 8\pi^2\hbar^2(\pi^{1/2}m_0 d_{||} d_\perp)^{-1} \sum_{i=1}^{4} a_i(b_i + 4\pi^2\varrho_\perp^2)^{-1/2} \times$$
$$\times \exp\left(-4\pi^2x^2(b_i + 4\pi^2\varrho_\perp^2)^{-1}\right];$$
$$a_i = \{2.13, 2.53, 0.80, 0.32\} \times 10^{-1}\,\text{nm} ,$$
$$b_i = \{37.8, 16.5, 2.87, 0.386\} \times [10^{-1}\,\text{nm}]^2 .$$

Here d_p is the distance between the atomic planes, $\varrho_\perp$ the root-mean square amplitude of the atomic thermal vibrations, $d_{||(\perp)}$ are the interatomic distances in the plane. The planar potentials were calculated with the use of atomic scattering form factors according to the Hartree-Fock model in relativistic approximation [6].

The band spectrum of transverse energies for channeled electrons was defined by solution of the dispersion equation

$$\cos(k_\perp d_p) = \frac{\psi^+\psi^{-\prime} + \psi^-\psi^{+\prime}}{\psi^+\psi^{-\prime} - \psi^-\psi^{+\prime}}, \tag{2}$$

where $k_\perp$ is the transverse wave vector, ψ^- and ψ^+ are the odd and even solutions of the integrated equation found in the point $x = d_p/2$. The Bloch wave is represented in the form

$$\psi_{nk_\perp}(x) = A_{nk_\perp}\psi^+_{nk_\perp}(x) + B_{nk_\perp}\psi^-_{nk_\perp}(x) , \tag{3}$$

where

$$A_{nk\perp} = \psi_{nk\perp}^{-}\left(\frac{d_p}{2}\right)\cos\left(\frac{k_\perp d_p}{2}\right) - \psi_{nk\perp}^{-\prime}\left(\frac{d_p}{2}\right)\sin\left(\frac{k_\perp d_p}{2}\right),$$

$$B_{nk\perp} = i\left(\psi_{nk\perp}^{+}\left(\frac{d_p}{2}\right)\sin\left(\frac{k_\perp d_p}{2}\right) + \psi_{nk\perp}^{+\prime}\left(\frac{d_p}{2}\right)\cos\left(\frac{k_\perp d_p}{2}\right)\right).$$

The orientation dependences $N_n(\varphi) = |C_n|^2$ of the electron capture in the channeling quantum states were determined from the calculation of the excitation amplitudes

$$C_n = \int_{-\infty}^{\infty} \exp\left(-i(k_\perp + q_\perp)\,x\right)\,\psi_{nk\perp}(x)\,\mathrm{d}x\,, \tag{4}$$

where $q_\perp = 2\pi l/d_p$ is the vector of the reciprocal one-dimensional lattice, $\varphi = (k_\perp + q_\perp)/k$ with $\boldsymbol{k}$ as the total wave vector.

For the purpose of checking our numerical procedures we have carried out a careful comparison with the data published in the literature on the problem of transverse energy band population in the MeV electron energy region [7, 8]. In these papers a Bloch wave $\psi_{nk\perp}(x)$ is represented as a linear combination of plane waves. We think that such a procedure is very complicated and requires to take into account a great number of plane waves for the representation of the bound state Bloch wave, and therefore to diagonalize a large matrix. It is convenient to integrate the one-dimensional Schrödinger equation numerically. In this case we should not be care of the plane wave set convergence.

Our results are in good agreement with those of Anderson et al. [7] for 4.51 MeV electron channeling along the (100) planes of Si. The comparison with results of [8] is presented in Fig. 1, where the energy dependence of the first three bottom and top populations are represented for electron channeling along (110) planes of Si. As seen, there is a little disagreement of results which can be explained by the different accuracy in our and Tulupov's calculations. For the Molière approximation of the (110) plane potential used in [8] $V_0 \approx 7$ eV, and thus we think 21 plane waves are not sufficient for satisfying the inequality $\hbar^2(ng)^2/2M \gg V_0$ for 1 to 20 MeV electrons. Also, in [8] it had been stated that even-number bands have zero population at $k_\perp = 0$. The band edge Bloch wave parity changes when the band crosses the potential barrier top [9]. For the (110) plane of Si this takes place at an electron energy of about 2.5 MeV. Thus, if the electron energy is $E_0 > 2.5$ MeV the even-number band Bloch wave parity is odd at $k_\perp = 0$, and the population of these bands is zero at zero incidence angle of electrons on the crystal. For electron energy $E_0 < 2.5$ MeV the parity is even, and the second-band population is not zero. This effect permits to control the band populations by changing the electron energy. As seen in Fig. 1 the decrease of the second-band

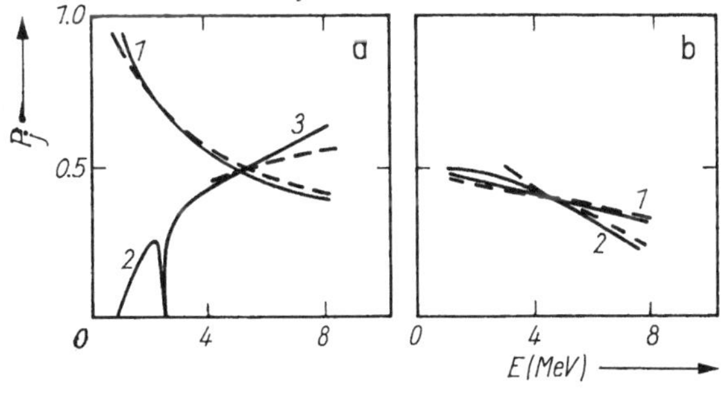

Fig. 1. First three band bottom and top populations P_j, at different electron energies. a) $k_\perp = 0$, b) $k_\perp = 2\pi/d_p$; solid lines our results, dashed lines from [8]

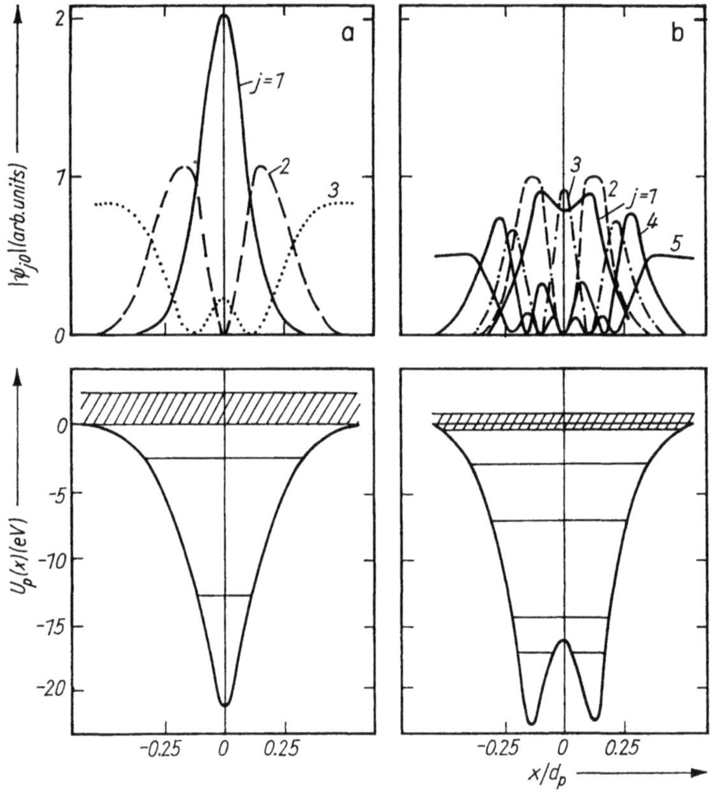

Fig. 2. Potential well and energy levels for 5.5 MeV electrons channeled along a) (110) and b) (111) planes in silicon. Density distribution of electron localization in different channeling states for a) (110) and b) (111) planes in silicon

population is accompanied by an increase of the third-band population. Many investigators have a similar program of calculating the transverse energy band spectrum, and its utilization for theoretical investigations became a common practice (see, e.g., [10 to 12]).

Our calculation results in the case of (110) atomic planes in Fig. 2a show two rather narrow sub-barrier bands and the wide bands of above-barrier transverse energies for channeled electrons. This calculation was done for electrons with the kinetic energy $E = 5.0$ MeV. The band population is defined by the dependence $N_n(\varphi) = |C_n|^2$, which is shown in Fig. 3a. Channeled electrons in the sub-barrier quantum states are strongly localized at the atomic plane positions in contrast to the electrons of the above-barrier bands as it follows from the calculation of the density distribution $\varrho(x) = |\psi_{nk_\perp}(x)|^2$, which is shown in Fig. 2a. However, the possibility of electron localization in the very close region near the atomic nuclei ($|x| < \varrho_\perp$, where $\varrho_\perp$ is the root-mean square amplitude of atomic thermal vibrations) is essential only for the first narrow band of the sub-barrier motion, the states of which will define all the processes of close interactions of channeled electrons with the crystal atoms. So, the channeled electrons of the first energetic band should be responsible for the scattering at large angles, the ionization of the inner shells of the atoms, and the generation of the lattice defects in the crystal structures.

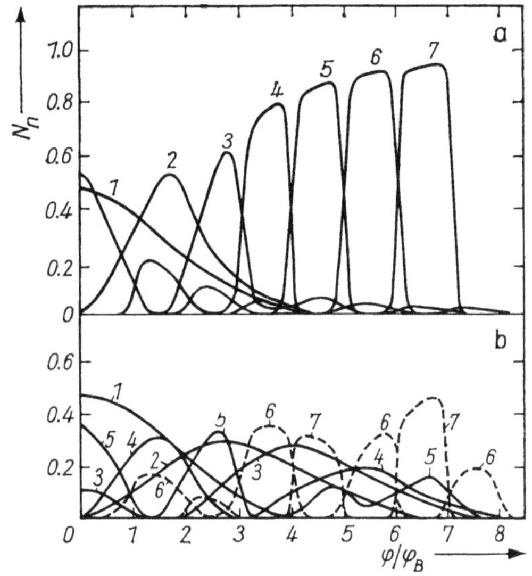

Fig. 3. Orientational population of different channeling states of 5.5 MeV electrons for a) (110) Si and b) (111) Si planes

The minimum electron density in the region $|x| < \varrho_\perp$ for the quantum states of the high-lying bands (and especially for the near-barrier ones) should define their appreciable contribution to the formation of the stable flux of channeled electrons transmitted through the crystal, which forms the observed angular distribution of scattered electrons in the present experiment. In the case of molcular-type channeling of 5.0 MeV electrons along (111) Si atoms planes there are four rather narrow subbarrier bands (see, Fig. 2b), the near-barrier band, and the wide bands for the above-barrier motion, which are populated according to the calculated $N_n(\varphi)$ dependences represented in Fig. 3b. As for the sub-barrier quantum states we should note that in the model of "strong coupling" they correspond to the "bonding" ($n = 1, 3, 5$) and "antibounding" ($n = 2, 4$) molecular orbitals of channeled electrons in the field of two potential centres. The channeled electrons transmitted through a crystal in molecular-type quantum states are localized preferentially near the potential barrier of a pair of atomic planes.

3. Experimental Results and Discussions

The experimental measurements were performed with the electron beam of the microtron-type accelerator of the Nuclear Physics Institute (Tomsk). The angular divergence of the electron beam was about 0.02° and its diameter at the target 0.5 mm. The recording of the scattered electron beam was done using a phototechnical film FT-31 (USSR). The accuracy of the crystal target orientation was about 0.01°.

3.1 Angular distributions of electrons transmitted through a (110) Si crystal

Fig. 4 shows the photos of the angular distributions of the electron beam and the curves of their intensity being obtained by photometry of the photographs for different angles φ of the beam direction relative to the (110) atomic planes of a Si crystal. At the crystal orientation $\varphi = 0°$ (4a) a three-peak structure of the angular distribution with the maxima of intensity at $\theta = 0°$ and $\pm 0.1°$ is observed in contrast to a two-peak distribution for electrons at smaller energies [13]. At $\varphi = 0.05°$ (4b) four maxima of intensity are observed at angles being different from the corresponding positions of the maxima in the previous case. For the crystal orientation $\varphi = 0.12°$ (4c) a picture of three maxima in the angular distribution is observed as for the orientation $\varphi = 0°$, but somewhat asymmetric. The case $\varphi = 0.18°$ (4d) demonstrates a four-peak intensity curve, but with the maxima in positions different from the case $\varphi = 0.05°$. This angular distribution of transmitted electrons is also asymmetric with respect to the

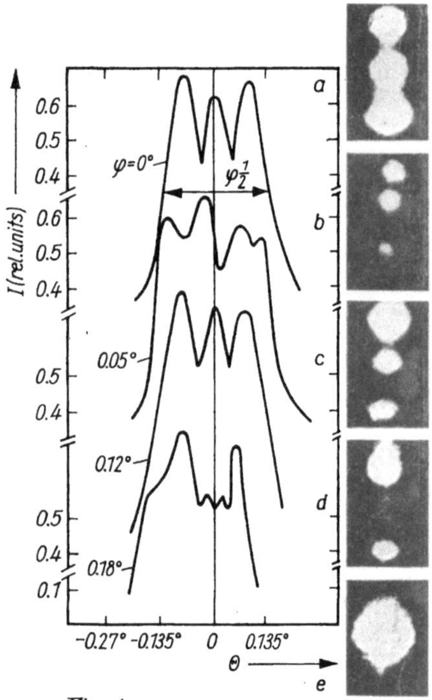

Fig. 4

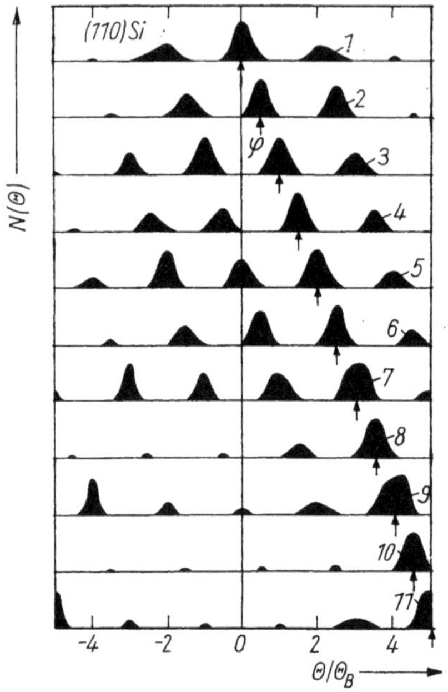

Fig. 5

Fig. 4. Photographs of the angular distribution of 5.6 MeV electrons transmitted through a (110) Si crystal of 2.0 μm thickness at different orientation angles a) $\varphi = 0°$, b) 0.05°, c) 0.12°, d) 0.18°, e) 0.29°

Fig. 5. Calculated shape of the angular distribution of 5.5 MeV electrons behind a (110) Si crystal at different orientation angles φ: (1) 0, (2) $0.5\varphi_B$, (3) φ_B, (4) $1.5\varphi_B$, (5) $2\varphi_B$, (6) $2.5\varphi_B$, (7) $3\varphi_B$, (8) $3.5\varphi_B$, (9) $4\varphi_B$, (10) $4.5\varphi_B$, (11) $5\varphi_B$

atomic plane direction. Finally, at crystal orientation $\varphi = 0.29°$ (4e) the electron flux along the directions close to $\theta = 0°$ is not observed, the distribution of electrons is very asymmetric and similar to the angular distribution of the electron beam at random transmission through the matter. One can suggest that the observed "switching" phenomenon of the angular distribution type of the transmitted electron beam is determined by changing the character of fast-electron interaction with the crystal lattice due to the preferential population of definite quantum states from the discrete spectrum of the energetic bands of transverse motion. At channeling of heavy particles and ultrarelativistic electrons such a "switching" phenomenon of the angular distribution is not observed due to the high density of energetic levels in the potential well [14] because of the quasiclassical character of transverse motion.

Fig. 5 shows the calculated evolution of the total angular distribution of the transmitted electron beam $N(\theta) = \sum_n N_n(\theta)$ which was summarized over all populated quantum states, in dependence on the angular orientation of the incident beam relative to the (110) atomic planes of the crystal target. In this case the angular divergence of the incident beam was taken $\Delta\varphi = 0.5\varphi_B$ (φ is the Bragg angle). The contribution to the angular distribution from electrons of the n-th state was obtained by calculating the momentum representation of the wave function of the transverse component for

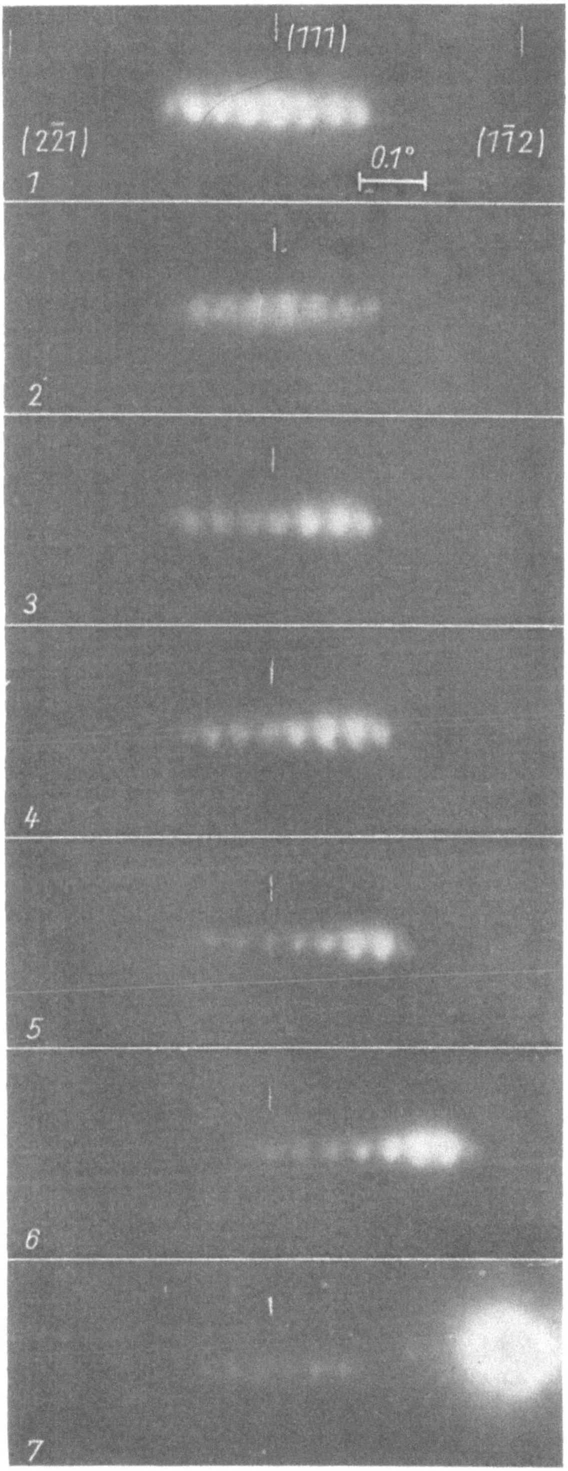

Fig. 6. Photographs of the angular distribution of 5.6 MeV electrons transmitted through a (111) Si crystal of 2.0 μm thickness at different orientation angles (1) $\varphi = 0°$, (2) 0.03°, (3) 0.06°, (4) 0.09°, (5) 0.135°, (6) 0.225°, (7) 0.38°

electron motion in a field of the atomic planes by analogy with the transformation (4). At the crystal orientation $\varphi = 0°$ the calculation shows, similar to our experimental data, three maxima of the intensity which in accordance with the calculated dependence $N_n(\varphi)$ in Fig. 3a are belonging to the third energy band of the channeled electrons. According to this calculation the fraction of the electron beam channeling in these states is about 56%.

When the crystal is tilted, the angular distribution of electrons is changed due to the channeled electron redistribution throughout the quantum states of the other energy bands. At the crystal orientation $\varphi = 1.5\varphi_B$ the second band of the bound motion is populated mainly, that determined the exhibition of two central peaks in the measured angular distribution of electrons at $\varphi = 0.05°$ in Fig. 4. The calculated fraction of the channeled electron beam belonging to the quantum states of this band is about 50%. The lateral maxima of the angular distribution are specified by some contribution of the quantum states from the fourth band of the transverse energy spectrum. With further crystal tilting, as it should follow from the orientation dependence $N_n(\varphi)$ in Fig. 3a, the main part of the channeled beam (about 60%) populates the third energy band again in accordance with experiment giving rise to a three-peak structure of the angular distribution of electrons (Fig. 4, $\varphi = 0.12°$). At crystal orientations $\varphi \geqq 4\varphi_B$ the only quantum states of the energy bands of the above-barrier motion with $n \geqq 4$ determine the shape of the angular distribution of the transmitted electrons, because the trapping of electrons in these quantum states is dominating. Let us note that in this case the maximum in the angular distribution of the transmitted electrons should be formed at the position of the incident beam, i.e. at $\theta = \varphi$ in accordance with the experimental data in Fig. 4.

3.2 Angular distributions of electrons transmitted through a (111) Si crystal

Fig. 6. shows the photos of the angular distributions of an electron beam transmitted through the crystal target at various angles φ relative to the (111) Si atomic planes. The angle between the ⟨110⟩ crystallographic axis and the incident beam in this case was $\varphi_a = 1.6°$. At the crystal orientation $\varphi = 0°$ (photo 1) the angular distribution shows some intensity maxima, the most pronounced ones are at $\theta = 0°$, $\pm 0.045°$, and $\pm 0.135°$. With crystal tilting up to $\varphi = 0.03°$ (photo 2) the three-peak structure of the angular distribution in the centre of the picture transforms into a two-peak one with maximum intensity of the transmitted electrons in directions close to $\theta = \pm 0.03°$. With further crystal tilting the number of intensity maxima in the angular distribution of electrons transmitted through the crystal is practically unchanged, but one can observe a displacement of the angular position for the most pronounced intensity maxima (photos 3, 4). At the crystal orientations $\varphi \geqq 0.09°$ the angular distribu-

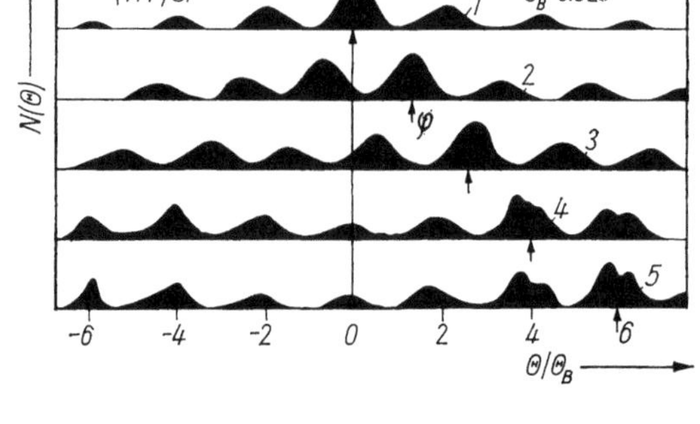

Fig. 7. Calculated shape of the angular distribution of 5.5 MeV electrons behind a (111) Si crystal at different orientation angles (1) $\varphi = 0°$, (2) 0.03°, (3) 0.06°, (4) 0.09°, (5) 0.135°

tions are asymmetric with the maximum intensity of the transmitted electrons near $\theta \approx \varphi$, i.e., in the directions close to the primary beam incidence. The enhancement of the crystal tilting angle φ leads to the gradual disappearance of the intensity maxima near the atomic plane direction, e.g., in photo 5, $\varphi = 0.135°$. At least, at $\varphi = 0.225°$ (photo 6) the angular distribution of the transmitted electrons assumes a special shape with a three-peak maximum in the position of the incident beam, the spreading of which is increasing with crystal tilting.

Fig. 7 shows the calculated evolution of the angular distribution of electrons, $N(\theta) = \sum_n N_n(\theta)$, behind the crystal in dependence on the tilting angle φ of (111) atomic planes relative to the incident beam with angular divergence $\Delta\varphi = 0.02°$. The kinetic energy of the incident electrons was taken $E = 5.0$ MeV. The contribution to the total angular distribution from electrons of each n-th state was obtained from the momentum representation of the wave function $\psi_{nk_\perp}(x)$ of the transverse motion with the use of a transformation analogous to (4). At beam incidence along the atomic plane direction $\varphi = 0°$, the calculated distribution demonstrates a few intensity maxima, the most pronounced three of them near the position of the incident beam, which are in accordance with the measured angular distributions (Fig. 5, photo 1). As it follows from the dependence $N_n(\varphi)$ in Fig. 3b, the angular distribution of the transmitted electrons in this case is determined mainly by the quantum states of the fifth energy band (and partially of the third one), since the electron trapping probability in the $n = 2$ and 4 quantum states is extremely small. The first band states ($n = 1$) should be not so stable because the strong localization of channeled electrons in this case at the atomic planes will give rise to large-angle scattering [15]. The fraction of the electron beam being captured into $n = 3$ and 5 quantum states of bound motion in the potential field of (111) atomic planes is about 50%.

With crystal tilting, e.g., at the angle $\varphi = 1.25\varphi_B \approx 0.03°$ ($\varphi_B = 0.023°$ in this case), on the contrary, the trapping probability in the quantum states with $n = 3$ and 5 is practically absent (see, Fig. 3b) and the angular distribution of the transmitted electrons is formed by the quantum states of the second, fourth, and sixth energetic bands. This phenomenon, as follows from the calculation (Fig. 7, distribution 2) and experimental data (Fig. 6, photo 2), leads to the formation of minimum intensity of the transmitted electron beam along the atomic plane directions, in contrast to the case $\varphi = 0°$. Further, at $\varphi = 2.5\varphi_B \approx 0.06°$ electron trapping in the states of the fifth near-barrier band becomes possible again in accordance with the $N_n(\varphi)$ dependence in Fig. 2b, that immediately leads to the appearance of the transmitted electron beam along the atomic plane direction (theoretical distribution 3 in Fig. 7, photo 3 in Fig. 6). In the case $\varphi = 4\varphi_B \approx 0.09°$ the above-barrier quantum states are dominating (with $n = 6, 7$), although in accordance with the calculation (Fig. 2b) a few sub-barrier states are excited, too. The odd sub-barrier states contribute to the intensity maxima near the direction of atomic planes and the above-barrier states with $n = 6$ and 7 determine the intensity distribution in the direction close to the incident beam (calculated distributions 4, 5 in Fig. 7 and photos 4, 5 in Fig. 6).

At large angles of crystal orientation $\varphi \gtrapprox 7\varphi_B \approx 0.16° \approx \varphi_c$, where $\varphi_c = \sqrt{2\,\Delta U_0/E}$ is the critical channeling angle [2], with ΔU_0 the depth of the potential well, the primary beam is completely captured into the above-barrier quantum states and the flux of scattered electrons is concentrated in the direction of the incident beam (see, photos 6, 7 in Fig. 6). The narrow angular distribution of electrons transmitted through the crystal is determined by the excitation of the quantum states of channeled electrons in two neighbouring energy bands due to the finite angular distribution of the incidence beam. For all these cases the channeling effect (strongly bound states of fast electrons [15]) is practically absent and one should observe electron diffraction at (111) atomic planes.

4. Concluding Remarks

The experimental investigations and theoretical analysis performed show the realization of the phenomenon of the preferential population for definite energetic bands of transverse energy at planar channeling of electrons in (110) Si crystals. This phenomenon can be used effectively in control and formation of a fast electron flux in a crystal, because the preferential population of definite quantum states determines the localization of the flux of channeled electrons in the crystal lattice. By crystal tilting this localization region may be set to a large or small distance from atomic rows or planes, which should determine the preferential yield of a definite type of collision processes for the channeled electron beam in the crystal.

The experimental data and theoretical considerations presented show also the real existence of molecular-type quantum states under channeling of fast electrons along pairs of (111) atomic planes in a Si crystal. The role of these channeling states at small crystal orientation angles, i.e., at $\varphi \lessapprox \varphi_c$, is a dominant one in the formation of the angular distributions of transmitted electrons. So, one can clearly distinguish the atomic-type and molecular-type quantum states at planar channeling of electrons with energy $E \lessapprox$ 10 to 20 MeV. Recently, the molecular-type channeling was discussed for the ultrarelativistic electron motion along pairs of atomic rows in a ⟨110⟩ diamond crystal [16]. In the case of ultrarelativistic electrons the molecular-type channeling also leads to marked peculiarities in the angular distributions of electrons transmitted through a crystal target and in the channeling radiation spectrum in comparison with the case of atomic-type motion.

References

[1] P. B. HIRSCH, A. HOWIE, R. B. NICHOLSON, D. W. PASHLEY, and M. J. WHELAN, Electron Microscopy of Thin Crystals, Butterworths, London 1965.
[2] J. LINDHARD, Kong. Danske Vid. Selsk., mat.-fys. Medd. **34**, No. 14 (1965).
[3] V. V. KAPLIN and S. A. VOROBIEV, Fiz. tverd. Tela **20**, 31 (1978).
[4] YU. M. ADISHCHEV, A. N. DIDENKO, V. V. KAPLIN, A. P. POTYLITSIN, and S. A. VOROBIEV, Phys. Letters A **83**, 337 (1981).
[5] J. U. ANDERSEN, S. DATZ, E. LAEGSGAARD, J. P. E. SELLSHOP, and A. H. SOERENSEN, Phys. Rev. Letters **49**, 215 (1982).
[6] P. A. DOYLE and P. S. TURNER, Acta cryst. **A24**, 390 (1968).
[7] J. U. ANDERSEN, K. R. ERIKSEN, and E. LAEGSGAARD, Physica Scripta **24**, 588 (1981).
[8] A. V. TULUPOV, Fiz. tverd. Tela **23**, 46 (1981).
[9] S. B. NURMAGAMBETOV, VINITI, N 2688-83 DEP, 1983.
[10] A. V. ANDREEV, S. A. AKHMANOV, V. A. VISLOUKH, and V. A. KUZNETSOV, Zh. eksper. teor. Fiz. **84**, 1743 (1983).
[11] K. KOMAKI and F. FUJIMOTO, Radiat. Eff. **56**, 13 (1981).
[12] N. P. KALASHNIKOV, M. N. MAMONOV, A. S. OLCHAK, and M. M. STRIKHANOV, Fiz. tverd. Tela **25**, 195 (1983).
[13] D. E. POPOV, V. V. KAPLIN, and S. A. VOROBIEV, phys. stat. sol. (b) **96**, 263 (1979).
[14] B. A. CHEVGANOV and J. D. FERANCHUK, J. Physique **43**, 1687 (1982).
[15] V. V. KAPLIN and S. A. VOROBIEV, Zh. exper. teor. Fiz. **73**, 28 (1977).
[16] E. G. VYATKIN, YU. M. FILIMONOV, A. M. TARATIN, and S. A. VOROBIEV, phys. stat. sol. (b) **117**, 41 (1983).

(Received August 14, 1984)

phys. stat. sol. (b) 126, 575 (1984)

Subject classification: 13.1; 22

Sektion Physik der Humboldt-Universität zu Berlin[1])

The Effect of Many-Electron Correlation on Photothresholds of Semiconductors and Valence Band Discontinuities at Heterojunctions

By

F. Bechstedt, R. Enderlein, and O. Heinrich

Photothresholds of semiconductors are known to deviate by 3 to 4 eV from the predictions of one-electron band theory. It is shown that these discrepancies are due to many-electron correlation effects, in particular, to electronic relaxation. The corrections to photothresholds due to electronic relaxation are calculated by means of Green's function technique applying the linear shielded potential approximation to the exchange-correlation self-energy. Within the uncertainties of LCAO theories the calculated photothresholds of group IV, III–V, and II–VI semiconductors agree with experimental data. A simple physical picture of the correlation effect is provided. Relaxation corrections to valence band discontinuities from one-electron band theories are also calculated.

Bekanntlich unterscheiden sich Photoemissions-Schwellenergien um 3 bis 4 eV von den Ergebnissen der Einelektronen-Bandtheorie. Es wird gezeigt, daß diese Diskrepanzen auf Vielteilchen-Korrelationseffekte, insbesondere auf die elektronische Relaxation, zurückzuführen sind. Die Korrekturen der Photoemissions-Schwellenergien infolge elektronischer Relaxation werden mit Hilfe der Methode der Green'schen Funktionen berechnet, wobei die Korrelations-Selbstenergie in linearer Näherung bezüglich der abgeschirmten Coulomb-Wechselwirkung genommen wird. Innerhalb der Ungenauigkeiten der LCAO-Ergebnisse stimmen die berechneten Photoemissions-Schwellenenergien für Halbleiter der Gruppe IV, für III–V- und II–VI-Halbleiter mit experimentellen Werten überein. Es wird eine einfache physikalische Interpretation für den Korrelationseffekt angegeben. Relaxations-Korrekturen an den Valenzband-Diskontinuitäten aus Einelektronen-Bandtheorien werden ebenfalls berechnet.

1. Introduction

Photoemission and Auger spectroscopy of core electrons [1, 2] have demonstrated the importance of many-electron correlation effects in solids, in particular of electronic relaxation, i.e. of the polarization or relaxation of solid state electrons during excitation and the related change of Coulomb energy. It was found that experimental values of core electron binding energies were up to 10 eV smaller than the corresponding theoretical values if relaxation self-energy corrections [3] were disregarded in the calculations. If, on the other hand, relaxation effects were considered, the discrepancies between theory and experiment practically could be removed [4]. These findings, by no means, were surprising. The same situation was met, and solved, many years earlier in the case of free atoms. Atomic ionization energies could only be explained theoretically if the ΔSCF-correction within the Hartree-Fock calculation scheme was made (see e.g. [5] and references therein) which accounts essentially for the same physical effect as electronic relaxation. On this background electronic relaxation and its importance for core level excitations of solids now is clearly established whereas

[1]) Invalidenstr. 110, DDR-1040 Berlin, GDR.

its relevance for other types of electronic elementary excitations involving valence or/and conduction states is still not fully recognized. To mention only some of such excitations: the transition of an electron from the top of the valence band of a semiconductor to infinity with zero kinetic energy which determines the photothreshold; the excitation of an electron at a semiconductor heterojunction from the top of the valence band of one semiconductor to the top of the valence band of the other semiconductor which determines the valence band discontinuity; the excitation of a core electron to the bottom of the conduction band. In this case the relaxation energy is different from relaxation energies involved in the two photoemission experiments which are used for an independent measurement of the transition energy between the core level and the bottom of the conduction band. The "core exciton binding energy" obtained by comparing the two transition energies, therefore, must be expected to be, at least, partially due to electronic relaxation [6]; finally, the direct excitation of an electron from the top of the valence band to the conduction band which determines the direct optical gap in tetrahedral semiconductors. Ab-initio Hartree-Fock calculations of energy bands give considerably larger values of optical gaps than are observed experimentally. This discrepancy was already predicted and attributed to electronic relaxation much earlier [7 to 9]. Later investigations of electronic correlation effects on band structure [10 to 15] confirmed these earlier predictions. Density-functional calculations of band structure were also shown to be very sensitive to the correct account of correlation [16, 17]. Nevertheless there remain unresolved problems concerning the correct treatment of relaxation in calculating optical gaps.

In the present paper we shall concentrate ourselves on the effect of electronic relaxation on photothresholds. Their changes at semiconductor heterojunctions are also considered and used for the calculation of relaxation corrections to valence band discontinuities. Like core electron binding energies the absolute values of photothresholds Φ are not reproduced correctly by one-electron band calculations [18, 19]. In typical cases experimental values of Φ are by 3 to 4 eV smaller than the calculated distances of valence band maxima from the vacuum level. Although these calculations are not very precise in predicting *absolute* positions of energy bands, their inaccuracy cannot amount to 4 eV. There must be other reasons for this remarkable discrepancy. Work functions of metals are strongly influenced by the dipole layer at the surface [20]. This layer makes the electrostatic potential of an electron jump by several eV if the electron moves through the surface. The effect is due to *free* electrons. The corresponding free electron effect in semiconductors is much smaller and has practically no influence on their photothresholds. There may be, however, another dipole layer at a semiconductor surface. It is due to the redistribution of *valence* electrons, and therefore, depends sensitively on the particular surface orientation, on surface treatment and coverage, and, in general, is less effective on photothresholds than in the case of metals. For these reasons surface dipole layers do not provide an universal explanation of the deviations of photothresholds of semiconductors from the predictions of one-electron band theory. Harrison [19] connects these deviations with the image force on the outgoing photoelectron and gives an estimate of about 2 eV of the effect. The image force concept is very close to the concept of electronic relaxation. Both rely on electric polarization of the solid due to an excess charge and related changes of Coloumb energies. Our approach is rigorous in the sense that, in calculating excitation energies of photoelectrons by means of Green's function technique or another formalism of many-body theory, the relaxation correction occurs automatically as the dominating contribution to the correlation self-energy in the final state. One finds that one-particle excitation energies of photoelectrons are lowered by the Coloumb energy which is gained if the remaining electrons of the solid are allowed to redistribute according to the influence of the created electron–hole pair. It is mainly the repelling force of the negative polarization charge around the positive hole which transfers the gained Coloumb energy to the outgoing photoelectron by giving it an additional acceleration. The repelling force makes that part of the excitation energy of the photoelectron is taken from the electronic system itself lowering its energy in the final state. We will demonstrate that the observed large discrepancies between photothresholds from one-electron band theory and experiment can indeed be removed if electronic relaxation is regarded. This does not mean, however, that other corrections do not play *any* role. Surface dipole layers, adatoms in the case of covered surfaces, or conditions of measurement have to be regarded within a complete theory. But we will not deal with such effects in the present paper. The grounding of the sample during measurement, as usual, will be accounted for by setting the zero point of the energy scale equal to the Fermi energy. The lattice relaxation energy is relatively small due to the small lattice susceptibility as compared to the electronic one in the case of semiconductors under consideration. It can be neglected.

The paper is organized as follows. In Section 2 we derive the connection between the photothreshold and the relaxation self-energy and give the general equations for the calculation of the latter. In Section 3 we specify the wave function of the hole and perform the explicit calculation of relaxation self-energies for various group IV, III–V, and II–VI semiconductors. In Section 4 we combine our calculated relaxation energies with results of LCAO theories on the position of valence band maxima to get theoretical estimates for photothresholds and band discontinuities. The results are discussed and compared with experiment.

2. Photothreshold and Relaxation Self-Energy

The photothreshold Φ of a semiconductor can be defined as the excitation energy of an electron–hole pair with the electron at the minimum of the continuum part of the energy spectrum and the hole at the maximum of the valence band. Due to the complete delocalization of the electron in this case excitonic effects are small and can be neglected. Then, using an energy scale where the continuous spectrum starts from zero, the photothreshold Φ can be directly expressed by the energy of the excited hole,

$$\Phi = E_{\text{hole}} \,. \tag{1}$$

Equation (1) relates the photothreshold to the corresponding pole of the one-particle Green's function

$$G(1, 1') = \frac{1}{i\hbar} \langle T(\psi(1)\, \psi^+(1')) \rangle \,. \tag{2}$$

In (2) we have used the abbreviation $1 = \boldsymbol{x}, t$ for the space-time coordinate of an electron (the spin is omitted). The Green's function G obeys the following Dyson equation

$$\left(-i\hbar \frac{\partial}{\partial t} - \frac{\hbar^2}{2m} \Delta + V(\boldsymbol{x})\right) G(1, 1') + \int \mathrm{d}2\Sigma(1, 2)\, G(2, 1') = -\delta(1 - 1') \tag{3}$$

with $V(\boldsymbol{x})$ as the periodic potential of lattice ions and $\Sigma(1, 2)$ as the full self-energy. Σ can be divided into the Hartree part Σ^{H}, the exchange part Σ^{ex}, and the correlation part Σ^{rel}, according to

$$\Sigma = \Sigma^{\text{H}} + \Sigma^{\text{ex}} + \Sigma^{\text{rel}} \,. \tag{4}$$

The correlation part of Σ has been denoted by Σ^{rel} to indicate that it is essentially due to electronic relaxation, as will become clear from the further calculations. Following [21 to 25], we expand $\Sigma^{\text{ex}} + \Sigma^{\text{rel}}$ in a power series with respect to the screened Coulomb potential

$$v_s(1, 1') = \int d2 v(1 - 2)\, \varepsilon^{-1}(1', 2)\,, \tag{5}$$

$$v(1 - 1') = \frac{e^2}{|\boldsymbol{x} - \boldsymbol{x}'|}\, \delta(t - t') \tag{6}$$

with $\varepsilon(1, 2)$ as the longitudinal dielectric function. The linear term of this expansion reads

$$\Sigma^{\text{ex}}(1, 1') + \Sigma^{\text{rel}}(1, 1') = i\hbar v_s(1, 1')\, G(1, 1')\,. \tag{7}$$

The exchange contribution Σ^{ex} to Σ is defined as

$$\Sigma^{\text{ex}}(1, 1') = i\hbar v(1, 1')\, G(1, 1')\,. \tag{8}$$

Combining (7) and (8) one gets

$$\Sigma^{\text{rel}}(1, 1') = i\hbar \int d2 v(1 - 2)\, (\varepsilon^{-1}(1', 2) - \delta(1' - 2))\, G(1, 2)\,. \tag{9}$$

In HF-calculations of energy spectra of crystals the relaxation self-energy Σ^{rel} is omitted whereas the Hartree-Fock part $\Sigma^{\text{H}} + \Sigma^{\text{ex}}$ of Σ is taken into account in addition to the periodic potential $V(\boldsymbol{x})$ of lattice ions. The corresponding HF-Bloch states $\psi_{l\boldsymbol{k}}(\boldsymbol{x})$ and energies $E_l(\boldsymbol{k})$ are given by the equations

$$-\frac{\hbar^2}{2m}\, \Delta\psi_{l\boldsymbol{k}}(\boldsymbol{x}) + \int d^3x'\; V(\boldsymbol{x}, \boldsymbol{x}')\, \psi_{l\boldsymbol{k}}(\boldsymbol{x}') = E_l(\boldsymbol{k})\, \psi_{l\boldsymbol{k}}(\boldsymbol{x})\,, \tag{10}$$

$$V(\boldsymbol{x}, \boldsymbol{x}') = V(\boldsymbol{x})\, \delta(\boldsymbol{x} - \boldsymbol{x}') + \int d(t - t')\, (\Sigma^{\text{H}}(1, 1') + \Sigma^{\text{ex}}(1, 1'))\,. \tag{11}$$

Now representing the Dyson equation (3) by means of the Bloch functions from (10) as basis set, a series of approximations can be introduced which make further calculations easy. We keep only the main terms. In particular, we omit the non-diagonal parts of G and Σ^{rel} with respect to the band index [24] and take account only of the diagonal parts G and Σ^{rel}. This approximation is valid if $|\Sigma_{ll'}^{\text{rel}}| \ll |E_l - E_{l'}|$ can be assumed [25]. The non-diagonal part of ε which accounts for local-field effects can be easily treated by means of perturbation theory. This was demonstrated in [4] in the case of core level excitations. In the present case of excitations from valence band maxima the corresponding corrections to Σ^{rel} are small compared to the diagonal contribution. They can be completely neglected. We stress that ε corresponds to the diagonal dielectric function of HF-*Bloch* electrons rather than *free* electrons, i.e. ε accounts in particular for the generation of electron–hole pairs. This was shown to be very important for the correct calculation of correlation effects in one-particle spectra [15, 17]. If in (9) G is approximately replaced by the Green's function of a HF-Bloch electron (again not of a *free* electron) an explicit expression for the self-energy Σ^{rel} is obtained. It depends still on energy due to the ω-dependence of the dielectric function. If, however, $|\Sigma_l^{\text{rel}}|$ is small compared to $|E_l|$ this dependence can be omitted. We note that this approximation does not apply to the self-energy of a free electron gas. In this case dynamical corrections are quite important [15, 17]. The latter ones are automatically included in our expression for the self-energy. With these approximations the diagonal part $G_l(\boldsymbol{k}, E)$ of the Fourier-transformed hole Green's function at zero temperature can be written as

$$G_{\text{v}}(\boldsymbol{k}, E) = (-E - E_{\text{v}}(\boldsymbol{k}) - \Sigma_{\text{v}}^{\text{rel}}(\boldsymbol{k}))^{-1}\,, \tag{12}$$

$$\Sigma_{\mathrm{v}}^{\mathrm{rel}}(\boldsymbol{k}) = \frac{1}{2} \int \frac{\mathrm{d}^3\boldsymbol{q}}{(2\pi)^3} \frac{4\pi e^2}{|\boldsymbol{q}|^2} \, |B_{\mathrm{vv}}^{\boldsymbol{kk}+\boldsymbol{q}}(\boldsymbol{q})|^2 \, (1 - \varepsilon^{-1}(\boldsymbol{q}, 0)) \,, \tag{13}$$

$$B_{ll'}^{\boldsymbol{kk}'}(\boldsymbol{q}) = \int_{\text{el.c.}} \mathrm{d}^3\boldsymbol{x} \, \mathrm{e}^{i\boldsymbol{qx}} \, \psi_{l\boldsymbol{k}}^*(\boldsymbol{x}) \, \psi_{l'\boldsymbol{k}'}(\boldsymbol{x}) \,. \tag{14}$$

The self-energy (13) is real and, therefore, immediately describes the energy shift of the HF valence band $E_{\mathrm{v}}(\boldsymbol{k})$ due to relaxation. Expression (13) is a slight generalization of the dielectric polarization energy E_{pol} of a certain charge distribution of total charge e, embedded in a polarizable medium of static $\boldsymbol{q}$-independent dielectric constant ε_0. For a homogeneous spherical distribution of radius r electrostatic yields

$$E_{\mathrm{pol}} = \frac{1}{2} \frac{e^2}{r} \left(1 - \frac{1}{\varepsilon_0}\right). \tag{15}$$

The same expression is obtained from (13) if spatial dispersion in ε is neglected and the Bloch distribution $|\psi_{\mathrm{v}\boldsymbol{k}}|^2$ is replaced by the spherical distribution. These observations make the physical meaning of $\Sigma_{\mathrm{v}}^{\mathrm{rel}}(\boldsymbol{k})$ quite obvious: $\Sigma_{\mathrm{v}}^{\mathrm{rel}}(\boldsymbol{k})$ represents the dielectric polarization energy due to a hole in the Bloch state $\psi_{\mathrm{v}\boldsymbol{k}}$.

For the explicit calculation of $\Sigma_{\mathrm{v}}^{\mathrm{rel}}(\boldsymbol{k})$ from (13) one needs the $\boldsymbol{q}$-dependent static dielectric function $\varepsilon(\boldsymbol{q}, 0)$ of the HF-Bloch electrons. We use an approximate expression of ε which was proposed by various authors (see, e.g., [25, 26] and references therein). It reads

$$\varepsilon(\boldsymbol{q}, \omega) = 1 + \frac{\omega_{\mathrm{p}}^2}{\omega_{\mathrm{g}}^2(\boldsymbol{q}) - (\omega + i0)^2} \tag{16}$$

with $\omega_{\mathrm{g}}(\boldsymbol{q})$ being an effective $\boldsymbol{q}$-dependent gap frequency and ω_{p} the plasma frequency of valence electrons. In [25] the following relation for $\omega_{\mathrm{g}}(\boldsymbol{q})$ was derived:

$$\omega_{\mathrm{g}}^2(\boldsymbol{q}) = \omega_{\mathrm{p}}^2 \left(\frac{1}{\varepsilon_0 - 1} + \left(\frac{q}{q_{\mathrm{TF}}}\right)^2\right) + \left(\frac{\hbar q^2}{2m}\right)^2 \tag{17}$$

(q_{TF} means the Thomas-Fermi wave vector of valence electrons). Despite of its simplicity (17) reproduces the essential features of the $\boldsymbol{q}$- and ω-dependent actual dielectric function. The poles of $\varepsilon(\boldsymbol{q}, \omega)$ occur at $\omega = \pm\omega_{\mathrm{g}}(\boldsymbol{q})$ and that of $\varepsilon^{-1}(\boldsymbol{q}, \omega)$ at $\omega = \pm\Omega(\boldsymbol{q})$ with $\Omega^2(\boldsymbol{q}) = \omega_{\mathrm{p}}^2 + \omega_{\mathrm{g}}^2(\boldsymbol{q})$, indicating that the dielectric function (16) accounts both for the generation of electron–hole pairs as well as for the generation of (modified) plasmons. In the static limit (16) corresponds to the Penn model of the dielectric function of tetrahedral semiconductors [27]. Further arguments for the validity of the approximate formula (16) were given in [25]. Expression (16) has already been successfully used for the calculation of relaxation energies in the case of core level excitations [4]. Since the relaxation energy relates to ε only in an integral way, fine structures not reproduced by the approximate formula (16), are averaged out in $\Sigma_{\mathrm{v}}^{\mathrm{rel}}$ to a certain extent. The above arguments give sufficient evidence of the applicability of the dielectric function (16) for an approximate calculation of relaxation energies of valence band excitations. Nevertheless one must be aware of the limited accuracy of the results due to the use of an approximate dielectric function.

We conclude this section by relating the relaxation energy explicitly to the photothreshold Φ. If the maximum of the valence band v is located at $\boldsymbol{k}_0$, from (1) and (12) it follows immediately

$$\Phi = -E_{\mathrm{v}}(\boldsymbol{k}_0) - \Sigma_{\mathrm{v}}^{\mathrm{rel}}(\boldsymbol{k}_0) \,. \tag{18}$$

3. Relaxation Self-Energies of Holes at Valence Band Maxima

In this section we will specify the general relations of Section 2 to the particular case of tetrahedrally coordinated group IV, III–V, and II–VI semiconductors. The valence band maximum is located at Γ for these materials. The threefold degenerate valence states at Γ, $\psi_{v\Gamma}$, transform according to the vector representation of the point group of Γ. In calculating $\Sigma_v^{rel}(\Gamma)$ from (13) we observe that the largest contributions to the integral over $\boldsymbol{q}$ result from q-vectors close to the centre of the first BZ. For such $\boldsymbol{q}$ the corresponding Bloch functions $\psi_{v\boldsymbol{q}}$ can be approximately replaced by $\exp(i\boldsymbol{q}\cdot\boldsymbol{x}) \times \times \psi_{v\Gamma}(\boldsymbol{x})$ neglecting the $\boldsymbol{q}$-dependence of the Bloch factor of $\psi_{v\boldsymbol{q}}(\boldsymbol{x})$. The explicit form of $\psi_{v\Gamma}(\boldsymbol{x})$ is taken from tight-binding theory. Within the bond orbital approximation it holds [18, 19, 28]

$$\psi_{v\Gamma}(\boldsymbol{x}) = \frac{1}{N^{1/2}} \sum_{\boldsymbol{R}} \left(\left(\frac{1-\alpha_p}{2}\right)^{1/2} \varphi_p^A(\boldsymbol{x}-\boldsymbol{R}) + \left(\frac{1+\alpha_p}{2}\right)^{1/2} \varphi_p^B(\boldsymbol{x}-\boldsymbol{R}-\boldsymbol{\tau})\right). \tag{19}$$

Here $\boldsymbol{R}$ denotes a lattice point, $\boldsymbol{\tau}$ is the vector between atoms A and B in the same elementary cell, N the number of elementary cells per unit volume, and α_p the polarity of bonds. The wave functions $\varphi_p^{A,B}$ mean atom-like p-orbitals at atoms A and B, respectively, obtained by a Löwdin-orthogonalization from the true atomic p-orbitals. With the wave functions from (19) the Bloch integral (14) at $\boldsymbol{k} = \Gamma$ becomes

$$B_{vv}^{\Gamma\Gamma+\boldsymbol{q}}(\boldsymbol{q}) = \left(\frac{1-\alpha_p}{2}\right) \varrho_p^A(\boldsymbol{q}) + \left(\frac{1+\alpha_p}{2}\right) \varrho_p^B(\boldsymbol{q})\,, \tag{20}$$

$$\varrho_p^{A,B}(\boldsymbol{q}) = \int d^3\boldsymbol{x}\; e^{i\boldsymbol{q}\cdot\boldsymbol{x}}\, |\varphi_p^{A,B}(\boldsymbol{x})|^2 \tag{21}$$

Table 1

Relaxation self-energies $\Sigma_v^{rel}(\Gamma)$ of holes at valence band maxima of tetrahedrally coordinated semiconductors. Material parameters used in the calculation are also shown: lattice constant a, static electronic dielectric constant ε_0, plasmon energy $\hbar\omega_p$ [4], effective gap energy $\hbar\omega_g(0)$, and bond polarity α_p [18]

semiconductor	a (at. units)	ε_0	$\hbar\omega_p$ (eV)	$\hbar\omega_g(0)$ (eV)	α_p	$\Sigma_v^{rel}(\Gamma)$ (eV)
C	6.740	5.7	31.2	14.38	0	4.32
Si	10.263	12.0	16.6	5.01	0	3.15
Ge	10.691	15.9	15.6	4.04	0	3.08
α-Sn	12.267	24.0	12.7	0.61	0	2.75
AlP	10.301	8.5	16.5	6.02	0.47	3.09
AlAs	10.620	10.3	15.8	5.18	0.44	3.06
AlSb	11.593	10.2	13.8	4.54	0.54	2.81
GaP	10.300	8.4	16.5	6.07	0.52	3.09
GaAs	10.684	10.9	15.6	4.96	0.50	3.06
GaSb	11.561	14.5	13.9	3.78	0.44	2.89
InP	11.090	9.6	14.8	5.05	0.58	2.93
InAs	11.406	12.3	14.2	4.22	0.53	2.91
InSb	12.242	15.7	12.7	3.31	0.51	2.75
ZnS	10.222	5.1	16.7	8.25	0.73	2.98
ZnSe	10.710	5.9	15.6	7.05	0.72	2.93
ZnTe	11.510	7.3	14.0	5.58	0.72	2.82
CdS	11.047	5.3	14.9	7.19	0.77	2.80
CdSe	11.489	6.1	14.0	6.20	0.77	2.76
CdTe	12.246	7.2	12.7	5.10	0.76	2.66

In (20) factors of the kind exp $(i\boldsymbol{q}\cdot\boldsymbol{r})$ have been replaced by 1. The reason is again that the most important $\boldsymbol{q}$-vectors for the calculation of $\Sigma_v^{rel}(\Gamma)$ are close to the centre of the first BZ. For the same reason the main contributions of the atom-like wave functions φ_p to $\varrho_p^{A,B}$ originate at large distances from the core region. Thus the φ_p can be taken as one-parameter Slater orbitals of type

$$\varphi_p(\boldsymbol{x}) \sim x_p r\, e^{-\beta r}, \qquad \beta = \frac{2\pi}{a}\lambda, \tag{22}$$

with x_p the component of $\boldsymbol{x}$ parallel to the p-orbital, a the lattice constant, and λ a dimensionless parameter. According to [29, 30] band calculations by means of such orbitals can be optimized with $\lambda = 1.7$ for group IV semiconductors and $\lambda^A = 1.6$, $\lambda^B = 1.8$ for III–V materials. The extrapolation of these values to the case of II–VI compounds yields $\lambda^A = 1.5$, $\lambda^B = 1.9$. For the calculation of relaxation self-energies $\Sigma_v^{rel}(\Gamma)$ one needs, in addition to λ, values of lattice constants, plasma frequencies of valence electrons, static electronic dielectric constants, and bond polarities of the materials under consideration. The effective gap frequency at $\boldsymbol{q} = \boldsymbol{0}$ and q_{TF} are calculated from ω_p and ε_0. The corresponding data (except for q_{TF}) are listed in Table 1 together with the calculated relaxation self-energies $\Sigma_v^{rel}(\Gamma)$. One notes from Table 1 that $\Sigma_v^{rel}(\Gamma)$ is close to 3 eV, nearly independent of the particular material. The simple electrostatic formula (15) yields a value of ≈ 2.8 eV with $\varepsilon_0 = 12$ and the hole radius equal to half the interatomic spacing of Si. The small changes of $\Sigma_v^{rel}(\Gamma)$ between different semiconductors are mainly due to the variations of the lattice constant and of the static dielectric constant. Of course, $\Sigma_v^{rel}(\Gamma)$ does not depend on the direction of the orbital in (22), i.e. on the particular choice of the hole state among the threefold degenerate valence states at Γ. For diamond correlation self-energy corrections to HF-valence band holes were calculated in [15]. The shift of 4.3 eV at Γ from [15] corresponds to our value of 4.32 eV for this material.

4. Photothresholds and Band Discontinuities. Comparison with Experiment

According to (18) the photothreshold Φ is obtained by subtracting the relaxation self-energy $\Sigma_v^{rel}(\Gamma)$ from the distance $-E_v(\Gamma)$ of the valence band maximum from the vacuum level. Theoretical values of $E_v(\Gamma)$ are rare in the literature. Usually, in band calculations $E_v(\Gamma)$ is taken as the reference point on the energy scale without specifying its position or with fixing it at zero. If only nearest-neighbour interactions are taken into account and wave function overlap of adjacent atoms is neglected, tight-binding theory provides a simple approximate expression for $E_v(\Gamma)$. It reads [31]

$$E_v(\Gamma) = \left(\frac{\varepsilon_p^A + \varepsilon_p^B}{2}\right) - \left(\left(\frac{\varepsilon_p^A - \varepsilon_p^B}{2}\right)^2 + V_{xx}^2\right)^{1/2} \tag{23}$$

with $\varepsilon_p^{A,B}$ the p-orbital energy of atoms A and B, respectively, and V_{xx} the interaction matrix element between nearest-neighbour atoms. There are various sets of tight-binding parameters, including ε_p^A, ε_p^B, and V_{xx}, in the literature [32 to 34]. In Table 2 we have listed two of them together with the calculated valence band maxima $E_v(\Gamma)$ according to (23). Columns 2, 3, 4, and 8 are from Harrison [32], columns 5, 6, 7, and 9 from Vogl et al. [33], but with the exclusion of ZnS, CdS, CdSe, and CdTe. For the latter II–VI compounds the ε_p-values were obtained as follows: for ZnS the data of ZnSe and ZnTe were linearely interpolated, for the Cd compounds the same values were used as for the corresponding Zn compounds. The V_{xx}-values were calculated from the ZnSe and ZnTe values by using the d^{-2}-scaling of V_{xx} [18]. For all

Table 2

Photothresholds calculated from equation (18) using valence band maxima from LCAO-calculations according to [32 to 34]. The two sets of tight-binding parameters from [32, 33] (for exceptions see text) are also listed. The experimental photothresholds Φ are shown for comparison. All energies are in eV

semi-con-ductor	[32]			[33]			$-E_v(\Gamma)$			$\Phi = -E_v(\Gamma) - \Sigma_v^{rel}(\Gamma)$			Φ exp.
	$-\varepsilon_p^A$	$-\varepsilon_p^B$	V_{xx}	$-\varepsilon_p^A$	$-\varepsilon_p^B$	V_{xx}	[32]	[33]	[34]	[32]	[33]	[34]	
Si	6.52	6.52	2.98	6.52	6.52	1.72	9.50	8.24	7.13	6.35	5.09	3.98	5.10
Ge	6.36	6.36	2.74	6.36	6.36	1.61	9.12	7.97	6.92	6.04	4.89	3.84	4.80
α-Sn	5.94	5.94	2.08	5.94	5.94	1.33	8.04	7.27	6.48	5.29	4.52	3.73	4.51
AlP	4.86	8.33	2.95	5.11	8.08	2.37	10.03	9.40	8.30	6.94	6.31	5.21	—
AlAs	4.86	7.91	2.78	5.08	7.69	1.88	9.57	8.67	7.88	6.51	5.61	4.82	—
AlSb	4.86	7.24	2.33	5.03	7.07	1.72	8.67	8.05	7.27	5.86	5.24	4.46	5.22
GaP	4.90	8.33	2.95	5.12	8.11	2.15	10.00	9.24	8.27	6.91	6.15	5.18	5.83
GaAs	4.90	7.91	2.75	5.09	7.72	1.95	9.53	8.76	7.86	6.47	5.70	4.80	5.47
GaSb	4.90	7.24	2.35	5.04	7.10	1.58	8.69	7.96	7.23	5.80	5.07	4.34	4.71
InP	4.69	8.33	2.55	4.92	8.10	1.88	9.64	8.97	8.46	6.71	6.04	5.53	5.69
InAs	4.69	7.91	2.41	4.89	7.71	1.84	9.21	8.62	8.01	6.30	5.71	5.10	5.31
InSb	4.69	7.29	2.09	4.84	7.09	1.40	8.41	7.76	7.34	5.66	5.01	4.59	4.77
ZnS	3.38	10.27	3.00	4.33	9.52	3.43	11.40	11.23	10.45	8.42	8.25	7.47	7.50
ZnSe	3.38	9.53	2.73	4.21	8.70	3.01	10.58	10.21	9.46	7.65	7.28	6.53	6.82
ZnTe	3.38	8.59	2.37	4.09	7.88	2.80	9.50	9.36	8.77	6.68	6.54	5.95	5.76
CdS	3.38	10.27	2.57	4.33	9.52	2.94	11.12	10.85	10.69	8.32	8.05	7.89	7.26
CdSe	3.38	9.53	2.37	4.21	8.70	2.72	10.35	9.98	9.84	7.59	7.22	7.08	6.62
CdTe	3.38	8.59	2.09	4.09	7.88	2.39	9.32	9.04	8.92	6.66	6.38	6.26	5.78

materials the absolute values for $\varepsilon_p^{A,B}$ [33] were shifted such that their centres of gravity, $\frac{1}{2}(\varepsilon_p^A + \varepsilon_p^B)$, are identical with those from [32]. The valence band maxima $E_v(\Gamma)$ in column 10 were obtained by taking account of wave function overlap in a way which follows Enderlein and Harrison [34]. The different sets of $E_v(\Gamma)$ lead to different sets of predicted photothresholds. They are listed in columns 11, 12, and 13 of Table 2. In column 14 we show the experimental values of Φ according to [35 to 38]. The comparison between theory and experiment gives clear evidence for the importance of electronic relaxation effects on photothresholds (see Fig. 1). In all cases experimental values of Φ are much closer to the theoretical values if the relaxation self-energy is regarded in addition to $E_v(\Gamma)$. The deviations between theoretical and experimental values of Φ are of the same magnitude as the deviations between the different sets of $E_v(\Gamma)$. This indicates that the main reason for the observed discrepancies between theoretical and experimental Φ in Table 1 are the uncertainties of band calculations on valence band maxima $E_v(\Gamma)$. Other corrections to Φ mentioned in Section 1 seem to be of minor influence. If one accepts this interpretation one can evaluate the accuracy of the different sets of $E_v(\Gamma)$-values and of the corresponding tight-binding parameters by inspecting the deviations between calculated and measured photothresholds. The Harrison-values of $E_v(\Gamma)$ turn out to be about 1 eV too large, perhaps due to an overestimation of the interatomic matrix element V_{xx} and to the use of atomic $\varepsilon_p^{A,B}$-values also in the case of the II–VI compounds with their large charge transfer between cations and anions. The correction of the Harrison-values by the regard of wave function overlap leads to an improvement of $E_v(\Gamma)$ for materials with polarities $\alpha_p > 0.5$, but the overlap seems to be overestimated for

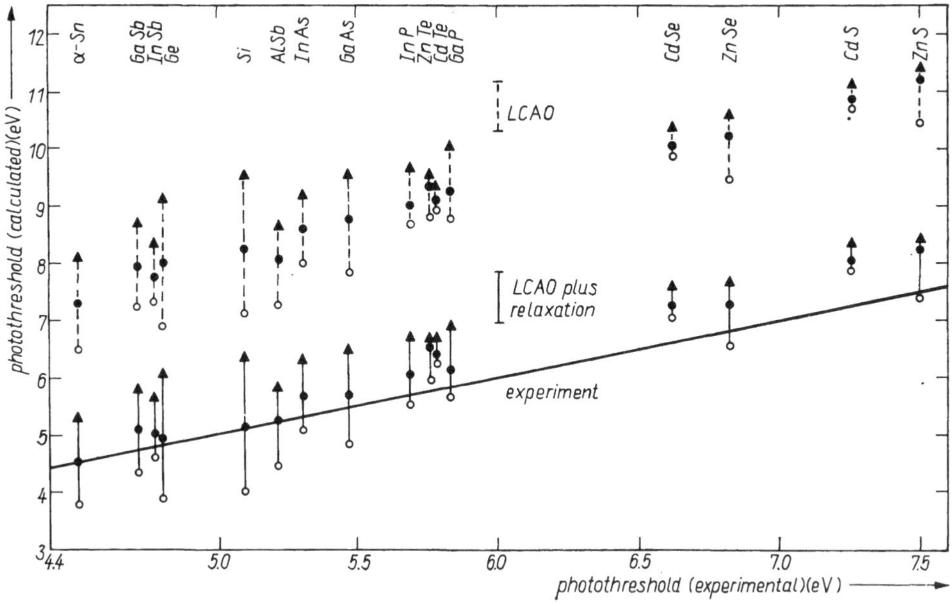

Fig. 1. Plot of calculated photothresholds against experimental data for various group IV, III–V, and II–VI-semiconductors. The vertical lines indicate the spread of the three different LCAO-calculations. Triangles corresponds to [32], full circles to [33], and open circles to [34]. The upper data (dashed vertical line) are without relaxation corrections, the lower ones with (full vertical line)

smaller polarities. The parameters of Vogl et al. turn out to be relatively precise for covalent semiconductors, but with increasing polarity the precision decreases.

The results on photothresholds are also relevant for valence band discontinuities Δ_v at heterojunctions between two semiconductors S and S′. By exciting an electron from the valence band maximum of the semiconductor S to the vacuum level and then taking it back to the valence band maximum of the semiconductor S′ one finds the following relation:

$$\Delta_v(\Gamma) = \Phi - \Phi' = -\big(E_v(\Gamma) - E'_v(\Gamma)\big) - \big(\Sigma_v^{rel}(\Gamma) - \Sigma_v^{rel'}(\Gamma)\big) . \tag{24}$$

We use this relation for an estimation of relaxation corrections to band discontinuities. The corrections are listed in Table 3 together with the predictions of tight-binding theory for $-(E_v(\Gamma) - E'_v(\Gamma))$ and experimental values of Δ_v measured mainly by means of photoemission. The relaxation corrections amount to several tenths up to one half of an eV. These corrections have to be taken into account necessarily if band discontinuities are to be calculated with the corresponding accuracy. At the present time band calculations of $E_v(\Gamma)$ obviously are much more inaccurate. This is the reason why the agreement between calculated values of Δ_v (Γ) with the inclusion of relaxation in columns 6, 7, and 8 of Table 3, and the experimental values in column 9 is still rather poor, although somewhat better than without relaxation corrections.

In conclusion, we have shown that photothresholds of semiconductors are considerably lowered due to the relaxation of valence electrons around the hole. If hole relaxation self-energies are subtracted from distances between valence band maxima and the vacuum level, the large discrepancies between one-electron LCAO predictions and experimental observations of photothresholds are essentially removed. There may be one reason why electronic relaxation for a long time has not been considered

Table 3

Valence band discontinuities at heterojunctions of semiconductors S–S′. The LCAO-results — $(E_v(\Gamma) - E'_v(\Gamma))$ according to [32 to 34] are represented first, followed by the relaxation corrections $\Sigma_v^{rel}(\Gamma) - \Sigma_v^{rel'}(\Gamma)$, the relaxation corrected discontinuities $\Delta_v(\Gamma)$, and the experimental data for $\Delta_v(\Gamma)$ from [39 to 41]. All energies are in eV

S–S′	$-(E_v(\Gamma) - E'_v(\Gamma))$			$\Sigma_v^{rel}(\Gamma) - \Sigma_v^{rel'}(\Gamma)$	$\Delta_v(\Gamma)$			$\Delta_v(\Gamma)$ exp.	ref.
	[32]	[33]	[34]		[32]	[33]	[34]		
Si–Ge	0.38	0.27	0.21	0.07	0.31	0.20	0.14	0.17	[39]
GaP–Ge	0.88	1.00	1.14	0.01	0.87	0.99	1.13	0.96	[39]
GaAs–Ge	0.41	0.79	0.94	−0.02	0.43	0.81	0.96	0.33	[39]
GaSb–Ge	−0.43	−0.01	0.31	−0.19	−0.24	0.18	0.50	0.21	[39]
InP–Ge	0.52	1.00	1.54	−0.15	0.67	1.15	1.69	0.69	[39]
InAs–Ge	0.09	0.65	1.09	−0.17	0.26	0.82	1.26	0.33	[39]
InSb–Ge	−0.71	−0.21	0.42	−0.33	−0.38	0.12	0.75	0.11	[39]
ZnSe–Ge	1.46	2.24	2.54	−0.15	1.61	2.39	2.69	1.41	[39]
ZnTe–Ge	0.38	1.39	1.85	−0.26	0.64	1.65	2.11	0.98	[39]
CdS–Ge	2.00	2.88	3.77	−0.28	2.28	3.16	4.05	1.73	[39]
CdSe–Ge	1.23	2.01	2.92	−0.32	1.55	2.33	3.24	1.33	[39]
CdTe–Ge	0.20	1.07	2.00	−0.42	0.62	1.49	2.42	0.88	[39]
CdS–InP	1.48	1.88	2.23	−0.13	1.61	2.01	2.36	1.63	[40]
AlAs–GaAs	0.04	−0.09	0.02	0.00	0.04	−0.09	0.02	0.15	[41]
GaAs–AlAs	−0.04	0.09	−0.02	0.00	−0.04	0.09	−0.02	−0.40	[41]
ZnSe–GaAs	1.05	1.45	1.60	−0.13	1.18	1.58	1.73	0.95	[42]
GaAs–InAs	0.32	0.14	−0.15	+0.15	0.17	−0.01	−0.30	0.15	[42]

to be relevant for the solution of the photothreshold problem of semiconductors. It might be hard to understand how the gain in Coloumb energy during relaxation is transferred to the photoelectron, lowering the minimum photon energy for electron emission. The answer has already been given at the beginning: A certain amount of negative polarization charge is attracted from the positive hole during relaxation. It repels the photoelectron and increases its (kinetic) energy by the same amount which the remaining electron system of total Coloumb energy loses in the final state as compared to the initial state.

References

[1] M. CARDONA and L. LEY, in: Photoemission in Solids, Vol. 1, Ed. M. CARDONA and L. LEY, Springer-Verlag, Berlin/Heidelberg/New York 1978 (p. 1).
D. A. SHIRLEY, ibid. (p. 165).
[2] L. YIN, T. TSANG, and J. ADLER, Phys. Letters A 57, 193 (1976); Phys. Rev. B **15**, 2974 (1977).
[3] F. BECHSTEDT, phys. stat. sol. (b) **112**, 9 (1982).
[4] F. BECHSTEDT, R. ENDERLEIN, and R. WISCHNEWSKI, phys. stat. sol. (b) **107**, 637 (1981).
[5] A. ROSÉN and J. LINDGREN, Phys. Rev. **176**, 114 (1968).
[6] A. ZUNGER, Phys. Rev. Letters **50**, 1215 (1983).
[7] N. F. MOTT and M. J. LITTLETON, Trans. Faraday Soc. **34**, 485 (1938).
[8] W. B. FOWLER, Phys. Rev. **151**, 657 (1966).
[9] W. BRINKMAN and B. GOODMAN, Phys. Rev. **149**, 597 (1966).
[10] N. O. LIPARI, phys. stat. sol. **40**, 691 (1970).
[11] A. B. KUNZ and D. J. MICKISH, Phys. Rev. B 8, 779 (1973).
[12] E. O. KANE, Proc. 12th Internat. Conf. Phys. Semicond., B. G. Teubner-Verlagsgesellschaft, Stuttgart 1974, (p. 169).

[13] N. E. BRENNER and T. C. COLLINS, ibid. (p. 174).
[14] F. GUINEA and C. TEJEDOR, J. Phys. C **13**, 5515 (1980).
[15] G. STRINATI, H. J. MATTAUSCH, and W. HANKE, J. Phys. Soc. Japan **49**, Suppl. A, 77 (1980).
[16] A. ZUNGER and A. J. FREEMAN, Phys. Rev. B **16**, 2901 (1977).
[17] C. S. WANG, and W. E. PICKET, Phys. Rev. Letters **51**, 597 (1983).
[18] W. A. HARRISON and S. CIRACI, Phys. Rev. B **10**, 1516 (1974).
[19] W. A. HARRISON, Electronic Structure and the Properties of Solids, W. H. Freeman and Co., San Francisco 1980.
[20] N. D. LANG and W. KOHN, Phys. Rev. B **1**, 4555 (1970).
[21] H. STOLZ, Einführung in die Vielelektronentheorie der Kristalle, Akademie-Verlag, Berlin 1974.
[22] L. P. KADANOFF and G. BAYM, Quantum Statistical Mechanics, W. A. Benjamin, Inc., New York 1962.
[23] L. HEDIN, Phys. Rev. **139**, A 796 (1963).
[24] L. HEDIN and S. LUNDQVIST, Solid State Phys. **23**, 1 (1969).
[25] F. BECHSTEDT and R. ENDERLEIN, phys. stat. sol. (b) **94**, 239 (1979); **95**, 185 (1979).
[26] J. C. INKSON, J. Phys. C **5**, 2599 (1972).
[27] D. R. PENN, Phys. Rev. **128**, 2093 (1962).
[28] E. ANTONČIK, in: Electrons in Crystalline Solids, IAEA, Vienna 1973 (p. 461).
[29] D. J. CHADI, Phys. Rev. B **16**, 3572 (1977).
[30] A.-B. CHEN and A. SHER, Phys. Rev. B **22**, 3886 (1980).
[31] D. J. CHADI and M. L. COHEN, phys. stat. sol. (b) **68**, 405 (1975).
[32] W. A. HARRISON, J. Vac. Sci. Technol. **14**, 1016 (1977).
[33] P. VOGL, H. P. HJALMARSON, and J. D. DOW, J. Phys. Chem. Solids **44**, 365 (1983).
[34] R. ENDERLEIN and W. A. HARRISON, Phys. Rev. B **30**, 1867 (1984).
[35] G. W. GOBELLI and F. G. ALLEN, Phys. Rev. **127**, 141 (1962); **137**, A 245 (1965).
[36] T. E. FISHER, Phys. Rev. **139**, A 1228 (1965); **142**, 519 (1966).
[37] R. K. SWANK, Phys. Rev. **153**, 844 (1967).
[38] V. S. FOMENKO, Emission Properties of Materials, Naukova Dumka, Kiev 1981 (in Russian).
[39] G. MARGARITONDO, Solid State Electronics **26**, 499 (1983).
[40] J. L. SHAY, S. WAGNER, and J. C. PHILLIP, Appl. Phys. Letters **28**, 31 (1976).
[41] S. P. KOWALCZYK, E. A. KRAUT, J. R. WALDROP, and R. W. GRANT, J. Vac. Sci. Technol. **19**, 573 (1981); **21**, 482 (1982); B **1**, 684 (1983).
[42] W. J. SCHAFFER, M. D. LIND, S. P. KOWALCZYK, and R. W. GRANT, J. Vac. Sci. Technol. B **1**, 689 (1983).

(Received September 7, 1984)

phys. stat. sol. (b) **126**, 587 (1984)

Subject classification: 13.1; 22.8

Department of Physics, Al-Azhar University, Cairo[1])

Temperature Dependence of the Electronic Energy Bands of $NaNO_2$Crystal from 20 to 245 °C

By

A. M. El-Dib and H. F. Hassan

The electronic band structure of $NaNO_2$ crystals is calculated. The band structure is studied in the temperature range from 20 to 245 °C in the principal directions as well as at the high symmetry points inside the Brillouin zone. The temperature dependence of the minimum energy gap is also investigated. The calculations are carried out using a semiempirical LCAO method. The general features of the electronic energy bands at various temperatures are similar to that at 20 °C. The optical energy gap calculated at 20 °C equals to 3.24 eV which is in good agreement with experimental results.

Die Struktur der Energiebänder eines $NaNO_2$-Kristalls, sowie das Minimum der verbotenen Zone für Temperaturen zwischen 20 und 245 °C werden berechnet. Die Berechnung wird mittels der halbempirischen LCAO-Methode für Punkte hoher Symmetrie und für die Hauptrichtungen der Brillouin-Zone durchgeführt. Das Ergebnis dieser Betrachtung ergibt eine gewisse Ähnlichkeit der allgemeinen Merkmale der bei verschiedenen Temperaturen berechneten Strukturen der Energiebänder, die sich kaum von denen bei 20 °C unterschieden, sowie, in Übereinstimmung mit den Experimenten, für den optischen Energieabstand bei 20 °C den Wert von 3,24 eV.

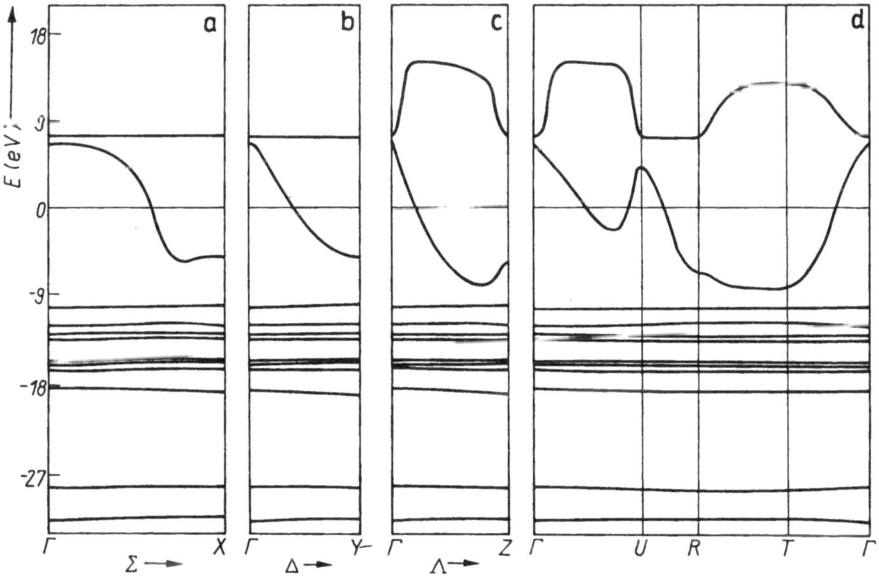

Fig. 1. The electronic energy bands of $NaNO_2$ at 20 °C in a) Σ direction, b) Δ direction, c) Λ direction, and d) Γ–U–R–T–Γ directions

[1]) Nasr City, Cairo, Egypt.

1. Introduction

After the discovery of the ferroelectricity in sodium nitrite by Sawada et al. [1], the crystal has initiated scientific investigations. A short review about these investigations is given by Lines and Glass [2]. Recently measurements of unit cell dimensions [3], infrared spectroscopy [4], and spontaneous polarization [5] were done. However, the electronic band structure of this crystal was studied in a few works [6, 7].

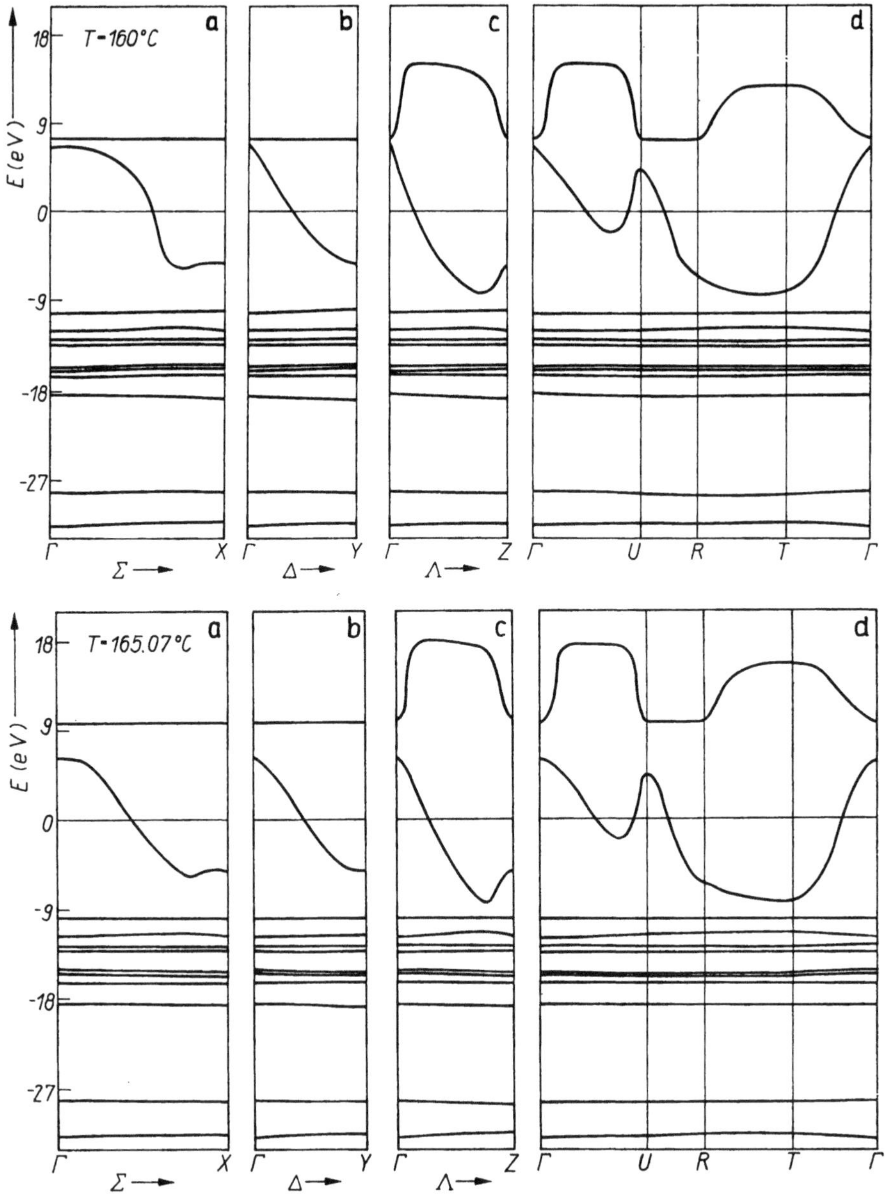

Fig. 2

In this work, the electronic energy bands of $NaNO_2$ are calculated. The temperature dependence of these energy bands is studied in the range from 20 to 245 °C. The temperature dependence of the minimum energy gap is also calculated in the mentioned temperature range.

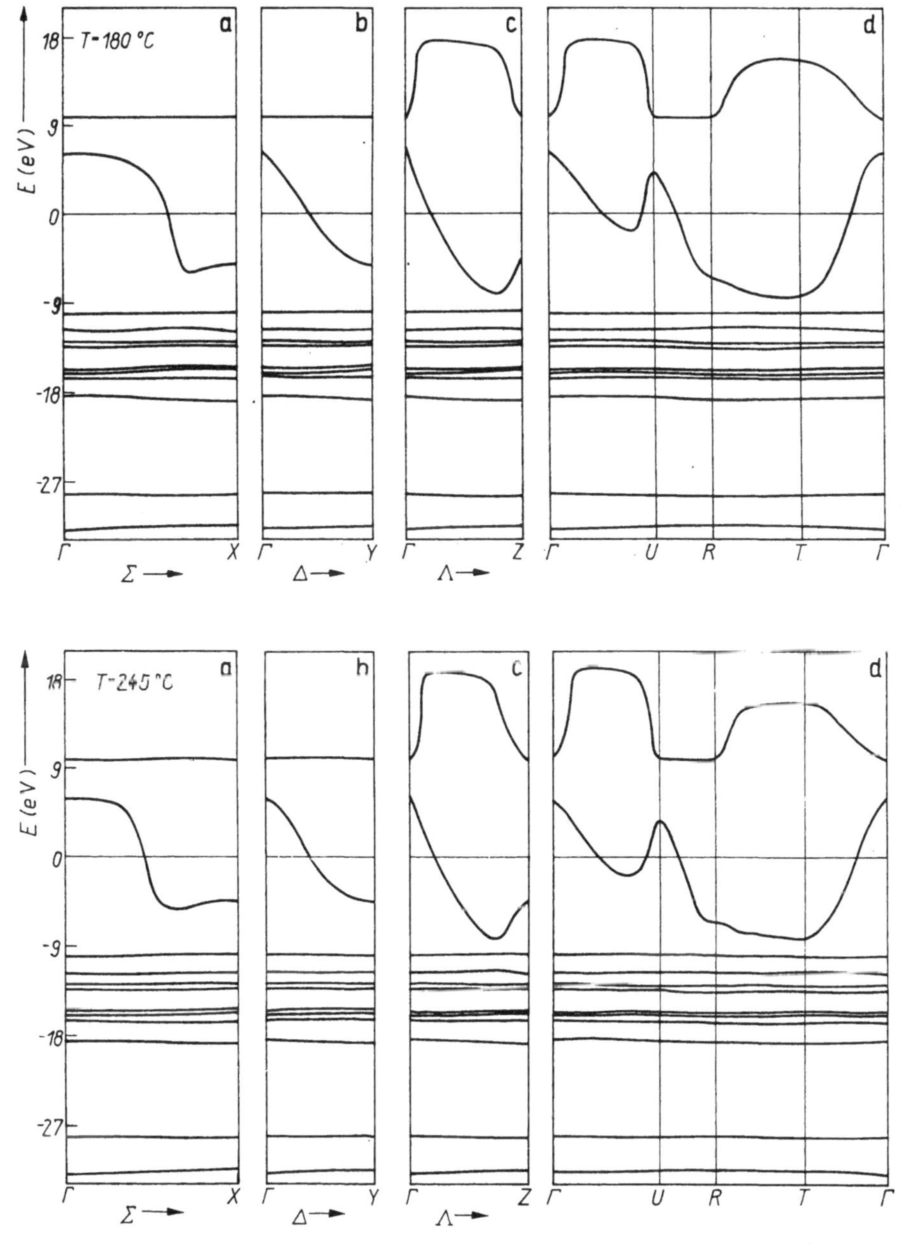

Fig. 2. Energy bands of $NaNO_2$ at T = 160, 165.07, 180, and 245 °C

2. Method

A semiempirical LCAO method is used in this calculation. This method has been already explained in our previous work [7, 8]. The basis functions considered are 3s for Na, 2s, 2p for N and O atoms. However, the overlap integrals of the type ($3s_{Na}$, X) where X is any orbit, are performed in the reciprocal space, while those which do not contain $3s_{Na}$ orbits are calculated in the direct space by the usual technique.

In order to obtain the parameters for calculations, the lattice constants measured by Kucharczyk et al. [3] at each temperature and the atomic coordinates given by Kay and Frazer [9, 10] in the corresponding phase are used. In addition, a charge self-consistent procedure [7, 8] is applied at the Γ point at each temperature in order to obtain the self-consistent values of the effective diagonal elements of the Hamiltonian matrix as well as the values of charges on atoms. The charge self-consistent values, therefore, are used to perform the calculations of the energy bands at other points inside the Brillouin zone.

3. Results

3.1 The energy bands

The electronic energy bands in Σ, Δ, and Λ directions at 20 °C in the ferroelectric phase are represented in Fig. 1a, b and c where the Σ, Δ, and Λ directions are parallel to the principal directions of the Brillouin zone. The energy bands at 20 °C in some other directions and high symmetry points (Γ–U), (U–R), (R–T), and (T–Γ) are represented in Fig. 1d. The electronic energy bands in all the mentioned directions are calculated again at 160 °C in the ferroelectric phase as well as at 165.07, 180, and 245 °C at which the crystal is in its paraelectric phase. Fig. 2 shows the energy bands at the mentioned temperatures. In Fig. 3 the 9th, 10th, and 11th bands in the Λ direction are represented at temperatures 20, 160, 180, and 245 °C, so that the effect of temperature on these bands can be observed. The 12th and 13th bands in Λ direction are plotted in Fig. 4 and 5, respectively.

3.1.1 Discussion and conclusion

1. The general features of the electronic band structure at various temperatures are similar to that at 20 °C.

2. The electronic energy bands from 1 to 10 have weak dependence on the wave vector. In particular, bands 2, 4, 6, 7, 8, and 10 appear to be approximately straight lines with negligible slope. This denotes that these bands are still belonging to the localized states of the NO_2^- group. However, the symmetry of these bands shows that the $3s_{Na}$ orbit is not represented in them.

3. The $\boldsymbol{K}$ dependence of the 11th and 13th bands in the Σ, Δ, and Λ directions, show that these bands have strong $\boldsymbol{K}$ dependence. The symmetry of these bands shows that they contain the $3s_{Na}$ orbit.

4. Since the dispersion in the 10th band is too weak, the electrons occupying this band will have considerably high effective mass and small mobility. The 10th band, thus, will play no role in the electric conduction.

5. By means of group-theoretical study and according to the Mott theory of transitions, one can show that the electric dipole transition from the 9th to the 10th levels is only allowed if the incident radiation has an X polarization component. But in this case, if an X polarization component exists, the absorption spectrum of this transition will lead to a sharp absorption line rather than a fundamental absorption edge and

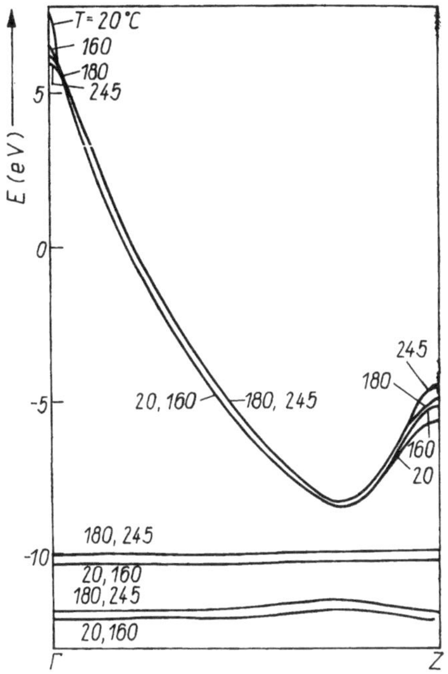

Fig. 3

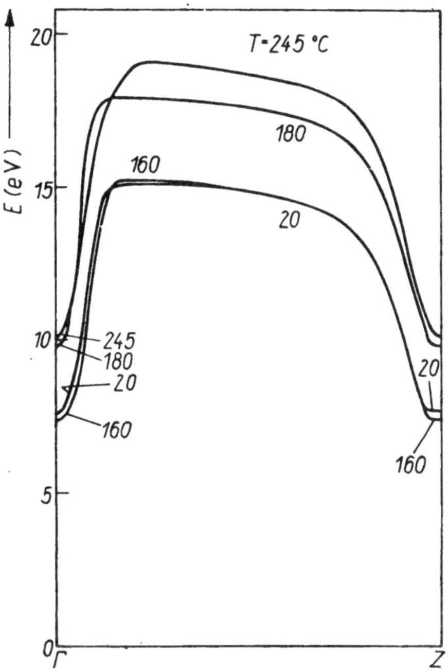

Fig. 4

Fig. 3. The structure of the 9th, 10th, and 11th bands in Λ direction at different temperatures

Fig. 4. The structure of the 12th band in Λ direction at different temperatures

may be interpreted, experimentally, as an impurity, or other types of absorption lines. Thus, it is predicted that the optical absorption edge will be between the top of the 9th band and the bottom of the 11th one. The energy difference between these bands, therefore, is termed the optical energy gap, while the energy difference between the 9th and the 10th bands is termed the thermal ionization energy gap.

6. Since the effective mass of the electrons and holes in the 10th and 9th bands is high, one may expect a formation of excitons with high binding energy. The exciton band may exist near the valence one, and thus will strongly reduce the number of free electron–hole pairs. This phenomenon may reduce the electric conductivity in the crystal.

7. It is shown from Fig. 3 that the phase transition is accompanied with approximately a constant energy shift for the 9th, 10th, and 11th bands. It is shown also that as the temperature increases the width of the allowed 11th band decreases. This behaviour is expected since the interaction between the atoms and molecules decreases as the distance between them increases as a result of the thermal expansion. On the contrary, the width of the allowed 12th band, as shown from Fig. 4, increases as the temperature increases. This is also expected since the dispersion of this band is mainly in Λ direction $(0, 0, K_z)$ where the dimension of the unit cell in z-direction decreases with increasing temperature. This figure shows also that the 12th band is shifted upwards in the paraelectric phase relative to the ferroelectric one. From Fig. 5 it is shown also that the width of the 13th band in Λ direction in the paraelectric phase is larger than that in the ferroelectric phase.

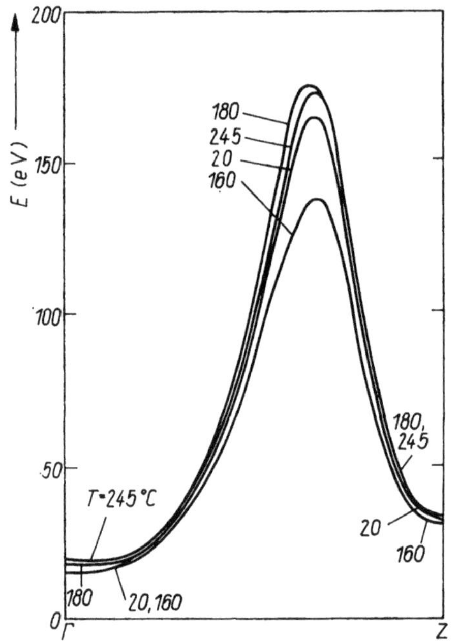

Fig. 5. The structure of the 13th band in Λ direction at different temperatures

8. It is noticed from the comparison of the energy bands in the ferro- and paraelectric phases that the points in the Brillouin zone at which the minimum energy gap occurs are approximately not affected by the phase transition.

3.2 The minimum energy gap (E_g)

It is seen from the results in the previous section that the minimum optical energy gap (E_g) for $NaNO_2$ occurs in Λ direction. The dependence of E_g on temperature is shown in Fig. 6.

3.2.1 Discussion and conclusion

The comparison of the results of this work and the experimental results of Verkhovskaya and Sonin [11] leads to the following remarks:

1. The change in the width of the forbidden gap, E_g, due to the phase transition according to Verkhovskaya is —0.02 eV while according to this work about —0.23 eV.

2. Due to the work of Verkhovskaya, the width of E_g in the ferro- and paraelectric regions changes linearly with the temperature coefficient

$$\frac{dE_g}{dT} \approx -(7.3 \pm 0.25) \times 10^{-4}\ \mathrm{eVK^{-1}}\,.$$

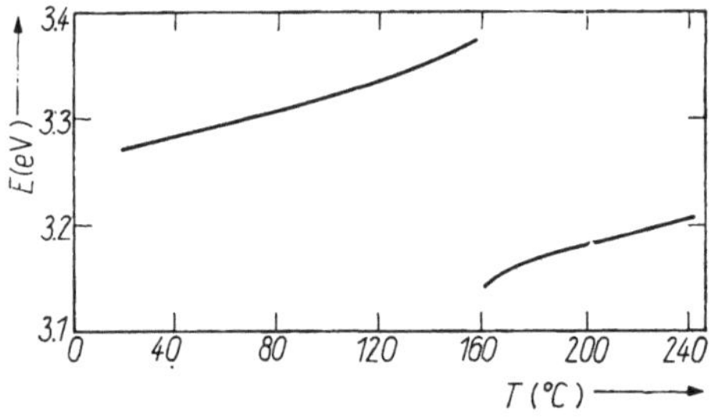

Fig. 6. Temperature dependence of energy gap of $NaNO_2$ crystal

Due to this work, it is shown that the width of the forbidden gap in the region far from the phase transition region changes linearly with the temperature coefficients

$$\frac{dE_g}{dT} \approx 6.06 \times 10^{-4}\,\text{eVK}^{-1}$$

in the ferroelectric phase and

$$\frac{dE_g}{dT} \approx 6.29 \times 10^{-4}\,\text{eVK}^{-1}$$

in the paraelectric phase.

However, the value of the optical energy gap given in this work (3.24 eV at 20 °C) is in good agreement with the experimental results of Verkhovskaya and Sonin [11] (3.14 eV at the same temperature).

3. The study of the temperature dependence of the thermal energy gap shows that the minimum thermal energy gap, E_{th}, equals ≈ 1.37 eV at 20 °C and occurs in the Σ direction. It shows also that E_{th} depends linearly on temperature with the coefficients

$$\frac{dE_{th}}{dT} \approx 5.07 \times 10^{-4}\,\text{eVK}^{-1}$$

in the ferroelectric phase and

$$\frac{dE_{th}}{dT} \approx 1.25 \times 10^{-3}\,\text{eVK}^{-1}$$

in the paraelectric phase.

Acknowledgements

The authors are grateful to Prof. A. A. Mohamed, in Ain Shams University and Dr. A. F. Hassan, in Helwan University, for their valuable discussion.

References

[1] S. Sawada, S. Nomura, S. Fujii, and I. Yoshida, Phys. Rev. Letters 1, 320 (1958).
[2] M. E. Lines and A. M. Glass, Principles and Applications of Ferroelectric and Related Materials, Claredon Press, Oxford 1977 (p. 181).
[3] D. Kucharczyk, A. Pietraszko, and K. Łukaszewicz, phys. stat. sol. (a) **37**, 287 (1976).
[4] E. A. Ivanova, and E. V. Chisler, Soviet Phys. — Solid State 17, 1919 (1976).
[5] W. Buchheit and J. Petersson, Solid State Commun. **34**, 649 (1980).
[6] K. Kam-Shing, and J. H. Henkel, Phys. Rev. B **17**, 1361 (1978).
[7] A. M. Eldib, Egypt J. Chem. **22**, 109 (1979).
[8] H. F. Hassan, Ph. D. Thesis, Ain Shams University, Cairo 1983.
[9] M. I. Kay, Ferroelectrics **4**, 235 (1972).
[10] M. I. Kay and B. C. Frazer, Acta Cryst. **14**, 56 (1961).
[11] A. K. Verkhovskaya and A. C. Sonin, Soviet Phys. — J. exper. theor. Phys. **25**, 249 (1967).

(Received August 1, 1984)

phys. stat. sol. (b) **126**, 595 (1984)

Subject classification: 4; 13; 22.1.2

Sektion Physik der Karl-Marx-Universität, Leipzig[1]),
Arbeitsgemeinschaft A^{III}–B^{V}-Halbleiter

Simulation fo Dynamical Bond Charge Transfer Properties in Elemental Semiconductors by Means of a Simple Quantum-Mechanical Model

By

U. PIETSCH

A simple model is presented, that constructs the valence charge density of monoatomic semiconductors by linear superposition of appropriate diatomic molecules including the bond charges between next neighbour atoms. By means of this model the study of dynamical density properties is possible being in general calculated by full self-consistent pseudopotential schemes. An essential feature of that density variation is the bond charge transfer, which will be analyzed with respect to the variation of the next neighbour distance and the bond angle. During stretching or compressing of the next neighbour distance the variation in the $|s\rangle$ portion of the valence orbitals has to be taken into account. Using the results for silicon, which are fitted to the self-consistent outcome and to experimental data, an estimation of the transfer constants of other group-IV elemental crystals is given. The model can be applied to the interpretation of dynamical bond charge transfer influenced experiments, like pressure and temperature dependence of the "forbidden" X-ray reflection power.

Es wird ein einfaches Model dargestellt, das die Valenzelektronendichte monoatomarer Halbleiter durch lineare Superposition zweiatomiger Moleküle konstruiert und die Bindungsladung zwischen nächsten Nachbarn berücksichtigt. Mit Hilfe dieses Modells ist die Untersuchung dynamischer Dichteeigenschaften möglich, die sonst nur mittels selbst-konsistenter Pseudopotentialansätze berechnet werden kann. Der wesentliche Anteil solcher Dichteänderungen, der Bindungsladungstransfer, wird bezüglich der Änderung des nächsten Nachbarabstandes und der Änderung der Bindungswinkel analysiert. Es ist zu beachten, daß sich während der Streckung bzw. Stauchung des nächsten Nachbarabstandes der $|s\rangle$-Anteil der Valenzorbitale ändert. Unter Verwendung der Ergebnisse für Silizium, die an die Resultate selbstkonsistenter Rechnungen und am Experiment angepaßt werden, werden die Transferkonstanten anderer Elementhalbleiter der vierten Hauptgruppe bestimmt. Das Modell kann zur Interpretation solcher Experimente herangezogen werden, die vom dynamischen Bindungsladungstransfer beeinflußt sind, wie z. B. die Druck- und Temperaturabhängigkeit der „verbotenen" Röntgenbeugungsreflexe.

1. Introduction

For the interpretation of the valence charge density in elemental semiconductors its maximum is essentially localized in the middle of the next neighbour distance. Because of the overlap of the valence orbitals (bond charge) due to the chemical bond this charge density accumulation is much greater than the simple superposition of spherical atomic charge densities. The bond charge (BC), on the other hand, can be considered as a consequence of the incomplete screening of the ionic potentials caused by the finiteness of the static dielectric constant of semiconductor.

The static valence charge density was determined by both the Fourier transformation of structure amplitudes measured by X-ray diffraction experiments [1] and by

[1]) Linnéstr. 5, DDR-7010 Leipzig, GDR.

the evaluation of structural properties applying self-consistent calculation methods [2 to 7]. Since pseudopotentials were used the computed valence charge densities are inexact near the core region. For the experimental determination of the BC special models were developed interpreting BC-influenced X-ray diffraction data [8 to 12]. Another paper will illustrate that the "experimental" and the calculated valence charge densities of silicon are in qualitative and quantitative agreement. In principal, the tendency of the decreasing BC amount from diamond to α-tin was confirmed, corresponding to the dielectric theory [14]. Hence the problem of the determination of the *static* valence charge density of elemental semiconductors seems to be solved. Further, *dynamical* properties of the valence charge density (VCD) will be attractive, for instance, for the polarization density of "frozen" phonons or the hydrostatic pressure induced charge response [15 to 18].

During the phonon or thermally induced atomic displacement in the crystal some valence charge is transferred from the stretched to the compressed bonds. This effect shown by Maschke and Baldereschi [17] must be divided into the bond charge transfer and the change in the spherical VCD distributed around the ions. The transfer amount of both parts can be described by transfer constants.

An experimental proof of the dynamical charge transfer was given by the temperature dependence of the mean square displacement of the BC in silicon, GaAs, and InSb [19, 20], the piezoelectrically influenced change in the X-ray diffraction intensity in GaAs [21] and the hydrostatic pressure induced variation of the "forbidden" (222) reflection power in silicon and InSb [22, 23]. For the interpretation of those effects the calculation by self-consistent pseudopotential methods is expensive and formidable. Moreover, details of the transfer process are difficult to extract from the final results of those calculations.

This situation suggests the introduction of a semiempirical model which involves the experience known. It will be employed for the qualitative and quantitative investigation of the bond charge transfer properties and for the interpretation of X-ray diffraction experiments. In this paper a simple quantum-mechanical model will be presented for the calculation of the BC transfer constants of elemental semiconductors. Both the total VCD and the BC are constructed by superposition of diatomic molecule densities. Each molecule contains an individual overlap charge. The coefficients of the Gaussian-like model valence functions are fitted to the VCD known of the unperturbed crystal. All calculations are applied to silicon. The transfer constants are fitted to the value calculated by Maschke and Baldereschi [17] and to the experiment [19]. The principal results will be expanded to the computation of the transfer constants of the other elemental semiconductors considering the varied $|s\rangle$ portion of the VCD and the BC amount.

2. Molecular-Theoretic Background

The VCD of a diatomic molecule is described by the population of bonding states

$$\varrho(\boldsymbol{r}) = \frac{1}{1 + S^n} \{\Psi_i^2 + \Psi_j^2 + 2S^{n-1}\Psi_i\Psi_j\} . \tag{1}$$

Depending on the approximation used the parameter n is $n = 1$ or 2, according to the molecular orbital method [24, 25] or to the valence bond theory [26]. Principally the total molecule density is divided into a spherical part around the ions

$$\varrho_{s,i}(\boldsymbol{r}) = \frac{1}{1 + S^n} \Psi_i^2(\boldsymbol{r}) \tag{2a}$$

and an overlap part called bond charge (BC)

$$\varrho_{BC}(\boldsymbol{r}) = \frac{2S^{n-1}}{1 + S^n} \Psi_i(\boldsymbol{r}) \Psi_j(\boldsymbol{R} - \boldsymbol{r}) . \tag{2b}$$

Choosing the overlap integral S

$$S = \int \Psi_i(\boldsymbol{r}) \Psi_j(\boldsymbol{R} - \boldsymbol{r}) \, d^3\boldsymbol{r} \tag{3}$$

both the total spherical part and the BC are calculated to be

$$\int \varrho_{s,i}(\boldsymbol{r}) \, d^3\boldsymbol{r} = \frac{1}{1 + S^n} \tag{4a}$$

and

$$\int \varrho_{BC}(\boldsymbol{r}) \, d^3\boldsymbol{r} = \frac{2S^n}{1 + S^n} . \tag{4b}$$

If the overlap integral between the orbitals Ψ_i and Ψ_j become great the interaction energy of both orbitals will be essential, too. Therefore the maximum of the molecule bonding energy correlates to the maximum overlap of bonding orbitals [25]. In the solid state that condition is incompletely fulfilled because of the complex interaction of the different molecule orbitals in the crystal. For the further procedure the use of Gaussian-like model valence orbitals is favourable owing to the simple mathematical management. Only $|s\rangle$ and $|p\rangle$ like orbitals will be taken into account

$$|s\rangle = N_s \, r^2 \exp(-\alpha r^2) , \tag{5a}$$

$$|p_x\rangle = N_p \, x^3 \exp(-\beta r^2) . \tag{5b}$$

For both the maximum overlap between two bonding orbitals is obtained for a next neighbour distance R greater than zero.

By means of the hybridization of atomic orbitals the chemical bond of covalent solids can be described by a sum of two-centre molecules. The hybridization of valence orbitals Ψ_i and Ψ_j with

$$\Psi_{i,j} = N_{i,j}^{-1/2} \{|s\rangle + \lambda_{i,j} |p_x\rangle\} = |sp_x\rangle \tag{6}$$

$(x = \lambda^2)$ is determined by the orthogonality condition

$$1 + \lambda_i \lambda_j \cos \theta_{ij} = 0 . \tag{7}$$

θ_{ij} means the angle between the hybrid orbitals centred at the same atom. The value of the overlap integral changes in dependence on the hybridization state. Therefore, S depends on the concrete coordination angle. For an increasing θ_{ij} the $|p\rangle$ contribution of a $\{|s\rangle + \lambda|p\rangle\}$ hybrid orbital $(1 < \lambda < \infty)$ must decrease in order to fulfil the orthogonality condition (7). It follows a decreasing overlap amount of bonding orbitals (cf. [25] Fig. 33).

3. Construction of a "Realistic" Valence Charge Density

Before the evalution of the dynamic bond charge transfer properties a "realistic" VCD of the crystal must be constructed. That means, all static VCD properties known and general experience should be taken into account:

1. The VCD is characterized by a single maximum in the middle of the next neighbour distance ($\varrho_{max} = \varrho(R/2)$).

2. The total BC amount between next neighbours is supposed to about $2e/\varepsilon_0$ (0.17 electrons for silicon) according to the dielectric theory (ε_0 means the static electronic dielectric constant [13, 14]).

3. The BC-dependent X-ray structure amplitudes describing the q-space dependence of the BC, should agree with the experiment.

Our model contains the following basic approximations:

(i) The total VCD is built by superposition of two-centre molecules. For a construction of the "realistic" VCD around a bond the superposition of diatomic molecule densities of the first coordination sphere should be sufficient. Our procedure corresponds with the description of the solid state properties by two-body interaction including angle-dependent forces.

(ii) The model valence orbitals are constructed from $|s\rangle$ and $|p\rangle$ like atomic orbitals. The exact energy and $\boldsymbol{k}$-space dependence of all valence bands are replaced by an averaged model wave function

$$\Psi = (\mu^2 + 1 + \lambda^2)^{-1/2} \{\mu|s\rangle + |sp_x\rangle\} . \tag{8}$$

The model wave function contains two different parts. The last one treats the covalent portion of the bond described by (6). Using this part alone a closed VCD can be obtained using a larger BC amount in comparison with $2e/\varepsilon_0$ only. Therefore the charge density near the centre of the bond has to diminish. That is allowed to be described by an additional $|\bar{s}\rangle$ like portion being dealt with the metal-like contribution of the chemical bond. The addition of $|\bar{s}\rangle$ like portion does not change the orthonormalization condition.

From the numerical point of view the metallic $|\bar{s}\rangle$ and the covalent basic function $|s\rangle$ in (6) are not different essentially. Therefore, both functions will be adapted further by the same screening constant. Fig. 1 shows the total VCD for silicon by VB approximation. The screening constants of the orbitals and the parameter μ are fitted adopting the above conditions (Table 1). All free parameters are fitted so that μ as small as possible with the VCD maximum being just closed. The fitted value of about $\mu = 1.8$ is somewhat greater than this parameter determined by a Mössbauer experiment [28]. Our calculated VCD is in agreement with general density plots using X-ray diffraction data [1] and pseudopotential calculations [2 to 7]. This is also true for the bond charge density (Figure 2).

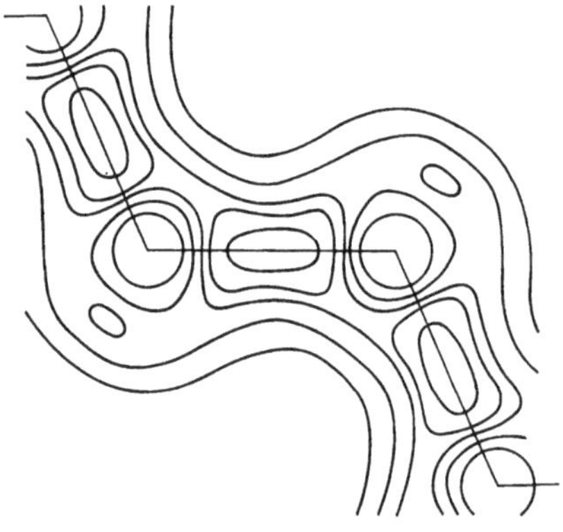

Fig. 1

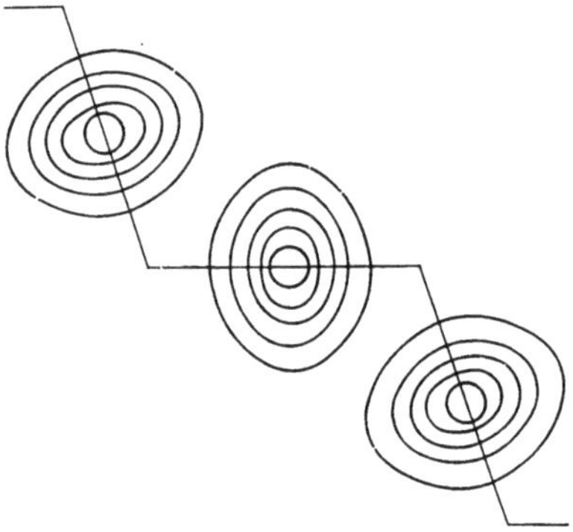

Fig. 2

Fig. 1. Valence charge density of silicon calculated in the (011) plane by linear superposition of diatomic molecule densities in VB approximation. The model parameters are listed in Table 1; contours at 0.21×10^{30} electrons m^{-3}

Fig. 2. BC density for silicon calculated by the same approximation as in Fig. 1; contours at 0.04×10^{30} electrons m^{-3}

Table 1
Parameters of the model wave functions following (8)

substance	μ	α	β	BC_0 (electrons)	R_0 (at. units)
C (diamond)	0.0	1.5	0.80	0.35	2.92
Si	1.8	0.54	0.46	0.17	4.44
Ge	2.9	0.45	0.43	0.13	4.63
α-Sn	4.5	0.33	0.31	0.10	5.31

Both the (222) and (442) "forbidden" X-ray scattering factors are measurable, they are caused by the existence of a BC. Their calculation provides

$$f_{BC}(q) = \frac{2S}{1+S^2} \int \Psi_i(\boldsymbol{r}) \Psi_j(\boldsymbol{R} - \boldsymbol{r})\, e^{-i\boldsymbol{q}\boldsymbol{r}} d^3\boldsymbol{r} = \frac{2S^2}{1+S^2}\, e^{-q^2/8\alpha} \tag{9}$$

employing Gaussian-like orbitals (5). In our model the (222) scattering factor is reconstructed to about 70%, but the (442) reflection power turns out to be exact. The outcome is comparable with the result of the adaption of the experimentally determined X-ray structure amplitudes by the static BC model [11] and depends on the use of Gaussian-like basis functions.

4. Determination of the Bond Charge Transfer Constants of Silicon

The overlap integral depends on both the next neighbour distance R and the coordination angle θ_{ij}. Therefore, we expand the BC in terms of both dependences

$$\begin{aligned} BC &= BC_0 + \gamma_1 \Delta R + \gamma_2 \Delta R^2 + \delta_1 \Delta\theta + \delta_2 \Delta\theta^2 + \dots \\ &= BC_0 + \gamma_1 \Delta R + \gamma_2 \Delta R^2 \,. \end{aligned} \tag{10}$$

The value BC_0 stands for the unvaried state fitted in the manner of Section 3. Principally a variation of the bond angles is allowed to be expressed by means of a changing next neighbour distance.

Therefore, it is useful to prefer averaged transfer constants in comparison with the experiment. Fig. 3 shows the principal slope of the ΔR-dependent parts of the transfer constants. The full description of the transfer properties requires the second deviation in (10). For maximum $\langle p|p\rangle$ overlap γ_1, is equal to zero, but γ_2 increases with expanding lattice. For pure $\langle s|s\rangle$ overlap γ_2 diminishes in the same direction.

Choosing bonded $|sp_3\rangle$ hybrides different transfer parts must be taken into account. For the hydrostatic pressure γ_i terms are essential, only. Angle-dependent parts δ_i are

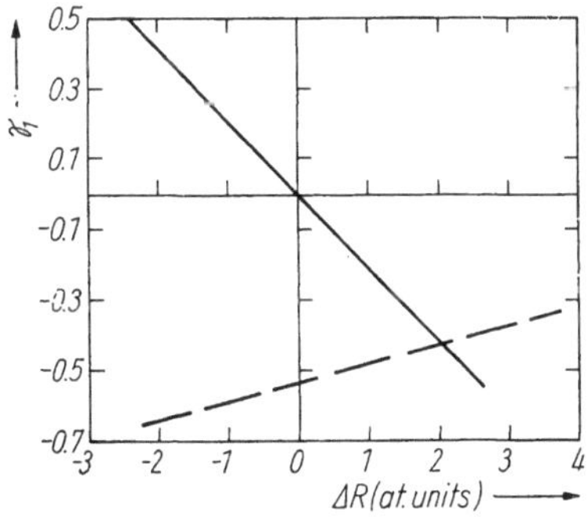

Fig. 3. The principal dependence of the BC transfer constant $\gamma_1 = \partial BC/\partial R|_0$ versus the next neighbour distance ΔR for $\langle s \mid s\rangle$ (dashed line) and $\langle p \mid p\rangle$ bonding (solid line). ΔR is positive used for expanding lattice. The slope of the $\langle p \mid p\rangle$ overlap is multiplied by a factor 10

required for a displacement of some atoms in an otherwise rigid lattice, like thermal displacement or uniaxial strain, for instance. Phonon induced displacements corresponding to optic or acoustic modes have to be distinguished. For the separation of the pure angle-dependent part both the total transfer and the only θ_{ij} dependent transfer constant contribution are calculated and subtracted from each other.

At once Table 2 demonstrates the different transfer constants of silicon using full $|sp_3\rangle$ hybridized valence orbitals in maximum overlap. In the first line one of the atoms has been replaced in an almost rigid lattice. This case corresponds with a thermal displacement. The $|p\rangle$ like portion dominates in the quadratic term. The linear contribution is determined by the angle dependence alone.
In the second line a collective displacement of both sublattices is simulated, a "frozen" TO(Γ) phonon displacement, for example. The angle dependent part is twice that due to thermal displacement. In the third line the transfer coefficients are listed employing the model valence orbitals given in (8). Using a fixed metallic parameter μ the total transfer constants are about 100% greater in comparison with self-consistent pseudopotential results [7] and the experimental determination being 0.2 and 0.16 electrons/m [19], respectively. On the other hand, we must know that the covalent bond becomes more metallic under hydrostatic pressure. At the phase transition point the elemental semiconductors transform into a metallic state. It was demonstrate that the $|s\rangle$ like behaviour of the VCD increases also for small displacements [27]. That dynamical property of the VCD shall be included into our model by the following empirical relation:

$$\Delta\mu = -\frac{\Delta s}{\Delta r}\,\Delta R\,. \tag{11}$$

The reconstruction of the exact theoretically and experimentally determined transfer constants by means of our model requires $\Delta s/\Delta r = 1.50 \pm 0.05 \times 10^{10}\ \mathrm{m}^{-1}$ using the mean parameter $\bar{\gamma}_1$. Using (11) the transfer constants are smaller than shown in the third line. That is demonstrated in the fourth line. The quadratic contribution changes sign. For a mean-square displaced silicon atom in a rigid lattice the BC amount of the stretched bond decreases by about 9%. The quadratic contribution is about 0.5%

Table 2

Transfer constants of elemental semiconductors following (10)

substance	γ_1	γ_2	δ_1	δ_2	$\bar{\gamma}_1$	$\bar{\gamma}_2$ f)	
Si	0.00	−0.51	−0.16	−0.06	−0.16	−0.57	a)
	0.00	−0.51	−0.32	−0.10	−0.32	−0.16	b)
	−0.59	+0.55	+0.18	−0.33	−0.41	+0.22	c)
	−0.37	+0.24	+0.18	−0.33	−0.19	−0.09	d)
C	−0.51	+0.25	+0.25	−0.45	−0.26	−0.20	e)
Ge	−0.33	+0.31	+0.14	−0.24	−0.19	+0.07	e)
α-Sn	−0.23	+0.24	+0.09	−0.13	−0.14	+0.11	e)

a) $|sp_3\rangle$ hybrid orbitals, thermal displacement.
b) $|sp_3\rangle$ hybrid orbitals, "frozen" TO(Γ) phonon displacement.
c) Model valence orbitals (8), the coefficients remain constant during the thermal displacement.
d) Model valence orbitals (8), the coefficients change during the thermal displacement according to (11).
e) Model valence orbitals (8), transfer model as in d).
f) γ_1, δ_1, and $\bar{\gamma}_1$ at 10^{10} electron m^{-1}, γ_2, δ_2, and $\bar{\gamma}_2$ at 10^{20} electrons m^{-1}.

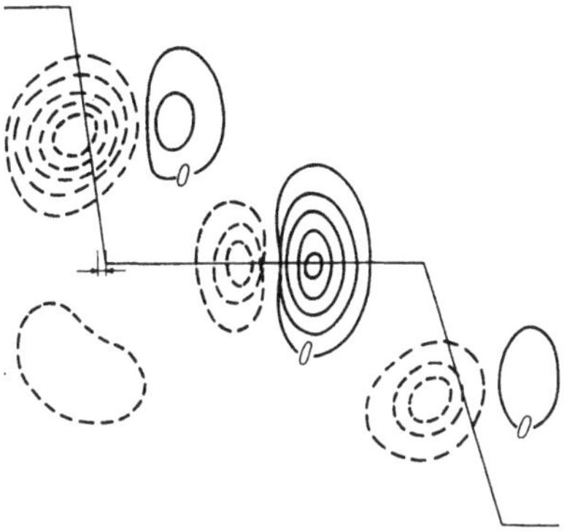

Fig. 4

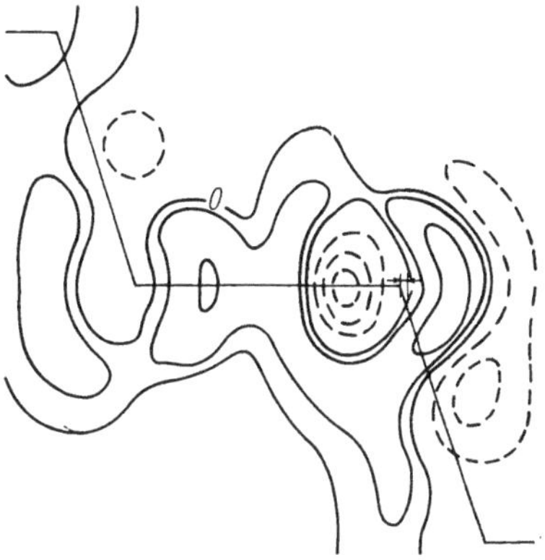

Fig. 5

Fig. 4. BC polarization density for silicon using the model approximation as in Fig. 1. The displacement corresponds to the "frozen" TO(Γ) phonon excitation. The zero density line is marked; negative contours are drawn dashed; contours at 0.00135×10^{30} electrons m^{-3}

Fig. 5. Valence charge polarization density for silicon using the model approximation as in Fig. 1 and 4, contours at 0.0076×10^{30} electron $\cdot m^{-3}$

only. Therefore, for a practicable handling the linear term is sufficient for the description of the BC transfer.

Fig. 4 shows the polarization density of silicon calculated by our model. The next neighbour distance of the central bond is compressed by a displacement of the right-hand atom of about 0.8×10^{-10} m towards the almost rigid lattice. The BC is transferred from the stretched to the compressed bonds. Note that the spherically distributed valence charge amount is changed during the ionic displacement if the quadratic terms of the transfer constants are not equal to zero.

The polarization density of the total VCD caused by the above excitation is plotted in Fig. 5. The behaviour is comparable with that given by Harmon et al. [16] and Maschke et al. [17] in a qualitative manner. The density maximum of the right-hand side can be explained by the changed amount of superimposed valence orbitals caused by the reduced coordination angle between the model valence orbitals.

5. Calculation of the BC Transfer Constants of the Group IV Elemental Semiconductors

After the determination of the transfer properties for silicon the transfer constants of the other group IV elemental crystals will be estimated. The fitting conditions given in Section 3 are to transform to the Phillips BC amount and the right next neighbour distance. The fitted model orbital parameters are listed in Table 1. The enhanced $|s\rangle$ character of the model valence orbitals follows the well-known increasing metallic portion of the chemical bond from diamond to α-tin. This is documented by the increasing dielectric constant in the same direction, too. Using the isomer shift from Mössbauer spectroscopy measuring the $|s\rangle$ character in the neighbourhood of a substituent, the $|s\rangle$ portion is determined to be 1.34, 1.50, and 1.56 for silicon, germanium, and α-tin, respectively [28].

Compared to those values our fitted μ parameter is greater being caused by the relatively rigorous model conditions. Whereas for diamond the full $|sp_3\rangle$ hybridization seems to be fulfilled we found a relatively strong deviation for the other elemental crystals. The change in the |s⟩ character during the ionic displacement is unknown. Therefore, the simplest approximation employs a constant $\Delta s/\Delta r$ parameter fitted for silicon. The transfer constants are given in Table 2. The total transfer diminishes from diamond to α-tin, due to the increasing μ the angle-dependent part decreases from diamond to α-tin. The pure ΔR part rises in the same direction. Considering the reduced BC amount the absolute transfer turns out to be about 4%, 12%, and 20% at a mean thermal displacement of about $\sqrt{\bar{u}^2} = 0.05$, 0.076, 0.083, and 0.126×10^{-10} m [8, 9] for diamond, silicon, germanium, and α-tin, respectively. The portion of the quadratic terms rises from 0.2 to 2.0% in the same direction.

6. Application to X-Ray Diffraction Experiments

The BC transfer is an important factor for the interpretation of all dynamic properties of semiconductors, which depends on the real VCD, like the phonon induced effect and experiments using external fields or pressure. Especially for X-ray diffraction experiments some applications will be given in what follows.

The hydrostatic pressure dependence of the silicon (222) reflection power has been measured by Yoder-Short et al. [22]. They found an increased reflection intensity up to a pressure of about 10 GPa. Then a strong decrease in the intensity was measured up to the phase transition point. This experiment can be interpreted by means of the BC transfer model. Up to 10 GPa the rising reflection power is caused by the increased overlap of the valence orbitals described by the ΔR dependent transfer constant [23]. Using the linear term only, the hydrostatic pressure dependence of the BC comes from the lattice parameter a_0 and the bulk modulus B_s [29],

$$\frac{\Delta \mathrm{BC}}{\Delta p} = -\frac{a_0 \bar{\gamma}_1}{3 B_s}. \tag{12}$$

These values for the group IV crystals are listed in Table 3. The absolute variation in the BC amount under pressure increases from diamond to α-tin. In view of the experimental possibilities the greatest effect is expected for germanium.

For uniaxial stress both the next neighbour dependent and the angle-dependent part of the transfer constants determine the transfer properties. Some calculations for silicon are presented in [23]. Relative to silicon we expect for germanium a smaller influence of the angle-dependent parts on the absolute variation of the (222) reflection power, which depends on the smaller |p⟩ like portion of the model valence orbitals. Therefore, the (222) reflection power using [001] stress must increase less for germanium than for silicon. However, for [001] stress we expect a decreasing slope for ger-

Table 3

Variation of the BC amount by different excitations and measurable by X-ray diffraction experiments

substance	$\frac{\partial \mathrm{BC}}{\partial T}$ (10^{-6} electrons/K)	$\frac{\partial \mathrm{BC}}{\partial p}$ (10^{-3} electrons/GPa)	$\frac{\bar{v}^2_{\mathrm{BC}}}{\bar{u}^2_{\mathrm{ion}}}$
C	− 2.1	− 1.4	1.4
Si	− 7.5	− 6.9	4.0
Ge	−11.4	− 8.6	6.2
α-Sn	−31.1	−11.7	8.1

manium and α-tin. Another BC transfer influenced X-ray experiment is the investigation of the temperature dependence of the (222) reflection power. Some authors found the same temperature dependence for the BC as well as for the strong reflections for silicon and germanium above room temperature [30, 31]. The measurement of α-tin up to 250 K yielded the same result [9]. In contrast to these experiments we obtained a greater mean square displacement of the BC between 92 and 293 K compared to the strong reflections for silicon [19].

The interpretation of the temperature dependence of the BC scattering factor is very complex. First we must consider the linear thermal expansion coefficient $\alpha(T)$ of the lattice. The absolute variation of the BC,

$$\frac{\Delta \mathrm{BC}}{\Delta T} = \alpha(T)\, a_0 \bar{\gamma}_1 \,, \tag{13}$$

depends on the $\bar{\gamma}_1$ transfer constant and the lattice parameter, only. However, the numerical values are small (Table 3). Second, we must consider the dynamic charge transfer during the phonon induced oscillation of lattice. In a former publication we have employed a simple model for the calculation of the mean square displacement of BC. Using a mean transfer constant and an equal portion of acoustic and optical mode excitation the ratio of the mean square displacement of the BC and the atoms is determined to be

$$\frac{\bar{v}^2_{\mathrm{BC}}}{\bar{u}^2_{\mathrm{ion}}} = \frac{1}{6}\left\{2 + \left(1 + \frac{2\bar{\gamma}_1}{\mathrm{BC}_0}\,\frac{a_0}{6}\sqrt{3}\right)^2\right\} \tag{14}$$

for the diamond structure (Table 3) [19].

A more exact calculation of the ratio $\bar{v}^2/\bar{u}^2$ must take into account the real phonon excitation in dependence on temperature. The proportion of the acoustic and optical modes and its temperature dependence was calculated by Soma et al. [32]. At temperatures near 0 K optical and acoustic zone centre modes are excited in nearly equal number. The transfer constants of the optical modes are greater than the acoustic ones. The absolute displacement and therefore the absolute BC transfer is small. At increasing temperature the absolute displacement increases and the measureable transfer amount must rise. But the contribution of the optical modes to the Debye-Waller factor decreases compared with the acoustic ones. At the Debye temperature all phonon modes are excited. The averaged displacement of zone edge phonons is smaller than the zone centre phonons, too. Therefore, the measureable BC variation must decrease up to the Debye temperature. We conclude that the effect of the BC transfer should be investigated near nitrogen temperature for silicon and germanium.

7. Discussion

Our model calculation of the density response caused by atomic displacement starts from well-known physical density properties and contains a few model approximations, only. Consequently, the BC transfer properties are determined without additional approaches during the calculation procedure. These input properties are

1. the closed VCD maximum in the middle of the covalent bond,
2. the BC amount fitted at the known (222) X-ray reflection power,
3. the mean BC transfer constant of about 0.16 to 0.20 × 10^{10} electrons/m for silicon, and
4. the change in the $|s\rangle$ character of the valence orbitals during the variation of the next neighbour distance.

For diamond the condition of only one VCD maximum seems to be questionable. New theoretic calculations yielded a double-hump feature [3]. On the other hand, for germanium and α-tin the limitation of $|s\rangle$ and $|p\rangle$ like valence orbitals represents a stronger restriction than for silicon and diamond.

In comparison with full self-consistent calculations of the VCD dependent properties our model offers some approximations. The complicated many-body interaction is replaced by a sum of two-body interactions comprising angle-dependent forces. But with regard to usual empirical models our scheme includes the bond charge as an essential consequence of the covalent bond. The overlap charges between more than next neighbours are neglected. Considering a BC between next nearest neighbours of about 0.17 electron the BC between second nearest neighbours is determined to be smaller than 0.005 electron only. The assumption that according to the diagonal screening of ionic pseudopotentials the BC is about $2e/\varepsilon_0$ is fulfilled for the elemental semiconductors approximately [18, 33] and can be confirmed by X-ray diffraction data [13].

The chemical bond of our model VCD is determined by metallic and purely covalent parts. The covalent part corresponds to a full $|sp_3\rangle$ hybridization of valence orbitals. The metallic contribution is replaced by the additional $|s\rangle$ portion ($\mu > 0$). But the coefficients seem to be somewhat overdetermined caused by the rigorous model conditions.

On the other hand, the restriction of the valence bond theory has no qualitative influence on the calculated transfer constants. Using the molecular–orbital theory the screening constants must be varied to fulfil the fitting conditions. Taking into account the validity regions of both approximations the BC is overdetermined by the VB theory and smaller according to the MO concept.

In our model the variation in the $|s\rangle$ character of orbitals at changed next neighbour distance is obtained by means of a simple empirical formula. The $\Delta s/\Delta r$ parameter fitted to silicon is applied to the other elemental semiconductors. The results are in agreement with the tendency of the transformation of the diamond structure to the metallic β-tin structure under pressure [4].

The radial part of the used Gaussian wave function goes down strong near the bond region in comparison with Slater like functions. Therefore our $\Delta s/\Delta r$ parameter is somewhat overdetermined. A better model for the determination of the $\Delta s/\Delta r$ parameter should involve the trend of closed shells in the direction from elemental semiconductor to the ionic compounds. However, we expect that more sophisticated models will not vary the change of the transfer constants in a qualitative manner. A decrease in the $\Delta s/\Delta r$ parameter of about 50% causes an increase of the transfer constant of about 25% for germanium.

Considering all listed approximations our model cannot correctly replace self-consistent calculations of the VCD including internal and external excitations. In our mind its advantage consists in the convenient handling of the transfer parameter for the interpretation of VCD dependent experiments. Moreover, new experimental results can be incorporated in a simple manner.

Acknolwedgements

The author would like to thank Dr. K. Maschke (EPF Lausanne) and Prof. Dr. K Unger for helpful and critical discussions.

References

[1] Y. W. YANG and P. COPPENS, Solid State Commun. 15, 1555 (1974).
[2] J. R. CHELIKOWSKY and M. L. COHEN, Phys. Rev. B 14, 556 (1976).

[3] M. T. YIN and M. L. COHEN, Phys. Rev. Letters **50**, 2006 (1983).
[4] M. T. YIN and M. L. COHEN, Phys. Rev. B **26**, 5668 (1982).
[5] A. ZUNGER and M. L. COHEN, Phys. Rev. B **20**, 4082 (1979).
[6] A. ZUNGER and A. J. FREEMAN, Phys. Rev. B **15**, 5049 (1977).
[7] A. BALDERESCHI, K. MASCHKE, A. MILCHEV, R. PICKENHAIN, and K. UNGER, phys. stat. sol. (b) **108**, 511 (1981).
[8] B. DAWSON, Proc. Roy. Soc. **A298**, 264, 298, 379, and 395 (1967).
[9] D. H. BILDERBACK and R. COLELLA, Phys. Rev. B **11**, 793 (1975).
[10] D. H. BILDERBACK and R. COLELLA, Phys. Rev. B **13**, 2476 (1976).
[11] U. PIETSCH, phys. stat. sol. (b) **102**, 127 (1980).
[12] U. PIETSCH, phys. stat. sol. (b) **106**, K9 (1981).
[13] U. PIETSCH and K. UNGER, to be published.
[14] J. C. PHILLIPS, Phys. Rev. **166**, 832 (1968).
[15] H. WENDEL and R. M. MARTIN, Festkörperprobleme **19**, 21 (1979).
[16] B. N. HARMON, W. WEBER, and D. R. HAMANN, Phys. Rev. B **25**, 1109 (1982).
[17] K. MASCHKE and A. BALDERESCHI, Proc. Internat. Conf. Semiconductors, Edinbourgh 1978.
[18] J. BAUR, K. MASCHKE, and A. BALDERESCHI, Phys. Rev. B **27**, 3720 (1983).
[19] U. PIETSCH and K. UNGER, phys. stat. sol. (b) **102**, 503 (1980); **107**, K19 (1981).
[20] U. PIETSCH and K. UNGER, phys. stat. sol. (b) **104**, 253 (1981).
[21] I. FUJIMOTO, Japan J. appl. Phys. **19**, L345 (1980).
[22] D. R. YODER-SHORT, R. COLELLA, and B. A. WEINSTEIN, Phys. Rev. Letters **149**, 1438 (1982).
[23] U. PIETSCH, phys. stat. sol (b) **120**, 183 (1983).
[24] J. C. SLATER, Quantum Theory of Molecules and Solids, McGraw-Hill Publ. Co., New York 1963.
[25] C. A. COULSON, Valence, Oxford University Press, 1961.
[26] L. PAULING, Die Natur der chemischen Bindung, Verlag Chemie, Weinheim 1964.
[27] K. MASCHKE, private communication.
[28] E. ANTONCIK and B. L. GU, Physica Sripta **25**, 835 (1982).
[29] LANDOLDT/BÖRNSTEIN, Vol. 17a, Ed. O. MADELUNG, Springer-Verlag, 1982.
[30] J. B. ROBERTO, B. W. BATTERMAN, and D. KEATING, Phys. Rev. B **9**, 2590 (1974).
[31] I. FUJIMOTO, Phys. Rev. B **9**, 591 (1974).
[32] T. SOMA and H. MATSUO, phys. stat. sol (b) **111**, K93 (1982).
[33] J. P. WALTER and M. L. COHEN, Phys. Rev. B **4**, 1877 (1971).

(Received June 26, 1984)

phys. stat. sol. (b) **126**, 607 (1984)

Subject classification: 13 and 20; 22.2.1

Zentralinstitut für Elektronenphysik der Akademie der Wissenschaften der DDR, Berlin[1]) *(a) and Institute of Physics, Academy of Sciences of the Ukrainian SSR, Kiev (b)*

The Role of Electron–Hole Interaction in the Cooling Process of Highly Excited Carriers

By

M. Asche (a) and O. G. Sarbei (b)

When the energy exchange by electron–hole interaction is properly taken into account, the holes after optical excitation never achieve such a high temperature as the electrons for the concentrations usually envolved, and the cooling rate of the electrons therefore is significantly lower than in the case of an equal plasma temperature assumed until now. Theoretical expressions are obtained in Born approximation using the Thomas-Fermi model for screening and calculations are performed for carrier relaxation in GaAs after instantaneous excitation.

Если точно учесть обмен энергии за счет электронно дырочного взаимодействия, то после оптического возбуждения температура дырок не может достичь такой высокой величины как у электронов для обычно реализуемой концентрации носителей, и вследствие этого охолождение электронов происходит значительно медленнее, чем в случае единой температуры плазмы, как это предполагалось до сих пор. Получены выражения для обмена энергии в борновском приближении с использованием модели Томаса-ферми для экранирования, и проведены расчеты релаксации носителей в GaAs после мгновенного возбуждения.

1. Introduction

As well known the cooling process of an optically excited electron–hole plasma can be examined by time-resolved ps spectroscopy of transmission, reflection, or luminescence. For GaAs as a typical semiconductor for these investigations initially the electrons have much more energy than the holes on account of the band structure, however, in all calculations of the cooling process it is assumed that beginning from a certain short time after excitation the electrons and holes can be described by Fermi or Maxwell distribution functions with the same temperature. Then the main energy loss of the plasma is due to the phonon emission by the holes. But equal carrier temperature means that in the beginning of the cooling process the holes have to gain much more energy from the electrons than they deliver to the lattice in order to increase their temperature so significantly. As the effective masses of electrons and holes in GaAs differ by an order of magnitude, this initially quick heating of the holes seems not very reasonable from a physical point of view except the concentration of the plasma being very high. Therefore, the too fast cooling rate obtained in the earlier calculations [1] is not surprising. This discrepancy between theoretically and experimentally determined cooling rates was reduced by including screening of carrier–phonon interaction [2 to 8] and departure from phonon equilibrium [7, 4, 6]. The purpose of this paper is to estimate the electron–hole interaction and to show that the more convincing assumption of different electron and hole temperatures leads to a significantly slower cooling down of the electrons than obtained for equal carrier temperature.

[1]) Hausvogteiplatz 5–7, DDR-1086 Berlin, GDR.

2. Theoretical Considerations

The mean change of carrier energy by carrier–carrier collision per one carrier is given by

$$\left\langle\frac{\partial E_1}{\partial t}\right\rangle_{ij} = \frac{1}{n_1(2\pi)^6} \iiiint [E_{K_1} - E_{K_1'}]\, w(K_1, K_1', K_2, K_2')\, f(K_1)\, [1 - f(K_1')] \times \times f(K_2)\, [1 - f(K_2')]\, \mathrm{d}K_1\, \mathrm{d}K_1'\, \mathrm{d}K_2\, \mathrm{d}K_2' \tag{1}$$

with the scattering probability for two Coulomb interacting particles

$$w(K_1, K_1', K_2, K_2') = \frac{2\pi}{\hbar}\left(\frac{4\pi e_1 e_2}{\varepsilon}\right)^2 \frac{\delta_{K_1+K_2, K_1'+K_2'}\, \delta(E_{K_1'} + E_{K_2'} - E_{K_1} - E_{K_2})}{[(K_1 - K_1')^2 + q_D^2]}, \tag{2}$$

where q_D is the screening parameter and $\int f(K_i)\, \mathrm{d}K_i = n_i$. Supposing $f(K_2)$ to be a Maxwellian distribution as realized for the holes of GaAs in most experiments the integration over K_2' using $\delta_{K_1+K_2, K_1'+K_2'}$ is trivial.

In order to perform the integration over K_1' with $f(K_1')$ as a Fermi distribution for the electrons analytically we assume that the energy exchange of an electron with a hole at each scattering event is small in comparison to the energy of the electron on account of the different effective masses of both particles. Therefore, $f(E_{K_1'})$ can be expanded in powers of $\Delta E = E_{K_1} - E_{K_1'}$. Introducing the variable $q = K_1' - K_1$ instead of K_1' we get

$$\left\langle\frac{\partial E_1}{\partial t}\right\rangle_{ij} = \frac{4e^4}{\varepsilon^2 n_1 \hbar} \iiiint \Delta E\, f(E_{K_1})\left[1 - f(E_{K_1}) + \frac{\partial f(E_{K_1})}{\partial E_{K_1}}\Delta E\right] f(E_{K_2}) \times \times \delta(E_{K_1+q} + E_{K_2-q} - E_{K_1} - E_{K_2})\, [q^2 + q_D^2]^{-2}\, \mathrm{d}q\, \mathrm{d}K_1\, \mathrm{d}K_2\,, \tag{3}$$

where $\Delta E = -\hbar^2 K_1 q(\cos\alpha + q/2K_1)/m_1$ with α being the angle between q and K_1'. Furtheron we introduce the relative wave vector of the two colliding particles $p = 2m(K_1/m_1 - K_2/m_2)$ with $1/m = (1/m_1 + 1/m_2)$.

The integration over q then leads to

$$\left\langle\frac{\partial E_1}{\partial t}\right\rangle_{ij} = -\frac{8\pi e^4 m}{\varepsilon^2 m_1 n_1 \hbar}\Bigg\{\iint f(E_{K_1})\, [1 - f(E_{K_1})]\, f(E_{K_2})\, \frac{K_1}{p_2} \times \times \left[\frac{p}{2K_1} - \cos\theta\right] \Phi\left(\frac{p}{q_D}\right) \mathrm{d}K_1\, \mathrm{d}K_2 - -\frac{\hbar^2}{m_1}\iint f(E_{K_1})\, \frac{\mathrm{d}f(E_{K_1})}{\mathrm{d}E_{K_1}}\, f(E_{K_2})\, \frac{K_1^2}{p}\left\{\left[2\left(\frac{p}{2K_1} - \cos\theta\right)^2 - \sin^2\theta\right] \times \times \left[\frac{p^2}{2(p^2 + q_D^2)} - \frac{\Phi(p/q_D)}{(p/q_D)^2}\right] + \frac{\sin^2\theta}{2}\, \Phi\left(\frac{p}{q_D}\right) \mathrm{d}K_1\, \mathrm{d}K_2\right\}. \tag{4}$$

In this equation

$$\Phi\left(\frac{p}{q_D}\right) = \ln\left(1 + \frac{p^2}{q_D^2}\right) - \frac{p^2}{p^2 + q_D^2}$$

has been introduced and θ is the angle between K_1 and p. The angular integration of

$\boldsymbol{K}_2$ can be performed using

$$\int \dots \mathrm{d}\boldsymbol{K}_2 = -\left(\frac{m_2}{2m}\right)^3 \iint \dots \delta(E_{\boldsymbol{K}_2} - E_2)\, \mathrm{d}\boldsymbol{p}\, \mathrm{d}E_2\,,$$

where the argument of the δ-function has to be expressed by $\boldsymbol{K}_1$ and $\boldsymbol{p}$

$$E_{\boldsymbol{K}_2} - E_2 = \frac{\hbar^2 m_2}{2}\left(\frac{K_1^2}{m_1^2} - \frac{K_1 p}{m_1 m}\cos\theta + \frac{p^2}{4m^2}\right) - E_2\,.$$

The angular integration over $\boldsymbol{K}_1$ is trivial again. Some tedious calculations lead to rather voluminous expressions not presented here in full extension but only for the zeroth order,

$$\left\langle\frac{\partial E_1}{\partial t}\right\rangle_{ij} = \frac{\sqrt{\pi}\, e^4 (m_1 + m_2)\, n_2}{\varepsilon^2 (k_\mathrm{B} T_1 k_\mathrm{B} T_2 m_1 m_2)^{1/2} \int f(x_1) \sqrt{x_1}\, \mathrm{d}x_1} \iint f(x_1)\,[1 - f(x_1)]\, \mathrm{e}^{-x_2} \times$$
$$\times \left\{\frac{m_2 - m_1}{2(m_2 + m_1)}\left[y \ln\left(\frac{y^2}{y_\mathrm{D}^2} + 1\right) - 3y + 3y_\mathrm{D} \operatorname{arctg}\frac{y}{y_{\mathrm{D}_1}}\right] + \right.$$
$$+ \left(\frac{k_\mathrm{B} T_1}{m_1} x_1 - \frac{k_\mathrm{B} T_2}{m_2} x_2\right) \times$$
$$\times \left.\left[\frac{1}{y} \ln\left(1 + \frac{y^2}{y_\mathrm{D}^2}\right) - \frac{1}{y_\mathrm{D}} \operatorname{arctg}\frac{y}{y_\mathrm{D}}\right]\right\}_{\pm(\sqrt{2k_\mathrm{B}T_1x_1/m_1} - \sqrt{2k_\mathrm{B}T_2x_2/m_2})}^{\sqrt{2k_\mathrm{B}T_1x_1/m_1} + \sqrt{2k_\mathrm{B}T_2x_2/m_2}} \mathrm{d}x_1\, \mathrm{d}x_2 \tag{5}$$

with $y_\mathrm{D} = \hbar q_\mathrm{D}(m_1 + m_2)/2m_1 m_2$ and $x_i = E_i/k_\mathrm{B}T_i$. In this expression the limits for y are indicated, the positive sign of the lower limit corresponds to $K_1/m_1 > K_2/m_2$ and the negative sign to $K_1/m_1 < K_2/m_2$, respectively.

If $K_2/m_2 \ll K_1/m_1$ the full expression for $\langle \partial E_1/\partial t\rangle_{ij}$ can be expanded in powers of $K_2 m_1/K_1 m_2 = \sqrt{m_1 T_2 x_2/m_2 T_1 x_1}$ and the integration over x_2 can be analytically performed. In this case the remaining integral over x_1 has the form

$$\left\langle\frac{\partial E_1}{\partial t}\right\rangle_{ij} = \frac{\sqrt{2m_1}\, \pi e^4 n_2}{\sqrt{k_\mathrm{B} T_1}\, \varepsilon^2 m_2 \int f(x_1)\sqrt{x_1}\, \mathrm{d}x_1} \left\{\int f(x_1)\,[1 - f(x_1)] \times \right.$$
$$\times \left[\ln\left(1 + \frac{x_1}{x_\mathrm{D}}\right) - \frac{x_1}{x_1 + x_\mathrm{D}}\right] \mathrm{d}x_1 + \frac{4m_1^2 m_2}{(m_1 + m_2)^2} \int x_1 f(x_1) \frac{\mathrm{d}f(x_1)}{\mathrm{d}x_1} \times$$
$$\times \left.\left[1 + \frac{x_\mathrm{D}}{x_1 + x_\mathrm{D}} - \frac{2x_\mathrm{D}}{x_1} \ln\left(1 + \frac{x_1}{x_\mathrm{D}}\right)\right] \mathrm{d}x_1\right\} \tag{6}$$

with $x_\mathrm{D} = m_1 y_\mathrm{D}^2/2k_\mathrm{B}T_1 = m_1 \hbar^2 q_\mathrm{D}^2/8m^2 k_\mathrm{B} T_1$ and can be numerically calculated.

3. Results of Numerical Calculations

For the numerical calculations a simple model is used. It is supposed that the duration of the laser pulse is so short that during this time the carrier cannot cool down essentially. Corresponding to this condition for GaAs the laser pulse should be about a tenth of a ps. As shown in the Appendix for the monoenergetically excited carrier concentrations of about 10^{17} to 10^{18} cm^{-3} the electron–electron and hole–hole collisions are too weak to determine the shape of the distribution function, but the short duration of the laser pulse leads to an excitation of the carriers within significant energy regions which

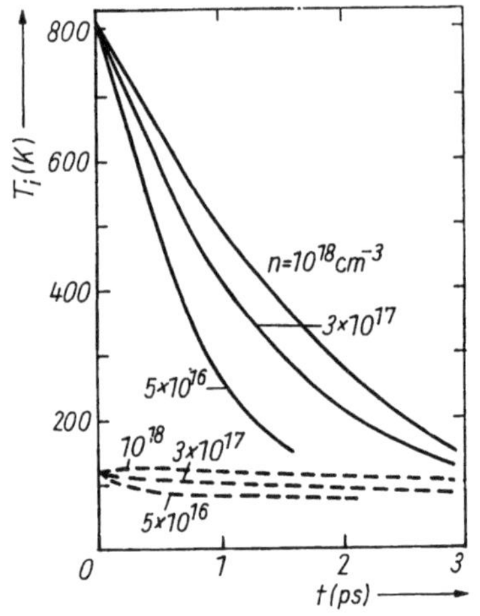

Fig. 1. Electron (1) and hole (2) temperatures in dependence on time after short pulse excitation for $T_1(0) = 810$ K, $T_2(0) = 120$ K for indicated carrier concentrations

can be determined by the uncertainty principle. This spread of the initial distribution of the carriers is of the order of the energy of optical phonons. In spite of this region being only small in comparison to the mean value of the initial electron energy Fermi and Maxwell functions are assumed as approximations in order to estimate the effect of electron–hole interaction. The ratio of the initial energies of electrons and holes for an excitation with $h\nu$ is determined by the relation

$$h\nu - E_{\text{gap}} = E_1 + E_2 = \frac{\hbar^2 K^2}{2m_0}\Big[1 + AK^2 \pm \pm\sqrt{B^2K^4 + C^2(K_x^2K_y^2 + K_y^2K_z^2 + K_z^2K_x^2)}\Big] + \frac{E_{\text{gap}}}{2}\left[\left(1 + \frac{2E_p\hbar^2K^2}{m_0E_{\text{gap}}^2}\right)^{1/2} - 1\right]$$

with $2E_p/m_0 = 21.3$ eV [8], $A = -7.65$, $B = -4.82$, and $|C| = 7.71$ [9]. The corresponding temperatures of electrons and holes are given by

$$E_1 = \frac{3}{2}\,k_BT_1\,\frac{F_{3/2}(\eta, T_1)}{F_{1/2}(\eta, T_1)} \quad \text{and} \quad E_2 = \frac{3}{2}\,k_BT_2\,,$$

respectively, with $F_{i/2}(\eta, T)$ for the Fermi functions.

The screening parameter used for electron–hole collisions as well as electron–phonon interaction is determined by the contributions of electrons and holes and has the form

$$q_D^2 = \frac{4\pi ne^2}{\varepsilon}\left[\frac{F_{-1/2}(\eta, T_1)/F_{1/2}(\eta, T_1)}{k_BT_1} + \frac{1}{k_BT_2}\right].$$

Besides $\langle\partial E/\partial t\rangle_{eh}$ the other expressions for the energy changes are obtained by integration of (A5) to (A8) over $f(E_1)\,[1 - f(E_1')]\,E_1^{1/2}\,dE_1$ as usual.

In Fig. 1 the calculated data for the time dependence of the electron and hole temperatures are shown for an excitation by $h\nu - E_{\text{gap}} = 0.13$ eV, i.e. $T_1(0) \approx 810$ and $T_2(0) \approx 120$ K, with carrier concentration as parameter. The curves demonstrate that an initial heating of the holes up to the electron temperature as accepted in [1, 4 to 6, 10] is lacking. On the contrary T_1 and T_2 significantly differ throughout the whole relaxation process. Even for the highest concentration regarded here only a small increase of T_2 can be remarked. It is to be noticed that for $T_1 \approx 2T_2$ the first-order contribu-

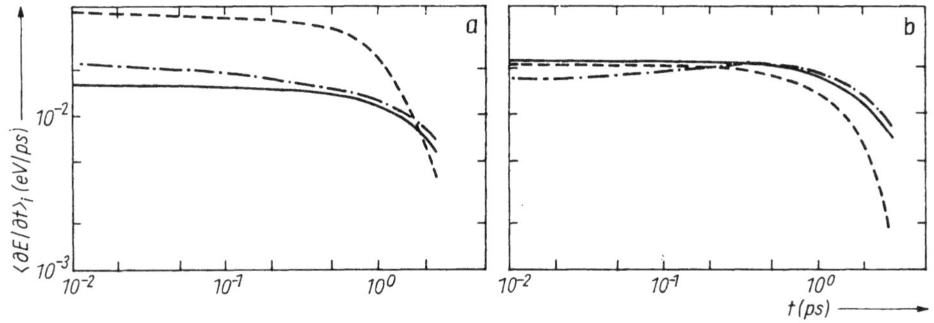

Fig. 2. Dependence of $\langle \partial E/\partial t\rangle_i$ on time for the interactions of electrons and holes (———), electrons and optical phonons (– – – –), holes and optical phonons (—·—·—) for a) $n = 3 \times 10^{17}$ and b) 1×10^{18} cm^{-3} carriers

tion grows up to 25% of the whole term for $\langle \partial E/\partial t\rangle_{eh}$ and in the region of $T_1 < 2T_2$ (6) is not valid anymore.

The changes of the electron cooling rates with increasing carrier concentration are determined by the balance of the growing screening of the electron–phonon interaction processes and the rising energy delivered to the holes. Fig. 2 a and b represent the values of $\langle \partial E/\partial t\rangle_{eh}$ together with $\langle \partial E/\partial t\rangle_{opt}$ for polar and deformation potential scattering for $n = 3 \times 10^{17}$ and 10^{18} cm^{-3}, respectively. The energy delivered to acoustic phonons by piezoelectric and deformation potential interactions is not remarkable for this region considered. For 3×10^{17} cm^{-3} electrons and holes the energy delivered by the electrons to polar phonons is the dominating loss mechanism down to about 1 ps, in contrast to the case of equal electron and hole temperatures, when the holes determine the energy loss of the plasma. The energy obtained by the holes from the electrons is of the order of the energy transfer of the holes to optical phonons both by polar and deformation potential scattering, but always remains lower and therefore a slow but steady cooling of the holes results. For 10^{18} cm^{-3} electrons and holes, on the other hand, the energy transfer from the electrons to the holes is already equal or higher than the energy loss of the electrons to the phonons as the polar interaction is screened very strongly. With regard to the holes the energy gained from the electrons slightly surmounts the loss to optical phonons below 0.4 ps and leads to the small hole heating depicted in Fig. 1 for this concentration. For this case the behaviour is investigated in

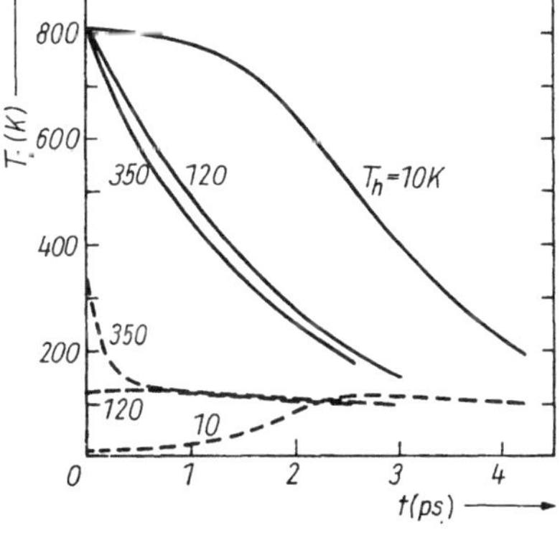

Fig. 3. Carrier temperatures versus time in dependence on $T_2(0)$ as indicated for 1×10^{18} cm^{-3} carriers (upper curves for electrons, lower curves for holes)

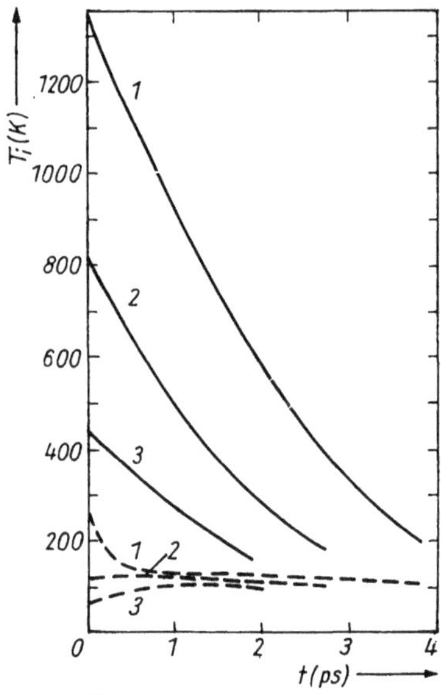

Fig. 4

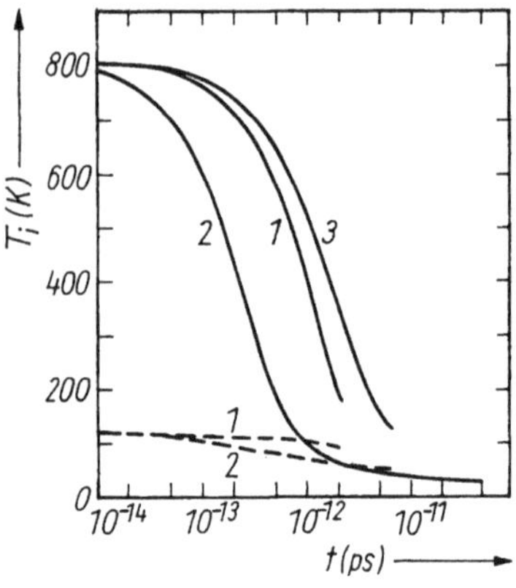

Fig. 5

Fig. 4. Carrier temperatures in dependence on time for different excitation energies $h\nu - E_{gap}$: (1) 0.04, (2) 0.13, (3) 0.2 eV (upper curves for electrons, lower curves for holes)

Fig. 5. Carrier temperatures versus time (log scale) after short pulse excitation for 3×10^{17} cm^{-3} electrons and holes. (1) Electron–hole interaction taken properly into account, (2) unique plasma temperatures, (3) decoupled electron and hole relaxation (upper curves for electrons, lower curves for holes)

more detail changing the initial hole temperatures. Fig. 3 demonstrates that this influence leads to no qualitative differences, because the holes either quickly cool down from 350 to about 120 K or are slowly heated up from 10 to 130 K in accordance with the results for an initial temperature of 120 K. Of course, the cooling rate of the electrons exhibits a strong influence of the hole temperature on the screening effect. On the other hand, Fig. 4 illustrates the cooling process according to different excitation energies $h\nu - E_{gap}$ equal to 0.04, 0.13, and 0.2 eV corresponding to pairs of electron and hole initial temperatures of about 430 and 64 K, 810 and 120 K, 1370 and 270 K, respectively. Qualitatively the features do not differ and again demonstrate that it is not realistic to assume equal electron and hole temperatures, but that the interaction of both types of particles has to be taken into account properly.

In order to compare the presented numerical results with the corresponding data obtained for cooling calculated with equal electron and hole temperatures and for decoupled energy relaxation of electrons and holes (but taking the simultaneous screening by both types of carriers into acccount furtheron) Fig. 5 exhibits the behaviour calculated for the same initial condition in the case of 3×10^{17} cm^{-3} carrier pairs. It is evident that the cooling rate is strongly reduced for the electron–hole interaction calculated according to the formulas presented in this paper in comparison to a proposed ab initio rise of the hole temperature to the values of the electron temperature. On the other hand, for the independent cooling process of both types of carriers the cooling rate of the electrons is about two times slower for this example, however, the difference increases with rising carrier concentration, of course.

4. Conclusion

The experimental data published up to now are related to much longer laser pulses than assumed here and the measurements were performed only after an already significant relaxation of the excited plasma. In order to compare the calculated data with the experimentally determined temporal change of the electron temperature one has to depart from the simple model used in the present paper. Instead of an instantaneous excitation the energy change already during the excitation pulse has to be included. Furtheron the electron–hole interaction in the region of $T_{el} \gtrapprox T_h$ has to be enumerated more precisely. Besides the reabsorption of the nonequilibrium phonons has to be taken into account, as in [6] its influence is clearly demonstrated, but its role will be reduced by a correct treatment of electron–hole interaction instead of an equal plasma temperature.

Appendix

The aim of the appendix is to compare the roles of emission of phonons and of carrier–carrier collisions for carriers monoenergetically excited and isotropically distributed in $\boldsymbol{K}$ space. Of course, for such a distribution $\langle \partial E/\partial t \rangle_{ii}$ equals zero for electron–electron

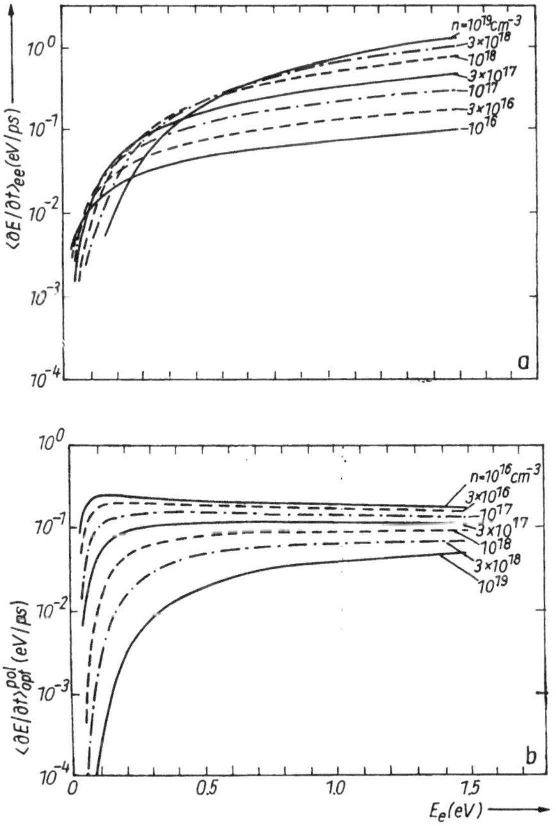

Fig. 6. Dependence of a) $\langle \partial E/\partial t \rangle_{ee}$ and b) $\langle \partial E/\partial t \rangle^{pol}_{opt}$ on energy for electron concentrations as indicated

and hole–hole scattering, because a carrier experiences the same amount of scattering events with energy losses as with energy gains. Therefore we use

$$\left\langle\frac{\partial E_1}{\partial t}\right\rangle_{ii} = \frac{1}{2(2\pi)^6} \iiiint |E_{K_1} - E_{K_1'}|\, w(K_1, K_1', K_2, K_2')\, \delta(E_{K_1} - E_1) \times \times \delta(E_{K_2} - E_2)\, dK_1\, dK_1'\, dK_2\, dK_2' \times \times \left[\int \delta(E_{K_1} - E_1)\, dK_1 \int \delta(E_{K_2} - E_2)\, dK_2\right]^{-1} \quad \text{(A1)}$$

instead of (1) in the case of electron–electron and hole–hole interaction. Trivial but rather tiresome calculations lead to

$$\left\langle\frac{\partial E_1}{\partial t}\right\rangle_{ii} = \frac{\pi e^4 n}{2\varepsilon^2 \sqrt{2m_i E_{D_i}}} \left\{\pi - 2 \arcsin \frac{1}{\sqrt{4E/E_{D_i} + 1}} + \right. \left. + \sqrt{\frac{E_{D_i}}{E}} \left[3 - 2 \ln\left(1 + \frac{4E}{E_{D_i}}\right) - \frac{3E_{D_i}}{4E}\left(\pi - 2 \arcsin \frac{1}{\sqrt{4E/E_{D_i} + 1}}\right)\right]\right\}. \quad \text{(A2)}$$

For electron–hole collision the change for $v_1 > v_2$ is

$$\left\langle\frac{\partial E_1}{\partial t}\right\rangle_{12} = \frac{\pi e^4 n_2}{\varepsilon^2 m_2 v_1} \left\{\ln\left[\frac{(v_1 + v_2)^2 + v_0^2}{v_0^2}\, \frac{(v_1 - v_2)^2 + v_0^2}{v_0^2}\right] + \right. + 3\, \frac{m_2 - m_1}{m_1} - \frac{m_2 v_1}{m_1 v_2} \ln\left[\frac{(v_1 + v_2)^2 + v_0^2}{(v_1 - v_2)^2 + v_0^2}\right] - \left. - \frac{1}{2m_1 v_2}\left[3v_0(m_2 - m_1) + \frac{v_2^2 - v_1^2}{v_0}(m_2 + m_1)\right] \operatorname{arctg} \frac{2v_2 v_0}{v_1^2 - v_2^2 + v_0^2}\right\} \quad \text{(A3)}$$

or for $v_1 < v_2$

$$\left\langle\frac{\partial E_1}{\partial t}\right\rangle_{12} = \frac{\pi e^4 n_2}{\varepsilon^2 m_2 v_1} \left\{\ln\left[\frac{(v_1 + v_2)^2 + v_0^2}{(v_1 - v_2)^2 + v_0^2}\right] + 3\, \frac{(m_2 - m_1)\, v_1}{m_1\, v_2} - \right. - \frac{m_2 v_1}{m_1 v_2} \ln\left[\frac{(v_1 + v_2)^2 + v_0^2}{v_0^2}\, \frac{(v_1 - v_2)^2 + v_0^2}{v_0^2}\right] - \left. - \frac{1}{2m_1 v_2}\left[3v_0(m_2 - m_1) + \frac{v_2^2 - v_1^2}{v_0}(m_2 + m_1)\right] \operatorname{arctg} \frac{2v_1 v_0}{v_2^2 - v_1^2 + v_0^2}\right\} \quad \text{(A4)}$$

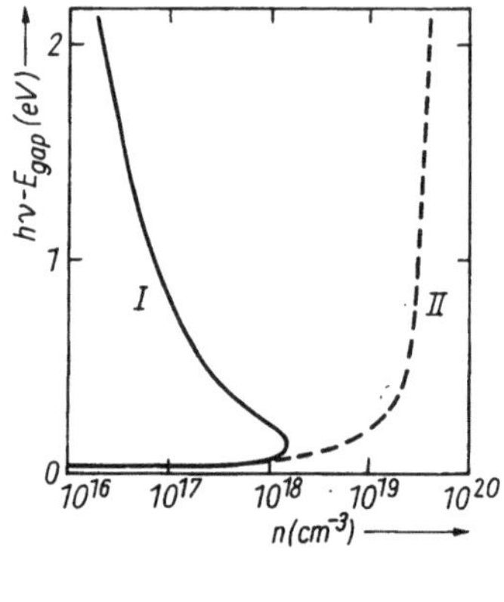

Fig. 7. Excitation energies $h\nu - E_{gap}$ for equal $\hbar\omega/\langle\partial E/\partial t\rangle_{opt}$ and $\hbar\omega/\langle\partial E/\partial t\rangle_{ee}$ versus electron concentration (———) and of equal $\hbar\omega/\langle\partial E/\partial t\rangle_{opt}$ and $E/\langle\partial E/\partial t\rangle_{ee}$ (– – –)

with $v_0 = (m_1 + m_2)\,\hbar/2m_1m_2\lambda_D$. This equation is written with v_i instead of E_i in order to obtain a more transparent form. For the other scattering processes the energy changes read

$$\left\langle\frac{\partial E_i}{\partial t}\right\rangle_{\rm opt}^{\rm pol} = \frac{\sqrt{m_i}\,e^2\omega^2}{2\sqrt{2}}\left(\frac{1}{\varepsilon_\infty}-\frac{1}{\varepsilon_0}\right)\frac{1}{\sqrt{E}}\left\{(N+1)\left[\ln\frac{(\sqrt{E}+\sqrt{E-\hbar\omega})^2+E_{D_i}}{(\sqrt{E}-\sqrt{E-\hbar\omega})^2+E_{D_i}}+\right.\right.$$
$$\left.+E_{D_i}\left(\frac{1}{(\sqrt{E}+\sqrt{E-\hbar\omega})^2+E_{D_i}}-\frac{1}{(\sqrt{E}-\sqrt{E-\hbar\omega})^2+E_{D_i}}\right)\right]-$$
$$-N\left[\ln\frac{(\sqrt{E}+\sqrt{E+\hbar\omega})^2+E_{D_i}}{(\sqrt{E}-\sqrt{E+\hbar\omega})^2+E_{D_i}}+\right.$$
$$\left.\left.+E_{D_i}\left(\frac{1}{(\sqrt{E}+\sqrt{E+\hbar\omega})^2+E_{D_i}}-\frac{1}{(\sqrt{E}-\sqrt{E+\hbar\omega})^2+E_{D_i}}\right)\right]\right\}, \tag{A5}$$

$$\left\langle\frac{\partial E_i}{\partial t}\right\rangle_{\rm opt}^{\rm def} = \frac{m_i^{3/2}D_i^2}{\sqrt{2}\,\pi\hbar^2\varrho}\left\{(N_0+1)\sqrt{E-(\hbar\omega_0+\Delta_i)}-N_0\sqrt{E+(\hbar\omega_0+\Delta_i)}\right\} \tag{A6}$$

with $\Delta_e = \Delta_{\Gamma L}$, $\Delta_h = 0$,

$$\left\langle\frac{\partial E_i}{\partial t}\right\rangle_{\rm ac}^{\rm def} = \frac{2\sqrt{2}\,m_i^{5/2}\,\Xi_i^2}{\pi\varrho\hbar^4}\,E^{3/2}\left(1-2\,\frac{k_BT_0}{E}\right), \tag{A7}$$

$$\left\langle\frac{\partial E_i}{\partial t}\right\rangle_{\rm ac}^{\rm piezo} = \frac{32\sqrt{2}\,\pi m_i^{3/2}}{15\varrho\hbar^2}\left(\frac{\beta e}{\varepsilon_0}\right)^2 E^{1/2}\left[\frac{1+\dfrac{E_{D_i}}{2E}}{1+\dfrac{E_{D_i}}{4E}}-\frac{E_{D_i}}{2E}\ln\left(1+\frac{4E}{E_{D_i}}\right)\right] \tag{A8}$$

with $E_{D_i} = \hbar^2q_D^2/2m_i$.

For the screening parameter the expression [11]

$$q_D^2 = -4\pi e^2\sum_{\boldsymbol K}\left\{\frac{1}{E_{\boldsymbol K}-E_{\boldsymbol K+\boldsymbol Q}+\hbar\omega-i\hbar\alpha}+\frac{1}{E_{\boldsymbol K}-E_{\boldsymbol K-\boldsymbol Q}-\hbar\omega+i\hbar\alpha}\right\}f(K)$$

was used. For long-range static scattering, i.e. $Q\to 0$, and with $f(\boldsymbol K_i) = C_i\,\delta(E_{\boldsymbol K_i}-E_i)$ for both types of carriers we get

$$q_D^2 = \frac{2\pi e^2}{\varepsilon}\left(\frac{n}{E_e}+\frac{p}{E_h}\right).$$

Fig. 6 a and b demonstrate the dependences of $\langle\partial E/\partial t\rangle_{ee}$ and $\langle\partial E/\partial t\rangle_{\rm opt}^{\rm pol}$ on the energy of the excited electrons with their concentration as parameter. As to be seen for both scattering mechanisms the behaviour is determined by the pronounced influence of screening superimposed on the explicit dependence on electron energy. It is to be remembered that the screening is preponderantly determined by the holes with their much smaller energies in comparison to the electrons. The calculated values for the other processes are much lower and not shown here.

In Fig. 7 those carrier concentrations and energies of exciting irradiation are shown by a solid line, for which the energy changes in unit time by both processes specified in

Fig. 6 are the same. For $h\nu - E_{gap} = E_e + E_h < \hbar\omega + E_h$, of course, no emission of optical phonons is possible, but for $E_e \geqq \hbar\omega$ this process becomes dominating for $1/\langle\partial E/\partial t\rangle$ up to carrier concentrations of the order of 10^{18} cm^{-3} and the relaxation of electrons is cascade-like in the region I, where the smearing of the electrons over the states is small in comparison to energy losses by emission. It seems to be astonishing that for high excitation energies the e–e collision again becomes dominating over the phonon emission, but this is caused by the stronger decrease of screening with rising energy for e–e collisions in comparison to polar interaction of electrons and phonons.

By the dashed curve the limiting condition is indicated for which the time for the smearing over the whole electron energy $E/\langle\partial E/\partial t\rangle_{ee}$ is equal to $\hbar\omega/\langle\partial E/\partial t\rangle^{pol}_{opt}$. Within region II a Fermi distribution can be used as approximation. In the region between I and II the distribution is more complicated and should be determined by the time-dependent Boltzmann equation.

References

[1] J. SHAH, J. Physique C **7**, 445 (1981).
[2] H. EHRENREICH, J. Phys. Chem. Solids 8, 130 (1959).
[3] E. J. YOFFA, Phys. Rev. B **23**, 1909 (1981).
[4] M. PUGNET, J. COLLET, and A. CORNET, Solid State Commun. **38**, 531 (1981).
[5] W. GRAUDSZUS and E. O. GÖBEL, Physica (Utrecht) **117B**, 555 (1983).
[6] W. PÖTZ and P. KOCEVAR, Phys. Rev. B **28**, 7040 (1983).
[7] F. SEITZ, Phys. Rev. **73**, 549 (1948).
[8] R. ULBRICH, Phys. Rev. B 8, 5719 (1978).
[9] P. LAWAETZ, Phys. Rev. B **4**, 3460 (1971).
[10] D. VAN DER LINDE and R. LAMBRICH, Phys. Rev. Letters **42**, 1090 (1978).
[11] J. M. ZIMAN, Principles of the Theory of Solids, Chap. 5, Cambridge University Press, 1964.

(Received September 12, 1984)

phys. stat. sol. (b) **126**, 617 (1984)
Subject classification: 17; 13.3

Fachbereich Physik der Universität Osnabrück[1])

Relativistic Theory of Photoemission from Solids

I. Key Aspects of the Model

By

G. THÖRNER and G. BORSTEL

The key aspects known to be important in photoemission: matrix elements, scattering of the outgoing wave, band structure, and surface effects are formulated within a relativistic single-particle theory with the objective of developing a workable relativistic model for photoemission from magnetic and nonmagnetic solids.

Die für Photoemission wichtigen Einzelaspekte: Matrixelemente, Streuung der auslaufenden Welle, Bandstruktur und Oberflächeneffekte werden im Rahmen einer relativistischen Einteilchentheorie behandelt. Ziel ist die Entwicklung eines relativistischen Modells für Photoemission aus magnetischen und unmagnetischen Festkörpern.

1. Introduction

Angle-resolved ultraviolet photoelectron spectroscopy (ARUPS) within the last ten years has become one of the most powerful experimental methods for investigating the electronic properties of clean and adsorbate-covered surfaces [1 to 4]. The success of this technique may primarily be attributed to its high surface sensitivity and to the fact that photoemission data may be interpreted surprisingly well in the framework of single-particle theories. Recently ARUPS has been supplemented by the new technique of ultraviolet bremsstrahlung isochromat spectroscopy (UBIS) [5, 6] which utilizes the inverse photoemission process and allows for probing the electronic structure in an energy range which is not accessible in ARUPS.

In order to interpret experimental ARUPS or UBIS data it is necessary to have a reliable theoretical model to calculate the emission spectra. For clean crystals such interpretations in the past have been given mostly within the three-step model of photoemission [7, 8]. In this model the photoemission process is divided into three independent steps, namely (i) excitation in the bulk crystal, (ii) transport of the photoelectron to the surface, and (iii) transmission through the surface into the vacuum. The conceptual simplicity of this model is obtained by breaking the total coherent quantum process of photoemission from solids into three incoherent separate parts which then are calculated independently of each other. Such a separation into independent processes can be valid only if the lifetime of the photoexcited electron is sufficiently long. Furthermore, due to the assumed independence of the different steps, in particular the subsequent steps (ii) and (iii) of the model may be calculated only on the basis of rather crude approximations. These severe restrictions have led to the development of more sophisticated one-step theories of photoemission from solids [9, 10]. Here photoemission is treated as a single coherent quantum process. Energies and wave functions are determined taking into account explicitly the presence of the surface. From these wave functions of the semi-infinite crystal the photocurrent is then calculated utilizing Fermi's golden-rule formula. For clean sur-

[1]) Postfach 4469, D-4500 Osnabrück, FRG.

faces the first calculations within a one-step model were published by Pendry and Titterington [11] and by Jepsen [12].

These one-step theories so far are all based on the Schrödinger equation, i.e. relativistic corrections are neglected. This restriction has severe consequences for the applicability of the model: (i) ARUPS measurements have demonstrated that even for nonmagnetic materials with low atomic number Z like copper ($Z = 29$) effects due to spin–orbit interaction are present in the photoemission spectra [13]. As the influence of relativistic corrections increases with Z, theoretical ARUPS spectra, calculated within the Schrödinger theory, will become more and more inadequate for interpreting photoemission data. The same holds for magnetic materials like nickel ($Z = 28$) where the exchange splitting of the energy bands is comparable to the spin–orbit splitting. (ii) Emission of spin-polarized photoelectrons from nonmagnetic crystals following the excitation with circularly polarized light is entirely dependent on spin–orbit interaction in the solid [14]. This process, which has led to the development of a powerful source for spin-polarized electrons (GaAs) [15], may be described approximately, just as ordinary ARUPS, by a three-step model, which incorporates the spin dependence within the Pauli formalism [16]. But due to its conceptual limitations and especially due to the impossibility to describe the spin dependence of the transmission step (iii) properly within this model, its virtue for a quantitative interpretation of the coming generation of spin-resolved ARUPS experiments [17] is rather restricted.

The aim of the present contribution is therefore to develop a relativistic one-step model of photoemission. This first part of the theory is devoted to those key aspects which are characteristic of the one-step model. The coupling of these key steps within one coherent process, which will then result in the final formula for calculating the photocurrent, will be described in the forthcoming second part of this contribution. The following Section 2 deals with the description of the excitation process for a single atom within the Dirac formalism. The matrix elements for electric dipole transitions are calculated for two equivalent representations of the interaction operator, namely the "$\boldsymbol{\alpha}$-form", which is inherently relativistic, and the "$\boldsymbol{r}$-form", which in the nonrelativistic limit gives just the familiar length form for the transition matrix elements. In Section 3 we give an account of relativistic scattering theory for the excited electron in the solid. This theory originally has been developed for calculating spin-dependent low-energy electron diffraction (LEED) [18] at solid surfaces and is therefore not new. With respect to photoemission it is thus only necessary to reformulate this relativistic LEED formalism in such a way that its connection with a one-step model of photoemission becomes more transparent. Section 4 is devoted to the transmission of the photoelectron through the surface barrier. In previous calculations within the nonrelativistic one-step model this part of the potential usually has been assumed to exhibit a step-like behaviour. The actual form of this barrier is in fact not crucial as long as one is primarily interested in bulk emission. For emission from surface states, however, this point is clearly of large importance. In view of very recent UBIS experiments [19, 20] which prove the existence of unoccupied surface states forming a Rydberg series just below the vacuum level of the solid, we will discuss in this section more realistic barriers which in particular exhibit the $1/4z$ image potential behaviour for large distances z from the surface.

2. Optical Excitation in a Single Crystal Atom

This section deals with the basic problem of photoemission, namely the calculation of optical transition matrix elements. The physical importance of this process is reflected in the following two points: Firstly, the selection rules for optical excitation of electrons have shown to be extremely useful for interpreting experimental ARUPS and UBIS data [4, 21, 22]. Secondly, for excitation with circularly polarized light spin-orbit interaction may give rise to a highly selective population in the spin states of the photoelectron even in nonmagnetic systems [14], which leads to the possibility to develop effective sources for spin-polarized electrons.

In the relativistic theory the interaction of a single electron with classical monochromatic electromagnetic radiation is described (the Coulomb gauge and relativistic units $m = \hbar = c = 1$ are used) by the operator [23]

$$H = -e\boldsymbol{\alpha}\boldsymbol{A} \exp(i\boldsymbol{q} \cdot \boldsymbol{r} - i\omega t) . \tag{1}$$

Herein $e < 0$ denotes the electron charge and $\boldsymbol{A}$ the amplitude of the electromagnetic vector potential with wave vector $\boldsymbol{q}$ and frequency ω. The three components α_k of the vector $\boldsymbol{\alpha}$ are (4×4) Dirac matrices which are defined through the tensor product

$$\alpha_k = \sigma_1 \otimes \sigma_k ; \qquad k = 1, 2, 3 , \tag{2}$$

where

$$\sigma_1 = \begin{pmatrix} 0 & 1 \\ 1 & 0 \end{pmatrix}, \qquad \sigma_2 = \begin{pmatrix} 0 & -i \\ i & 0 \end{pmatrix}, \qquad \sigma_3 = \begin{pmatrix} 1 & 0 \\ 0 & -1 \end{pmatrix} \tag{3}$$

denote the (2×2) Pauli spin matrices. As H refers to a single electron, our description will proceed in the framework of a relativistic single-particle theory for the many-electron system (see below). In what follows we assume that the vector potential amplitude $\boldsymbol{A}$ inside the solid is fully determined by classical macroscopic dielectric theory, i.e. local-field corrections due to the inhomogeneity of the electronic system are neglected. In the visible and ultraviolet spectral region, where $\boldsymbol{q} \cdot \boldsymbol{r} \ll 1$, one then can use the electric dipole approximation and from Fermi's golden-rule formula the transition probability density per unit time P_{fi} for a transition $|\mathrm{i}\rangle \to |\mathrm{f}\rangle$ is then obtained in the form ($\boldsymbol{\alpha}$-form)

$$P_{\mathrm{fi}} = 2\pi e^2 \, |\langle \varphi_{\mathrm{f}} | \, \boldsymbol{\alpha}\boldsymbol{A} \, | \varphi_{\mathrm{i}} \rangle|^2 \, \delta(E_{\mathrm{f}} - E_{\mathrm{i}} - \omega) \, \varrho(E_{\mathrm{f}}) . \tag{4}$$

Herein φ_{f} and φ_{i} denote the four-component eigenspinors of the unperturbed relativistic Dirac Hamiltonian H_{D} corresponding to the final state energy E_{f} and the initial state energy E_{i}, respectively. $\varrho(E_{\mathrm{f}})$ is the density of final states.

The reduction of the quantum-mechanical many-electron problem to a set of single-particle equations can be done in the spirit of relativistic density functional theory. Following Rajagopal [24] and MacDonald and Vosko [25], Strange et al. [26] have recently shown that the relevant Kohn-Sham-Dirac single-particle equations for the ground state of an inhomogeneous relativistic electron system in the presence of external fields may be written, neglecting diamagnetic effects, as

$$H_{\mathrm{D}}\varphi_j = (\boldsymbol{\alpha} \cdot \boldsymbol{p} + \beta + V^{\mathrm{eff}}[n, \boldsymbol{m}] \, 1_4 + \beta\tilde{\boldsymbol{\sigma}} \cdot \boldsymbol{B}^{\mathrm{eff}}[n, \boldsymbol{m}]) \, \varphi_j = \varepsilon_j\varphi_j \tag{5}$$

with $\boldsymbol{p} = -i\nabla$, $\beta = \sigma_3 \otimes 1_2$, $\tilde{\sigma}_k = 1_2 \otimes \sigma_k$. The term $\tilde{\boldsymbol{\sigma}} \cdot \boldsymbol{B}^{\mathrm{eff}}$ may be thought as arising from the coupling of the electron spin to an external magnetic field and to an effective space-dependent internal magnetic field due to spin correlations in the electron system. It depends on the number density $n(\boldsymbol{r})$ of the electrons and on the spin density vector $\boldsymbol{m}(\boldsymbol{r})$. For nonmagnetic systems in zero external magnetic field this term vanishes. The electrostatic effective potential V^{eff}, which also depends on $n(\boldsymbol{r})$ and $\boldsymbol{m}(\boldsymbol{r})$, describes the Coulomb interaction with the nuclei and all electrons plus the exchange-correlation interaction in the inhomogeneous polarized electron system.

Equation (5) is strictly valid only for the ground state of the many-electron system and the correct physical interpretation of the energy parameters ε_j, which enter the theory as Lagrange multipliers, is still an open question [27]. If, like in photoemission, one is primarily interested in quasi-particle properties, (5) must be modified by contributions due to the complex nonlocal energy-dependent self-energy. In a local approximation (5) may then be maintained by assuming that both V^{eff} and $\boldsymbol{B}^{\text{eff}}$ are complex and energy-dependent and by identifying the ε_j with the quasi-particle energies E_j [27 to 29].

Coming back now to the transition matrix element $\langle\varphi_{\text{f}}|\, \boldsymbol{\alpha} \cdot \boldsymbol{A}\, |\varphi_{\text{i}}\rangle$ in (4), we notice that one may derive an equivalent expression which is just the relativistic analogue of the well-known nonrelativistic length form for the matrix element ($\boldsymbol{r}$-form): Since both φ_{f} and φ_{i} are eigenstates of H_{D}, the Heisenberg equation of motion

$$\boldsymbol{\alpha} = -i[\boldsymbol{r}, H_{\text{D}}] \tag{6}$$

for the relativistic velocity operators $\boldsymbol{\alpha}$ results in

$$\langle\varphi_{\text{f}}|\, \boldsymbol{\alpha}\boldsymbol{A}\, |\varphi_{\text{i}}\rangle = i(E_{\text{f}} - E_{\text{i}})\, \langle\varphi_{\text{f}}|\, \boldsymbol{A}\boldsymbol{r}\, |\varphi_{\text{i}}\rangle\,. \tag{7}$$

As the operator $\boldsymbol{r}$ has no matrix elements linking the small and large components of a Dirac spinor, in that representation the correspondence to nonrelativistic theory is most clearly seen. There exists, however, a caveat for a straight-forward application of (7) in actual calculations: If the effects of the energy-dependent finite lifetime of the quasiparticles are included in (5) via an energy-dependent absorptive part in V^{eff} and $\boldsymbol{B}^{\text{eff}}$, φ_{f} and φ_{i} are no more eigenfunctions belonging to the same Hamiltonian and (7) therefore is not valid. In actual calculations the original $\boldsymbol{\alpha}$-form for the matrix elements is thus preferable.

If $\boldsymbol{B}^{\text{eff}}$ in (5) is assumed to be along the z-axis, the term $\tilde{\boldsymbol{\sigma}} \cdot \boldsymbol{B}^{\text{eff}}$ becomes $\tilde{\sigma}_3 B^{\text{eff}}$. The general solution of (5) for an atom then has the form [30]

$$\varphi(\boldsymbol{r}) = \sum_{\varkappa,\mu} \frac{1}{r} \begin{pmatrix} f_\varkappa^\mu(r) & \chi_\varkappa^\mu(\hat{r}) \\ i g_\varkappa^\mu(r) & \chi_{-\varkappa}^\mu(\hat{r}) \end{pmatrix}. \tag{8}$$

Herein $\chi_\varkappa^\mu(\hat{r})$ are the two-component spin orbitals [31],

$$\chi_\varkappa^\mu(\hat{r}) = \sum_{\sigma=\pm 1/2} C(l\,\tfrac{1}{2}\,j;\, \mu - \sigma, \sigma)\; Y_l^{\mu-\sigma}(\hat{r})\, \chi^\sigma\,, \tag{9a}$$

with χ^σ denoting the two Pauli basis spinors. The quantum number $\varkappa$ is related to the orbital angular momentum quantum number l by

$$l = \begin{cases} \varkappa & \text{for } \varkappa > 0\,, \\ -\varkappa - 1 & \text{for } \varkappa < 0\,. \end{cases} \tag{10a}$$

For fixed value of $\varkappa$, the quantum number μ takes one of the values $-j$, $-j + 1, \dots$ $\dots, j - 1, j$, where j is the total angular momentum number. The dependence of the radial wave functions $f(r)$ and $g(r)$ on μ is a result of the fact that the term $\tilde{\sigma}_3 B^{\text{eff}}$ breaks the spherical symmetry of H_{D}, and is therefore characteristic of magnetic systems.

Using for φ_{f} and φ_{i} the ansatz according to (8), the matrix elements in $\boldsymbol{\alpha}$-form become

$$\langle\varphi_{\text{f}}|\, \boldsymbol{\alpha}\boldsymbol{A}\, |\varphi_{\text{i}}\rangle = \sum_{\varkappa\mu} \sum_{\varkappa'\mu'} (M_{\varkappa\mu,\varkappa'\mu'} + N_{\varkappa\mu,\varkappa'\mu'})\,, \tag{11}$$

where

$$M_{\varkappa\mu,\varkappa'\mu'} = \int \mathrm{d}r\, f_{\varkappa,\text{f}}^\mu(r)\, i g_{\varkappa',\text{i}}^{\mu'}(r) \int \mathrm{d}\Omega\, \chi_\varkappa^{\mu *} \boldsymbol{A}\boldsymbol{\sigma} \chi_{-\varkappa'}^{\mu'}\,, \tag{12a}$$

$$N_{\varkappa\mu,\varkappa'\mu'} = -\int \mathrm{d}r\, i g_{\varkappa,\text{f}}^\mu(r)\, f_{\varkappa',\text{i}}^{\mu'}(r) \int \mathrm{d}\Omega\, \chi_{-\varkappa}^{\mu *} \boldsymbol{A}\boldsymbol{\sigma} \chi_{\varkappa'}^{\mu'}\,. \tag{12b}$$

The integrations over the solid angle Ω in (12) may all be done analytically. Using (9a) for $\chi_\varkappa^\mu$ and the corresponding relationship for $\chi_{-\varkappa}^\mu$,

$$\chi_{-\varkappa}^\mu(\hat{r}) = \sum_{\sigma=\pm 1/2} C(\bar{l}\tfrac{1}{2}j;\, \mu-\sigma, \sigma)\, Y_{\bar{l}}^{\mu-\sigma}(\hat{r})\, \chi^\sigma\,, \tag{9b}$$

where $\bar{l}$ belongs to $-\varkappa$ according to

$$\bar{l} = \begin{cases} \varkappa - 1 & \text{for } \varkappa > 0\,, \\ -\varkappa & \text{for } \varkappa < 0\,, \end{cases} \tag{10b}$$

one obtains by means of the identities [31]

$$\sigma_1\chi^\sigma = \chi^{-\sigma}\,, \qquad \sigma_2\chi^\sigma = i(-1)^{1/2-\sigma}\,\chi^{-\sigma}\,, \qquad \sigma_3\chi^\sigma = (-1)^{1/2-\sigma}\,\chi^\sigma \tag{13}$$

the following results:

$$\begin{aligned}\int d\Omega\, \chi_\varkappa^{\mu *}\sigma_1\chi_{-\varkappa'}^{\mu'} = {} & C(l\tfrac{1}{2}j;\, \mu-\tfrac{1}{2}, \tfrac{1}{2})\, C(\bar{l}'\tfrac{1}{2}j';\, \mu'+\tfrac{1}{2}, -\tfrac{1}{2})\, \delta_{l\bar{l}'}\delta_{\mu-1,\mu'} + \\ & + C(l\tfrac{1}{2}j;\, \mu+\tfrac{1}{2}, -\tfrac{1}{2})\, C(\bar{l}'\tfrac{1}{2}j';\, \mu'-\tfrac{1}{2}, \tfrac{1}{2})\, \delta_{l\bar{l}'}\delta_{\mu+1,\mu'}\,,\end{aligned} \tag{14a}$$

$$\begin{aligned}\int d\Omega\, \chi_\varkappa^{\mu *}\sigma_2\chi_{-\varkappa'}^{\mu'} = {} & -iC(l\tfrac{1}{2}j;\, \mu-\tfrac{1}{2}, \tfrac{1}{2})\, C(\bar{l}'\tfrac{1}{2}j';\, \mu'+\tfrac{1}{2}, -\tfrac{1}{2})\, \delta_{l\bar{l}'}\delta_{\mu-1,\mu'} + \\ & + iC(l\tfrac{1}{2}j;\, \mu+\tfrac{1}{2}, -\tfrac{1}{2})\, C(\bar{l}'\tfrac{1}{2}j';\, \mu'-\tfrac{1}{2}, \tfrac{1}{2})\, \delta_{l\bar{l}'}\delta_{\mu+1,\mu'}\,,\end{aligned} \tag{14b}$$

$$\int d\Omega\, \chi_\varkappa^{\mu *}\sigma_3\chi_{-\varkappa'}^{\mu'} = \sum_{\sigma=\pm 1/2} (-1)^{1/2-\sigma}\, C(l\tfrac{1}{2}j;\, \mu-\sigma, \sigma)\, C(\bar{l}'\tfrac{1}{2}j';\, \mu'-\sigma, \sigma)\, \delta_{l\bar{l}'}\delta_{\mu\mu'}\,. \tag{14c}$$

The results for the integrals of type $\int d\Omega\, \chi_{-\varkappa}^{\mu *}\sigma_k\chi_{\varkappa'}^{\mu'}$ ($k = 1, 2, 3$) are obtained from those in (14) by interchanging the quantum numbers l and $\bar{l}$.

The derivation of the matrix elements in the $\boldsymbol{r}$-form (length form) proceeds along similar lines and is summarized in the Appendix.

3. Relativistic Multiple-Scattering Formalism

In this section a short outline of relativistic multiple scattering theory for electrons in ferromagnets is given. It is based on a relativistic layer-KKR formalism which has established itself as a very efficient scheme to calculate spin-dependent LEED intensities for semi-infinite crystals [18]. The KKR method makes use of the muffin-tin approximation, i.e. $V(\boldsymbol{r})$ and $B^{\text{eff}}(\boldsymbol{r})$ are assumed to be spherically symmetric inside a muffin-tin sphere and constant in the interstitial region. The total scattering process is then separated into three steps: (i) scattering by a single muffin-tin sphere, (ii) multiple scattering within a monoatomic crystal layer in the xy plane parallel to the surface (intra-layer scattering), and (iii) multiple scattering between layers, including the surface layer (inter-layer scattering). With the exception of step (i) which only recently has been extended to ferromagnetic crystal atoms [26, 30], the relativistic scattering formalism is well known from the theory of spin-dependent LEED. We thus may restrict ourselves to describe the scheme in a compact form, which allows for a straightforward incorporation into the relativistic photoemission theory.

If the general ansatz for $\varphi(\boldsymbol{r})$ according to (8) is substituted into the Dirac equation, one obtains for each value of μ and $\varkappa$ a system of coupled differential equations for the radial functions $f_\varkappa^\mu$, $g_\varkappa^\mu$, $f_{-\varkappa-1}^\mu$, and $g_{-\varkappa+1}^\mu$. This coupling of $f_\varkappa^\mu$, $g_\varkappa^\mu$ with $f_{-\varkappa-1}^\mu$ and $g_{-\varkappa+1}^\mu$ is due to the presence of the effective magnetic field $B^{\text{eff}}(r)$ and leads, via the small component $g_{-\varkappa+1}^\mu$ and the associated coupling of type $l \pm 2$, to a coupling of an infinite number of states, which all exhibit the same parity $(-1)^l$. Fortunately in actual magnetic systems effects originating from the coupling to the small component $g_{-\varkappa+1}^\mu$ turn out to be at least two orders of magnitude smaller than those originating from the coupling to the large component $f^{\mu}_{-\varkappa-1}$[30, 32]. Coupling to $g^{\mu}_{-\varkappa+1}$ therefore can safely be neglected. For each parity the infinite system of coupled differential equations then decomposes into subsystems of four coupled equations for the functions $f^{\mu}_{\varkappa}$, $g^{\mu}_{\varkappa}$, $f^{\mu}_{-\varkappa-1}$, $g^{\mu}_{-\varkappa-1}$, each subsystem thus being associated with one definite value of l. This means that for scattering from ferromagnetic crystal atoms the usual relativistic partial wave analysis [31] can be maintained and need only be modified by taking into account the coupling of $\varkappa$ to $-\varkappa - 1$, i.e. the mixing of the two states $j = l \pm \frac{1}{2}$.

The relativistic scattering calculation for a single ferromagnetic crystal atom thus proceeds as follows: Within the muffin-tin sphere there exist for every μ and every $l \neq 0$ two linearly independent radial solutions $R^{\mu i}_{l}(r)$ which behave regularly at the origin $r = 0$,

$$R^{\mu i}_{l}(r) = \frac{1}{r}\begin{bmatrix} f^{\mu i}_{\varkappa} \\ i g^{\mu i}_{\varkappa} \\ f^{\mu i}_{-\varkappa-1} \\ i g^{\mu i}_{-\varkappa-1} \end{bmatrix}; \qquad i = 1, 2\,. \tag{14}$$

For a given energy E and fixed μ and $l \neq 0$ the general solution inside the sphere

$$R^{\mu}_{l}(r) = a^{\mu 1}_{l} R^{\mu 1}_{l}(r) + a^{\mu 2}_{l} R^{\mu 2}_{l}(r) \tag{15}$$

must be matched at the sphere radius r_0 to the corresponding general radial solution in the interstitial region, which has the form

$$\begin{bmatrix} A^{\mu}_{\varkappa} \mathrm{j}_l(kr) + B^{\mu}_{\varkappa} \mathrm{h}^{(1)}_{l}(kr) \\ i\varepsilon k S_{\varkappa}[A^{\mu}_{\varkappa} \mathrm{j}_{\bar{l}}(kr) + B^{\mu}_{\varkappa} \mathrm{h}^{(1)}_{\bar{l}}(kr)] \\ A^{\mu}_{-\varkappa-1} \mathrm{j}_l(kr) + B^{\mu}_{-\varkappa-1} \mathrm{h}^{(1)}_{l}(kr) \\ i\varepsilon k S_{-\varkappa-1}[A^{\mu}_{-\varkappa-1} \mathrm{j}_{\bar{l}}(kr) + B^{\mu}_{-\varkappa-1} \mathrm{h}^{(1)}_{\bar{l}}(kr)] \end{bmatrix} \tag{16}$$

with $E = (k^2 + 1)^{1/2}$, $\varepsilon = 1/(E + 1)$, $S_{\varkappa} = \varkappa/|\varkappa|$, $\bar{l} = l - S_{\varkappa}$, where j_l denotes the spherical Bessel functions and $\mathrm{h}^{(1)}_{l}$ the spherical Hankel functions of the first kind. Matching (15) and (16) for $r = r_0$ then results in the desired relationship between the incident partial wave amplitudes $A^{\mu}_{\varkappa}$ and the scattered partial wave amplitudes $B^{\mu}_{\varkappa}$. By defining

$$\begin{aligned} &\boldsymbol{a} = \begin{pmatrix} a^{\mu 1}_{l} \\ a^{\mu 2}_{l} \end{pmatrix}, \quad \boldsymbol{A} = \begin{pmatrix} A^{\mu}_{\varkappa} \\ A^{\mu}_{-\varkappa-1} \end{pmatrix}, \quad \boldsymbol{B} = \begin{pmatrix} B^{\mu}_{\varkappa} \\ B^{\mu}_{-\varkappa-1} \end{pmatrix}, \\ &F = \begin{pmatrix} f^{\mu 1}_{\varkappa} & f^{\mu 2}_{\varkappa} \\ f^{\mu 1}_{-\varkappa-1} & f^{\mu 2}_{-\varkappa-1} \end{pmatrix}, \quad G = \begin{pmatrix} g^{\mu 1}_{\varkappa} & g^{\mu 2}_{\varkappa} \\ g^{\mu 1}_{-\varkappa-1} & g^{\mu 2}_{-\varkappa-1} \end{pmatrix}, \\ &J = \begin{pmatrix} \mathrm{j}_l & 0 \\ 0 & \mathrm{j}_l \end{pmatrix}, \quad \bar{J} = \begin{pmatrix} \mathrm{j}_{\bar{l}} & 0 \\ 0 & -\mathrm{j}_{l+S_{\varkappa}} \end{pmatrix}, \\ &H = \begin{pmatrix} \mathrm{h}^{(1)}_{l} & 0 \\ 0 & \mathrm{h}^{(1)}_{l} \end{pmatrix}, \quad \bar{H} = \begin{pmatrix} \mathrm{h}^{(1)}_{\bar{l}} & 0 \\ 0 & -\mathrm{h}^{(1)}_{l+S_{\varkappa}} \end{pmatrix}, \end{aligned} \tag{17}$$

and taking the argument of all functions at $r = r_0$, the matching condition is equivalent to the following two matrix equations:

$$\begin{aligned} &\frac{1}{r_0} F\boldsymbol{a} = J\boldsymbol{A} + H\boldsymbol{B}\,, \\ &\frac{1}{r_0} G\boldsymbol{a} = \varepsilon k S_{\varkappa}[\bar{J}\boldsymbol{A} + \bar{H} B]\,. \end{aligned} \tag{18}$$

These can be solved to yield

$$\boldsymbol{B} = T\boldsymbol{A}\,, \tag{19}$$

where the (2×2) scattering matrix T is given by

$$T = \begin{pmatrix} t_{\varkappa}^{\mu 1} & t_{\varkappa}^{\mu 2} \\ t_{-\varkappa-1}^{\mu 1} & t_{-\varkappa-1}^{\mu 2} \end{pmatrix} = [GF^{-1}H - \varepsilon k S_\varkappa \bar{H}]^{-1} [\varepsilon k S_\varkappa \bar{J} - GF^{-1}J]\,. \tag{20}$$

In the special case $l = 0$ (i.e. $\varkappa = -1$, $\mu = +\frac{1}{2}$ or $\mu = -\frac{1}{2}$) the scattering matrix T breaks down to a single number

$$t_{-1}^{\mu} = \frac{\mathrm{j}_0(kr_0)\, g_{-1}^{\mu}(r_0) + \varepsilon k \mathrm{j}_1(kr_0)\, f_{-1}^{\mu}(r_0)}{\mathrm{h}_0^{(1)}(kr_0) g_{-1}^{\mu}(r_0) + \varepsilon k \mathrm{h}_1^{(1)}(kr_0)\, f_{-1}^{\mu}(r_0)}\,. \tag{21}$$

For nonmagnetic systems ($B^{\text{eff}} = 0$) and $l \neq 0$ the states $j = l + \frac{1}{2}$ and $j = l - \frac{1}{2}$ are decoupled. The matrices F, G and the scattering matrix T then become diagonal and the diagonal elements of T are usually expressed in terms of the scattering phase shifts δ_l^+ $(j = l + \frac{1}{2})$ and δ_l^- $(j = l - \frac{1}{2})$ [31].

Once the single-site scattering matrix T has been calculated, the treatment of intra-layer and inter-layer scattering proceeds, due to the assumption that the effective potential is constant in the interstitial region, within the usual spin-dependent LEED formalism. For a crystal with one atom per unit cell intra-layer scattering is described by a scattering matrix M which connects the wave field incident from both sides on a single layer (chosen in the xy plane centred at $z = 0$) with the outgoing wave field. In a plane-wave representation one has within the Pauli representation (i.e. neglecting the small components) for the incident wave field the form

$$\sum_{\boldsymbol{g},s} \left[u_{\boldsymbol{g}s}^+ \, \mathrm{e}^{i\boldsymbol{k}_{\boldsymbol{g}}^+ \cdot \boldsymbol{r}} + u_{\boldsymbol{g}s}^- \, \mathrm{e}^{i\boldsymbol{k}_{\boldsymbol{g}}^- \cdot \boldsymbol{r}}\right] \chi^s; \qquad s = \pm\tfrac{1}{2} \tag{22}$$

with $\boldsymbol{k}_{\boldsymbol{g}}^{\pm} = (\boldsymbol{k}_{\boldsymbol{g}}^{\parallel}, \pm |k_{\boldsymbol{g}}^z|)$. Herein $\boldsymbol{g}$ refers to the surface reciprocal lattice vectors and $u_{\boldsymbol{g}s}^+$ and $u_{\boldsymbol{g}s}^-$ denote the plane-wave amplitudes for plane waves incident from the left-hand $(z < 0)$ and the right-hand $(z > 0)$ side, respectively. The outgoing wave field is written similiarly as

$$\sum_{\boldsymbol{g},s} \left[v_{\boldsymbol{g}s}^- \, \mathrm{e}^{i\boldsymbol{k}_{\boldsymbol{g}}^- \cdot \boldsymbol{r}} + v_{\boldsymbol{g}s}^+ \, \mathrm{e}^{i\boldsymbol{k}_{\boldsymbol{g}}^+ \cdot \boldsymbol{r}}\right] \chi^s\,, \tag{23}$$

where now $v_{\boldsymbol{g}s}^-$ and $v_{\boldsymbol{g}s}^+$ refer to waves outgoing to the left $(z < 0)$ and to the right $(z > 0)$, respectively. The connection between incoming and outgoing amplitudes is then given by

$$v_{\boldsymbol{g}s}^{\tau} = \sum_{\tau'\boldsymbol{g}'s'} M_{\boldsymbol{g}\boldsymbol{g}'ss'}^{\tau\tau'} u_{\boldsymbol{g}'s'}^{\tau'} \qquad (\tau = +, -)\,. \tag{24}$$

Herein the matrix elements $M_{\boldsymbol{g}\boldsymbol{g}'ss'}^{\tau\tau'}$ may be expressed completely in terms of structural parameters for the layer and by the single-site scattering matrix T for every l and μ, cf. (27) in [30]. In view of inter-layer scattering, the intra-layer scattering matrix M is transformed into a transfer matrix Q which relates the plane wave amplitudes $u_{\boldsymbol{g}s}^+$, $v_{\boldsymbol{g}s}^-$ on the left-hand side of the layer to the amplitudes $u_{\boldsymbol{g}s}^-$, $v_{\boldsymbol{g}s}^+$ on the right-hand side. With the conventions set up by Pendry (cf. (4.81) in [33]) one has

$$\begin{aligned}
Q^{\mathrm{I}}_{gg'ss'} &= M^{++}_{gg'ss'} \exp\left(i\boldsymbol{k}^+_g \frac{\boldsymbol{c}}{2} + i\boldsymbol{k}^+_{g'} \frac{\boldsymbol{c}}{2}\right),\\
Q^{\mathrm{II}}_{gg'ss'} &= M^{+-}_{gg'ss'} \exp\left(i\boldsymbol{k}^+_g \frac{\boldsymbol{c}}{2} - i\boldsymbol{k}^-_{g'} \frac{\boldsymbol{c}}{2}\right),\\
Q^{\mathrm{III}}_{gg'ss'} &= M^{-+}_{gg'ss'} \exp\left(-i\boldsymbol{k}^-_g \frac{\boldsymbol{c}}{2} + i\boldsymbol{k}^+_{g'} \frac{\boldsymbol{c}}{2}\right),\\
Q^{\mathrm{IV}}_{gg'ss'} &= M^{--}_{gg'ss'} \exp\left(-i\boldsymbol{k}^-_g \frac{\boldsymbol{c}}{2} - i\boldsymbol{k}^-_{g'} \frac{\boldsymbol{c}}{2}\right),
\end{aligned} \tag{25}$$

where $\boldsymbol{c}$ connects two equivalent atoms in consecutive layers and the origin is changed now from the centre of the layer to points half-way between layers. Besides being the key quantity in inter-layer scattering, the transfer matrix Q allows also for finding the band structure and Bloch waves of the ideal crystal. As shown by Pendry (cf. (4.6) in [33]), the eigenvalues λ of the matrix

$$\begin{bmatrix} Q^{\mathrm{I}}, & Q^{\mathrm{II}} \\ -[Q^{\mathrm{IV}}]^{-1} Q^{\mathrm{III}} Q^{\mathrm{I}}, & [Q^{\mathrm{IV}}]^{-1}[1 - Q^{\mathrm{III}} Q^{\mathrm{II}}] \end{bmatrix} \tag{26}$$

occur in the form $\lambda = \exp(i\boldsymbol{k} \cdot \boldsymbol{c})$, i.e. for given values of E and $\boldsymbol{k}^{\|}$ they determine $k^z(E)$ and thus give the band structure.

Inter-layer scattering is usually calculated with the layer-doubling method [33]: Let $Q(1)$ and $Q(2)$ denote the transfer matrices of two layers. Transmission and reflection of both layers together are then described by transfer matrices $Q(12)$ according to

$$\begin{aligned}
Q^{\mathrm{I}}(12) &= Q^{\mathrm{I}}(2)\,[1 - Q^{\mathrm{II}}(1)\, Q^{\mathrm{III}}(2)]^{-1}\, Q^{\mathrm{I}}(1)\,,\\
Q^{\mathrm{II}}(12) &= Q^{\mathrm{II}}(2) + Q^{\mathrm{I}}(2)\,[1 - Q^{\mathrm{II}}(1)\, Q^{\mathrm{III}}(2)]^{-1}\, Q^{\mathrm{II}}(1)\, Q^{\mathrm{IV}}(2)\,,\\
Q^{\mathrm{III}}(12) &= Q^{\mathrm{III}}(1) + Q^{\mathrm{IV}}(1)\,[1 - Q^{\mathrm{III}}(2)\, Q^{\mathrm{II}}(1)]^{-1}\, Q^{\mathrm{III}}(2)\, Q^{\mathrm{I}}(1)\,,\\
Q^{\mathrm{IV}}(12) &= Q^{\mathrm{IV}}(1)\,[1 - Q^{\mathrm{III}}(2)\, Q^{\mathrm{II}}(1)]^{-1}\, Q^{\mathrm{IV}}(2)\,.
\end{aligned} \tag{27}$$

By repeating this procedure one can obviously calculate the transfer matrices for a system of 2, 4, 8, ... layers. In order to obtain a convergent result, it is then only required that finite absorption is present in the interstitial region of constant effective potential.

4. Surface Barrier in Metals

The effective surface barrier potential for an photoelectron travelling from the crystal into the vacuum gives rise to the following effects: (i) The photoelectrons are diffracted at the surface, i.e. the surface barrier contributes to the multiple-scattering processes described in the last section; (ii) near the surface barrier there may exist electronic surface states, both occupied and unoccupied, and transitions from or into these states may contribute to the photocurrent; (iii) the wave functions of the bulk states will more or less be modified near the surface. Although not all these effects are equally sensitive to the particular form of the surface barrier, a detailed knowledge of the barrier is usually essential for a complete analysis of ARUPS and UBIS data.

From the discussion in Section 2 it is clear that within a local approximation to the self-energy the electrostatic interaction V^{eff} and the magnetic interaction $\tilde{\sigma}_3 B^{\mathrm{eff}}$ may be combined to a single spin- and energy-dependent potential $V^{\sigma}(E, \boldsymbol{r})$. In this context it must be mentioned that in particular for the surface barrier part V_s^σ of V^σ a question mark hangs over the validity of a local optical potential as assumed here. Harris and Jones [34] and Echenique and Pendry [35] have shown that the surface barrier potential is strongly velocity-dependent, but so far it is not known how important this and other departures from locality might be in actual calculations.

The surface barrier potential $V_s^\sigma(E, \boldsymbol{r})$ must show periodicity parallel to the surface, i.e.

$$V_s^\sigma(E, \boldsymbol{r}) = \sum_{\boldsymbol{g}} V_{s\boldsymbol{g}}^\sigma(E, z)\, e^{i\boldsymbol{g}\cdot\boldsymbol{r}_\parallel}\,. \tag{28}$$

Although self-consistent electronic structure calculations for solid films have demonstrated that V_s^σ has a distinctly three-dimensional character, at the present level of accuracy in LEED and photoemission calculations three-dimensional effects seem to be of relative insignificance [36]. Thus all terms with $\boldsymbol{g} \neq \boldsymbol{0}$ are usually disregarded in (28) and V_s^σ is approximated in one-dimensional form,

$$V_s^\sigma(E, \boldsymbol{r}) = V_s^\sigma(E, z)\,. \tag{29}$$

Let us now assume that the crystal occupies the halfspace $z > 0$ extending inwards from half an inter-layer spacing outside the topmost atomic layer. The asymptotic form of the real part of V_s^σ for $z \to -\infty$ must be that of an image potential [34, 37]

$$\text{Re}\; V_s^\sigma(E, z) = \frac{e^2}{4z}\,. \tag{30}$$

Just as dynamical corrections to the image behaviour of Re V_s^σ, the imaginary part of V_s^σ is strongly velocity-dependent due to the nonlocality of the self-energy [35, 37]. In a local description one thus has to be content with phenomenological forms for Im V_s^σ. The decay processes responsible for a nonzero Im V_s^σ fall rapidly in strength outside the surface [35, 37]. For this reason the asymptotic behaviour of Im V_s^σ is usually described by a rapidly decreasing function of type $-z^{-2}$ [38], $-\exp(-\lambda z^2)$ [39], $-\exp(-\lambda\,|z|)$ [36].

The asymptotic image behaviour of Re V_s^σ gives rise to barrier-induced unoccupied surface states in an energy range just below the vacuum level, if the latter falls into a gap of the bulk band structure. For these surface states the image potential combined with a repulsive barrier at the surface results in a Rydberg series of bound states with energies

$$E_n = -\frac{27.2}{32(n + a)^2}\,\text{eV}\,. \tag{31}$$

The quantum defect a ranges from -0.5 to $+0.5$ and depends on the particular form of Re V_s^σ in the non-asymptotic region near the surface. Since Im V_s^σ decays rapidly with increasing $|z|$, this Rydberg series is in principle fully detectable within the limitations of experimental resolution [40]. Indirect evidence for these image potential surface states on metals has been collected from the observation of resonances in elastic and inelastic electron emission [41]. Utilizing the technique of UBIS it has very recently become possible to probe these states by direct measurements [19, 20].

Near $z = 0$ the barrier potential $V_s^\sigma(E, z)$ must join smoothly to the inner potential $V_0^\sigma(E)$ in the interstitial regions of the bulk crystal. Within the muffin-tin approximation V_0^σ behaves essentially like the self-energy of a homogeneous spin-polarized electron gas. The spin dependence of Re V_0^σ for the static case is of the order of 0.1 to 0.2 eV for ferromagnets like Fe, Co, Ni [42]. Probably more important in actual calculations is the spin dependence of Im V_0^σ, which for energies below 40 eV has been shown to become quite pronounced [43, 44]. Since the exchange-correlation interaction weakens with increasing electron energy, $|\text{Re } V_0^\sigma|$ will decrease with increasing energy. Typical variations in Re V_0^σ are 1 to 2 eV in a range up to $E = 40$ eV [45]. The imaginary part of V_0^σ is strictly zero at the Fermi energy E_F and just above E_F behaves like $-(E - E_F)^r$ where $1 < r \lessapprox 2$ [39]. For higher energies in the range $E_F + 25$ eV to $E_F + 150$ eV the analysis of experimental LEED data shows that Im V_0^σ is essentially a constant of order -4 eV [46]. The limiting form for high energies is a slow fall with increasing energy as $-E^{-1/2} \ln E$ [47].

Due to the energy and spin dependence of the inner potential in ferromagnets the barrier potential V_s^σ will show these dependences, too. There is, however, so far no clear prescription for the barrier potential in the intermediate non-asymptotic region near the surface. In view of the fact that V_s^σ in that region is complicated by dynamical and three-dimensional effects, which in a form like $V_s^\sigma(E, z)$ are neglected, one might even ask, if a unique prescription for $V_s^\sigma(E, z)$ may exist at all. On the other hand, recent interpretations of high-resolution electron scattering data have led to more and more refined empiric forms of the barrier, the most recent prescriptions being those of Rundgren and Malmström [48], Dietz et al. [49], and Jones et al. [36]. The recent observation of surface states of different physical origin in UBIS [50], namely a barrier induced (Shockley-type) image-potential state and crystal-induced (Tamm-type) surface states at a Cu(001) surface, combined with a detailed experimental knowledge of the corresponding energy versus momentum dispersions will probably allow for further refinements of the barrier potential.

Appendix

The transition matrix elements in the $\boldsymbol{r}$-form (cf. (7)) are given by

$$\langle \varphi_f | \boldsymbol{A}\boldsymbol{r} | \varphi_i \rangle = \sum_{\varkappa\mu} \sum_{\varkappa'\mu'} (P_{\varkappa\mu,\varkappa'\mu'} + Q_{\varkappa\mu,\varkappa'\mu'}) , \tag{A1}$$

where

$$P_{\varkappa\mu,\varkappa'\mu'} = \int \mathrm{d}r\, f^{\mu}_{\varkappa,f}(r)\, f^{\mu'}_{\varkappa',i}(r) \int \mathrm{d}\Omega\, \chi^{\mu*}_{\varkappa} \boldsymbol{A}\boldsymbol{r} \chi^{\mu'}_{\varkappa'} , \tag{A2a}$$

$$Q_{\varkappa\mu,\varkappa'\mu'} = \int \mathrm{d}r\, g^{\mu}_{\varkappa,f}(r)\, g^{\mu'}_{\varkappa',i}(r) \int \mathrm{d}\Omega\, \chi^{\mu*}_{-\varkappa} \boldsymbol{A}\boldsymbol{r}\, \chi^{\mu'}_{-\varkappa'} . \tag{A2b}$$

The angular parts of the integrals can be transformed into a more appropriate form by expressing the components of the vector $\boldsymbol{r}$ in terms of the spherical harmonics of order $l = 1$. Adding the additionally appearing factor r to the radial part of the integral and using Gaunt's formula [51], one obtains for the angular integrals

$$\begin{aligned}\int \mathrm{d}\Omega\, \chi^{\mu*}_{\varkappa} \begin{bmatrix} x \\ y \end{bmatrix} \chi^{\mu'}_{\varkappa'} &= \left(\frac{2l'+1}{2(2l+1)}\right)^{1/2} C(l'1l;\, 000)\, \times \\ &\times \begin{bmatrix} 1 \\ i \end{bmatrix} \{C(l\tfrac{1}{2}j;\, \mu - \tfrac{1}{2}, \tfrac{1}{2})\, C(l'\tfrac{1}{2}j';\, \mu' - \tfrac{1}{2}, \tfrac{1}{2})\, [C(l'1l;\, \mu' - \tfrac{1}{2}, -1, \mu - \tfrac{1}{2}) \mp \\ &\mp C(l'1l;\, \mu' - \tfrac{1}{2}, 1, \mu - \tfrac{1}{2})] + \\ &+ C(l\tfrac{1}{2}j;\, \mu + \tfrac{1}{2}, -\tfrac{1}{2})\, C(l'\tfrac{1}{2}j';\, \mu' + \tfrac{1}{2}, -\tfrac{1}{2})\, [C(l'1l;\, \mu' + \tfrac{1}{2}, -1, \mu + \tfrac{1}{2}) \mp \\ &\mp C(l'1l;\, \mu' + \tfrac{1}{2}, 1, \mu + \tfrac{1}{2})]\} ,\end{aligned} \tag{A3 a, b}$$

$$\begin{aligned}\int \mathrm{d}\Omega\, \chi^{\mu*}_{\varkappa} z \chi^{\mu'}_{\varkappa'} &= \left(\frac{2l'+1}{2l+1}\right)^{1/2} C(l'1l;\, 000)\, \times \\ &\times [C(l\tfrac{1}{2}j;\, \mu - \tfrac{1}{2}, \tfrac{1}{2})\, C(l'\tfrac{1}{2}j';\, \mu' - \tfrac{1}{2}, \tfrac{1}{2})\, C(l'\,1l;\, \mu' - \tfrac{1}{2}, 0, \mu - \tfrac{1}{2}) + \\ &+ C(l\tfrac{1}{2}j;\, \mu + \tfrac{1}{2}, -\tfrac{1}{2})\, C(l'\tfrac{1}{2}j';\, \mu' + \tfrac{1}{2}, -\tfrac{1}{2})\, C(l'1l;\, \mu' + \tfrac{1}{2}, 0, \mu + \tfrac{1}{2})] .\end{aligned} \tag{A3c}$$

The results for the integrals of type $\int d\Omega\, \chi^{\mu*}_{-\varkappa} r \chi^{\mu'}_{-\varkappa'}$ can be calculated from those in (A3) by interchanging the quantum numbers l and $\bar{l}$. Note that in (A3) the selection rules for the quantum numbers l and μ are implicitly contained in the Clebsch-Gordan coefficients.

Acknowledgements

One of the authors (G.T.) wishes to thank H. Ibach and J. Kirschner for their hospitality at the Institut für Grenzflächenforschung und Vakuumphysik (KFA Jülich), where parts of this work have been obtained. The authors are much obliged to the Deutsche Forschungsgemeinschaft for financial support.

References

[1] B. Feuerbacher, B. Fitton, and R. F. Willis, Eds., Photoemission and the Electronic Properties of Surfaces, Wiley, New York 1978.
[2] M. Cardona and L. Ley, Eds., Photoemission in Solids, Vol. 1, Springer-Verlag, Berlin 1978.
[3] R. H. Williams, G. P. Srivastava, and I. T. McGovern, Rep. Progr. Phys. **43**, 87 (1980).
[4] F. J. Himpsel, Adv. Phys. **32**, 1 (1983).
[5] V. Dose, Prog. Surface Sci. **13**, 225 (1983).
[6] D. P. Woodruff, P. D. Johnson, and N. V. Smith, J. Vacuum Sci. Technol. **A1**, 1104 (1983).
[7] W. E. Spicer, Phys. Rev. **112**, 114 (1958).
[8] C. N. Berglund and W. E. Spicer, Phys. Rev. **136**, A 1030, A 1044 (1964).
[9] A. Liebsch, Phys. Rev. Letters **32**, 1202 (1974).
[10] J. B. Pendry, Surface Sci. **57**, 679 (1976).
[11] J. B. Pendry and D. J. Titterington, Commun. Phys. **2**, 31 (1977).
[12] D. W. Jepsen, Phys. Rev. B **20**, 402 (1979).
[13] H. Przybylski, A. Baalmann, G. Borstel, and M. Neumann, Phys. Rev. B **27**, 6669 (1983).
[14] S. F. Alvarado, W. Eib, F. Meier, H. C. Siegmann, and P. Zürcher, see [1] (p. 437).
[15] D. T. Pierce, R. J. Celotta, G. C. Wang, W. N. Unertl, A. Galejs, C. E. Kuyatt, and S. R. Mielczarek, Rev. Sci. Instrum. **51**, 478 (1980).
[16] G. Borstel and M. Wöhlecke, Phys. Rev. B **26**, 1148 (1982).
[17] A. Eyers, F. Schäfers, G. Schönhense, U. Heinzmann, H. P. Oepen, K. Hünlich, J. Kirschner, and G. Borstel, Phys. Rev. Letters **52**, 1559 (1984).
[18] R. Feder, J. Phys. C **14**, 2049 (1981).
[19] V. Dose, W. Altmann, A. Goldmann, U. Kolac, and J. Rogozik, Phys. Rev. Letters **52**, 1919 (1984).
[20] D. Straub and F. J. Himpsel, Phys. Rev. Letters **52**, 1922 (1984).
[21] W. Altmann, V. Dose, A. Goldmann, U. Kolac, and J. Rogozik, Phys. Rev. B **29**, 3015 (1984).
[22] G. Thörner and G. Borstel, Solid State Commun. **49**, 997 (1984).
[23] A. Messiah, Quantum Mechanics, Vol. 2, North-Holland Publ. Co., Amsterdam 1969.
[24] A. K. Rajagopal, Adv. Chem. Phys. **41**, 59 (1980).
[25] A. H. MacDonald and S. H. Vosko, J. Phys. C **12**, 2977 (1979).
[26] P. Strange, J. Staunton, and B. L. Gyorffy, J. Phys. C **17**, 3355 (1984).
[27] S. Lundqvist and N. H. March, Eds., Theory of the Inhomogeneous Electron Gas, Physics of Solids and Liquids Series, Plenum Press, New York 1983.
[28] L. J. Sham and W. Kohn, Phys. Rev. **145**, 561 (1966).
[29] L. Hedin and S. Lundqvist, Solid State Phys. **23**, 1 (1969).
[30] R. Feder, F. Rosicky, and B. Ackermann, Z. Phys. **B52**, 31 (1983).
[31] M. E. Rose, Relativistic Electron Theory, Wiley, New York 1961.
[32] B. Ackermann and R. Feder, Solid State Commun. **49**, 489 (1984).
[33] J. B. Pendry, Low-Energy Electron Diffraction, Academic Press, London 1974.
[34] J. Harris and R. O. Jones, J. Phys. C **6**, 3585 (1973).
[35] P. M. Echenique and J. B. Pendry, J. Phys. C **8**, 2936 (1975).

[36] R. O. Jones, P. J. Jennings, and O. Jepsen, Phys. Rev. B **29**, 6474 (1984).
[37] J. C. Inkson, J. Phys. F **3**, 2143 (1973).
[38] G. Malmström and J. Rundgren, J. Phys. C **13**, L 61 (1980).
[39] E. G. McRae and C. W. Caldwell, Surface Sci. **57**, 77 (1976).
[40] P. M. Echenique and J. B. Pendry, J. Phys. C **11**, 2065 (1978).
[41] E. G. McRae, Rev. mod. Phys. **51**, 541 (1979).
[42] V. L. Moruzzi, J. F. Janak, and A. R. Williams, Calculated Electronic Properties of Metals, Pergamon Press, New York 1978.
[43] R. W. Rendell and D. R. Penn, Phys. Rev. Letters **45**, 2057 (1980).
[44] G. Tréglia, F. Ducastelle, and D. Spanjaard, J. Physique **43**, 341 (1982).
[45] P. J. Jennings and S. M. Thurgate, Surface Sci. **104**, L 210 (1981).
[46] J. B. Pendry, see [1] (p. 87).
[47] B. S. Ing and J. B. Pendry, J. Phys. C 8, 1087 (1975).
[48] J. Rundgren and G. Malmström, J. Phys. C **10**, 4671 (1977).
[49] R. E. Dietz, E. G. McRae, and R. L. Campbell, Phys. Rev. Letters **45**, 1280 (1980).
[50] V. Dose, U. Kolac, G. Borstel, and G. Thörner, Phys. Rev. B **29**, 7030 (1984).
[51] M. E. Rose, Elementary Theory of Angular Momentum, Wiley, New York (1957) (p. 62).

(Received September 19, 1984)

phys. stat. sol. (b) **126**, 629 (1984)

Subject classification: 17; 13.5.2; 21

Zentralinstitut für Elektronenphysik der Akademie der Wissenschaften der DDR, Berlin[1])

On the Theory of Ion-Induced Electron Emission

By

M. Rösler and W. Brauer

A microscopic theory of H^+ ion-induced electron emission is developed in close analogy to the theory of secondary electron emission and applied to polycrystalline aluminum. The formation of excited electrons by screened Coulomb interaction with H^+ ions of primary energy E_0 between 20 and 800 keV is considered. Both, elastic and inelastic collisions of these electrons are taken into account. As far as possible the results calculated are compared with recent experimental data. Energy and angular distributions of ejected electrons prove to be in accordance with measurements. The calculated electronic yield γ shows a maximum in the primary energy dependence at $E_0 \approx \approx 55$ keV with a maximum value $\gamma_{max} = 1.1$. With increasing primary energy there are larger deviations from the measured yield values.

In enger Analogie zur Theorie der Sekundärelektronenemission wird eine mikroskopische Theorie der durch H^+-Ionen induzierten Elektronenemission entwickelt und auf polykristallines Aluminium angewendet. Dazu wird die Bildung angeregter Elektronen untersucht, die durch H^+-Ionen der Energie E_0 zwischen 20 und 800 keV über das abgeschirmte Coulomb-Potential entstehen. Elastische und inelastische Stöße dieser Elektronen werden berücksichtigt. So weit als möglich werden die berechneten Ergebnisse mit neuen experimentellen Daten verglichen. Energie- und Winkelverteilungen der ausgelösten Elektronen stehen in Übereinstimmung mit den Meßresultaten. Die berechnete elektronische Ausbeute γ zeigt ein Maximum in der Abhängigkeit von der Primärenergie bei $E_0 \approx 55$ keV mit einem Maximalwert $\gamma_{max} = 1{,}1$. Mit wachsender Primärenergie gibt es größere Abweichungen von den gemessenen Ausbeutewerten.

1. Introduction

The "kinetic" emission of electrons from a solid surface under the influence of ion bombardment by particles of sufficient velocity ($v_0 > 2 \times 10^7$ cm s^{-1}) plays a fundamental role in the physics of ion–solid interactions [1, 2]. Especially, the ejection is of great importance in connection with problems of plasma–wall interactions in thermonuclear fusion reactors [3, 4]. Up to the present no quantitative theory of the phenomenon does exist.

The electron yield γ in ion-induced electron emission (IEE) is defined as the number of ejected electrons per incident ion. More detailed information is provided by the energy distribution $j(E)$ which fulfils the relation $\gamma = \int j(E)\, dE$. Experimental results for Al are shown in Fig. 1 and 2 exhibiting the dependence of γ on the projectile energy E_0 and velocity v_0.

In 1982 the energy distribution $j(E)$ of sputter-cleaned Al for IEE was measured by Hasselkamp and Scharmann at an energy of $E_0 = 500$ keV [6]. They observed a main maximum at 2 eV and two weak shoulders at 11 and 6 eV for projectiles H^+, He^+, and Ar^+ [7] (Fig. 3). The complete agreement with corresponding curves for secondary electron emission (SEE) in Al [8] represents a hint for developing a theory of IEE in close analogy to existing theories of SEE [9].

---

[1]) Hausvogteiplatz 5—7, DDR-1086 Berlin, GDR.

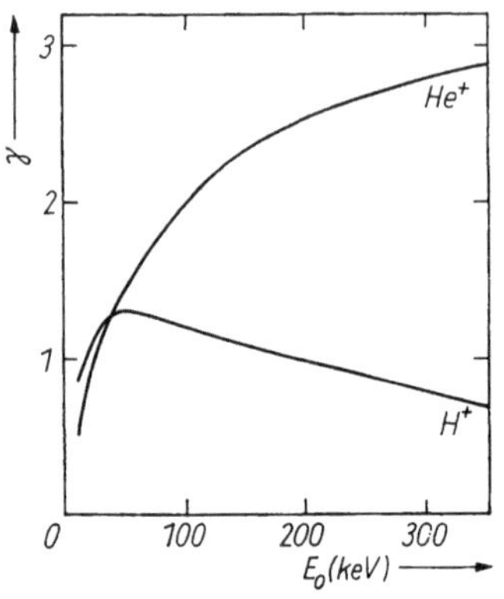

Fig. 1

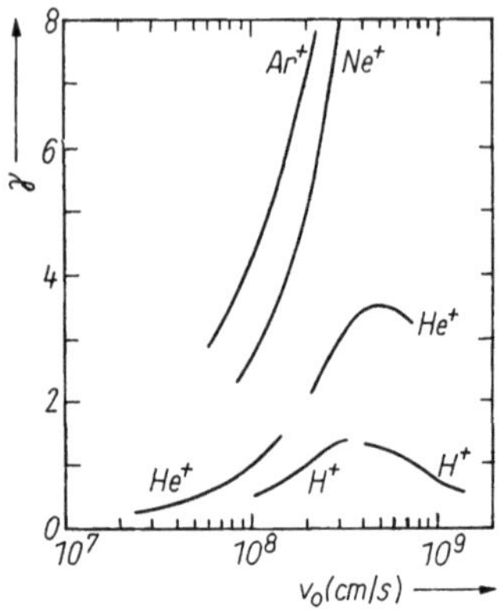

Fig. 2

Fig. 1. Yield γ as a function of ion energy E_0 in Al (after [5])

Fig. 2. Yield γ as a function of ion velocity v_0 in Al (after [2])

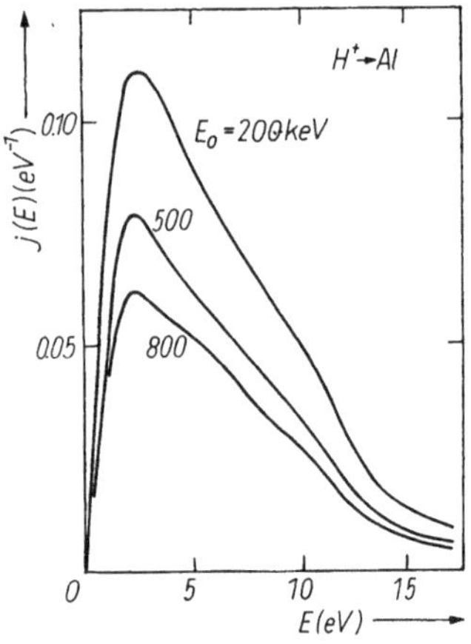

Fig. 3. Energy distribution of ejected electrons in Al (after [7])

2. Semiempirical Considerations

Before formulating a theory of IEE in such a manner we want to study the relation between IEE and SEE by the following rough consideration. The yield $\gamma(E_0)$ can be described by an expression of the form

$$\gamma(E_0) = A \int_0^{R(E_0)} S(E_0, x)\, p(x)\, dx\,. \tag{1}$$

Here the excitation function $S(E_0, x)$ gives the number of electrons at x per cm^3 thrown into states above the Fermi surface by incident ions or primary electrons of energy E_0 and unit current density. $p(x)$ is the probability to escape from the surface at $x = 0$, $R(E_0)$ denotes the range of the projectiles of energy E_0, A is a material dependent constant. The function $p(x)$ obeys the relations

$$p(0) = 1 \quad \text{and} \quad p(x) = 0 \quad \text{for} \quad x > L$$

with L characterizing the maximum escape depth of the order 5 nm (Fig. 4).

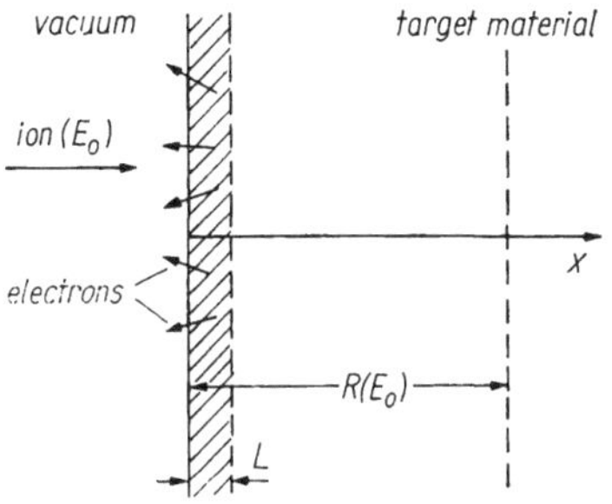

Fig. 4. Particle-induced electron emission process (schematic)

Defining the energy $E_{0,\max}$ by $\gamma'(E_{0,\max}) = 0$ we obtain from (1) the following equation:

$$[F(E_0) + G(E_0)]|_{E_0=E_{0,\max}} = 0 \tag{2}$$

with the abbreviations

$$F(E_0) = R'(E_0)\, S(E_0, R(E_0))\, p(R(E_0)) ,$$

$$G(E_0) = \int_0^{R(E_0)} \frac{\partial S(E_0, x)}{\partial E_0} p(x)\, dx .$$

Because of $F \geqq 0$, one obtains from (2) either the relation $F(E_{0,\max}) = G(E_{0,\max}) = 0$ or the compensation of both terms in (2) with $\partial S(E_0, x)/\partial E_0|_{E_{0,\max}} < 0$ for a sufficiently extended x-interval.

Obviously, the first case corresponds to IEE. For $H^+ \rightarrow Al$ one has (Fig. 1) $E_{0,\max} \approx \approx 55$ keV and R (55 keV) ≈ 500 nm [10] which is two orders of magnitude larger than the escape depth L. Therefore it follows $p(R(E_{0,\max})) = 0$. The second case describes SEE. Because of $E_{0,\max} = 300$ eV (for Al) one obtains $R(E_{0,\max}) \approx L$ and $\partial S(E_0, 0)/\partial E_0|_{E_{0,\max}} < 0$ [11]. Thus, there are different reasons for the appearance of yield maxima in IEE and SEE.

In analogy to the semiempirical theory of SEE [12] nearly all theoretical considerations of the IEE are based on the two following assumptions [13, 1, 14]:

$$S(E_0, x) = \frac{1}{\overline{\Delta E_0}} \frac{dE_0}{dx}$$

and

$$p(x) = e^{-x/L} . \tag{3}$$

$\overline{\Delta E_0}$ stands for the averaged energy exchange per collision. The application of both equations to the SEE seems very questionable. According to [9] the excitation function is represented by the sum of three terms

$$S = S_e + S_p + S_c . \tag{4}$$

The first term describes excitation of single conduction electrons, the second term pertains to electron excitation by plasmon decay, and the third term represents the excitation of core electrons. In the same manner one can write for the stopping power [15]

$$\frac{dE_0}{dx} = \left.\frac{dE_0}{dx}\right|_e + \left.\frac{dE_0}{dx}\right|_p + \left.\frac{dE_0}{dx}\right|_c . \tag{5}$$

At $E_0 = 2$ keV, for example, the last term in (5) is the prevailing one (in the case of Al) [11], contrary to (4) in which the term S_c is smaller than the first two terms.

Monte-Carlo calculations have shown [16] that the electron cascade [17] in transport of secondary electrons is not in accordance with the use of the exponential approximation (3). Therefore, the transport process of the IEE has to be developed in close analogy to the theory of SEE with both the elastic and inelastic scattering processes taken properly into account.

3. General Theory of H⁺-Induced Electron Emission

In the following we restrict ourselves to perpendicular incidence of H^+ ions with energies between 20 and 800 keV on an Al target. Thereby we eliminate additional complications, for instance, recoiling target atoms [5] and molecular effects [18]. The aim of the theory is the calculation of the energy angular distribution $j(E, \boldsymbol{\Omega})$ of emitted electrons, i.e. the current density of ejected $(E, \boldsymbol{\Omega})$ electrons related to the unit of primary current. This quantity is calculated according to (8) of [9] (referred to as RB hereafter) from the density $\hat{f}(E', \boldsymbol{\Omega}')$ of excited electrons $(E' > E_F)$ in the metal. This distribution function is the result of excitation and interaction processes of the metal electrons and can be calculated within the framework of the Boltzmann equation.

The general expression for the transition probability (neglecting exchange) between two point charges screened according to $\varepsilon(q, \omega)$ can be written as

$$W_{\boldsymbol{k}_1\boldsymbol{k}_2\to\boldsymbol{k}_1'\boldsymbol{k}_2'} = \frac{32\pi^3 e^4}{\hbar} \frac{1}{\Omega} \sum_{\boldsymbol{q}} \frac{1}{q^4 \left|\varepsilon(q, E_{\boldsymbol{k}_1} - E_{\boldsymbol{k}_1'})\right|^2} \left|\langle \boldsymbol{k}_1' | \, \mathrm{e}^{i\boldsymbol{q}\boldsymbol{r}} \, | \boldsymbol{k}_1 \rangle\right|^2 \times$$
$$\times \left|\langle \boldsymbol{k}_2' | \, \mathrm{e}^{-i\boldsymbol{q}\boldsymbol{r}} \, | \boldsymbol{k}_2 \rangle\right|^2 \delta(E_{\boldsymbol{k}_1} + E_{\boldsymbol{k}_2} - E_{\boldsymbol{k}_1'} - E_{\boldsymbol{k}_2'}) \,. \tag{6}$$

In the description of the transport process all particle states in (6) are related to electrons. The summation over one of the final states (besides those over $(\boldsymbol{k}_2 < k_F)$) results in the quantities

$$W_{\boldsymbol{k}_1\to\boldsymbol{k}_1'} = \sum_{\substack{\boldsymbol{k}_2(<k_F)\\ \boldsymbol{k}_2'(>k_F)}} W_{\boldsymbol{k}_1\boldsymbol{k}_2\to\boldsymbol{k}_1'\boldsymbol{k}_2'}$$

and

$$W^s_{\boldsymbol{k}_1,\boldsymbol{k}_2'} = \sum_{\substack{\boldsymbol{k}_2(<k_F)\\ \boldsymbol{k}_1'(>k_F)}} W_{\boldsymbol{k}_1\boldsymbol{k}_2\to\boldsymbol{k}_1'\boldsymbol{k}_2'} \,.$$

With the definitions

$$\frac{1}{\tau(\boldsymbol{k})} = \sum_{\boldsymbol{k}'(>k_F)} W_{\boldsymbol{k}\to\boldsymbol{k}'}$$

and

$$K^\sigma(\boldsymbol{k}', \boldsymbol{k}) = \tau(\boldsymbol{k}') \, W^\sigma_{\boldsymbol{k}',\boldsymbol{k}}, \qquad W^\sigma_{\boldsymbol{k}',\boldsymbol{k}} = W_{\boldsymbol{k}'\to\boldsymbol{k}} + W^s_{\boldsymbol{k}',\boldsymbol{k}}$$

one gets for

$$\psi(\boldsymbol{k}) = \frac{1}{\tau(\boldsymbol{k})} \hat{f}(\boldsymbol{k})$$

the Boltzmann equation (see (13) of RB)

$$\psi(\boldsymbol{k}) = S(\boldsymbol{k}_0, \boldsymbol{k}) + \sum_{\boldsymbol{k}'} \psi(\boldsymbol{k}') \, K^\sigma(\boldsymbol{k}', \boldsymbol{k}) \,. \tag{7}$$

Taking the following equations into account

$$\boldsymbol{k}_1 \equiv \boldsymbol{k}_0 \quad \text{and} \quad \boldsymbol{k}_1' \equiv \boldsymbol{k}_0' \quad \text{with}$$

$$E_{\boldsymbol{k}_1} \equiv E^{\mathrm{i}}_{\boldsymbol{k}_0} = \frac{\hbar^2 k_0^2}{2M_\mathrm{p}} \quad \text{and} \quad E_{\boldsymbol{k}_1'} \equiv E^{\mathrm{i}}_{\boldsymbol{k}_0'} = \frac{\hbar^2 k_0'^2}{2M_\mathrm{p}}$$

we obtain for the excitation function $S(\boldsymbol{k}_0, \boldsymbol{k})$ from (6)

$$S(\boldsymbol{k}_0, \boldsymbol{k}) = \frac{1}{v_0} \sum_{\boldsymbol{k}_0', \boldsymbol{k}_2 (< k_\mathrm{F})} W_{\boldsymbol{k}_0 \boldsymbol{k}_2 \to \boldsymbol{k}_0' \boldsymbol{k}} \,; \qquad v_0 = \frac{\hbar k_0}{M_\mathrm{p}} \tag{8}$$

with M_p denoting the mass of the H^+ ion.

As mentioned above the excitation function $S(\boldsymbol{k}_0, \boldsymbol{k})$ consists of three terms

$$S(\boldsymbol{k}_0, \boldsymbol{k}) = \sum_i S_i(\boldsymbol{k}_0, \boldsymbol{k}) \,; \qquad i = \mathrm{e}, \mathrm{p}, \mathrm{c} \,. \tag{9}$$

Correspondingly, one has

$$K^\sigma(\boldsymbol{k}', \boldsymbol{k}) = K^{\sigma(\mathrm{inel})}(\boldsymbol{k}', \boldsymbol{k}) + K^{(\mathrm{el})}(\boldsymbol{k}', \boldsymbol{k})$$

with

$$K^{\sigma(\mathrm{inel})}(\boldsymbol{k}', \boldsymbol{k}) = \sum_i K_i^{\sigma(\mathrm{inel})}(\boldsymbol{k}', \boldsymbol{k}) \tag{10}$$

describing the inelastic part of the transport scattering process. Finally, one obtains the electron mean free path

$$\frac{1}{l(\boldsymbol{k})} = \frac{m}{\hbar k} \frac{1}{\tau(\boldsymbol{k})} = \frac{m}{\hbar k} \left(\frac{1}{\tau^{(\mathrm{inel})}(\boldsymbol{k})} + \frac{1}{\tau^{(\mathrm{el})}(\boldsymbol{k})} \right)$$

with

$$\frac{1}{\tau^{(\mathrm{inel})}(\boldsymbol{k})} = \sum_i \frac{1}{\tau_i(\boldsymbol{k})} \,.$$

4. Excitation Function

Inserting (6) into (8) we get a general expression for the excitation function. In analogy to (17) of RB we have

$$S(\boldsymbol{k}_0, \boldsymbol{k}) = \frac{64\pi^3 e^4 M_\mathrm{p}}{\hbar^2 k_0} \frac{1}{\Omega^2} \sum_{\boldsymbol{k}', \boldsymbol{q}, \boldsymbol{k}_0'} \frac{1}{q^4 \, |\varepsilon(q, E^{\mathrm{i}}_{\boldsymbol{k}_0} - E^{\mathrm{i}}_{\boldsymbol{k}_0'})|^2} \, \theta[E_\mathrm{F} - E_{\boldsymbol{k}'}] \times$$
$$\times \, \delta_{\boldsymbol{k}_0', \boldsymbol{k}_0 + \boldsymbol{q}} \, |\langle \boldsymbol{k}| \, e^{-i\boldsymbol{q}\boldsymbol{r}} \, |\boldsymbol{k}'\rangle|^2 \, \delta(E^{\mathrm{i}}_{\boldsymbol{k}_0} + E_{\boldsymbol{k}'} - E^{\mathrm{i}}_{\boldsymbol{k}_0'} - E_{\boldsymbol{k}}) \,. \tag{11}$$

Regarding the dielectric function, the electronic states, and Bloch energies different approximations are necessary in arriving at explicit expressions for the excitation mechanisms.

The angular dependence of the transport problem is handled by expansion into Legendre polynomials. We define the $S^{(\Lambda)}$-approximation of the excitation function by ($\boldsymbol{k} \to E$, $\cos\theta$, $\theta = \sphericalangle\,(\boldsymbol{k}_0, \boldsymbol{k})$)

$$S(E_0, E, \cos\theta) \approx S^{(\Lambda)}(E_0, E, \cos\theta) = \sum_{\lambda=0}^{\Lambda} S_\lambda(E_0, E) \, P_\lambda(\cos\theta) \,.$$

Thus, the Boltzmann equation (7) is split into $\Lambda + 1$ independent integral equations. The treatment of the angular dependence of the phenomenon by expansion into Legendre polynomials is only suitable if Λ represents a sufficiently small number.

4.1 Excitation of single conduction electrons

In this case we describe the electrons by the free-electron-gas model. Therefore, we obtain from (11) (see (19) of RB)

$$S_e(\boldsymbol{k}_0, \boldsymbol{k}) = \frac{64\pi^3 e^4 M_p}{\hbar^2 k_0} \frac{1}{\Omega^2} \sum_{\boldsymbol{k}'} \frac{1}{|\boldsymbol{k}' - \boldsymbol{k}|^4 \, |\varepsilon^L(|\boldsymbol{k}' - \boldsymbol{k}|, E_{\boldsymbol{k}} - E_{\boldsymbol{k}'})|^2} \times$$
$$\times \, \theta[E_F - E_{\boldsymbol{k}'}] \, \delta(E^i_{\boldsymbol{k}_0 + \boldsymbol{k}' - \boldsymbol{k}} + E_{\boldsymbol{k}} - E^i_{\boldsymbol{k}_0} - E_{\boldsymbol{k}'}) \, ,$$

where $\varepsilon^L(q, \omega)$ is the well-known Lindhard function. One angular integral may be carried out with the help of the δ-function. This leads to restrictions for the excitation angle

$$\cos \theta_1 \leqq \cos \theta \leqq \text{Min} \, [\cos \theta_2, 1] \tag{12}$$

with

$$\cos \theta_{1,2} = \frac{1}{2k_0 k} \left[(k^2 - k_F^2) \left(\frac{M_p}{m} + 1 \right) \mp 2k_F \sqrt{k_0^2 - \frac{M_p}{m} (k^2 - k_F^2)} \right].$$

Owing to the large mass difference of colliding particles there does exist, in contrast to SEE, an excitation in direction of the primary beam, too. In order to illustrate this fact, the angular distribution of the excitation is shown in Fig. 5 for two primary energies ($E_0 = 40$ and 500 keV) at 20 and 100 eV, respectively.

The condition for the occurrence of excitation is expressed by $\cos \theta_1 < 1$ leading to an upper boundary E_m for the excitation from the conduction band

$$E_m = \frac{\hbar^2}{2m} \left\{ \frac{2k_0 + k_F \left(\dfrac{M_p}{m} - 1 \right)}{\dfrac{M_p}{m} + 1} \right\}^2 .$$

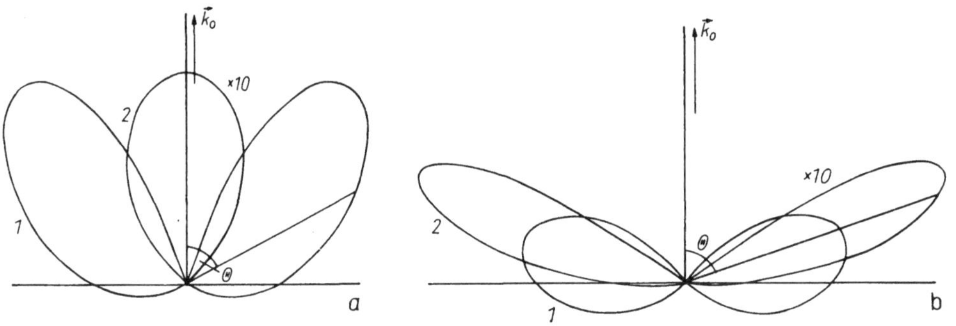

Fig. 5. Angular dependence of excitation by dynamically screened ion–electron scattering for (1) $E = 20$ and (2) 100 eV, respectively. a) $E_0 = 40$ and b) 500 keV

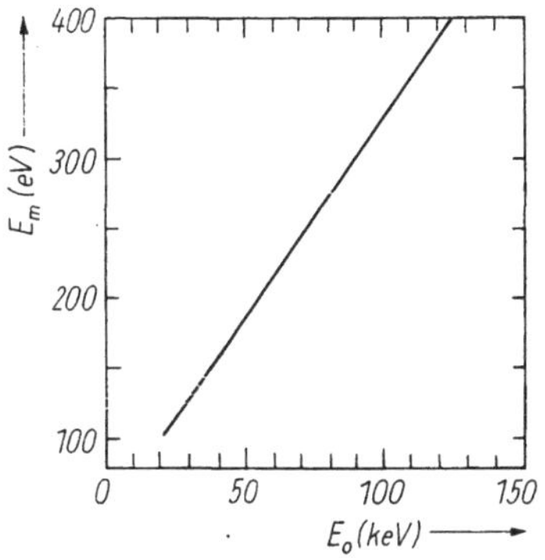

Fig. 6

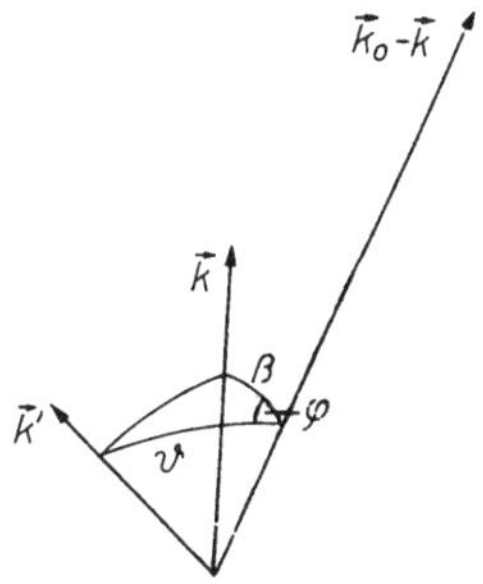

Fig. 7

Fig. 6. Upper boundary for the excitation of single electrons from the conduction band as a function of ion energy E_0

Fig. 7. Momentum diagram defining the different variables in the excitation function S_e

In Fig. 6 the primary energy dependence of this quantity is shown. The restriction to excitation energies smaller than E_m plays an important role only for small primary energies.[2])

In order to obtain explicit results, we must calculate numerically the following double integral (keeping in mind the condition (12)):

$$S_e(E_0, E, \cos\theta) = \frac{k}{\pi^3 e^2 a_B^3 k_0 |\boldsymbol{k}_0 - \boldsymbol{k}|} \left(\frac{M_p}{m}\right)^2 \int\limits_{k_{\min}}^{k_F} k' \, dk' \int\limits_0^{2\pi} \frac{d\varphi}{|\boldsymbol{k}' - \boldsymbol{k}|^4 \, |\varepsilon^L(|\boldsymbol{k}' - \boldsymbol{k}|, E_{\boldsymbol{k}} - E_{\boldsymbol{k}'})|^2}$$

with

$$k_{\min} = \frac{\left| |\boldsymbol{k}_0 - \boldsymbol{k}| - \sqrt{k_0^2 + k^2 \left(\frac{M_p}{m}\right)^2 - 2k_0 k \frac{M_p}{m} \cos\theta} \right|}{\frac{M_p}{m} - 1}$$

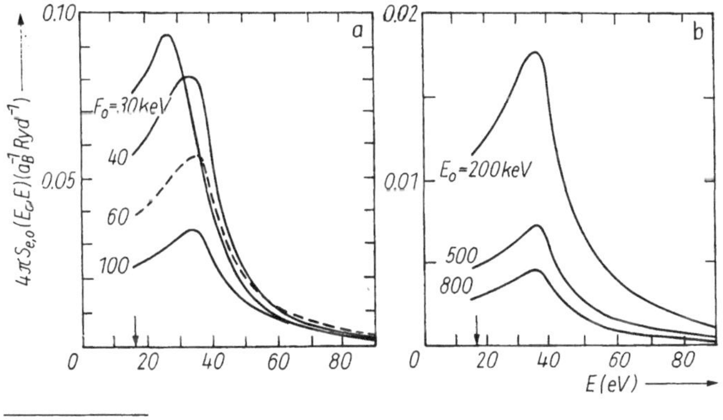

Fig. 8. Energy-dependent excitation function by dynamically screened ion–electron scattering at a) low and b) high primary energies. The arrow indicates the vacuum level $W = E_F + \Phi$. Φ is the work function

[2]) Here and hereafter the terms small primary energies and high primary energies pertain to $E_0 \lessapprox 60$ and $\gtrapprox 500$ keV, respectively.

and (see Fig. 7)

$$|\boldsymbol{k}' - \boldsymbol{k}|^2 = k^2 + k'^2 - 2kk' (\cos \vartheta \cos \beta + \sin \vartheta \sin \beta \cos \varphi) ,$$

$$\cos \beta = \frac{k_0 \cos \theta - k}{|\boldsymbol{k}_0 - \boldsymbol{k}|} ,$$

where the angle ϑ is fixed by the δ-function condition.

The energy-dependent excitation function is shown in Fig. 8 for different primary energies.

In terms of the $S^{(\Lambda)}$-approximation, we obtain satisfactory results by using $\Lambda = 4$.

4.2 *Excitation by decay of plasmons*

In this section we consider the excitation of electrons by decay of plasmons with the latter generated by incident H^+ ions. As to Al the plasmon damping $\Gamma(q)$ is primarily determined by interband transitions. However, with increasing wave number q there are important contributions from other processes (decay into two electron–hole pairs, decay into one electron–hole pair and a plasmon with lower energy [19]). Obviously, the interband processes being responsible for the decay of plasmons are identical to the excitation processes. With regard to our calculation of the excitation function S_p it is necessary to know the quantity $\Gamma(q)$. Especially, for small primary energies we need $\Gamma(q)$ for large wave numbers (smaller than the plasmon cut-off wave number q_c), only. However, a calculation of $\Gamma(q)$ for all wave numbers including all essential decay processes is out of the scope of this work. Therefore, an interpolation formula for $\Gamma(q)$ was used in evaluating the excitation function. The experimental results [20] suggest a parabolic expression

$$\Gamma(q) = \Gamma_0 + \Gamma_2 \left(\frac{q}{k_F}\right)^2$$

with $\Gamma_0 = 0.5$ eV and $\Gamma_2 = 3$ eV.

Using the extended zone scheme in describing the interband transitions we obtain for the corresponding matrix element a representation by means of the well-known Bloch integrals $B^{\boldsymbol{K}}(\boldsymbol{k}, \boldsymbol{k}')$ ((26) of RB), where $\boldsymbol{K}$ denotes reciprocal lattice vectors. As to the nearly-free-electron metal Al the perturbation theory for nearly degenerated states leads both to square-root expressions for the Bloch energies $\hat{E}_{\boldsymbol{k}}$ and to the following formula for the Bloch integral (see (28) of RB):

$$|B^{\boldsymbol{K}}(\boldsymbol{k}, \boldsymbol{k}')|^2 = |V_{\boldsymbol{K}}|^2 \frac{(D_{\boldsymbol{k}} + D_{\boldsymbol{k}'})^2}{(D_{\boldsymbol{k}}^2 + |V_{\boldsymbol{K}}|^2)(D_{\boldsymbol{k}'}^2 + |V_{\boldsymbol{K}}|^2)} \tag{13}$$

with $D_{\boldsymbol{k}} = E_{\boldsymbol{k}} - \hat{E}_{\boldsymbol{k}}$. The $V_{\boldsymbol{K}}$ are the Fourier coefficients of the local model potential of Heine, Abarenkov, and Animalu (see [21]).

In evaluating (11), we must take into account the plasmon-pole contribution of $1/|\varepsilon^L(q, \omega)|^2$. Using (13) we have

$$S_p(\boldsymbol{k}_0, \boldsymbol{k}) = \frac{64\pi^3 e^4 M_p}{\hbar^2 k_0} \frac{1}{\Omega^2} \sum_{\boldsymbol{K}, \boldsymbol{q}(<q_c)} |V_{\boldsymbol{K}}|^2 \frac{1}{q^4 |\varepsilon^L(q, E^i_{\boldsymbol{k}_0} - E^i_{\boldsymbol{k}_0+\boldsymbol{q}})|^2} \times$$
$$\times \frac{(D_{\boldsymbol{k}} + D_{\boldsymbol{k}+\boldsymbol{q}+\boldsymbol{K}})^2}{(D_{\boldsymbol{k}}^2 + |V_{\boldsymbol{K}}|^2)(D_{\boldsymbol{k}+\boldsymbol{q}+\boldsymbol{K}}^2 + |V_{\boldsymbol{K}}|^2)} \times$$
$$\times\, \theta[E_F - (E^i_{\boldsymbol{k}_0+\boldsymbol{q}} - E^i_{\boldsymbol{k}_0} + \hat{E}_{\boldsymbol{k}})]\, \delta(E^i_{\boldsymbol{k}_0+\boldsymbol{q}} + \hat{E}_{\boldsymbol{k}} - E^i_{\boldsymbol{k}_0} - \hat{E}_{\boldsymbol{k}+\boldsymbol{q}+\boldsymbol{K}}) .$$

The azimuthal angle integration in $\boldsymbol{q}$-space ($d^3q = q^2\, dq \sin\beta\, d\beta\, d\varphi$) is transformed to an integral over the energy loss $\Delta = E^{\mathrm{i}}_{\boldsymbol{k}_0} - E^{\mathrm{i}}_{\boldsymbol{k}_0+\boldsymbol{q}}$. Using the representation

$$\frac{1}{|\varepsilon^{\mathrm{L}}(q,\omega)|^2} = \frac{1}{\left(\left.\dfrac{\partial\varepsilon_1^{\mathrm{L}}(q,\omega)}{\partial\omega}\right|_{\omega_{\mathrm{p}}(q)}\right)^2} \frac{1}{(\omega-\omega_{\mathrm{p}}(q))^2 + \left(\dfrac{\Gamma(q)}{2\hbar}\right)^2},$$

being valid in the frequency region about the plasma frequency $\omega_{\mathrm{p}}(q)$, the Δ-integration can be carried out. $\varepsilon_1^{\mathrm{L}}$ is the real part of the Lindhard function.

In analogy to (32) of RB we define a function

$$N_{[\boldsymbol{K}]}(q, \Delta, \cos\theta) = $$
$$= \frac{E_{\mathrm{F}}^3 k_{\mathrm{F}}^3}{8\pi^2 q^3} \int d\varphi \int d\Omega_{\boldsymbol{K}} \frac{(D_{\boldsymbol{k}} + D_{\boldsymbol{k}+\boldsymbol{q}+\boldsymbol{K}})^2}{(D_{\boldsymbol{k}}^2 + |V_{\boldsymbol{K}}|^2)\,(D_{\boldsymbol{k}+\boldsymbol{q}+\boldsymbol{K}}^2 + |V_{\boldsymbol{K}}|^2)}\, \delta(\hat{E}_{\boldsymbol{k}+\boldsymbol{q}+\boldsymbol{K}} - \hat{E}_{\boldsymbol{k}} + \Delta)$$

which contains an averaging over all $\boldsymbol{K}$-directions together with the polar angle integration in $\boldsymbol{q}$-space. Thus, we can write the excitation function in the following form:

$$S_{\mathrm{p}}(E_0, E, \cos\theta) = \frac{4mk}{\pi^2\hbar^2 e^2 a_{\mathrm{B}}^3 k_{\mathrm{F}}^5 k_0^2} \left(\frac{M_{\mathrm{p}}}{m}\right)^2 \sum_{[\boldsymbol{K}]} n_{[\boldsymbol{K}]} \left(\frac{|V_{\boldsymbol{K}}|}{E_{\mathrm{F}}}\right)^2 \times$$
$$\times \int_{q_{\min}}^{q_{\mathrm{c}}} dq \int_{\mathrm{Max}\left\{E_{\boldsymbol{k}}-E_{\mathrm{F}},\, \hbar\omega_{\mathrm{p}}(q)-\frac{\Gamma(q)}{2}\right\}}^{\mathrm{Min}\,\{E_{\boldsymbol{k}},\,\Delta_{\mathrm{q}}\}} d\Delta\, \frac{1}{|\varepsilon^{\mathrm{L}}(q,\Delta)|^2} N_{[\boldsymbol{K}]}(q, \Delta, \cos\theta); \quad (14)$$

$\sum_{[\boldsymbol{K}]}$ denotes the summation over sets of reciprocal lattice vectors of equal lengths. Evaluating (14) we took into account only the interband processes described by $\boldsymbol{K}_1$ [111] and $\boldsymbol{K}_2$ [200].[3]) The corresponding values for $n_{[\boldsymbol{K}]}$ and $|V_{\boldsymbol{K}}|$ are: $n_{[\boldsymbol{K}_1]} = 8$, $n_{[\boldsymbol{K}_2]} = 6$ and $|V_{\boldsymbol{K}_1}| = 0.24$ eV, $|V_{\boldsymbol{K}_2}| = 0.79$ eV.

In Fig. 9 we plotted the RPA excitation spectrum together with the plasmon damping $\Gamma(q)$ and the upper limit $\Delta_{\mathrm{q}}(E_0) = \hbar^2 q(2k_0 - q)/2M_{\mathrm{p}}$ of Δ-integration. The relevant energy range for plasmon excitation is the region of width $\Gamma(q)$ about $\hbar\omega_{\mathrm{p}}(q)$ (shaded area in Fig. 9). In determining the lower limit of q-integration $q_{\min}$ in (14), we must allow for the relative positions of the different limits ($E_{\boldsymbol{k}}$, $\Delta_{\mathrm{q}}(E_0)$, $E_{\boldsymbol{k}} - E_{\mathrm{F}}$) with regard to this energy range, especially for small primary energies.

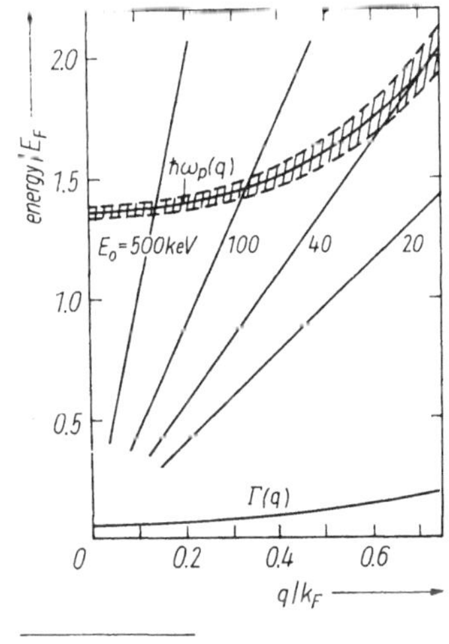

Fig. 9. Plasmon energy, plasmon damping, and upper limit $\Delta_q(E_0)$ for different primary energies, as a function of wave number

[3]) As in the case of SEE inclusion of the $\boldsymbol{K}_3$ [220]-contribution in the calculation of the excitation function leads to an unrealistic peak in the energy distribution $j(E)$ in contradiction to the experiment. The restriction to the $\boldsymbol{K}_1$- and $\boldsymbol{K}_2$-contributions can be justified possibly if we take into account the realistic band structure in the relevant energy range [22].

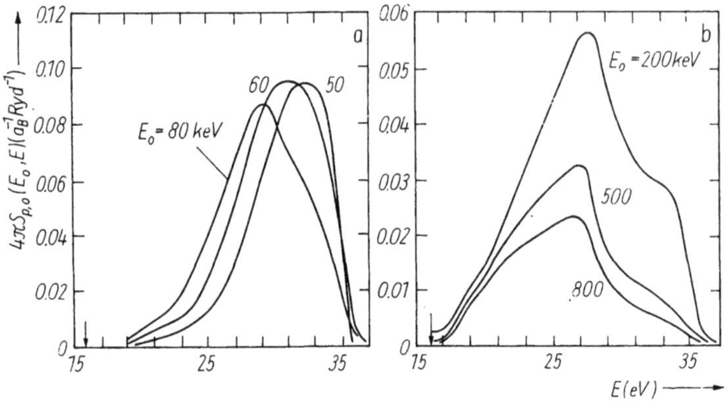

Fig. 10. Energy-dependent excitation function by plasmon decay at a) low and b) high primary energies. The arrow indicates the vacuum level

Additionally, we obtain a minimum primary energy $E_0^{\min} \approx 35$ keV, corresponding to $v_0^{\min} \approx 1.19\ e^2/\hbar = 2.6 \times 10^8$ cm s^{-1}. For lower values of E_0 (see the position of $\Delta_q(E_0)$ for $E_0 = 20$ keV compared with the plasmon energy in Fig. 9) the excitation of plasmons is impossible [6].

At high primary energies the excitation turns out to be nearly isotropic but for low primary energies we obtain an excitation preferably in forward direction (in the direction of the primary beam). We found that in the range of small primary energies a $S^{(6)}$-approximation is sufficient.

In Fig. 10 the energy dependence of the excitation function is shown for different primary energies. There are remarkable qualitative differences for excitation at low and high primary energies. At high values of E_0 the q-integration in (14) starts with small $q_{\min}$-values. Then for the excitation process only the region of small wave numbers is important (the function $N_{[\boldsymbol{K}]}$ decreases rapidly with increasing q). As in the case of SEE the excitation function begins to drop at the energy $E_F + \hbar\omega_p(0)$. At low values of E_0 the q-integration starts with higher $q_{\min}$-values ($\lessgtr q_c$). Then the possible excitation energies are determined by the dispersion of plasmon energy and damping.

4.3 Excitation of core electrons

In this case we obtain from (11) an expression corresponding to (18) of RB

$$S_c(E_0, E, \cos\theta) = \frac{k}{\pi^3 e^2 a_B^3 k_0^2} \left(\frac{M_p}{m}\right)^2 \sum_\nu \int_{k_0-k_0'}^{k_0+k_0'} \frac{dq}{q^3} \int_0^{2\pi} d\varphi\, |B_\nu(\boldsymbol{k}, \boldsymbol{q})|^2\,; \tag{15}$$

φ is the polar angle of $\boldsymbol{k}_0'$ with respect to $\boldsymbol{k}_0$, $\boldsymbol{q} = \boldsymbol{k}_0 - \boldsymbol{k}_0'$, where $k_0' = k_0 \sqrt{1 + (E_\nu - E)/E_0}$. $B_\nu(\boldsymbol{k}, \boldsymbol{q})$ denotes the matrix element of $e^{i\boldsymbol{qr}}$ calculated with the atomic wave function $\Phi_\nu(\boldsymbol{r})$ and the excited state described by an orthogonalized plane wave. On evaluating (15), we will take into account the excitation of L-shell electrons, only. As to the radial part of the atomic wave functions we will use the Herman-Skillman functions.

In Fig. 11 the energy-dependent excitation function for different primary energies is shown.

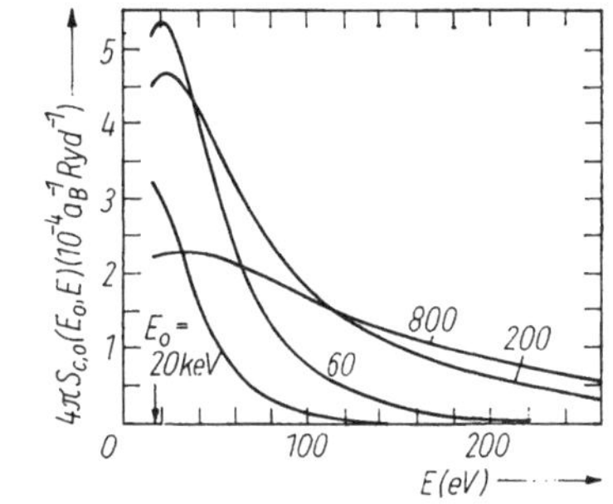

Fig. 11. Energy-dependent excitation function of core excitations from 2s and 2p core states at different primary energies. The arrow indicates the vacuum level

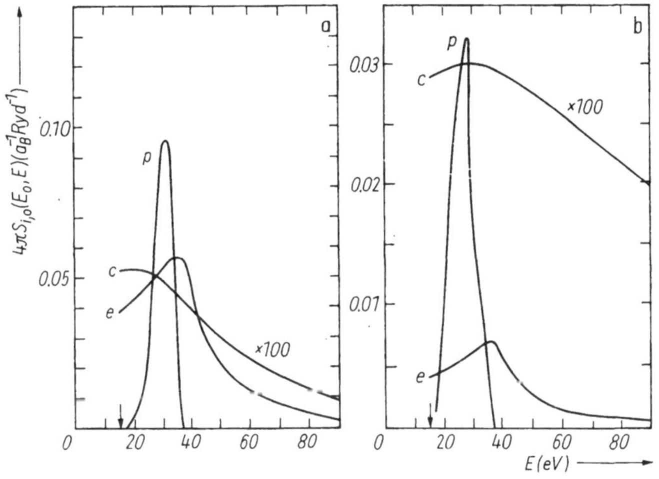

Fig. 12. Energy dependence of different excitation functions S_i (i = e, p, c) at a) $E_0 = 60$ and b) 500 keV. The arrow indicates the vacuum level

The calculation shows that for the essential energies the excitation is nearly isotropic. In terms of the $S^{(\Lambda)}$-approximation we found that the excitation function is very well represented by $\Lambda = 3$.

It seems to be interesting to compare the different excitation functions S_i. In Fig. 12 the energy dependences are plotted for $E_0 = 60$ and 500 keV.

5. Results

As mentioned above the correct excitation functions were replaced by suitable $S^{(\Lambda)}$-approximations. Then (7) splits into the equations (see (51) of RB)

$$\psi_\lambda(E) = (-1)^\lambda S_\lambda(E_0, E) + \int_E dE' \, \psi_\lambda(E') \, K_\lambda^\sigma(E', E); \qquad \lambda = 0, 1, \dots, \Lambda \,, \quad (16)$$

where ψ_λ and K_λ^σ are the expansion coefficients of the distribution function and the scattering function, respectively. The details for solving (16) are described in [9]. Explicit expressions for the different contributions to the mean free path and the scattering function are also given in this paper.

The interesting measurable quantities $j(E)$ and γ are calculated according to

$$j(E) = \theta[E' - E_F - \Phi] \, 2\pi l(E') \sum_\lambda A_\lambda(E') \, \psi_\lambda(E')$$

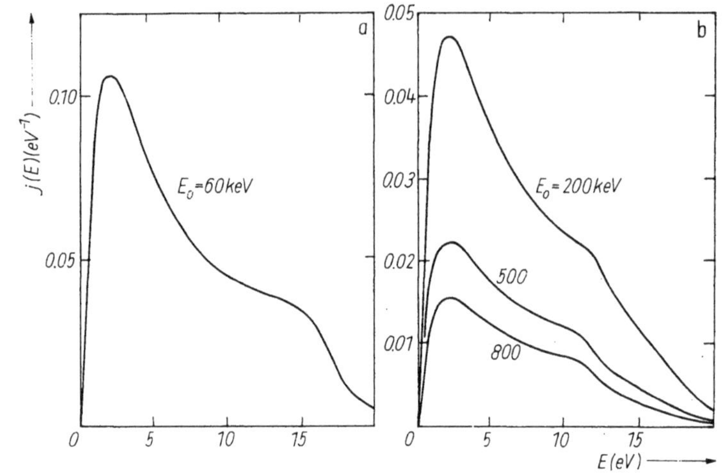

Fig. 13. Energy distribution of ejected electrons for a) $E_0 = 60$ keV and b) different high primary energies

and

$$\gamma = 2\pi \sum_{\lambda} \int_{E_F+\Phi} \mathrm{d}E' \, l(E') \, A_\lambda(E') \, \psi_\lambda(E') \, ,$$

where

$$A_\lambda(E') = \int_{\sqrt{(E_F+\Phi)/E'}}^{1} \mathrm{d}x \, x P_\lambda(x) \, .$$

Φ is the work function. The transformation between outer $(E, \boldsymbol{\Omega})$ and inner $(E', \boldsymbol{\Omega}')$ variables is given by (5) of RB.

In order to obtain the total energy distribution function of escaping electrons we must sum up the contributions of the different excitation mechanisms. In Fig. 13 we plotted this distribution function for different primary energies.

For all primary energies considered we found as a common feature a maximum at ≈ 2 eV. Differences are obvious in the behaviour at higher electron energies. These are attributed to the contribution from excitation by plasmon decay. For high primary energies we can see the plasmon shoulder at 11 to 12 eV being approximately equal to

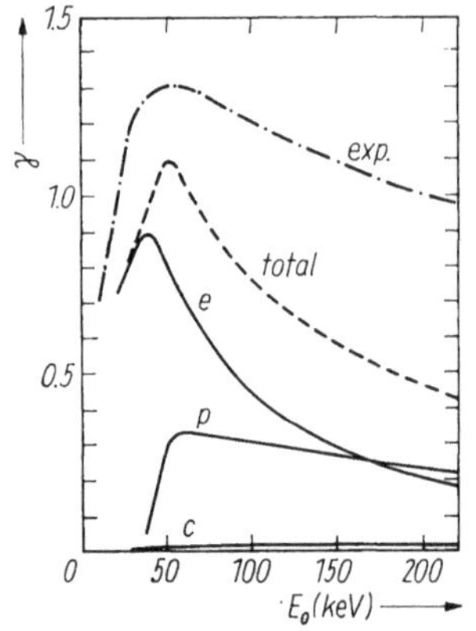

Fig. 14. Primary energy dependence of yield γ

Table 1

Comparison of theoretical and experimental values of γ for high primary energies

theory					experiment [7]
E_0 (keV)	contributions to γ			γ	γ
	e	p	c		
200	0.21	0.23	0.018	0.46	1.1
500	0.08	0.12	0.016	0.22	0.8
800	0.05	0.09	0.013	0.15	0.66

$\hbar\omega_p(q \approx 0) - \Phi$. For low primary energies (see e.g. Fig. 13a for $E_0 = 60$ keV) the shoulder is more pronounced, the energetic position of which is roughly determined by $\hbar\omega_p(q_{\min}(E_0)) - \Phi$.

In Fig. 14 the dependence of γ on E_0 is shown as well as the contributions of the different excitation mechanisms. The results for high primary energies are given in Table 1. At low primary energies the excitation of single conduction electrons is the prevailing mechanism. However, at high primary energies the contribution from the excitation by plasmon decay dominates over the excitation of single conduction electrons.

6. Conclusions

In close analogy to the theory of SEE here a microscopic theory of IEE is developed based on the excitation of conduction electrons, plasmons, and core electrons by H^+ ions in Al.

Several differences result in the excitation functions S_e and S_p compared with the case of SEE, especially for low primary energies $E_0 \lesssim E_{0,\max}$. Plasmon excitation comes only into being for H^+ ions with a primary energy of $E_0 \gtrsim 35$ keV, corresponding to a critical velocity $v_0^{\min} = 2.6 \times 10^8$ cm s^{-1}, which is dependent on the emitting metal only.

But, because Ar^+ ions with an energy $E_0 = 500$ keV (corresponding to $v_0 = 1.53 \times 10^8$ cm s$^{-1} < v_0^{\min}$) give rise to the appearance of plasmon shoulders also [6], one must regard the terms K_p^s in (10) as being responsible for this phenomenon. Owing to the unknown and certainly complicated interaction potential, there does exist up to the present no possibility for developing a quantitative theory of Ar^+-induced electron emission.

As a consequence of collisions of excited electrons with ion cores and metal electrons there comes out a nearly isotropic distribution of excited electrons resulting in a cosine-like angular distribution function of emitted electrons. Of course, this corresponds to the analogous result in SEE.

The theoretical energy distribution $j(E)$ of ejected electrons shows the well-known cascade maximum at ≈ 2 eV, with a halfwidth of ≈ 9 eV, and the plasmon shoulder at 11 to 12 eV resulting from the bulk plasmon decay process in exactly the same manner as in SEE. Due to the infinite medium used in the theoretical treatment of the excitation and transport process, surface plasmons were neglected in the course of calculations.

The electronic yield $\gamma(E_0)$ is in qualitative agreement with existing experimental data (Fig. 14 and Table 1). Especially the value of $E_{0,\max} = 55$ keV (corresponding to a velocity $v_{0,\max} = 1.5e^2/\hbar$) equals the experimental one, thus confirming the considerations of Section 2. The theoretical maximum value $\gamma_{\max} = 1.1$ is to be compared with the experimental value of 1.3. With increasing primary energy the theoretical yield curve lowers too rapid and differs e.g. at 800 keV from the experimental curve by a factor four.

The contributions of conduction electrons and electron excitation by plasmon decay play an overwhelming role compared to the contribution of core electrons, which are much more important in the stopping power of Al for H^+ ions.

References

[1] R. A. Baragiola, E. V. Alonso, and A. Oliva Florio, Phys. Rev. B **19**, 121 (1979).
[2] D. Hasselkamp, K. G. Lang, A. Scharmann, and N. Stiller, Nuclear Instrum. and Methods **180**, 349 (1981).
[3] L. S. Hall, Nuclear Fusion **174**, 681 (1977).
[4] P. G. Coakley and N. Hershkowitz, Phys. Letters A **78**, 145 (1980).
[5] B. Svensson and G. Holmén, J. appl. Phys. **52**, 6928 (1981).
[6] D. Hasselkamp and A. Scharmann, Surface Sci. **119**, L 388 (1982).
[7] D. Hasselkamp and A. Scharmann, Vakuum-Techn. **32**, 9 (1983).
[8] T. E. Everhart, N. Saeki, R. Shimizu, and T. Koshikawa, J. appl. Phys. **47**, 2941 (1976).
[9] M. Rösler and W. Brauer, phys. stat. sol. (b) **104**, 161, 575 (1981).
[10] H. A. Andersen and J. F. Ziegler, Hydrogen Stopping Powers and Ranges in all Elements, Pergamon Press, New York 1977.
[11] J. C. Ashley, C. J. Tung, and R. H. Ritchie, IEEE Trans. Nuclear Sci. **22**, 2533 (1975).
[12] H. Salow, Phys. Z. **41**, 434 (1940).
[13] E. J. Sternglass, Phys. Rev. **108**, 1 (1957).
[14] K. H. Krebs, Vacuum **33**, 555 (1983).
[15] R. Shimizu and T. E. Everhart, Appl. Phys. Letters **33**, 784 (1978).
[16] T. Koshikawa and R. Shimizu, J. Phys. D **7**, 1303 (1974).
[17] P. A. Wolff, Phys. Rev. **95**, 56 (1954).
[18] D. Hasselkamp and A. Scharmann, Phys. Letters A **94**, 259 (1983).
[19] K. Sturm, Adv. Phys. **31**, 1 (1982).
[20] H. Raether, Excitation of Plasmons and Interband Transitions by Electrons, Springer-Verlag, 1980.
[21] A. O. E. Animalu and V. Heine, Phil. Mag. **12**, 1249 (1965).
[22] H. J. Levinson, F. Greuter, and E. W. Plummer, Phys. Rev. B **27**, 727 (1983).

(Received October 4, 1984)

phys. stat. sol. (b) **126**, 643 (1984)

Subject classification: 13.4; 21; 21.1.1

Department of Materials Science and Engineering, Tokyo Institute of Technology, Yokohama[1])

Tight-Binding Electronic Theory for Lattice Defects in Transition Metals

Correlation Effects

By

K. MASUDA-JINDO[2])

A tight-binding type electronic theory is used to calculate the atomic relaxation (δR_i), formation and binding energies (E_{fv} and E_{bin}) of vacancy-type lattice defects in transition metals. The short-range repulsive energies between atomic sites i and j are simulated by the Born-Mayer potential. An electronic correlation contribution is taken into account using a second-order perturbation theory within the Hubbard model. It is shown that correlation effects play a significant role in determining the physical properties of the lattice defects (δR_i, E_{fv}, and E_{bin}) in transition metals. This tendency is observed for the lattice defects on the surface as well.

Eine elektronische "tight-binding"-Theorie wird benutzt, um die atomare Relaxation (δR_i), die Bildungs- und Bindungsenergien (E_{fv} und E_{bin}) von Leerstellen-Gitterdefekten in Übergangsmetallen zu berechnen. Die kurzreichweitigen Abstoßungsenergien zwischen den atomaren Plätzen i und j werden mit dem Born-Mayer-Potential simuliert. Unter Benutzung einer Störungstheorie zweiter Ordnung im Hubbardmodell wird ein elektronischer Korrelationsbeitrag berücksichtigt. Es wird gezeigt, daß Korrelationseffekte eine signifikante Rolle bei der Bestimmung der physikalischen Eigenschaften der Gitterdefekte (δR_i, E_{fv} und E_{bin}) in Übergangsmetallen spielen. Diese Tendenz wird auch für die Gitterdefekte an der Oberfläche beobachtet.

1. Introduction

It is well known that electron correlation of partially filled d-bands plays an important role in determining the cohesive properties of transition metals such as cohesive energy, bulk modulus, surface tension, and elastic constants [1]. In particular, the deviation from the parabolic variation of these quantities (when presented in sequence of the Periodic Table) is considered to arise mostly from the electron–electron interactions in d-bands. On the other hand, it gives a quite important influence on the optical excitation [2, 3], photoemission [4], and Auger spectra [5]. Friedel and Sayers [6], Kajzar and Friedel [7] have calculated the cohesive energy of a narrow band up to the second order in the Coulomb correlation ratio U/W using various approximation schemes (Hubbard, CPA, Gutzwiller), where $U(W)$ denotes the Coulomb integral (bandwidth), and demonstrated the importance of the correlation effects. Similar theoretical works have also been performed by several authors [8, 9].

It is the purpose of the present paper to investigate the effects of Coulomb correlation on the lattice defect properties of metals. Specifically, we calculate the atomic relaxation δR_i, formation and binding energies (E_{fv} and E_{bin}) of vacancy-type lattice defects in b.c.c. transition metals including the correlation effects. This kind of calculation would also be useful and informative for further studies on the lattice defects

[1]) Nagatsuta, Midori-ku, Yokohama 227, Japan.

[2]) K. Masuda-Jindo was formally known as K. Masuda.

of narrow band materials such as V_2O_3 or Sm compounds (SmB_6, SmS) and of the dense Kondo systems (CeB_6, $CeAl_3$, $CeCu_2Si_2$, etc.), where the correlation effects play a central role in discussing various physical properties [10 to 12].

We use the tight-binding (TB) type electronic theory to describe the transition metal d-band. To simulate the short-range repulsive energies between atomic sites i and j, arising mainly from the increase in sp electron kinetic energy upon compression, we use the Born-Mayer type repulsive potential [13]. Coulomb correlation effects are taken into account using the second-order perturbation theory within the Hubbard model [1 to 3]. We will show that the inclusion of correlation effects is important for the calculation of defect properties of transition metals.

2. Method of Calculations

Within the TB approximation and second-order perturbation treatment for the electron–electron interaction energy $E_{\text{el-el}}$, the cohesive energy E_c of transition metals can be expressed as [6]

$$E_c = 2 \sum_\lambda \int^{E_F} E\varrho_0^\lambda(E)\, dE + E_{\text{rep}} + \frac{U}{20} N_d(10 - N_d) - U^2C(N_d)\,, \tag{1}$$

where the factor 2 arises from the spin degeneracy and λ denotes five d-orbitals (xy, yz, zx, $x^2 - y^2$, and $3z^2 - r^2$), $U^2C(N_d)$ is the first correction due to deviations from the Hartree-Fock (HF) term $\Delta E_{HF} = (U/20)\, N_d(10 - N_d)$, owing to inter-atomic Coulomb correlations. The HF term ΔE_{HF} arises from the difference in the electron–electron interaction energy $E_{\text{el-el}}$ between the free atom and crystalline states. In atomic states, the Coulomb interaction energy is $(U/2)\, N_d(N_d - 1)$, since each of the N_d electrons sees $(N_d - 1)$ other electrons and interaction energy must not be counted twice. In a paramagnetic metal, the corresponding term is $(9U/20)\, N_d^2$, since each of the N_d electrons will see $(9/10)\, N_d$ electrons of different spin orbitals. Then, the difference in the $E_{\text{el-el}}$ between the two states becomes equal to ΔE_{HF}, just described above. This HF term reduces the amplitude of the cohesive energy without altering its form [6].

On the other hand, the Coulomb correlation term $U^2C(N_d)$ acts in a different manner. It restricts the atomic charge fluctuations compared to the system with uncorrelated d-electrons. The explicit expression of $C(N_d)$ can be obtained for the rectangular d-band (with bandwidth W_{eff}) and given as

$$U^2C(N_d) = 45 \left\{\frac{N_d}{10}\left(1 - \frac{N_d}{10}\right) U\right\}^2 W_{\text{eff}}^{-1}\,. \tag{2}$$

The second-order correlation term of the above equation (2) arises due to the virtual excitations of each $(N_d/10)$ electron per atom in a given spin orbital, into the empty portions $(1 - N_d/10)$ of orbitals with the same spin direction. As will be shown later, this term plays an important role in determining the lattice defect properties of transition metals. The more general form of $U^2C(N_d)$ has been given by Treglia et al. [3].

In the present study the band energy E_B (first term of (1)) is calculated in the TB model assuming a Gaussian shape [13 to 15] for the local density of states (DOS) $\varrho_i(E)$. This approximation scheme reproduces exactly the first and second moments μ_{1i} and μ_{2i}, and works well for the energetics of transition metals [14 to 16]. We take the free-atom level as the origin of the energies and take into account the diagonal matrix element V_i $(= \mu_{1i})$ of the self-consistent potential resulting from charge oscillations near the lattice defects. In the present calculation, V_i is determined from the condition that each atom is electrically neutral [14, 15],

$$V_i = (1 - \sqrt{\mu'_{2i}/\mu_2})\, E_F\,, \qquad \mu'_{2i} = \mu_{2i} - (\mu_{1i})^2\,. \tag{3}$$

For the calculation of the second moments μ_{2i}, we assume that hopping integrals h_{ij} between atomic sites i and j vary exponentially with the interatomic distance R_{ij} as [13 to 16]

$$h_{ij} = h_0 \exp(-qR_{ij})\,, \tag{4}$$

where the parameter q can be chosen so as to fit the elastic constants or phonon dispersion curve of the given transition metal. We take into account the correlation terms for each atomic site R_i by using a rectangular DOS of width W^i_{eff}, fitted to give the exact second moment on the atom R_i. This is known to be an excellent approximation [17]. To account for the repulsive core–core interaction energies at small distances, mainly due to the increase in sp kinetic energy upon compression, we use a Born-Mayer potential $C_0 \exp(-pR_{ij})$. The parameters appearing in the Born-Mayer potential are fitted to the experimental cohesive energy E_c and bulk modulus K (or elastic constants C_{ijkl}) [14 to 16, 20].

It is now possible to calculate the change in the total (electronic and repulsive) energies due to introduction of lattice defects. To calculate the atomic configurations around the vacancy-type lattice defects, we have relaxed the atoms in a spherical crystal region with radius $4(2R_0/\sqrt{3})$ (R_0 being the equilibrium interatomic distance), using a direct energy minimization procedure [16], and taking into account the first- and second-nearest-neighbour interactions. The choice of the relaxation region is made by considering the size dependence of the defect formation energies. Even for the larger crystallite with radius $6(2R_0/\sqrt{3})$, we have obtained almost the same defect formation energies (the difference between the results is less than about 2%).

3. Numerical Results and Discussions

The present numerical calculations are performed using parameter values presented in Table 1. We have calculated atomic relaxation δR_i, formation and binding energies (E_{fv} and E_{bln}) of single and divacancies in b.c.c. transition metals α-Fe, Mo, and W. The values chosen for U are such that $U/W_{\text{eff}} = 0.15$ to 0.20, which give confidence in the use of second-order perturbation theory, and are similar to those used by Desjonqueres and Spanjaard [18] for the adatom binding and migration energy calculations of Mo and W: The values of U in transition metals are generally estimated to be $U \approx 1$ to 3 eV (taking into account the screening effect) [19].

Table 1

Parameters qR_0 and pR_0, and calculated vacancy formation energies E_{fv} (eV). Values in parentheses are obtained for unrelaxed atomic configurations

	qR_0	pR_0	E (eV/atom)	U/W_{eff}	$E_{\text{fv}}^{\text{cal}}$ (eV)	$E_{\text{fv}}^{\text{exp.}}$ (eV)
α-Fe	2.6	12.8	4.2	0.00	1.386 (1.604)	1.3 [37]
				0.15	1.471 (1.746)	2.0 ± 0.2 [38]
				0.20	1.436 (1.706)	
Mo	2.4	12.8	7.2	0.00	2.435 (2.835)	2.5 to 3.0 [39]
				0.15	2.593 (3.036)	
W	2.4	13.0	8.2	0.00	2.801 (3.247)	3.3 [39]
				0.15	2.961 (3.428)	

Table 2
Lattice relaxation around a bulk vacancy*)

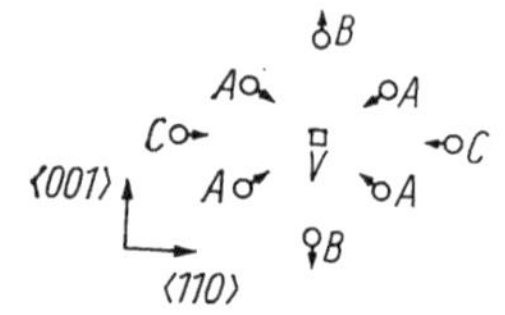

	α-Fe		Mo		W	
U/W_{eff}	0.0	0.15	0.0	0.15	0.0	0.15
$\delta_A R_0$	0.0368	0.0483	0.0404	0.0479	0.0391	0.0484
δ_B/R_0	−0.0251	−0.0286	−0.0273	−0.0293	−0.0267	−0.0285
δ_C/R_0	0.0096	0.0120	0.0110	0.0117	0.0105	0.0119

*) The direction and length of the arrows are schematic. Positive (negative) δ_i value indicates inward (outward) atomic relaxation.

The calculated formation energy E_{fv} and atomic relaxation δR_i of a single vacancy are presented in Tables 1 and 2, respectively: Energies are given in units of eV. In order to see the effects of Coulomb correlation, we also present the results obtained by using $U = 0$ [20]. As shown in Table 1, E_{fv} values with $U = 0$ are systematically smaller than the corresponding experimental results. However, we are able to obtain larger vacancy formation energies (better agreement with experimental results) by using values $U \neq 0$, i.e., by explicitly including the electron–electron interaction terms. This fact can be explained as follows: In the calculation of E_{fv}, most of the positive contribution comes from the change in the band energy term E_B while that of the negative contribution from the Born-Mayer repulsive term. One obtains the positive vacancy formation energy since the former contribution is larger than the latter. Furthermore, in order to reproduce the experimental cohesive energy with $U \neq 0$, one must take larger (but with negative sign) energy gain E_B due to d-band broadening, compared to the calculation with $U = 0$. This gives rise to the larger values of hopping integrals and enhances the resulting E_{fv} values.[3])

The magnitude of atomic relaxation around a vacancy is small and always less than 5% (in units of the nearest-neighbour distance R_0 of the perfect lattice), both for the calculations with $U = 0$ and with $U/W_{\text{eff}} = 0.15$ to 0.20 (as shown in Table 2). This is consistent with the previous theoretical calculations of lattice relaxations around a single vacancy [14, 15] or the relaxations near the surface of d-band metals [21]. However, it is worth pointing out that the considerably different (larger) atomic relaxations are obtained for the calculations with $U/W_{\text{eff}} \neq 0$.

Similar (and more pronounced) differences are also seen in the E_{bin} calculations of divacancies in transition metals (Table 3). In general, when the correlation terms are taken into account, E_{bin} increases, the effect being most pronounced for second-neighbour divacancy pairs. Here, it is interesting to note that next-nearest-neighbour divacancies are more stable than the first-neighbour divacancies in b.c.c. transition

[3]) However, the situation is not so simple: As shown in Table 1, E_{fv} calculated with $U/W_{\text{eff}} = 0.20$ for α-Fe is slightly smaller than that with $U/W_{\text{eff}} = 0.15$. This is due to the fact that E_{fv} is in fact determined from the two competitive, U-dependent contributions. The similar unexpected results are also observed for the E_{bin} calculations of the unrelaxed divacancy pairs (W) in Table 3.

Table 3

Calculated binding energies E_{bin} of divacancies both for the relaxed (a) and unrelaxed (b) atomic configurations. Energies are given in units of eV*)

configuration	α-Fe		Mo		W	
	U/W_{eff}					
	0.0	0.15	0.0	0.15	0.0	0.15
(a) 1	0.044	0.025	0.077	0.082	0.082	0.094
2	0.054	0.138	0.104	0.288	0.129	0.315
3	−0.181	−0.171	−0.360	−0.266	−0.388	−0.309
(b) 1	0.224	0.247	0.393	0.428	0.451	0.396
2	0.150	0.154	0.254	0.285	0.291	0.232
3	−0.032	−0.031	−0.053	−0.049	−0.060	−0.151

*) Configurations 1, 2, and 3 denote the first-nearest-, second-nearest-, and third-nearest-neighbour divacancy pairs, respectively. Negative (positive) E_{bin} represents the repulsive (attractive) interaction.

metals. This conclusion is consistent with the earlier calculations by Johnson [22] as well as with the recent experimental observations using the perturbed angular correlation (PAC) [23] or Mössbauer spectroscopy [24] techniques. These experimental techniques (though impurities are associated) have enabled us to get information on the configuration of vacancy-type lattice defects in metals, not only for divacancies but also for larger vacancy clusters.

The above-mentioned correlation effects on the defect properties can be considered to arise from the site dependence of the effective bandwidth W^i_{eff}, which can be estimated by the relation

$$W^i_{eff} = 2\sqrt{3\mu_{2i}} \, . \tag{5}$$

From the expression of the second-order perturbation energy of (2), one can see that Coulomb correlation terms play a more significant role for the systems with larger U/W_{eff} values. This means that the correlation effects become more important for atoms with larger change in the effective bandwidth W^i_{eff} (for the present case smaller coordination number Z_i): Accordingly, the atomic relaxation around the defects is significantly altered by them. In this respect, it is noted that there are four atomic sites having six nearest neighbours (compared to eight in the perfect lattice) around the next-nearest-neighbour divacancy in b.c.c. metals. In contrast, for the nearest-neighbour divacancies in b.c.c. metals such atomic sites do not exist. For this reason, binding energies obtained for the next-nearest-neighbour divacancy pairs with $U/W_{eff} \neq 0$ are markedly different from the corresponding values with $U/W_{eff} = 0$.

In addition to the importance of the correlation effects, we would like to point out the importance of the atomic relaxations in the E_{bin} calculations. As can be seen in Table 3, E_{bin} for the relaxed atomic configurations are significantly different from those of the unrelaxed atomic configurations. For divacancies in b.c.c. transition metals, unrelaxed binding energies E_{bin} of the first-neighbour pairs are greater than those of the next-nearest-neighbour pairs, whilst the reversed results are obtained for the relaxed divacancy configurations. This behaviour is common to the calculations with $U \neq 0$ and with $U = 0$ (independent of the inclusion of correlation effects), and indicates that the atomic relaxations around the vacancy-type lattice defects play a dominant role in determining the binding energies. Therefore, it should be

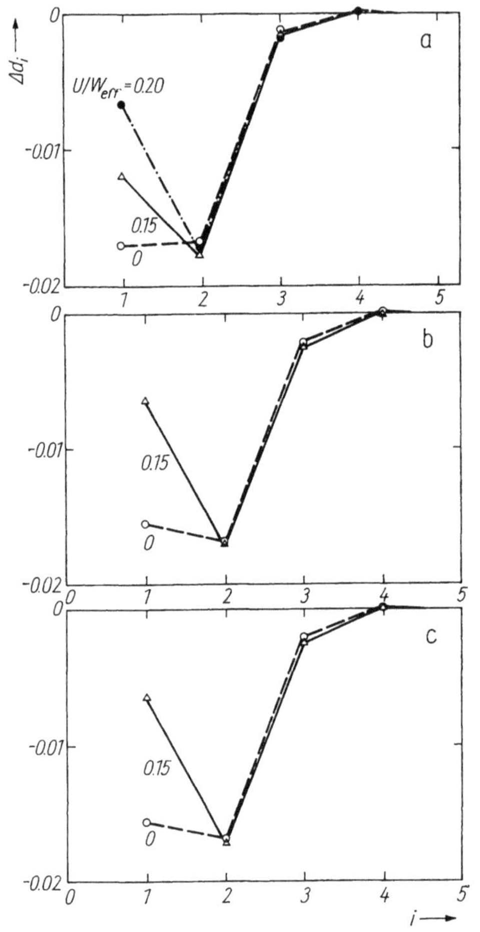

Fig. 1. Surface atomic relaxation Δd_i, between i-th and $(i + 1)$-th atomic planes, calculated for a) Fe(001), b) Mo(001), and c) W(001). Both $U/W_{eff} = 0$ and 0.15 (0.20 for α-Fe) are used. (For details of Δd_i see the text)

emphasized that satisfactory results might not be expected from the binding energy calculations of the unrelaxed vacancy–impurity, impurity–impurity, or vacancy–vacancy pairs [25 to 27].

In the present work, we have further investigated the effects of Coulomb correlation on the property of surface lattice defects. Using the same calculational method as mentioned above, we have also calculated the atomic relaxations of cleaved surface and of the surface vacancies of b.c.c. transition metals α-Fe, Mo, and W. Surface relaxation calculations are performed for the six atomic layers (first surface layer and five subsurface layers) of the Fe(001), Mo(001), and W(001) surfaces. This choice of relaxation region is quite sufficient for the present surface relaxation calculations. The calculated atomic relaxations of these surfaces with $U/W_{eff} = 0$, 0.15, and 0.20 are presented in Fig. 1a, b, and c, respectively: The changes in the interatomic layer spacing Δd_i, between i-th and $(i + 1)$-th atomic planes, are presented in units of the equilibrium bulk interlayer spacing d_0. In general, Δd_i for the first three surface layers are important and visible in the figures. One can see in Fig. 1 that the inclusion of correlation effects reduces significantly the first interlayer spacing compared to the calculations with $U/W_{eff} = 0$. This is due to the fact that the inward surface relaxation increases the local d-bandwidth W_{eff} of the surface atoms and thereby increases the Coulomb correlation energies. For this reason, the surface relaxation (first surface layer) becomes smaller when $U/W_{eff} \neq 0$ is used.

The calculated atomic relaxation around a surface vacancy is summarized in Table 4. One can see in Table 4 that the Coulomb correlation plays a significant role in determining the atomic relaxation around a surface vacancy: The magnitude of atomic relaxation depends sensitively on the value of U as in the case of the bulk vacancy. One can also see that the relaxation around a surface vacancy is considerably different in nature from that around a vacancy in bulk crystal. For instance, the second-

Table 4
Lattice relaxation around a surface vacancy*)

first layer

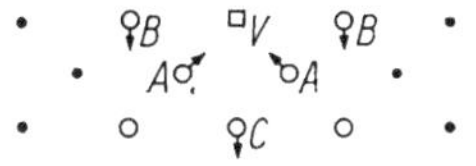

	α-Fe		Mo		W	
U/W_{eff}	0.0	0.15	0.0	0.15	0.0	0.15
δ_A/R_0	0.0067	0.0068	0.0074	0.0078	0.0069	0.0073
δ_B/R_0	−0.0041	−0.0020	−0.0043	−0.0023	−0.0021	−0.0012
δ_C/R_0	−0.0114	−0.0112	−0.0117	−0.0127	−0.0107	−0.0108

*) The direction and length of the arrows are schematic. Positive (negative) δ_i value indicates the atomic relaxation toward the surface vacancy (subsurface direction). Displacement of B atom parallel to the surface plane is almost zero and not included.

nearest-neighbour atom around a surface vacancy, denoted by B in Table 4, moves during the relaxation calculation towards the subsurface direction, and the relaxation parallel to the surface plane is quite small (practically equal to zero). This atomic relaxation is in marked contrast to that observed for the bulk vacancy. This means that the information on the atomic relaxation around a bulk vacancy cannot be used directly for the surface vacancies. The similar effects can also be expected for vacancies in the dislocation core region [28] or for vacancies near grain boundaries and near solid–solid interfaces [29].

In order to investigate the detailed electronic structure of the surface vacancy, we have calculated the local DOS $\varrho_i(E)$ on atoms near the surface vacancy and presented the results for Mo(001) in Fig. 2: The d-band parameters ddσ, ddπ, and ddδ of Mo are taken from [25]. The one-particle spectrum $n_i(E)$ in the presence of correlations

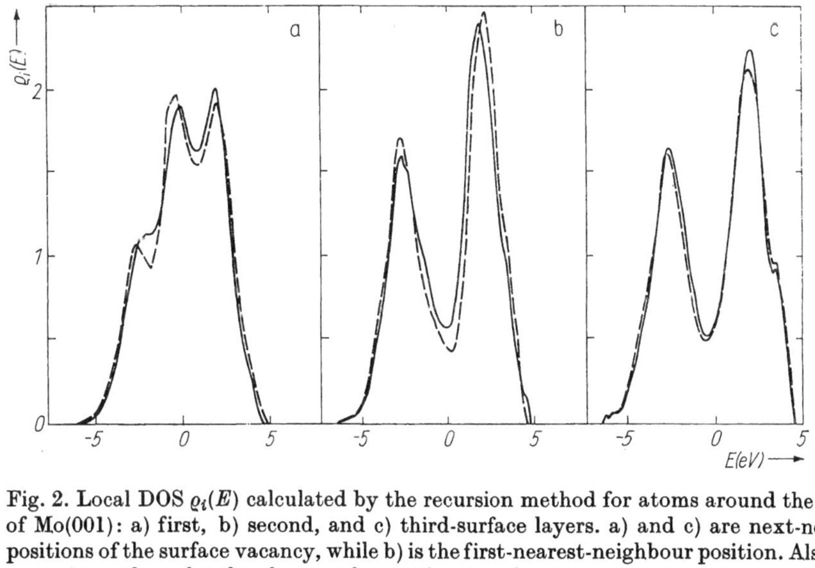

Fig. 2. Local DOS $\varrho_i(E)$ calculated by the recursion method for atoms around the surface vacancy of Mo(001): a) first, b) second, and c) third-surface layers. a) and c) are next-nearest-neighbour positions of the surface vacancy, while b) is the first-nearest-neighbour position. Also shown (dashed curves) are those for the clean surface without surface vacancy

has also been calculated from

$$n_i(E) = -(1/5\pi) \, \mathrm{Im} \sum_\lambda g_{ii}^{\lambda\lambda} \left(E - \Sigma_s(E)\right) , \tag{6}$$

with $U/W = 0.15$ and shown in Fig. 3. In the above equation (6), the single-particle self-energy $\Sigma_s(E)$ has been calculated using a so-called local approximation [3, 17, 18]. The validity of using this approximation scheme for transition metal surface has been discussed by Treglia et al. [17, 31]. In general, one can see in Fig. 2 that the local DOS $\varrho_i(E)$ (or one-particle spectra $n_i(E)$) around a surface vacancy are considerably different from the corresponding results for the clean surface (without the surface vacancy): These changes in the electronic structure due to the surface vacancy (Fig. 2 and 3) are larger than the DOS changes between the reconstructed and unreconstructed Mo(001) surfaces [30]. This indicates that the catalytic activity of surface vacancies is also considerably different from that of the clean and flat surface. In this respect, we would like to point out that the photo-electrochemical activities of the reduced (bombarded by the Ar beam) TiO_2 surface is much stronger than that of the stoichiometric surface [32]. The defect surface state induced by the oxygen vacancy can be considered to give rise to the pronounced surface chemical activity [33, 34].

In conclusion, we have calculated the atomic relaxation, formation and binding energies of vacancy-type lattice defects in b.c.c. transition metals using the TB-type electronic theory and including the Coulomb correlation effects. We are able to investigate systematically the correlation effects on the lattice defect properties of transition metals. It has been shown that the correlation effects play an important role in determining the lattice defect properties of transition metals (though similar effects have already been pointed out for cohesive properties of transition metals

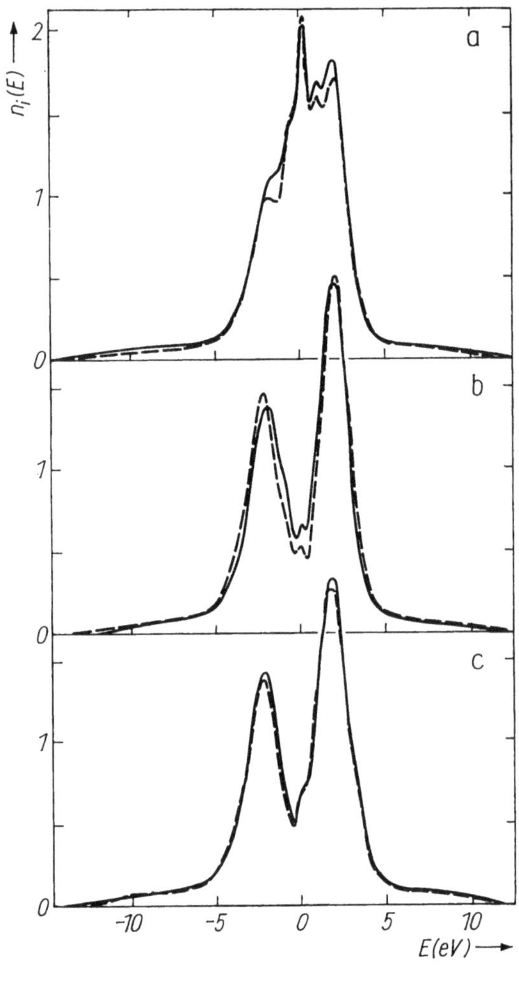

Fig. 3. One-particle spectra $n_i(E)$ for atoms around the surface vacancy of Mo(001): a), b), and c) are the corresponding results of Fig. 2a, 2b, and 2c. $U/W = 0.15$ is used. Also shown (dashed curves) are those for the clean surface without surface vacancy

[1 to 9]): The atomic relaxation around the defects are considerably altered due to the correlation effects and larger E_{fv} and E_{bin} (especially for the next-nearest-neighbour divacancies) values are obtained for vacancies in b.c.c. transition metals. In addition, the detailed electronic structures around the surface vacancy have been investigated using the calculated atomic configurations, in relation to the surface catalytic activities.

Finally, we note that the correlation effects are expected to play a significant role in other lattice defect problems [35, 36]. One of the most important unresolved problems on lattice defects in metals is concerned with the calculation of Peierls stress τ_P. For instance, calculated τ_P values of b.c.c. transition metals are usually one order of magnitude larger than the corresponding experimentals results [36]. This discrepancy might be removed to some extent by including the correlation effects: As mentioned before, the correlation term becomes more important for atomic sites with larger changes in the effective d-bandwidth W^i_{eff}. In fact, there exist considerable differences of the second moment $\mu_{2i}(W^i_{eff})$ in the local DOS between low- and high- (saddle point) energy configurations of screw dislocation [36]. However, in order to understand the properties of various lattice defects more quantitatively, more accurate and material-dependent U-values are required for specific transition metals [17]. Such studies would also be very useful for the further studies on the lattice defect properties of narrow-band materials.

References

[1] J. Friedel, Physics of Metals, Ed. J. M. Ziman, Cambridge University Press, 1969.
J. Friedel and C. M. Sayers, J. Physique Lettres **38**, L-263 (1977).

[2] G. Treglia, M. C. Desjonqueres, F. Ducastelle, and D. Spanjaard, Solid State Commun. **39**, 1113 (1981).

[3] G. Treglia, M. C. Desjonqueres, F. Ducastelle, and D. Spanjaard, J. Phys. C **13**, 6063 (1980).

[4] D. R. Penn, Phys. Rev. Letters **42**, 921 (1979).

[5] P. A. Bennett, J. C. Fuggle, F. U. Hillebrecht, A. Lenselink, and G. A. Sawatzky, Phys. Rev. B **27**, 2194 (1983).

[6] J. Friedel and C. M. Sayers, J. Physique **38**, 697 (1977).

[7] F. Kajzar and J. Friedel, J. Physique **39**, 397 (1978).

[8] A. M. Oles, Phys. Rev. B **23**, 271 (1981).
G. Allan, J. Friedel, and C. M. Sayers, J. Physique Lettres **41**, L-287 (1980).

[9] C. M. Sayers and F. Kajzar, J. Phys. F **11**, 1055 (1981).

[10] K. Yamada and K. Yoshida, Electron Correlation and Magnetism in Narrow-Band Systems, Ed. T. Moriya, Springer-Verlag, 1981 (p. 210).

[11] A. S. Edelstein, C. J. Tranchita, O. D. McMasters, and K. A. Gschneider, Solid State Commun. **15**, 81 (1974).

[12] B. H. Brandow, Adv. Phys. **26**, 651 (1977).
T. Oguchi, K. Terakura, and A. R. Williams, Phys. Rev. B **28**, 6443 (1983).

[13] F. Ducastelle, J. Physique **31**, 1055 (1970).

[14] G. Allan and M. Lannoo, J. Phys. Chem. Solids **37**, 699 (1976).

[15] G. Allan and M. Lannoo, phys. stat. sol. (b) **74**, 409 (1976).

[16] K. Masuda and A. Sato, Phil. Mag. **37B**, 531 (1978).

[17] G. Treglia, F. Ducastelle, and D. Spanjaard, J. Physique **41**, 281 (1980).

[18] M. C. Desjonqueres and D. Spanjaard, J. Phys. C **15**, 4007 (1982).

[19] G. Treglia, F. Ducastelle, and D. Spanjaard, J. Physique **43**, 341 (1982).

[20] K. Masuda, phys. stat. sol. (b) **112**, 609 (1982).

[21] K. Masuda, Phys. Rev. B **26**, 5968 (1982).

[22] R. A. Johnson, Phys. Rev. **134**, A1329 (1964).

[23] A. Weidinger, R. Wessner, E. Rechnagel, and Th. Wichert, Nuclear Instrum. and Methods **182/183**, 509 (1981).

[24] S. R. REINTSEMA, E. VERVIEST, J. ODEURS, and H. PATTYN, J. Phys. F 9, 1511 (1979).
[25] K. V. SATHYARAJ, D. ABLITZER, and C. DEMANGEAT, Phil. Mag. A 40, 541 (1979).
[26] C. DEMANGEAT, M. A. KHAN, G. MORAITOS, and J. C. PARLEBAS, J. Physique 41, 1001 (1980).
[27] E. KAUFFER and M. GEHL, J. Physique 35, 839 (1974).
[28] K. MASUDA, phys. stat. sol. (b) 99, 593 (1980).
[29] A. YANIV, Phys. Rev. B 22, 4776 (1980).
[30] I. TERAKURA, K. TERAKURA, and N. HAMADA, Surface Sci. 111, 479 (1981).
K. MASUDA, N. HAMADA, and K. TERAKURA, J. Phys. F 14, 47 (1984).
[31] G. TREGLIA, F. DUCASTELLE, and D. SPANJAARD, J. Magnetism magnetic Mater. 15/18, 1151 (1980).
[32] W. J. LO, Y. W. CHUNG, and G. A. SOMORJAI, Surface Sci. 71, 199 (1978).
V. E. HENRICH, G. DRESSELSHAUS, and H. J. ZEIGER, J. Vacuum Sci. Technol. 15, 534 (1978).
[33] M. TSUKADA, N. TSUDA, and F. MINAMI, J. Phys. Soc. Japan 49, 1115 (1980).
[34] M. TSUKADA, C. SATOKO, and H. ADACHI, J. Phys. Soc. Japan 47, 1610 (1979).
[35] J. R. BEELER, Radiation Effects Computer Experiments, Series on Defects in Solids, Vol. 13, North-Holland Publ. Co., Amsterdam 1983.
[36] A. SATO and K. MASUDA, Phil. Mag. B 43, 1 (1981).
[37] H. KIMURA, Progress in the Study of Point Defects, Ed. M. DOYAMA and S. YOSHIDA, University of Tokyo Press, Tokyo 1977 (p. 199).
[38] L. DE SCHEPPER, D. SEGERS, L. DORIKENS-VANPRAET, M. DORIKENS, G. KNUYT, L. M. STALS, and P. MOSER, Phys. Rev. B 27, 5257 (1983).
[39] L. KORNBLIT, Physica (Utrecht) 106B, 351 (1981).

(Received June 29, 1984)

phys. stat. sol. (b) 126, 653 (1984)

Subject classification: 13.4 and 20.3; 22.5.2

Institute of Physics, Academy of Sciences of the Estonian SSR, Tartu[1])

Decay Kinetics and Polarization of the A_X Emission of KI:In Crystal

By

V. NAGIRNYI, T. SOOVIK, and S. ZAZUBOVICH

The decay kinetics and polarization of the A_X emission are investigated for In^+ centres in KI:In at 0.35 to 80 K. The parameters of the relaxed A_X state are determined. Good agreement is obtained between the experimental results and theory, which suggests a strong interaction of impurity optical electrons with non-totally symmetric vibrations of the crystal lattice.

При 0,35 до 80 K исследована кинетика затухания и поляризация A_X-излучения In^+-центров в KI:In. Определены параметры релаксированного A_X-состояния. Получено хорошее согласие экспериментальных результатов с теорией, учитывающей сильное взаимодействие оптических электронов In^+ с неполносиметричными колебаниями кристаллической решетки.

1. Introduction

The emission spectrum of In^+ centres in KI:In on excitation in the A absorption band consists of two bands peaking at 2.81 eV (A_T) and 2.2 eV (A_X). They are caused by electronic transitions from two kinds of minima (T and X) coexisting on the adiabatic potential energy surface (APES) of the triplet excited state of the In^+ centre [1].

The A_T emission is caused by electronic transitions from tetragonal minima [2]. As stated in [3, 4], the properties of A_T emission and the structure of the A_T relaxed excited state (RES) of KI:In are described adequately in terms of the model [5, 6]. The trigonal symmetry of the minima responsible for the A_X emission is proposed in [7 to 9] where the effect of magnetic field and hydrostatic pressure on the A_X band has been investigated. More directly the symmetry of the A_X RES is determined in [10], where the polarization of the A_X emission has been found and studied.

The structure of the A_X RES of mercury-like centres in alkali halides and the processes occurring in this state are not sufficiently studied. It should be of special interest to investigate the applicability of the model [5, 6] to the description of the A_X luminescence properties of these systems. Such an investigation has been performed by us in [11] for Ga^+ centres in KCl:Ga and KBr:Ga crystals. Here we continue to study the A_X emission properties on the example of KI:In crystal.

2. Experimental Procedure

A KI:In single crystal was grown by the Stockbarger method from zone-refined KI salt. The method of preparation was described in [4]. The In^+ concentration was 0.9×10^{17} cm^{-3}. The experimental set-up and procedure were the same as described in [4, 12]. The polarization was observed head-on to the exciting light through a polaroid and a system of optical filters. The electric vector of exciting light $\boldsymbol{E}_{exc}$ was oriented along the [011] axis of the crystal ($\alpha = 45°$) when the spectral and temperature dependences of the polarization degree were measured. The azimuthal depend-

---

[1]) Riia 142, 202400 Tartu, USSR.

ences of the polarization degree were obtained by varying the angle α between $\boldsymbol{E}_{exc}$ and the [001] direction. The degree of polarization was defined by

$$P = \frac{I_{\parallel} - I_{\perp}}{I_{\parallel} + I_{\perp}},$$

where $I_{\parallel}$ and $I_{\perp}$ are the intensities of the emission polarized parallel or perpendicular to the direction of $\boldsymbol{E}_{exc}$. Due to such experimental conditions and the small intensity of A_X emission at $T < 50$ K, a careful separation from the emission of $(In^+)_2$ centres[2]) as well as from the much more intensive A_T emission was necessary. In our experiments the intensity of the disturbing emissions was not more than some per cents of the intensity of A_X emission.

3. Experimental Results

3.1 Polarization of emission

Spectral and polarization characteristics of the stationary A_X emission of KI:In at 4.2 K are shown in Fig. 1. A_X emission is polarized only at excitation in B and A absorption bands, the polarization degree being positive and growing as the energy of exciting photons decreases (curve 3). It is quite similar to the polarization spectrum of the A_T emission of KI:In [3, 4] as well as to that of the A_X emission of Ga^+ centres [11]. The maximum value of the polarization degree is small[3]) — about 6%. Probably at low temperatures the population of the X minima occurs mainly through the radiationless (tunnel) transitions from T minima preferentially populated optically. It leads to the strong depolarization of A_X emission. In [13] such T → X tunnel transitions have been observed in KI:Tl crystal.

The azimuthal dependence of the polarization degree (curve 4) is analogous to that observed in [10] for KI:In and in [11] for KCl:Ga and KBr:Ga, and it points to the trigonal symmetry of the X minima.

The polarization degree does not change as the temperature rises from 0.35 to 50 K (curve 5). The intensity of the A_X emission (curve 6) is also almost temperature

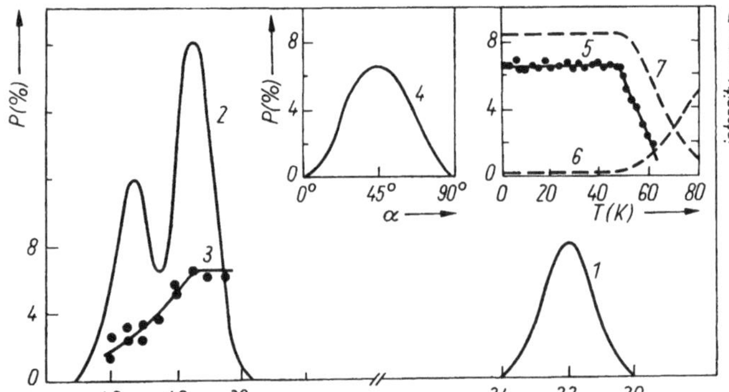

Fig. 1. Emission (1), excitation (2), and polarization (3) spectra as well as the azimuthal dependence of the polarization degree (4) of the A_X emission of KI:In at 4.2 K. Temperature dependence of the polarization degree (5) and intensity (6) of A_X emission, and the intensity of A_T emission (7)

---

[2]) The low-energy emission band of $(In^+)_2$ centres peaking at 2.27 eV is excited preferentially at 4.25 and 3.77 eV absorption bands.

[3]) The polarization spectrum obtained for A_X emission in [10] differs from curve 3 especially on exciting by photons with $h\nu > 3.9$ eV. The polarization degree (+30%) exceeds the limit value for the slow component of the A_X emission (+25% [6]).

independent in this region. The thermally stimulated $T \rightarrow X$ transitions occur at $T > 50$ K (curve 7) and lead to a thorough depolarization of A_X emission (curve 5).

The fact that the A_X emission is polarized at $T < 2.8$ K, when it is caused by the transitions from metastable APES minima, is very important. It points to the low (trigonal) symmetry of metastable minima as predicted by the theory [5, 6].

3.2 Decay kinetics of emission

Being analogous to the A_T emission [3, 4] the fast decay component is not observable in the A_X emission of KI:In even at 0.35 K.

At very low temperatures rather interesting peculiarities appear in the decay kinetics of the A_X emission as well as of the A_T emission of KI:In. Namely, two slow components can be observed in the decay of both emissions having the decay times $\tau_1 = 650$ μs, $\tau_1' = 1600$ μs for A_X band[4]) (Fig. 2) and $\tau_1 = 350$ μs, $\tau_1' = 950$ μs for A_T band. Their decay times and light sums do not change practically as the temperature rises up to 2 to 3 K. It means that both components are caused by transitions from metastable APES minima. The light sum of the slowest component of A_X emission is by about eight times larger, but that of the A_T emission is about two times smaller than the light sum of the faster component. For clarifying the main features of these components, the A_T emission has been studied in detail at low temperatures because of its larger intensity. The emission and excitation spectra of both decay components of the A_T emission coincide. The values of τ_1 and τ_1' are independent of the energy of emitted and exciting photons. The polarization characteristics of both components are qualitatively similar. The behaviour of the faster component of the A_T emission ($\tau_1 = 350$ μs) in the magnetic field can be described in detail in terms of the model [5, 6] (cf. also [3]). The value of τ_1' depends much more weakly on the magnetic field.

The reason for the existence of these two components in the decay of both A_T and A_X emissions occurring from the metastable minima of RES is not clear. We are sure that these two components do not arise from centres of different geometrical structure. The similar azimuthal dependences for both components of each emission band eliminate the possibility of explaining them by the tunnel splitting of metastable minima (see [6]). We are inclined to think that this effect may be explained by the strong hyperfine interaction in the In^+ centre (nuclear spin $I = 9/2$).

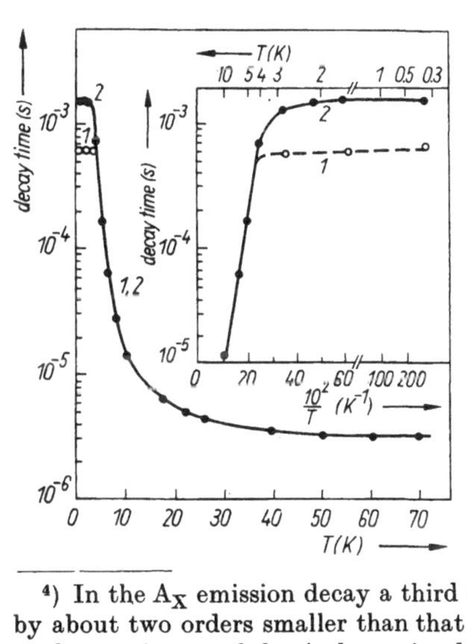

Fig. 2. Temperature dependence of the decay times $\tau(1)$ and $\tau'(2)$ of the A_X emission of KI:In

[4]) In the A_X emission decay a third component is also observed with $\tau = 20$ μs. Its light sum is by about two orders smaller than that of the slowest component. It is preferentially excited not in the maximum of the A absorption band of In^+ but in the 4.25 and 3.77 eV absorption bands. Probably, this component belongs to $(In^+)_2$ centres.

At $T > 4.2$ K both decay components of A_X emission (as well as those of A_T emission) cannot be separated, and the decay time shortens sharply as the temperature rises (Fig. 2). Above 50 K the decay time of A_X emission becomes independent of temperature again and is $\tau_2 = 3 \times 10^{-6}$ s (see also [1, 14]). A similar $\tau(T)$ dependence has been observed for the A_T emission of $Sn^{2+}v_c^-$ centres in alkali halides [15].

4. Discussion

4.1 Parameters of the A_X relaxed excited state

In model [5, 6] below the radiative doublet trigonal minima there are metastable singlet trigonal minima. The transition probabilities from radiative and metastable minima are denoted by γ_2 and γ_1, respectively. The probability of non-radiative transitions from radiative to metastable minima of the same orientation is p, and the energy distance between them is denoted by δ. Then the probability of thermal transitions from metastable to radiative minima is $p \exp(-\delta/kT)$. Using the value of $\tau = \tau_1$ in the low-temperature region, we may determine $\gamma_1 : \gamma_1 = \tau_1^{-1}$. Using the value of $\tau = \tau_2$ in the high-temperature region, we get $\gamma_2 = \frac{3}{2}\tau_2^{-1}$.

From the general equation [12] for $\lambda_s = \tau_s^{-1}$ (τ_s is the slow-component decay time at certain temperature T)

$$\lambda_s = \tfrac{1}{2}[\gamma_2 + \gamma_1 + p(1 + 2\,e^{-\delta/kT}) - \\ - \sqrt{p^2(1 + 2\,e^{-\delta/kT})^2 + 2(\gamma_2 - \gamma_1)\,p(1 - 2e^{-\delta/kT}) - (\gamma_2 - \gamma_1)^2}]\,; \tag{1}$$

taking into account that for In^+ centre $\gamma_2 \gg \gamma_1$, we obtain

$$p = \frac{\lambda_s(\lambda_s - \gamma_2)}{\lambda_s + 2\,e^{-\delta/kT}\,(\lambda_s - \gamma_2)}\,. \tag{2}$$

Assuming a weak dependence of p on temperature, one can get the value of δ by equating (2) at two close temperatures T_1 and T_2 (5 and 8 K) and the corresponding values of λ_{s1} and λ_{s2}. When δ is known, we can obtain the value of p from (2).

The parameters of the A_X RES are shown in Table 1. For a comparison the corresponding parameters for A_T emission are represented [3]. (In both cases p is given for $T = 5$ K.)

It is evident from Table 1 that in KI:In the probabilities of the radiative decay of metastable (γ_1, γ_1') as well as of the radiative (γ_2) minima of the A_X state are smaller than those of A_T state. It may be caused by the stronger interaction of In^+ optical electrons with trigonal vibrations.

In [16] the dependence of electronic wave functions on the coordinate of non-totally symmetric vibrations and the hyperfine interaction have been considered as the possible reasons of the radiative decay of metastable minima. On the basis of [16] we may conclude for KI:In that the radiative decay of the metastable A_T and A_X minima of KI:In (the nuclear spin $I = 9/2$) is essentially caused by hyperfine interaction. Really, in this case

$$\gamma_1 = \frac{2}{3}\left(\frac{A}{\delta}\right)^2 I(I + 1)\,\gamma_2\,. \tag{3}$$

Taking into account that $A = 2.4 \times 10^{-5}$ eV [17], we obtain from (3) $\gamma_1 \approx 10^{-3}\gamma_2$, which is in agreement with experiment.

As is evident from Table 1, the probabilities of radiative (γ_2) and non-radiative (p) transitions from the doublet A_X and A_T minima of KI:In are of the same order. It may lead to the existence of the fast decay component of both emissions. Using the

Table 1
Values of parameters of the A_X and A_T relaxed excited states

state	γ_1 (10^3 s^{-1})	γ_1' (10^3 s^{-1})	γ_2 (10^6 s^{-1})	δ (10^{-3} eV)	p (10^6 s^{-1})
A_X	1.6	0.6	0.5	2.1	1.2
A_T	2.86	1.05	1.6	2.2	1.1

general formula for λ_f [12] and the δ- and p-values from Table 1, one can calculate the decay time expected for the fast component of A_X emission: $\tau_f = \lambda_f^{-1} \approx 6 \times 10^{-7}$ s at 5 K. However, the fast component is not observable even at 0.35 K, i.e. its light sum (S_f) may be at least by two to three orders smaller than that of the main slow component (S_s). The fast decay component is also absent in the A_T emission of $Sn^{2+}v_c^-$ centres.

The S_f/S_s ratio depends (i) on the process occurring in the non-relaxed radiative excited state, in particular, on the probability ratio of non-radiative transition to the metastable state and relaxation to the minimum of the radiative state; these processes lead to a certain ratio of the initial populations (N^0_{rad}/N^0_{met}) of these states just after the excitation; (ii) on the processes in the relaxed radiative state, i.e. on the relation $\gamma_2/(\gamma_2 + p_0)$ of the probabilities of radiative decay (γ_2) and of one-phonon non-radiative transition (p_0) from the relaxed radiative state to the metastable one. According to [6]

$$p_0 \approx \frac{3\pi\zeta^2 c^2 \delta^3}{2\Delta^2 \omega_D^5} (\bar{n} + 1) . \tag{1}$$

Here c is the interaction constant with trigonal vibrations (in the case of A_T state) and with tetragonal ones (in the case of A_X state), ω_D is the Debye frequency, $\bar{n} = [\exp(\delta/kT) - 1]^{-1}$.

If $N^0_{rad} \gg N^0_{met}$ and the lifetime of the excited state is shorter than the time needed for the transition from radiative to metastable minima, the fast component is dominating (such situation may occur in KI:Tl, KBr:Pb, $\gamma_2^{-1} \approx 10^{-8}$ s). The decrease of spin–orbit interaction ζ leads to an increase of the lifetime of the radiative state (up to 10^{-4} s for Ga^+). The increase of the electron–phonon interaction Δ results in the growth of p_0. Therefore. Ga^+ centres ($N^0_{rad} \gtrapprox N^0_{met}$) have both fast and slow components in the decay of the A emission.

If $N^0_{met} \gg N^0_{rad}$, then only the slow component may dominate in the emission. This is the case of In^+ and $Sn^{2+}v_c^-$ centres. It means that the transitions to the metastable state from the non-relaxed radiative state occur much more quickly than the relaxation to the minima of the radiative state. Probably, it is caused by rather strong electron–phonon interaction in In^+ and $Sn^{2+}v_c^-$ centres and considerable values of ζ (0.3 to 0.5 eV).

Thus, the kinetics and polarization characteristics of the A_X emission of In^+ centres in KI:In (as well as of Ga^+ centres in KCl:Ga and KBr:Ga [11]) can be adequately described in the model [5, 6], which suggests strong interaction of impurity optical electrons with non-totally symmetric (trigonal) vibrations of the crystal lattice and considers spin–orbit interaction as a perturbation.

Acknowledgements

The authors would like to thank Prof. Dr. V. Hizhnyakov for his interest in this work and useful discussions and A. Stolovich for assistance in experiments at 0.35 K.

References

[1] A. FUKUDA, Phys. Rev. B 1, 4161 (1970).
[2] S. ZAZUBOVICH, Optika i Spektroskopiya **26**, 235 (1969).
[3] L. BRAGINA, V. HIZHNYAKOV, G. LIIDYA, V. NAGIRNYI, T. SOOVIK, S. ZAZUBOVICH, and N. JANSON, phys. stat. sol. (b) **120**, 463 (1983).
[4] V. NAGIRNYI, T. SOOVIK, S. ZAZUBOVICH, and N. JANSON, phys. stat. sol. (b) **120**, 117 (1983).
[5] V. V. HIZHNYAKOV, Preprint FI–36, Tartu 1975.
[6] V. HIZHNYAKOV and N. KRISTOFFEL, in: The Dynamical Jahn-Teller Effect in Localized Systems, Ed. Ju. E. PERLIN and M. WAGNER, Elsevier Publ. Co., Amsterdam/New York 1984 (p. 383).
[7] A. FUKUDA, A. MATSUSHIMA, and S. MASUNAGA, J. Lum. **12/13**, 139 (1976).
[8] A. FUKUDA, J. Phys. Soc. Japan **40**, 776 (1976).
[9] S. MASUNAGA, N. GOTO, A. MATSUSHIMA, and A. FUKUDA, J. Phys. Soc. Japan **43**, 2013 (1977).
[10] M. F. TRINKLER and I. S. ZOLOVKINA, phys. stat. sol. (b) **79**, 49 (1977).
[11] S. ZAZUBOVICH, V. NAGIRNYI, and T. SOOVIK, Fiz. tverd. Tela, in the press.
[12] V. HIZHNYAKOV, G. LIIDYA, V. NAGIRNYI, T. SOOVIK, and S. ZAZUBOVICH, phys. stat. sol. (b) **120**, 105 (1983).
[13] G. ZAVT, V. PLEKHANOV, V. HIZHNYAKOV, and V. SHEPELEV, Proc. Internat. Symp. Synergetics and Cooperative Phenomena in Solids and Macromolecules, Ed. A. SHERMAN, Valgus, Tallinn 1983 (p. 158).
[14] A. K. TALE, in: Radiatsionnaya fizika, Vol. VI, Zinatne, Riga 1970 (p. 49).
[15] V. HIZHNYAKOV, S. ZAZUBOVICH, and T. SOOVIK, phys. stat. sol. (b) **66**, 727 (1974).
[16] V. HIZHNYAKOV, T. SOOVIK, and S. ZAZUBOVICH, phys. stat. sol. (b) **106**, 581 (1981).
[17] P. G. BARANOV, Izv. Akad. Nauk SSSR, Ser. fiz. **45**, 253 (1981).

(Received July 26, 1984)

phys. stat. sol. (b) **126**, 659 (1984)

Subject classification: 20.1 and 20.3; 13.4; 22.5

Institute of Physics, Academy of Sciences of the Belorussian SSR, Minsk[1]) *(a) and Institute of Crystallography, Academy of Sciences of the USSR, Moscow (b)*

One- and Two-Photon Luminescence Excitation of Eu^{3+} in a LaF_3 Monocrystal

By

L. E. Kholodenkov (a), A. G. Makhanek (a), and A. A. Kaminskii (b)

For the first time results are given of a comparative study of the $^5D_2$ manifold Stark components of Eu^{3+} in a LaF_3 monocrystal by the polarization diagram method under one- and two-photon luminescence excitation. As a whole, the experimental results are in agreement with the theoretical approach under consideration and comply with the case of $C_{2v} \rightarrow C_2$ symmetry distortion. Stark components are identified by one- and two-photon polarization diagrams. It is shown that two-photon diagrams are much more informative in this respect.

Впервые представлены результаты сравнительного исследования штарковских компонент $^5D_2$-уровня Eu^{3+} в одноосном монокристалле LaF_3 методом поляризационных диаграмм при одно- и двухфотонном возбуждении люминесценции. Экспериментальные результаты в целом находятся в согласии с рассматриваемым теоретическим подходом и соответствуют искажению симметрии $C_{2v} \rightarrow C_2$. По одно- и двухфотонным поляризационным диаграммам проведена идентификация штарковских компонент и показано, что в этом отношении двухфотонные диаграммы обладают существенно большей информативностю.

1. Introduction

The subject of the present paper is the investigation of the light polarization dependence of one- and two-photon luminescence excitation Stark spectra of trivalent europium and the interpretation of observed spectra according to a given type of local symmetry as applied to an optically uniaxial crystal. The presence of an optical axis leads to the division of the incident electromagnetic wave of arbitrary polarization into ordinary and extraordinary waves with mutually orthogonal directions of linear polarization. This fact a priori restricts the investigation of the spectra of one- and two-photon transitions (OPT and TPT) to the above linear polarization states and to the case of circular polarization wave propagation along the optical axis. In this connection, a comparative investigation of OPT and TPT Stark spectra by the methods of one- and two-photon luminescence excitation (OPLE and TPLE), respectively, concerning the relative advantages and/or disadvantages of these methods in interpreting the Stark spectrum would be of interest.

Proceeding from the above, we chose a LaF_3-matrix as the subject of investigation. The study of one-photon Stark spectra of $Eu^{3+}:LaF_3$ has been the subject of a number of works, in particular [1 to 3]. As is obvious from these works, the replacement of La^{3+} ions by Eu^{3+} ions leads to lowering of the D_{3h} symmetry at the site of the replacement to a symmetry not higher than C_{2v}. The change of symmetry from rhombic to monoclinic is not followed by a change in the number of the observed Stark com-

[1]) Leninskii prospekt 70, 220602 Minsk, USSR.

ponents. In this case, all the components are generally observed both in OPT and TPT. Therefore, a suitable method of analysing the symmetry may be that relying on the analysis of the Stark component intensity dependence upon the exciting light polarization.

2. Experimental

There are several facts that should be noted in connection with measurements in the polarized light in $Eu^{3+}:LaF_3$. The first is concerned with the choice of a quantization axis for the Eu^{3+} ion. The symmetry of the europium ion position being not higher than C_{2v}, it would be reasonable from the physical point of view to choose the quantization axis along the second-order axis which lies in the plane orthogonal to the optical C_0-axis. The specific features of the LaF_3 structure are so [2] that there may be three such axes chosen with an angle of 120° between them.

The second fact is connected with the conditions of ordinary and extraordinary wave propagation through the sample. As is known [4], the effect of the light wave in an optically uniaxial crystal is associated with the electric induction vectors $\bar{\boldsymbol{D}}_e$ and $\boldsymbol{D}_0$. The possible states of passing wave polarization have been mentioned in the Introduction. It must be added that the rotation of linear polarization of the wave incident upon the crystal at an angle to $\boldsymbol{C}_0$ will lead to the redistribution of the vectors $\bar{\boldsymbol{D}}_e$ and $\boldsymbol{D}_0$ amplitudes in accordance with the Malus rule on the harmonic principle without changing the directions of these vectors. The vector $\boldsymbol{D}_0$ is rotated with the wave vector $\boldsymbol{k}$ parallel to the C_0-axis, whereas the rotation of $\bar{\boldsymbol{D}}_e$ takes place in the main plane and is possible only at the cost of changing the vector $\boldsymbol{k}$ orientation relative to the C_0-axis in the same plane, which is a rather time-consuming task under conventional experimental conditions.

The third fact is concerned with the essence of choosing the type of polarization when taking polarization measurements associated with the Stark spectrum. The dependence of OPT and TPT intensity upon the linear polarization of the exciting light is generally determined by the expression

$$|\sum_q (-1)^q d_q^{(1)} u_{-q}|^2, \qquad |\sum_{q,q'} (-1)^{q+q'} d_{qq'}^{(2)} u_{-q} u_{-q'}|^2, \qquad q, q' = 0, \pm 1, \quad (1)$$

respectively. Here, $d_q^{(1)}$ and $d_{qq'}^{(2)}$ are the spherical components of the first- and second-rank tensors of the dipole transition, u_{-q} are the spherical components of the linear polarization vector of the light wave. The states of the plane light wave polarization is determined, on the one hand, in the basis $\boldsymbol{e}_0'$, $\boldsymbol{e}_{+1}'$, $\boldsymbol{e}_{-1}'$ coupled with the direction of propagation of this wave so that $\boldsymbol{e}_0'$ is parallel to the wave vector $\boldsymbol{k}$, i.e. irrespective of interaction with an atomic system. On the other hand, it is most reasonable to consider the interaction between the atomic system and the light in the basis $\boldsymbol{e}_0$, $\boldsymbol{e}_{+1}$, $\boldsymbol{e}_{-1}$ coupled with the chosen quantization axis so that $\boldsymbol{e}_0$ is parallel to this quantization axis. If the above bases do not coincide the polarization state specified with respect to $\boldsymbol{k}$ of the wave will be different in going to the basis coupled with the quantization axis in accordance with the original basis $\boldsymbol{e}_0'$, $\boldsymbol{e}_{+1}'$, $\boldsymbol{e}_{-1}'$ transformation. It is logical to determine the polarization state based on the interaction involved, i.e. according to (1). So, the presence in (1) of components with $q = \pm 1$ or $q = 0$ alone physically means an "elementary" interaction with right–left-circularly or linearly polarized light, respectively. These conditions are realized for the circular $\sigma^{\pm}$-light with $\boldsymbol{k}$ parallel to the quantization axis and for the linearly polarized π-light with $\boldsymbol{k}$ orthogonal to the quantization axis. In all the other cases, a particular state of the incident wave polarization should be treated in terms of the interaction under consideration as a linear combination of $\sigma^{\pm}$, π-polarizations. Proceeding from the above and taking into account the specific features of the position on the C_0-axis relative to the chosen quantization axis in the present work the dependences of OPLE and TPLE Stark component intensities are studied upon the rotation of the linear incident light wave of the with certain fixed directions of $\boldsymbol{k}$ relative to the C_0-axis.

The measurements were made at 77 K on a sample containing 0.8 wt% of trivalent europium. The TPLE was carried out in a single-beam version [5] in the pulse regime. The value to be measured was the luminescence intensity of the $^5D_0 \rightarrow {}^7F_1$ magnetic dipole transition normalized to the laser pulse energy at OPLE and to the energy squared at TPLE. All measurements on luminescence excitation were taken at the $^7F_0 \rightarrow {}^5D_2$ transition. The linear polarization of the laser beam was rotated in the limits of 0° to 180°.

2.1 OPLE Stark spectrum

As applied to rhombic and monoclinic symmetries, the Stark state wave functions of the $^5D_2$ manifold, with allowance for the absence of mixing between the basis wave functions with magnetic numbers $M = 0, \pm 2$ and those with $M = \pm 1$, can be briefly written as

$$\alpha_i(2)\,|22\rangle + \alpha_i(-2)\,|2-2\rangle + \alpha_i(0)\,|20\rangle\,; \qquad i = 1, 2, 3\,,$$

$$\alpha_j(1)\,|21\rangle + \alpha_j(-1)\,|2-1\rangle\,; \qquad j = 1, 2\,.$$

Here, $|JM\rangle$ are basis wave functions with $J = 2$, $M = 0, \pm 1, \pm 2$, $\alpha_{i(j)}(M)$ are coefficients of basis wave function mixing determined by the matrix elements of the crystal field parameters B_{tp}.

Let us express in a brief form according to [6, 7 (p. 242)] the light linear polarization dependence of the relative intensity of the $i(j)$-Stark component of the excited manifold $\sum \alpha(M')\,|\gamma'J'M'\rangle$ for a transition of the $\gamma'J'M' \leftarrow \gamma JM$ type when $J, M = 0$,

Table 1

Angular dependences for linearly polarized light of relative intensity in OPT of the $i(j)$-Stark component for the manifold with $J = 2$ in the field of rhombic and monoclinic symmetry, $a_{i(j)}$, $b_{i(j)}$ are coefficients depending on $\alpha(M')$ and odd B_{tp}; $B^+_{tp} = \frac{1}{2}(B_{t-p} + B_{tp})$, $B^-_{tp} = \frac{1}{2}(B_{t-p} - B_{tp})$

symmetry	A-states	B-states	B_{tp}-parameters
C_{2v}	$\cos^2\theta$	$\sin^2\theta \sin^2\varphi$; $\sin^2\theta \cos^2\varphi$	$B_{20}B^+_{22}B_{40}B^+_{42}B^+_{44}$ ... $B_{20}B_{30}B^+_{32}$ ...
C_2	$\cos^2\theta$	$\sin^2\theta(a_j \cos^2\varphi + b_j \sin^2\varphi)$ $j = 1, 2$	$B_{20}B_{22}B^-_{22}B_{40}B^+_{42}B^-_{42}B^+_{44}B^-_{44}$ ... $B_{10}B_{30}B^+_{32}B^-_{32}$ ...
C_s	$\sin^2\theta(a_i \cos^2\varphi + b_i \sin^2\varphi)$ $i = 1, 2, 3$	$\cos^2\theta$	even as in C_2 $B^+_{11}B^-_{11}B^+_{31}B^-_{31}B^+_{33}B^-_{33}$...

i.e. at $\alpha(M) \equiv 1$,

$$I_{i(j)}(JM \to J'M') = \text{const} \left| \sum_{\substack{p,q \\ t,M'}} (-1)^{p+q+J-M} (2\lambda + 1)\, \alpha_{i(j)} K_{\lambda t}(r)(M') \times \right.$$
$$\left. \times \begin{pmatrix} 1 & \lambda & t \\ q & -p-q & p \end{pmatrix} \begin{pmatrix} J & \lambda & J' \\ -M & p+q & M' \end{pmatrix} B_{tp} u_{-q} \right|^2 . \quad (2)$$

We have for the manifold under consideration $J' = 2$, then $\lambda = 2$ and $t \leqq 3$.

The above dependence, as applied to the C_{2v}, C_2, C_s types of symmetry, is given in spherical coordinates in Table 1.

As an illustration, Fig. 1 shows a diagram of the relative intensities of the OPLE Stark spectrum measured under different conditions. Thus, Fig. 1a, b are an illustration of the spectra which were obtained when heating the sample to room temperature between the measurements. Fig. 1c shows a spectrum measured when irradiating the sample with intense laser light. All this is an indication of the changeability of the one-photon spectrum of the transition in question depending on the external conditions.

Consider now the dependence of the relative intensities of the components of spectrum a) of Fig. 1 upon the linear polarization rotation of the incident wave with $\boldsymbol{k}$ orthogonal to the C_0-axis. Such measurement data are given in Fig. 2. According to Table 1, they comply with the case of small distortion of $C_{2v} \to C_2$ symmetry. Indeed, the relationship between the intensities of components 1, 2, 4, 5 at the positions $\boldsymbol{D}_e \parallel \boldsymbol{C}_0$ and $\boldsymbol{D}_0 \perp \boldsymbol{C}_0$ do not contradict the choice of C_{2v} and C_2 symmetry types. At

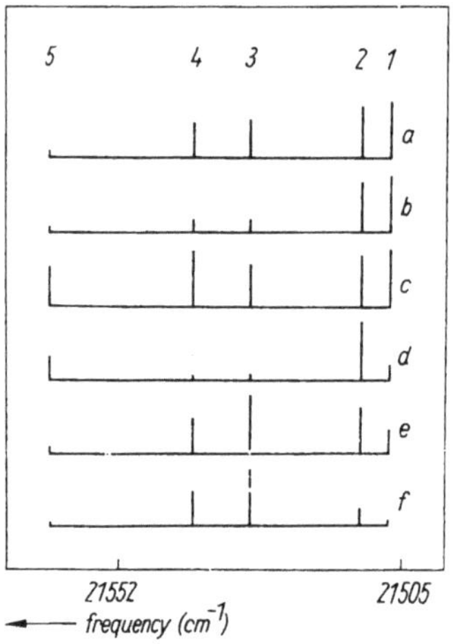

Fig. 1

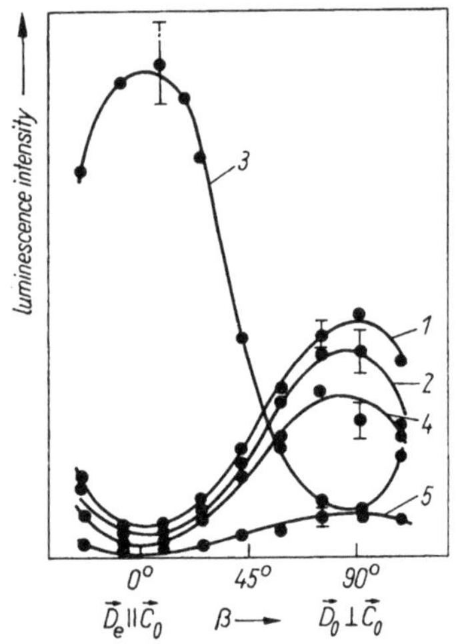

Fig. 2

Fig. 1. OPLE (a, b, c) and TPLE (d, e, f) Stark spectra with luminescence intensities normalized first to the laser pulse energy and squared energy, respectively, then to unity. a), b), c) $\boldsymbol{k} \perp \boldsymbol{C}_0$, $\boldsymbol{E}$ of the incident wave is at 60° with $\boldsymbol{C}_0$; d) $\boldsymbol{E}$ is at 25° with $\boldsymbol{C}_0$, $\boldsymbol{k} \perp \boldsymbol{C}_0$; e) $\boldsymbol{k}$ is at an angle other than 0° or 90° with $\boldsymbol{C}_0$; f) $\boldsymbol{k} \parallel \boldsymbol{C}_0$

Fig. 2. Polarization diagrams of the OPLE spectrum of the $^5D_2$ manifold as applied to the case a) in Fig. 1. Plotted on the ordinate in arbitrary units are relative values of the luminescence intensity normalized to the exciting laser radiation pulse energy. β is the rotation angle of the incident wave linear polarization. Mean errors are given for all curves the same time, the relationship between the intensities at the positions under consideration for component 3 indicates the predominance of C_{2v} symmetry, as follows from Table 1. The predominance of the above type of symmetry is also attested by the smaller value of the component 5 intensity when compared to other components (provided that this smallness is due to the fact that $B_{32} \approx B_{3-2}$). Let us dwell upon the identification of the components as applied to given conditions. Rhombic symmetry predominating, the components may be conditionally identified in terms of this symmetry. Directly identified from Fig. 2 can only be component 3 as basically belonging to one of the B-states. The particular B-state to which it belongs can be determined only by a unique choice of the $\boldsymbol{e}_x$, $\boldsymbol{e}_y$ basis vectors of the quantum system relative to the C_0-axis. Let us choose for definiteness $\boldsymbol{e}_x \parallel \boldsymbol{C}_0$. Then this component will belong to the B_1-state. Whatever the choice of orientation of the system of coordinates, component 5 may be basically considered as belonging to the A_2-state due to its small intensity. By this the identification of the components in a given spectrum is limited.

2.2 TPLE Stark spectrum

Let us find the expression for TPT describing the dependence of the relative intensity of the $i(j)$-Stark component of the manifold under consideration on the spherical components of linear polarization vectors in the case of two linearly polarized plane light waves of equal frequency. Then, (2) from the preceding section may be used as the original expression. The expression for TPT is obtained by formally replacing the odd operator of the crystal field potential by the operator of interaction between the dipole moment and an electromagnetic wave [7 (p. 288), 8]. At the same time the

$$p \to q\,, \quad B_{tp} \to u_{-2q'}\,, \quad u_{-q} \to u_{-1q}\,; \qquad t = 1, \;\; \lambda = 2 \tag{3}$$

substitution is made here, u_{-1q} and $u_{-2q'}$ are the spherical components of the linear polarization vectors of the two plane electromagnetic waves. In the case being discussed, the possibility arises of analyzing the Stark spectrum in the field of rhombic or monoclinic symmetry in terms of states with wave functions of the form

$$(2^+) = \frac{1}{\sqrt{2}}\,(|22\rangle + |2-2\rangle)\,, \qquad (2^-) = \frac{1}{\sqrt{2}}\,(|22\rangle - |2-2\rangle)\,,$$

$$(0) = |20\rangle\,, \quad (1^+) = \frac{1}{\sqrt{2}}\,(|21\rangle + |2-1\rangle)\,, \quad (1^-) = \frac{1}{\sqrt{2}}\,(|21\rangle - |2-1\rangle)\,.$$

Substituting values corresponding to the transition under consideration and making necessary calculations, it is possible to obtain an expression for the relative intensities of the $i(j)$-component in the form of a linear combination of angular parts of the states $(2^\pm)$, (0), $(1^\pm)$ with $\alpha(M')$-dependent coefficients. In Table 2, such an expression is given with analytical expressions for the angular parts $P(2^\pm)$, $P(0)$, $P(1^\pm)$, as well as with the coefficients $K\{\alpha(M')\}$ determining the relative contribution of each of the states under consideration. The data of Table 2 are given for the case of one axis of quantization and the value P depends on the mutual orientation of the vectors $\boldsymbol{D}_e$, $\boldsymbol{D}_0$, and the C_0-axis. In particular, it is interesting to see whether in a real crystal there is a dependence of the TPLE intensity on the rotation of $\boldsymbol{D}_0$ in the plane orthogonal to the C_0-axis, i.e. on the angle θ. Such measurements were made for component 2 and the measurement data are given in Fig. 3 where for comparison analytical dependences for the angular parts $P^2(2^+)$ and $P^2(0)$ are shown. As can be seen from Fig. 3, such a dependence does exist and this fact should be taken into account in analyzing the spectra taken at other orientations of $\boldsymbol{k}$ relative to the C_0-axis.

Table 2

Relative contribution of the states (2^+), (2^-), (0), and (1^+), (1^-) to the intensity of a TPT Stark component in a field of rhombic or monoclinic symmetry. $K_{i(j)}\{\alpha(M')\}$ are coefficients defined by $\alpha(M')$. The signs "+" and "−" in the right-hand column denote the presence or absence of a non-zero contribution, respectively

$K_{i(j)}\{\alpha(M')\}$	angular parts	P as a function of θ and φ	$k \perp C_0$		
			D_e^2	D_eD_0	D_0^2
$K_1\{\alpha(\pm 2)\}$	$1.2P^2(2^+)$	$\sin\theta_1 \sin\theta_2 \cos(\varphi_1 + \varphi_2)$	+	−	+
$K_2\{\alpha(\pm 2)\}$	$1.2P^2(2^-)$	$\sin\theta_1 \sin\theta_2 \sin(\varphi_1 + \varphi_2)$	−	+	−
$K_3\{\alpha(\pm 2)\}$	$1.2P(2^+)\,P(2^-)$		−	−	−
$K_4\{\alpha(0)\}$	$0.4P^2(0)$	$\sin\theta_1 \sin\theta_2 \cos(\varphi_1 - \varphi_2) - 2\cos\theta_1 \times \times \cos\theta_2$	+	−	+
$K_5\{\alpha(\pm 2, 0)\}$	$0.7P(2^+)\,P(0)$		+	−	+
$K_6\{\alpha(\pm 2, 0)\}$	$0.7P(2^-)\,P(0)$		−	−	−
$K_1\{\alpha(\pm 1)\}$	$1.2P^2(1^+)$	$\cos\theta_1 \sin\theta_2 \sin\varphi_2 + \cos\theta_2 \sin\theta_1 \sin\varphi_1$	−	−	+
$K_2\{\alpha(\pm 1)\}$	$1.2P^2(1^-)$	$\cos\theta_1 \sin\theta_2 \cos\varphi_2 + \cos\theta_2 \sin\theta_1 \cos\varphi_1$	−	+	−
$K_3\{\alpha(\pm 1)\}$	$1.2P(1^+)\,P(1^-)$		−	−	−

To illustrate the dependence of the relative distribution of the TPLE spectrum component intensities upon the orientation of $\boldsymbol{k}$ relative to the C_0-axis, spectra for three positions of $\boldsymbol{k}$ have been measured. They are illustrated in Fig. 1 (d, e, f).

Let us now consider the polarization diagrams of TPLE measured for the case (d) in Fig. 1. These diagrams are given in Fig. 4. Based on the data of Table 2 and Fig. 4, we analyze qualitatively the character of the Stark state mixing. So, it follows from Table 2 that for $\boldsymbol{D}_e \parallel \boldsymbol{C}_0$, the states (2^+) and (0) will contribute to the component intensities, in particular to the intensities of components 5,2. Then, since, unlike OPT, TPT is excited at the expense of $|\boldsymbol{D}_e| \cdot |\boldsymbol{D}_0|$ with maximum taking place at $\beta = 45°$ with the C_0-axis, this case is also to be considered. Thus, contributions to the intensities can be made this position only by the states (2^-) and (1^-). Taking into account that these states do not mix and that the states (2^+), (2^-), (0) mix, it can readily be determined that the state (2^-) contributes to the component 2 intensity, whereas the state (1^-) contributes to the intensity of component 1. Let us now consider the case of $\boldsymbol{D}_0 \perp \boldsymbol{C}_0$. As follows from Table 2, a contribution at this position can generally be made by the states (2^+), (0), (1^+). The presence of deep minima for curves 2, 5 in Fig. 4 is indicative of small values of $P^2(2^+)$ and $P^2(0)$ at this position. Thus, the maximum of curve 3 corresponds to the contribution of the state (1^+) to the intensity. As to component 4, it should be noted that its intensity for all three positions under consideration was much lower than the intensities of the other components (at the level of the accompanying background). In accordance with the present consideration, it should be assigned to the state (0). The smallness of this component intensity may be partially attributed, on the one hand, to the presence of the factor 0.4 in Table 2. On the other hand, it may be due to factors whose explanation is not to be found by the approach used in the present consideration.

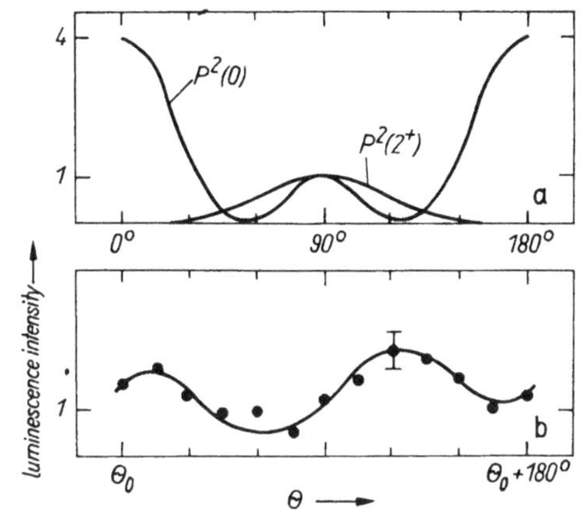

Fig. 3

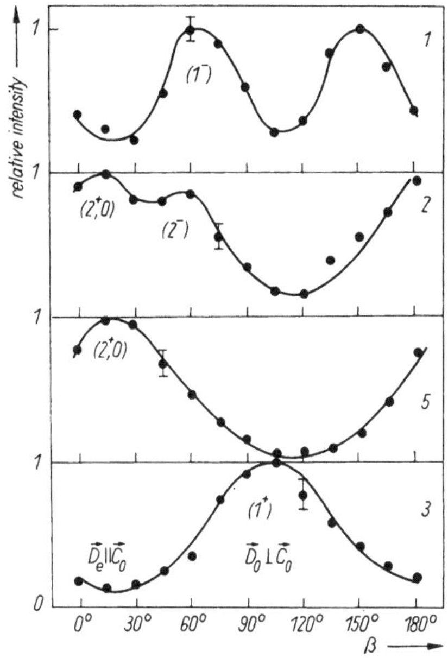

Fig. 4

Fig. 3. b) Luminescence intensity in arbitrary units normalized to squared energy of the exciting radiation pulse versus rotation $\boldsymbol{D}_0$ of the wave at $\boldsymbol{k} \parallel \boldsymbol{C}_0$ for component 2 of Fig. 1f. For comparison, Fig. 3a shows the angular parts of the states (2^+) and (0) versus the angle θ at $\boldsymbol{k} \parallel \boldsymbol{C}_0$ in the case of one axis of quantization

Fig. 4. Polarization diagrams of TPLE Stark components of the $^5D_2$ manifold. The relative intensities are normalized first to the laser pulse squared energy, then to unity, β is the angle of rotation of $\boldsymbol{E}$ at $\boldsymbol{k} \perp \boldsymbol{C}_0$. The ratio of maximum values of intensities 1, 2, 3, 4, 5 of the components is 1:1.4:0.9:0:0.6. Mean errors are given

As follows from the analysis, the Stark TPLE spectrum corresponds to the case of distortion of symmetry $C_{2v} \rightarrow C_2$. The distortion manifests itself largely as the presence of mixing of the states (2^+), (0), (2^-) in component 2. Taking into account the fact that the effectiveness of excitation for $\beta = 60°$ is approximately 0.25 of that for $\beta = 15°$, as well as the distribution of the maxima of curve 2 in Fig. 4, we find the proportion of the contributions of (2^+), (0), and (2^-) to be approximately 1:2.5. As with the OPLE spectrum, this proportion and the nature of the distribution of the contributions of the given states to the intensities of other components allow the components to be identified in terms of C_{2v} symmetry, namely, components 1 and 3 basically belong to the states B_1 and B_2, respectively, whereas components 2, 4, 5 basically belong to the states A_2, A_1, A_1, respectively.

3. Discussion

As noted in the Introduction, the interpretation of the Stark spectrum in a low symmetry field (C_{2v} and lower) on the basis of investigating the relative intensities of the components consists in choosing a correspondence between observed polarization and analytical dependences corresponding to a given type of symmetry and then in identifying the components according to the chosen type of symmetry. The analysis of OPLE and TPLE spectra reveals that they comply with the case of $C_{2v} \rightarrow C_2$ symmetry distortion. The distinguishing feature of these types of symmetry in terms of the parameters B_{tp} is the appearance of an imaginary part of these parameters in the case of C_2 symmetry. The degree of distortion is determined by the value of the ratio of imaginary and real parts for each parameter. This distortion must be different for OPT and TPT. As follows from (2), with allowance for (3), the crystal field affects OPT both directly through odd B_{tp} and the mixing coefficients $\alpha(M')$ which depend on even B_{tp}, whereas TPT is effected only through $\alpha(M')$. Therefore, it may be generally supposed that the relative changes in the OPT Stark component intensities caused by changing the crystal field parameters are more dramatic than in the case of TPT. The change of the parameters B_{tp} may be caused in particular by induced polarization of ligands due to intense laser radiation during excitation of the optical transition. Taking into account that a given transition [9] is supersensitive, even minor changes of the field parameters may significantly change the distribution of the OPLE component intensities. This is evidenced by spectrum (c) in Fig. 1. It should be noted that the TPLE polarization diagrams were measured in this work under intense laser radiation, whereas the OPLE spectrum diagrams were measured under weak laser radiation. The identification of the components considered in terms of C_{2v} symmetry yields, generally speaking, different results if OPT and TPT are compared. This fact indicates the susceptibility of the TPT spectrum to the effect of intense laser radiation. As can be seen, this effect manifests itself as changes in the nature of the Stark state mixing. Though the expressions determining the dependence of the relative intensity of $i(j)$-components of OPT and TPT on the light polarization do not directly contain the supersensitivity factor of the transition, it may be assumed that in the case of TPT this factor must also be taken into consideration through the coefficients of mixing $\alpha(M')$. This suggests that a correct comparison of the OPLE and TPLE spectrum components in terms of C_{2v} symmetry is possible only for spectra measured under weak laser radiation. This condition is not principal and enables experimental realization by measuring the TPLE spectrum in the presence of a luminescence registration system operating in the photon count regime with the signal being stored. Then, a comparison of the OPLE and TPLE polarization diagrams, where identification of components is possible, reveals the appreciably great information contained in the spectrum of the latter. So, in the case of OPLE only component 3 can be identified directly from the polarization diagrams, since it belongs to one of the states B_1, B_2, whereas all the other components may be identified from the TPLE polarization diagrams. Besides, there is a possibility of examining the changes in the nature of the Stark state mixing depending on the external conditions. More complete and versatile information of the TPLE spectrum about polarization is associated with the possibility of specifying and monitoring, in the general case, the polarization components of two electromagnetic waves rather than one as would be the case for OPLE. In a one-photon spectrum, to completely identify the components one has to resort to additional data associated with other optical transitions.

If in identifying the components a certain type of symmetry is strictly followed, then in our case one should operate in terms of C_2 symmetry irrespective of the predominance of C_{2v} symmetry. Then both OPLE and TPLE spectra provide a consistent picture where components 5, 4, 2 belong to three states A, while components 1, 3 belong to two states B.

It would be appropriate to note that the presence of two excitation waves, in particular ordinary and extraordinary, leads to TPLE excitation in the intermediate case where the incident wave polarization vector makes an angle of 45° with the C_0-axis. When this case is ignored loss of information may occur leading to erroneous results in interpreting the Stark spectrum. Therefore, measurements associated with rotation of the linear polarization vector must be also made in the case of a uniaxial crystal with a proper choice of the incident wave orientation $\boldsymbol{k}$.

References

[1] U. Vignanesvara Kumar, D. Ramchandra Rao, and Putcha Venkatesvarlu, J. chem. Phys. **66**, 2019 (1977).
[2] C. A. Morrison and R. P. Leavitt, J. chem. Phys. **71**, 2366 (1979).
[3] C. A. Morrison and R. P. Leavitt, J. chem. Phys. **73**, 749 (1980).
[4] M. Born and E. Wolf, Fundamentals of Optics, Izd. Nauka, Moskva 1973.
[5] L. E. Kholodenkov and A. G. Makhanek, phys. stat. sol. (b) **112**, K149 (1982).
[6] B. R. Judd, Phys. Rev. **127**, 750 (1962).
[7] A. G. Makhanek and V. S. Korolkov, Analytical Methods of the Quantum-Mechanical Theory of Perturbations, Izd. Nauka i Tekhnika, Minsk 1982.
[8] J. D. Axe, Phys. Rev. **136**, A42 (1964).
[9] B. R. Judd, J. chem. Phys. **70**, 4830 (1979).

(Received July 26, 1984)

phys. stat. sol. (b) **126**, **669** (1984)

Subject classification: 13.3; 22.2.2

Physics Department, University of Roorkee (a)
and Institut für Theoretische Physik, Freie Universität Berlin[1]) *(b)*

Tight-Binding Calculations for the Angle-Resolved Local Density of States of InP(110) Surfaces

By

S. JALAJ (a), I. SINGH[2]) (b), and S. K. JOSHI (a)

Angle-resolved electronic local density of states and layer density of states of indium phosphide(110) surfaces are calculated using empirical tight binding methods. The slab method is used for the purpose. The results are compared with recent angle-resolved photoemission spectroscopy (ARUPS) findings.

Die winkelaufgelöste lokale elektronische Zustandsdichte und die Zustandsdichte von Indiumphosphid(110)-Oberflächen werden mit empirischen „tight-binding“-Methoden berechnet. Zu diesem Zweck wird die Plattenmethode benutzt. Die Ergebnisse werden mit neueren Ergebnissen der winkelaufgelösten Photoemissionsspektroskopie (ARUPS) verglichen.

1. Introduction

The experimental and theoretical study of III–V zincblende semiconductor surfaces is becoming increasingly important both from a fundamental as well as technological point of view. Presently the theoretical studies of (110) surfaces of III–V semiconductors, their interfaces and superlattices are being made extensively because the recent advancement in experimental techniques, such as molecular beam epitaxy, have made it possible to fabricate remarkably perfect semiconductor interfaces and superlattices (for a recent review see Madhukar [1]). A complete knowledge of the electronic properties of clean surfaces is a pre-requisite to understand the properties of interfaces and superlattices.

In this paper we study theoretically the electronic properties of (110) surface (unrelaxed as well as relaxed) of InP. Recently the electronic properties of these surfaces have been studied experimentally using angle-resolved photoelectron spectroscopy by Williams and coworkers (McKinley et al. [2], Montgomery and Williams [3], Srivastava et al. [4], McKinley et al. [5]). Theoretically (110) surface of InP was studied recently using the empirical pseudopotential method (Srivastava and Singh [6]), self-consistent pseudopotential method (Manghi et al. [7]) and empirical tight binding method (Beres et al. [8], Srivastava et al. [4]). When these theoretical calculations were compared with the experimentally determined dispersion of energy bands in various symmetry directions, it was observed that the tight-binding results are relatively closer than pseudopotential ones. In the present work of calculating angle-resolved local density of states and layer density of states we use the same tight-binding parameters as used earlier (Srivastava et al. [4]). In Section 2 we discuss the

---

[1]) Arnimallee 14, D-1000 Berlin 33, Berlin (West).
[2]) On leave from Roorkee University, Roorkee, India.

basic ingredients of the theoretical method used in the present work. In Section 3 we present the ARLDS spectra of the relaxed (110) surface and the layer density of states of the unrelaxed as well as relaxed surface. The ARLDS results are compared with the available experimental data.

2. Method of Calculation

During the past one decade or so two very successful methods, (i) the slab method (Hirabayashi [9], Pandey and Phillips [10], Joannopoulos and Cohen [11], Chadi [12, 13]) and (ii) the scattering theoretic method (Pollmann and Pantelides [14, 15], Dandekar et al. [16], Chang [17], Schmeits et al. [18]) have been developed, to study surfaces and interfaces, within the tight-binding sp^3 scheme for semiconductors. The slab method has got the advantage that it can be easily extended to the study of superlattice structure of varying thickness of component materials. (Schullman and McGill [19], Jalaj et al. [20], Madhukar [1]). We shall make use of the slab method in these calculations.

In $\boldsymbol{r}$-representation the wave function for a slab of M layers is expressed as

$$\Psi_{n,\boldsymbol{k}}(\boldsymbol{r}) = \sum_{p=1}^{8} \sum_{m=1}^{M} C_{npm}(\boldsymbol{k})\, \chi_{pm,\boldsymbol{k}}(\boldsymbol{r}) \tag{1}$$

with

$$\chi_{pm,\boldsymbol{k}}(\boldsymbol{r}) = \frac{1}{\sqrt{N}} \sum_{l} \mathrm{e}^{i\boldsymbol{k}\cdot\boldsymbol{R}_l}\, \Phi_p(\boldsymbol{r} - \boldsymbol{R}_l - \boldsymbol{\tau}_m)\,. \tag{2}$$

Here $\boldsymbol{k}$ is a two-dimensional wave vector parallel to the surface. $\boldsymbol{R}_l$ is a lattice vector parallel to the surface and $\boldsymbol{\tau}_m$ a vector perpendicular to the surface. p is the orbital index. The electron eigenvalues and eigenfunctions are obtained by solving the secular equation

$$\sum_{p'=1}^{8} \sum_{m'=1}^{M} [H_{pm,p'm'}(\boldsymbol{k}) - E_n(\boldsymbol{k})\, \delta_{pp'}\delta_{mm'}]\, C_{np'm'}(\boldsymbol{k}) = 0\,, \tag{3}$$

where

$$H_{pm,p'm'}(\boldsymbol{k}) = \frac{1}{N} \sum_{ll'} \mathrm{e}^{i\boldsymbol{k}\cdot(\boldsymbol{R}_{l'} - \boldsymbol{R}_l)} \int \Phi_p(r - R_l - \boldsymbol{\tau}_m)\, H \Phi_{p'}(\boldsymbol{r} - \boldsymbol{R}_{l'} - \boldsymbol{\tau}_{m'})\, \mathrm{d}\tau\,. \tag{4}$$

The interaction integrals appearing in (4) are taken from the work of Chadi [13]. As mentioned in the introduction, these parameters give dispersion of surface energy bands in good agreement with the experimental results.

Now once one has obtained eigenvalues and eigenfunctions solving the secular equation (3) for $\boldsymbol{k}$-values spanning the Brillouin zone (Fig. 1), the total density of states $\varrho(E)$, the layer density of states $\varrho_m(E)$, and the ARLDS $\varrho_m(\boldsymbol{k}, E)$ may be calculated using the following formulae:

$$\varrho(E) = \sum_{m} \varrho_m(E) = \sum_{m} \{\sum_{\boldsymbol{k}} \varrho_m(\boldsymbol{k}, E)\}\,, \tag{5}$$

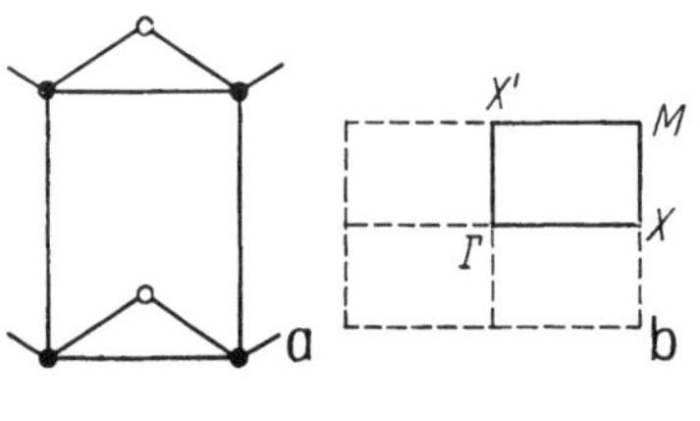

Fig. 1. a) Real-space unit cell and b) the Brillouin zone for the (110) surface

where

$$\varrho_m(\boldsymbol{k}, E) = \sum_{n,p} |\langle \Psi_{n,\boldsymbol{k}} \mid \chi_{pm,\boldsymbol{k}}\rangle|^2 \, \delta\big(E - E_n(\boldsymbol{k})\big)$$

$$= -\frac{1}{\pi} \operatorname{Im} \sum_{n,p} \frac{|C_{npm}(\boldsymbol{k})|^2}{E - E_n(\boldsymbol{k}) + i\Gamma}. \tag{6}$$

Now, obviously, the ARLDS spectra $\varrho_m(k, E)$ contain information about the surface band dispersion as well as the linewidth and shape of the surface resonances. The ARLDS results, therefore, may be directly compared with the ARUPS results. To simulate experimental situation, we incorporate the broadening effects in (5) and (6) by considering Γ in (6) as finite. In such a situation (6) becomes

$$\varrho_m(\boldsymbol{k}, E) = \frac{1}{\pi} \sum_{n,p} \frac{\Gamma\,|C_{npm}(\boldsymbol{k})|^2}{\{E - E_n(k)\}^2 + \Gamma^2}. \tag{7}$$

3. Results and Discussions

The indium- and phosphorus-derived local density of states in various layers for the ideal and relaxed (110) surfaces are shown in Fig. 2 a and b, respectively. The relaxation taken in these calculations is from Chadi [12]. We had also repeated the calculations taking more recent surface relaxation by dynamic LEED studies (Meyer et al. [21]) and did not find any significant change from the results with Chadi relaxation. That is why we did not think it worth to give the results for Meyer et al.'s relaxation.

We first consider the results for an ideal surface (Fig. 2a). It can be noticed that the LDS tend towards that of the bulk solid as one moves away from the surface layer towards the centre of the slab. We have seen that the total density of states of the sixth layer resembles perfectly the bulk density of states calculated within this tight-binding scheme (bulk results not shown). It can also be seen that the sixth layer density of states resembles with the pseudopotential calculations (Singh and Srivastava [22]) of bulk density of states. One observes that the lower valence band is mostly of P character, particularly in the s-like region (≈ -9.5 eV). In the upper valence band, the region around -5.0 eV is equally shared between In- and P-derived states, while near the top of the valence band, majority of states are P-derived. Near the bottom of the conduction band both the In- and P-derived states share, with In-derived states in majority. When we approach the surface, there is not much change in LDS up to the fourth layer. But in the third layer (not shown in the figure) some surface and resonant states appear. The new small peak at 0.0 eV has P-character, while a similar peak at 2.0 eV shares equally the In- and P-character. When we go to the second layer these peaks due to surface states become more pronounced. Peaks near energies -9.6 eV, -5.6 eV, and -1.8 eV also become pronounced.

Finally at the first layer, the surface state peaks around 0.0 eV and 2.0 eV become more pronounced, the former predominantly P-like and the latter In-like.

Let us now see what are the effects of surface relaxation on surface and resonant states. We observe that these effects are confined only up to second layer. The filled P-dangling bond derived state previously found near 0.0 eV moves down into the valence band near -0.8 eV and the empty In-dangling bond derived state moves up into the conduction band appearing now at 3.5 eV. Thus, the fundamental band gap is free of surface states. Amongst the resonance states, those near the bottom of the stomach region around -5.7 eV and near the top of the lower valence band around -9.5 eV are nearly unaffected. However, relaxation induces resonance states near the bottom of the lower valence band around -10.3 eV.

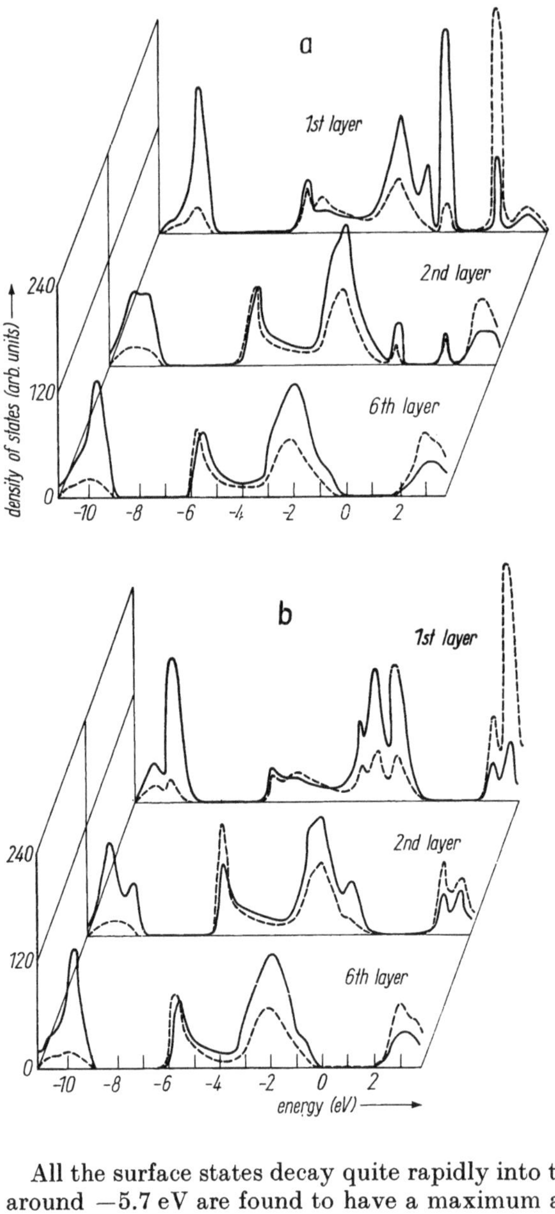

Fig. 2. a) In- and P-derived densities of states in the 1st, 2nd, and 6th layers of the unrelaxed InP (110) surface; ——— P, - - - - In. b) Same as in a) with Chadi relaxation

All the surface states decay quite rapidly into the bulk. However, the surface states around −5.7 eV are found to have a maximum amplitude at the second layer rather than at the top layer. This is understandable because this is a back-bonding state.

Now we turn to the discussion of ARLDS spectra calculated for four surface wave vectors on the boundary of the irreducible section of the surface Brillouin zone, i.e. the points Γ, X, M, and X′. These spectra are shown in Fig. 3a, 4a, 5a, and 6a for the points Γ, X, M, and X′, respectively. For comparison we show ARUPS results (Srivastava et al. [4]) in Fig. 3b, 4b, 5b, and 6b for these $\boldsymbol{k}$-points, respectively. As mentioned earlier for comparison purposes, we have broadened the whole ARLDS spectrum by a Lorentzian function (7) with halfwidth $\Gamma = 0.2$ eV. One thing should be borne in mind before studying the ARLDS and ARUPS curves that we have not taken into consideration the possible variation of the oscillator strength for states with different energies. Therefore we may not expect the relative peak heights in ARLDS to be matching with those in ARUPS spectra. Moreover, the growing behaviour of ARUPS spectra towards the low-energy side should be attributed to the secondary electron emission. It should be noted here that, in fact, even in the simplified picture the UPS intensity should depend on the initial (valence band) density of states, final density of states and $|M_{fi}|^2$, the matrix element (oscillator strength). (For a review see e.g. Williams et al. [23]). Therefore, in any photoemission calculation the matrix elements and the final-state effects are also important. In general, the initial state effects are believed to have dominant effects and therefore one may gain some insight by comparing ARUPS spectra with the ARLDS results (see e.g. Chang [17], Schmeits et al. [18]).

In Fig. 3a we show ARLDS for Γ-point for various layers. The uppermost curve in this figure shows the weighted sum of the lower six curves (for the six layers) with an exponential factor $e^{-z/d}$. In absence of any theoretical or experimental evidence for the value of escape depth d for InP, we have taken it to be 1.0 nm, the same as taken by Chang [17] for GaAs. Fig. 3b shows the ARUPS results for Γ-point, taken with fixed photon energy. In this curve the energy is measured below the Fermi energy E_F. It may be mentioned here that in these experiments (Srivastava et al. [4]) heavily doped n-type InP crystals were used and the Fermi energy was found roughly 1.0 eV above the valence band maximum (VBM). So while comparing various peak positions in Fig. 3a and b one should read 1 eV (instead of 0.0 eV) at E_F in Fig. 3b. (Similar is the case in Fig. 4b, 5b, and 6b). First peak in Fig. 3a (at around −2 eV) seems to be associated with the resonance states, as this peak is equally intense in all layers. Second and third peaks (respectively around −2.8 eV and −6.0 eV) seem entirely due to bulk states. In Fig. 3b the three peaks are having nearly the same positions.

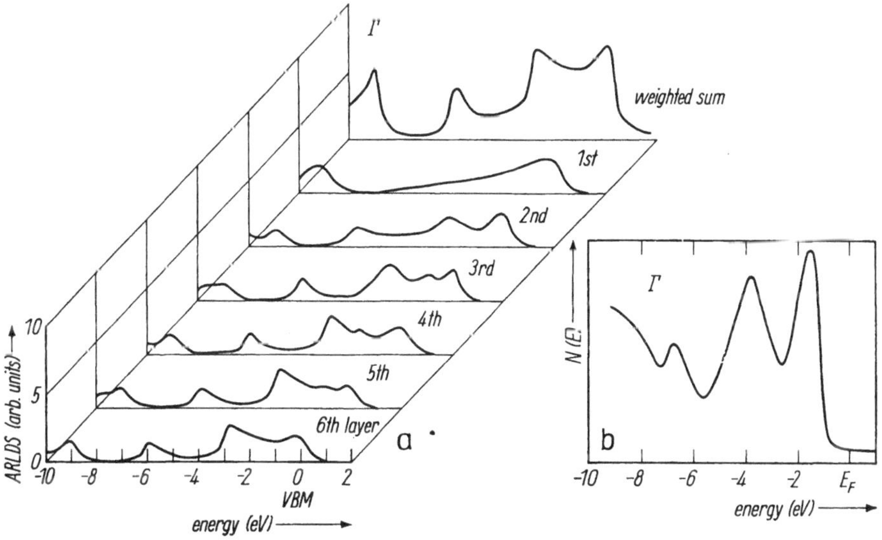

Fig. 3. a) Angle-resolved local density of states (ARLDS) for the InP (110) surface for the Γ-point; b) photoelectron spectra near Γ-point (from Srivastava et al. [4])

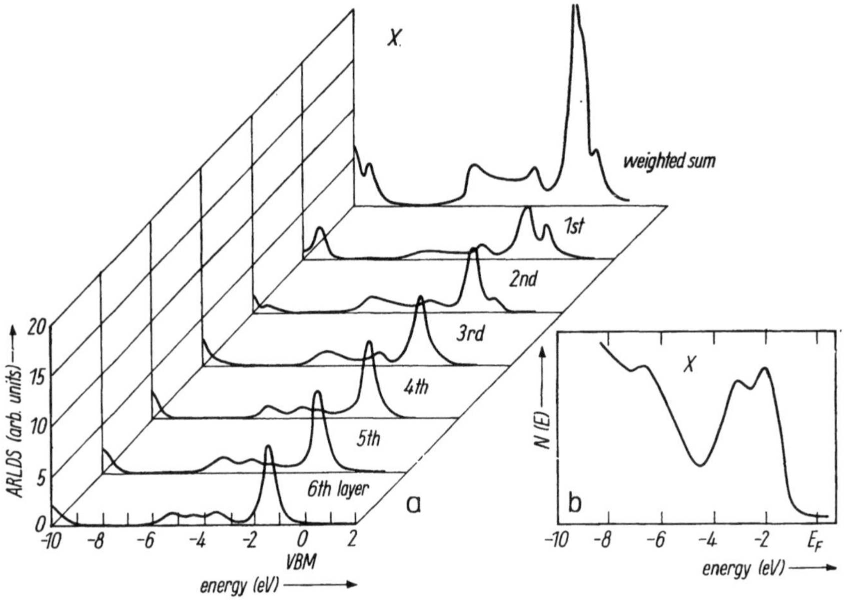

Fig. 4. a) Same as in Fig. 3a for X-point; b) same as in Fig. 3b near X-point

In Fig. 4a, for X point, the first peak (at around -0.6 eV) is entirely due to surface states, as this peak dies out rapidly as one moves from the first layer towards the bulk. The second and third peaks (at around -1.5 and -5.4 eV, respectively) seem to be due to bulk states. The spectrum in Fig. 4b shows peak positions at nearly these values.

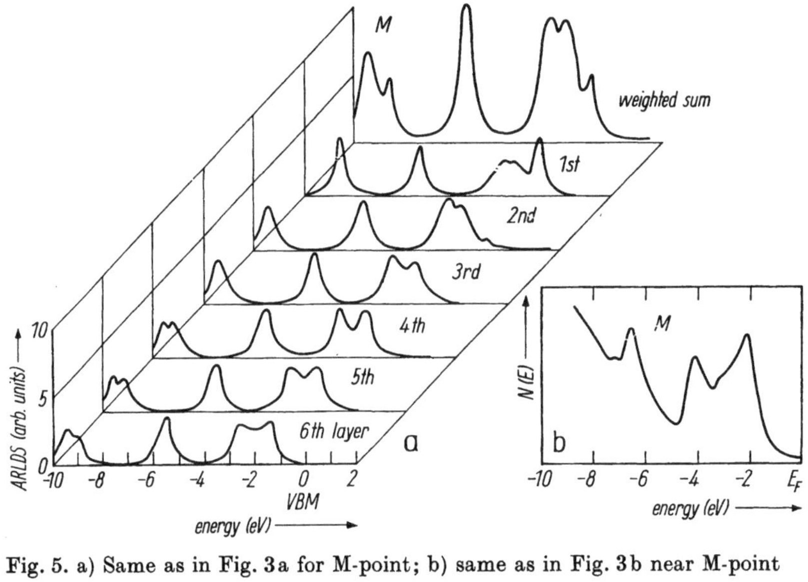

Fig. 5. a) Same as in Fig. 3a for M-point; b) same as in Fig. 3b near M-point

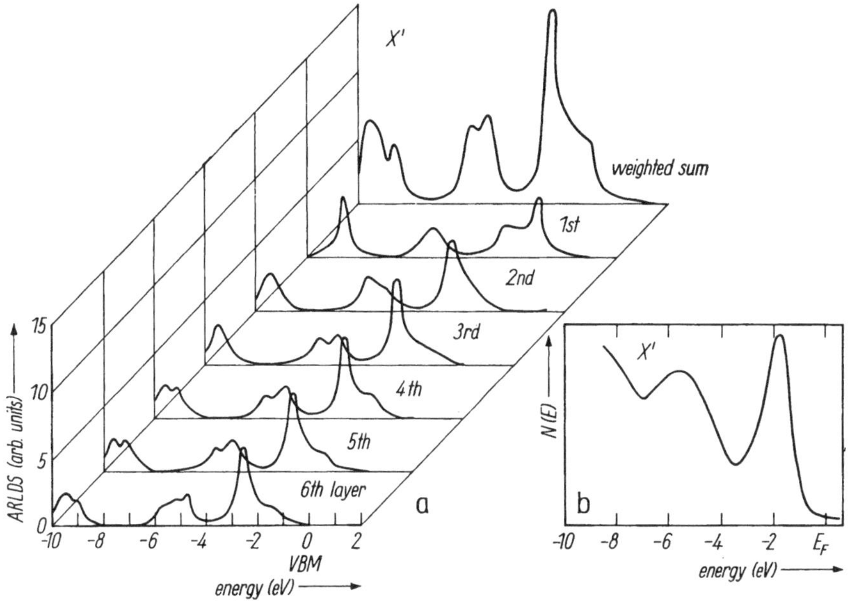

Fig. 6. a) Same as in Fig. 3a for X'-point; b) same as in Fig. 3b near X'-point

For the M point (Fig. 5a) we get a small peak in the weighted sum at energy around −0.8 eV. This should be associated, obviously, with the surface states as the contribution for this state is from the surface layer only. This surface state is in good agreement with the experimental results (Fig. 5b), where there is a corresponding peak around −1.0 eV. The second peak around −1.7 eV, which is associated with bulk states, is not present in Fig. 5b. Third and fourth (around −2.4 and −5.6 eV) have almost equal amplitude from all the six layers. There are corresponding peaks at approximately the same positions in Fig. 5b.

For the X' point (see Fig. 6) we see a shoulder at energy −1.0 eV which is due to surface states and is in reasonable agreement with the experiment. For this point one hardly finds an agreement for other peaks.

We have studied in this work the ARLDS and the layer density of states of the (110) surface of InP. We made a detailed comparison of the theoretical results with the available ARUPS results and found reasonably good agreement: it should be kept in mind that in these calculations we did not consider the possible variation of the oscillator strength for states with different energies. It seems logical to conclude that the empirical tight-binding sp^3 model together with the slab method may very well describe the electronic properties of (110) zincblende surfaces.

Acknowledgements

One of the authors (I.S.) is thankful to Prof. R. H. Williams for fruitful discussions on the subject during his (the former) stay at N.U.U. Coleraine. S.J. is thankful to C.S.I.R. (India) for financial support.

References

[1] A. Madhukar, J. Vacuum Sci. Technol. **20**, 149 (1982).
[2] A. McKinley, G. P. Srivastava, and R. H. Williams, J. Phys. C **13**, 1581 (1980).

[3] V. MONTGOMERY and R. H. WILLIAMS, J. Phys. C **15**, 5887 (1982).
[4] G. P. SRIVASTAVA, I. SINGH, V. MONTGOMERY, and R. H. WILLIAMS, J. Phys. C **16**, 3627 (1983).
[5] A. MCKINLEY, G. J. HUGES, and R. H. WILLIAMS, J. Phys. C **15**, 7049 (1982).
[6] G. P. SRIVASTAVA and I. SINGH, Vacuum **31**, 675 (1981).
[7] F. MANGHI, E. MALINAN, C. M. BERTONE, and C. CALANDRA, J. Phys. C **15**, 1106 (1982).
[8] R. P. BERES, R. E. ALLEN, and J. D. DOW, Phys. Rev. B **26**, 5702 (1982).
[9] K. HIRABAYASHI, J. Phys. Soc. Japan **27**, 1475 (1969).
[10] K. C. PANDEY and J. C. PHILLIPS, Phys. Rev. Letters **34**, 1450 (1975).
[11] J. D. JOANNOPOULOS and M. L. COHEN, Phys. Rev. B **10**, 5075 (1974).
[12] D. J. CHADI, Phys. Rev. B **18**, 1800 (1978).
[13] D. J. CHADI, Phys. Rev. B **19**, 2074 (1979).
[14] J. POLLMANN and S. T. PANTELIDES, Phys. Rev. B **18**, 5524 (1978).
[15] J. POLLMANN and S. T. PANTELIDES, Phys. Rev. B **21**, 709 (1980).
[16] N. V. DANDEKAR, A. MADHUKAR, and D. N. LOWY, Phys. Rev. B **21**, 5687 (1980).
[17] Y. C. CHANG, Phys. Rev. B **26**, 4400 (1982).
[18] M. SCHMEITS, A. MAZUR, and J. POLLMANN, Phys. Rev. B **27**, 5012 (1983).
[19] J. N. SCHULLMAN and T. C. MCGILL, Phys. Rev. B **19**, 6341 (1979).
[20] S. JALAJ, I. SINGH, and S. K. JOSHI, to be published.
[21] R. J. MEYER, C. B. DUKE, A. PATON, J. C. TSANG, J. L. YEH, A. KHAN, and P. MARK, Phys. Rev. B **22**, 6171 (1980).
[22] I. SINGH and G. P. SRIVASTAVA, phys. stat. sol. (b) **108**, 467 (1981).
[23] R. H. WILLIAMS, G. P. SRIVASTAVA, and I. T. MCGOVERN, Rep. Progr. Phys. **43**, 1357 (1980).

(Received August 15, 1984)

phys. stat. sol. (b) **126**, 677 (1984)

Subject classification: 13.4; 22.1.2

Department of Physics, Middle East Technical University, Ankara (a) and International Centre for Theoretical Physics, Trieste[1]) *(b)*

Ideal Vacancy States in Si

A Study of the Empirical Tight-Binding Method

By

C. TUNCAY (a) and M. TOMAK[2]) (b)

A series of calculations is performed to determine the sensitivity of ideal vacancy-induced deep levels on the empirical tight-binding parameters. A new parameter set is proposed for Si which yields a somewhat better band structure. The vacancy-induced bound states and resonances are found to depend on the parameters rather sensitively while the band structures do not show this sensitivity.

Es wird eine Serie von Rechnungen durchgeführt, um die Empfindlichkeit von idealen, durch Leerstellen induzierte Niveaus auf die empirischen „tight-binding"-Parameter zu untersuchen. Ein neuer Parametersatz für Si wird vorgeschlagen, der eine etwas bessere Bandstruktur ergibt. Es wird gefunden, daß die Leerstellen-induzierten gebundenen Zustände und Resonanzen ziemlich empfindlich von den Parametern abhängen, während die Bandstrukturen diese Empfindlichkeit nicht zeigen.

1. Introduction

In recent years, there has been an upswing of interest in understanding the electronic states introduced by vacancies in semiconductors [1]. A special emphasis is placed upon understanding the behaviour of the isolated vacancy in Si. This technologically extremely important and most extensively studied material provides a testing ground for theoretical models and calculations.

The single vacancy in Si was identified by Watkins [2] and was later studied by many workers in great detail [3]. In the theoretical studies of the ideal vacancy-induced deep levels the localized nature of the potential and the wave function is exploited by use of a basis set of localized orbitals. Early attempts along these lines were made by Koster and Slater [4] and by Callaway and Hughes [5] who used Wannier functions in Green's function method. The calculations turned out to be extremely complicated because of the difficulty involved in calculating the Wannier functions. However, Lannoo and Lenglart [6] observed that the application of the Green's function method with a basis set of localized atomic orbitals is much easier. The empirical tight-binding (ETB) as well as self-consistent pseudopotential methods with a localized basis set have been applied to the vacancy and impurity problem by a number of workers [7 to 11].

In the ETB method a set of localized orbitals on each atom is used and a band structure calculation, usually assuming nearest-neighbour interaction only, is performed. The Hamiltonian matrix elements are treated as adjustable parameters to be fitted to either the experimental band structure information or to the results of an exact calculation, at certain high-symmetry points in the Brillouin zone. One then

[1]) P.O.B. 586, 43100, Trieste, Italy.

[2]) Permanent address: Physics Department, Middle East Technical University, Inonu Bulvari, Ankara, Turkey.

calculates the defect-induced changes in the electronic structure in terms of this solution of the unperturbed system employing the Koster-Slater method in the language of Green's functions and scattering theory. Instead, in the self-consistent pseudopotential method the first step is to adopt a self-consistent description of the bulk solid and set up a scheme to find solutions for a given localized defect potential. The second step involves iteration to self-consistency for the defect potential.

The ETB method is rather qualitative in comparison with the more exact self-consistent pseudopotential methods. But it provides physical insight into more sophisticated calculational results and it could be used to treat many possible geometries and more complicated defects comparatively easily. The method has recently been applied to multivacancies, for example [12].

Thus, it is of interest to study the limitations of the ETB approach which may lead to a better understanding of the results obtained with it. In this work, we critically examine the ETB method as applied to the problem of an isolated, ideal vacancy in Si. In particular, we address ourselves to the question of sensitivity of the vacancy-induced deep levels to the tight-binding parameter sets used in the calculation.

The plan of the paper is as follows. In Section 2 the general formulation of the defect problem in ETB Green's function formalism is presented. Results are summarized in Section 3. Finally, in Section 4, a brief summary and discu ssion is given

2. Formulation of the Defect Problem in the ETB Method: Green's Function Formalism

If we let H^0 be the perfect crystal Hamiltonian and U the defect potential, the Schrödinger equation for the perturbed crystal becomes

$$(H^0 + U)\,\Psi = E\Psi\,. \tag{1}$$

First, let us concentrate on states within the band gaps. For these, the Green's operator for the perfect crystal is real and given by

$$G^0(E) = (E - H^0)^{-1}\,. \tag{2}$$

Equation (1) can then be written as

$$\Psi = (E - H^0)^{-1}\, U\Psi = G^0(E)\, U\Psi \tag{3}$$

or

$$[1 - G^0(E)\, U]\,\Psi = 0\,. \tag{4}$$

Thus, nontrivial solutions exist only when the determinant of the operator $1 - G_0U$ is zero. Expanding Ψ in any complete orthonormal set of states Φ_α, taking matrix elements of G_0 and U on the same basis and using (4) yields

$$\det \|\delta_{\alpha\alpha'} - \sum_{\alpha''} G^0_{\alpha\alpha''}(E)\, U_{\alpha''\alpha'}\| = 0\,, \tag{5}$$

where

$$G^0_{\alpha\alpha'}(E) = \langle\alpha|\,(E - H^0)^{-1}\,|\alpha'\rangle = \sum_{n\boldsymbol{k}} \frac{\langle\alpha \mid n\boldsymbol{k}\rangle\;\langle n\boldsymbol{k} \mid \alpha'\rangle}{E - E^0_{n\boldsymbol{k}}} \tag{6}$$

and where $\langle\alpha \mid n\boldsymbol{k}\rangle$ are expansion coefficients, $E^0_{n\boldsymbol{k}}$ are the eigenvalues of the perfect crystal Hamiltonian H^0, i.e., the energy bands. It is important to note that if the wave functions Φ_α are localized at atomic sites and the defect potential U has a finite range, the size of the determinant (5) reduces to the size of the nonzero part of the potential matrix.

Now, let us consider states within the energy bands of the perfect crystal. The definition of G^0 is in this case generalized to

$$G^0(E) = \lim_{\varepsilon \to 0} (E - H_0 + i\varepsilon)^{-1} \tag{7}$$

and (3) is replaced by the Lippmann-Schwinger equation

$$\Psi_{nk} = \Psi^0_{nk} + \lim_{\varepsilon \to 0^+} G^0(E + i\varepsilon)\, U\Psi_{nk}\,, \tag{8}$$

where Ψ^0_{nk} is a solution of

$$H^0\Psi^0_{nk} = E^0_{nk}\Psi^0_{nk}\,, \tag{9}$$

i.e. it is the perfect crystal Bloch function. Equation (4) is now replaced by

$$[1 - G^0(E)\, U]\, \Psi_{nk} = \Psi^0_{nk}\,. \tag{10}$$

If $D(E)$ denotes the determinant

$$D(E) = \det ||1 - G^0(E)\, U||\,, \tag{11}$$

the change in density of states can be written as

$$\Delta N(E) = \frac{1}{\pi} \frac{\mathrm{d}\delta(E)}{\mathrm{d}E}\,, \tag{12}$$

where $\delta(E)$ is the phase shift given by

$$\delta(E) = -\tan^{-1}\left[\frac{\operatorname{Im} D(E)}{\operatorname{Re} D(E)}\right]. \tag{13}$$

The rest of the useful relations can be found in [1].

The isolated ideal vacancy is considered to be the absence of the appropriate atom in the unit cell in an otherwise perfect crystal. In the tight-binding description U represents a matrix which annuls all the interactions between the atom removed and the rest of the crystal. An alternative way of this bond-cutting method of creating an ideal vacancy is to set the diagonal energies corresponding to the atom to be removed to infinity (i.e. a very large number in actual calculations). Thus, no electron is allowed to reach the site of the removed atom, a situation as if there is no atom there. The equivalence of these two methods is explicitly demonstrated in [13]. For a critique of the method see [14].

Since the defect potential matrix U in the s–p basis has the form

$$U = E_\infty I\,, \tag{14}$$

with I as the 4×4 unit matrix at the vacancy site, $D(E)$ itself becomes a 4×4 determinant. However, G^0 at the vacancy site is diagonal in atomic s–p basis because of the tetrahedral symmetry. Thus, (11) breaks up into sets of algebraic equations,

$$1 - G^0_{\alpha\alpha}(E)\, E_\infty = 0\,, \tag{15}$$

where α is s or any of the three p-orbitals at the vacancy site. In the limit $E_\infty \to \infty$, (15) becomes

$$G^0_{\alpha\alpha}(E) = 0\,, \tag{16}$$

the condition for vacancy-induced bound states. For α = s, we get the A_1 symmetric vacancy level and for α = p we get the T_2 symmetric vacancy level.

The condition (16) can be rewritten explicitly using (6) and (7),

$$G^0_{\alpha\alpha}(E) = \lim_{\varepsilon \to 0^+} \sum_{nk} \frac{|\langle \alpha \mid n\boldsymbol{k}\rangle|^2}{E - E^0_n(\boldsymbol{k}) + i\varepsilon} \tag{17}$$

which can be rewritten as

$$G^0_{\alpha\alpha}(E) = \lim_{\varepsilon \to 0^+} \int \frac{\sum_{nk} |\langle \alpha \mid n\boldsymbol{k}\rangle|^2 \, \delta(E' - E^0_{n\boldsymbol{k}})}{E - E' + i\varepsilon} \, \mathrm{d}E' \tag{18}$$

or

$$G^0_{\alpha\alpha}(E) = \mathrm{P} \int \frac{A_{\alpha\alpha}(E')}{E - E'} \, \mathrm{d}E' - i\pi A_{\alpha\alpha}(E) \, , \tag{19}$$

where P stands for "principal value", $A_{ss}(E)$ and $A_{pp}(E)$ are standard s- and p-partial density of states, respectively,

$$A_{\alpha\alpha}(E) = \sum_{n\boldsymbol{k}} |\langle \alpha \mid n\boldsymbol{k}\rangle|^2 \, \delta(E - E^0_{n\boldsymbol{k}}) \, . \tag{20}$$

Once $A_{\alpha\alpha}(E)$ is evaluated the real part of $G^0_{\alpha\alpha}(E)$ can be obtained by carrying out the principal value integrals. The condition (16) means that the bound states of A_1 and T_2 symmetries are given by the zeros of $\mathrm{Re}\, G^0_{ss}(E)$ and $\mathrm{Re}\, G^0_{pp}(E)$, respectively. The condition for bound states can be rewritten in such a way as to bring out the contribution of individual bands as

$$\int \frac{A_{\alpha\alpha}(E')}{E - E'} \, \mathrm{d}E' = 0 \tag{21}$$

or

$$\int\limits_{\text{valence}} \frac{A_{\alpha\alpha}(E')}{E - E'} \, \mathrm{d}E' = - \int\limits_{\text{conduction}} \frac{A_{\alpha\alpha}(E')}{E - E'} \, \mathrm{d}E' \, . \tag{22}$$

Equation (22) clearly reveals that, for the ideal vacancy, the bound state in the gap results from a delicate balance between valence and conduction bands, so that both must be known with equal accuracy.

3. Results

We have carried out a series of calculations employing the second-nearest-neighbour approximation with four atomic orbitals per atom basis set. Our main goal has been to determine the sensitivity of the vacancy-induced deep levels to the choice of ETB parameter sets and to the approximations made during the actual numerical calculations. In the following, we summarize our findings.

3.1 Band structure and density of states of pure crystal

To obtain s- and p-partial density of states, $A_{ss}(E)$ and $A_{pp}(E)$, we need to calculate first the band structure of pure Si. There have been many attempts within the ETB method differing from each other by varying degrees of empiricism and the type of the basis set used. Although often neglected some even include overlap terms between different orbitals. The common result is that though all these calculations give a reasonable valence band, they fail to produce an accurate enough conduction band. It is well established that within the nearest-neighbour approximation the single s and three p orbital basis set does not yield a good description of the conduction bands. The basic reason is that the antibonding basis corresponding to conduction bands is much less localized than the bonding basis and the matrix elements between different sites decay much more slowly with site separation because of this delocalization. There have been some attempts to improve the situation. In addition to the generalization of extending interaction parameters to more distant neighbour interactions with

Table 1

The ETB parameters employed in the calculations of ideal vacancy-induced deep levels in Si

parameter (eV)	[15]	[11]	present work
E_s	−4.204	−4.254	−4.265
E_p	0.186	0.219	0.183
$(ss\sigma)_1$	−2.080	−2.045	−2.069
$(sp\sigma)_1$	2.120	1.871	1.896
$(ss\sigma)_2$	0.000	0.070	0.020
$(sp\sigma)_2$	0.000	0.011	0.001
$(pp\sigma)_1$	2.321	2.150	2.220
$(pp\pi)_1$	−0.520	−0.550	−0.554
$(pp\sigma)_2$	0.580	0.482	0.509
$(pp\pi)_2$	−0.100	−0.094	−0.099

fixed orbital basis set there exists a second approach which remains in the nearest-neighbour approximation but enlarges the atomic orbital basis set. This second approach is used by the Illinois group under Dow [9] in a diversity of problems. A total of five atomic orbitals, two s and three p, are employed. This description evidently improves the conduction bands considerably.

The parameters used in this work are shown in Table 1 along with those of Pandey and Phillips [15] and Kauffer et al. [11]. The band structure calculated with our parameters is shown in Fig. 1. As can easily be noticed the valence bands are reproduced well but the conduction bands are rather flat. This is the general feature of the tight-binding method as discussed above. The results obtained with other parameter sets are similar. But the band gap is reproduced somewhat better with our parameters.

The second important step in the calculation of the vacancy-induced deep levels is the determination of the density of states (DOS) which is the imaginary part of the Green's function. We have employed all three techniques commonly in use in the calculation of DOS; namely, histogram, tetrahedron, and Gilat-Raubenheimer techniques. There is a slight dependence of the levels on the numerical technique used. One should be careful to carry highly accurate calculations of DOS, otherwise it is easy to get spurious results. The total density of states calculated by the histogram tech-

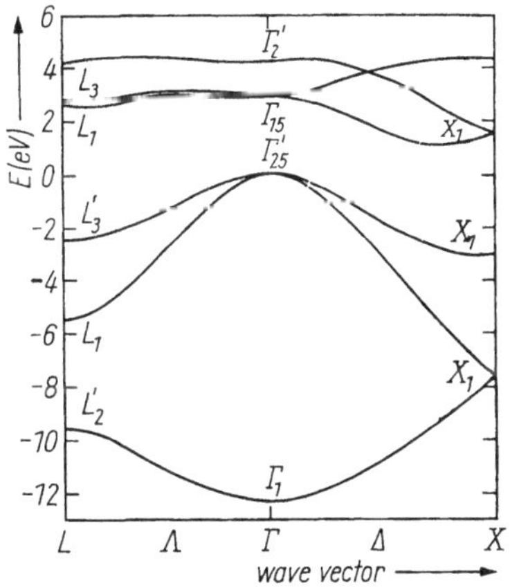

Fig. 1. Band structure of Si

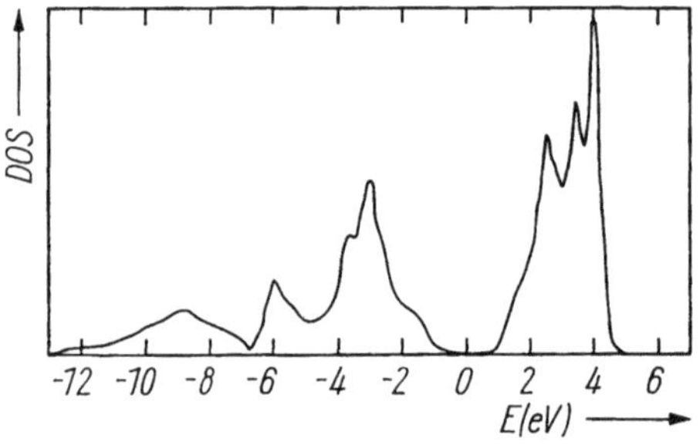

Fig. 2. The total density of states of Si

nique, which is the simplest, is shown in Fig. 2. s- and p-partial density of states are also calculated explicitly and agree very well with the established s and p character of the valence and conduction bands.

3.2 Vacancy-induced changes in DOS; phase shifts, bound states, and resonances

The vacancy-induced changes in DOS, phase shifts, bound states, and resonances for s- and p-like states are calculated using the fundamental relations given in Section 2. The starting point is the Hilbert transform of the s- and p- partial density of states to get the real parts of the corresponding Green's functions. The results are shown in Fig. 3 a, b and 4.

A critical review of the literature will be helpful to compare the results with the earlier ones. It has been established especially by self-consistent pseudopotential calculations that a vacancy in Si introduces a bound state of T_2 symmetry (p-like) in the energy gap and a sharp resonance of A_1 symmetry (s-like) just below the top of the valence band. The wave functions of these states exhibit a dangling bond character and are linear combinations of hybrid orbitals pointing from the neighbouring atoms towards the vacant site. When one studies the literature carefully, one immediately discovers that various ETB parameter sets represent comparably good fits to the bulk band structure and yet the deep levels obtained by employing these sets do indeed scan the whole range of the band gap. For example, Bernholc at al. [7] using the ETB parametrization of Pandey and Phillips [15] find a bound state 0.27 eV above the valence band edge (VBE). Kauffer et al. [11] employing a different set, find a bound state 0.12 eV above VBE. Papaconstantopoulos and Economou [16], who employed 20 parameters including up to third-neighbour interactions,

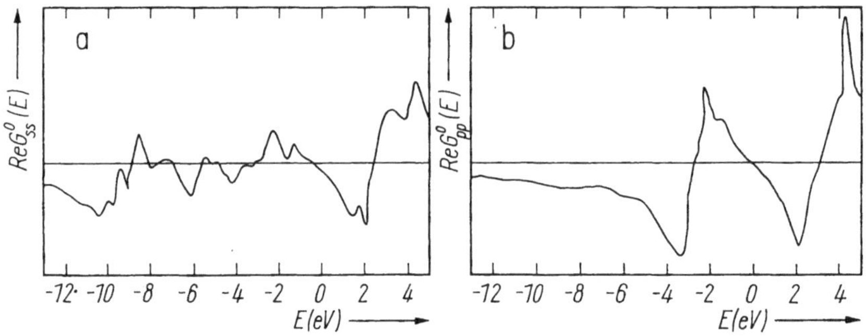

Fig. 3. The real part of the a) s-like and b) p-like Green function

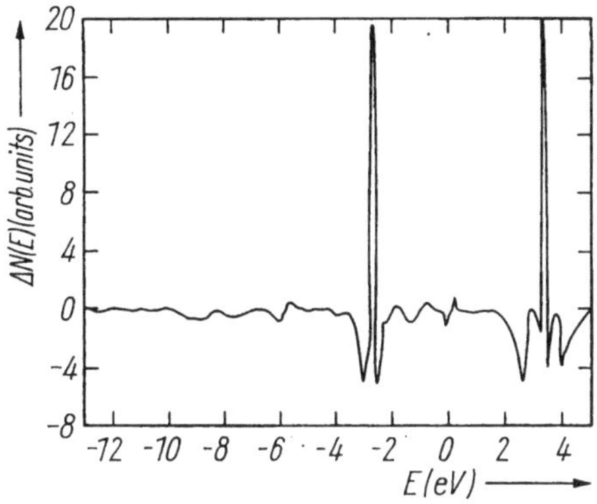

Fig. 4. Change in DOS induced by an ideal vacancy in Si

found a bound state 0.75 eV above the VBE. Several resonant and antiresonant structures within the bands are also reported. The A_1 resonance at about -8 eV occurs around the point where the bulk DOS has a cusp, which is a critical point in the band structure. This sharp resonance corresponds also to a noticeable drop in the A_1 phase shift. Broad resonances lie in a region of higher density of Bloch states.

Table 2

Band structure energies

band structure energy (eV) at high symmetry points	[15]	[11]	present work
$L_{2'}$	− 9.53	− 9.29	− 9.34
L_1	− 4.27	− 4.01	− 3.99
$L_{3'}$	− 3.92	− 3.43	− 3.57
L_1	1.58	1.57	1.56
$L_{2'}$	4.20	3.48	3.60
L_3	7.01	6.38	6.55
Γ_1	−12.52	−11.59	−12.51
$\Gamma_{25'}$	0.0	0.0	0.0
Γ_{15}	3.42	2.80	2.91
$\Gamma_{2'}$	4.12	4.77	4.04
X_1	− 8.17	− 7.66	− 7.57
X_4	− 3.20	− 3.00	− 3.06
X_1	1.84	1.43	1.51
X_3	4.38	4.20	4.28

Table 3

Vacancy energy levels

energies (eV)	[15]	[11]	present work
	−7.74	−7.27	−7.26
A_1	−0.98	−0.81	−0.88
	2.74	2.22	2.34
T_2	0.23	0.19	0.18

Thus, it is of interest to check, in an internally consistent manner, the sensitivity of the vacancy-induced deep levels to the ETB parameters used. This we have done by repeating all steps of the calculation, with the same numerical procedure and same accuracy, for different parameters. We have found that both the bound state in the fundamental gap and resonances in the bands shift sensitively in energy depending on the parameters. For example, in a calculation which employs the histogram method for DOS, we find that only the results of Kauffer et al. remain essentially the same, while the bound state is pushed down to the tip of the VBE for Pandey and Phillips and our parameters shown in Table 1. We also seem to find more resonances and antiresonances within the bands.

4. Summary and Conclusions

We have performed a series of calculations which clearly demonstrated that the vacancy-induced deep levels depend quite sensitively on which ETB parameter set one uses, although the band structures do not show this sensitivity. We thus conclude that the choice of ETB parameters should not be made solely in regard to band structure fitting. More stringent criteria are needed to fix these parameters. Clearly, an effort should be made towards a better understanding of the physics involved in each of the ETB parameters, instead of using them as mere numbers obtained through an empirical fitting procedure.

Similar conclusions were drawn by Das Sarma and Madhukar [13] who have carried out similar calculations for a number of III–V semiconductors. To obtain a physically meaningful set of parameters they have used, in addition to fitting to band structure at certain symmetry points, more stringent conditions such as the reproduction of the photothreshold of the material. They have also imposed the additional condition that the diagonal energies employed for a given atom be the same when measured with respect to vacuum, regardless of the solid in which the atom appears. The main motivation behind the additional requirements is to correctly reproduce both the energy bands and charge distribution of the perfect solid.

A possible improvement of the empirical parametrization scheme discussed so far is to parametrize the localized orbitals themselves. This approach has been developed by a number of workers [17 to 20]. The basic idea is to expand crystal wave functions in terms of Bloch sums of parametrized localized orbitals, generally taken as Gaussian or Slater orbitals. The expansion leads to an energy functional which can be minimized with respect to the orbital parameters. The resulting lower conduction bands are considerably improved over those obtained by the fitting procedure described above.

Acknowledgements

One of the authors (M.T.) would like to thank Prof. Abduš Salam, the International Atomic Energy Agency, and UNESCO for hospitality at the International Centre for Theoretical Physics, Trieste.

References

[1] For an extensive review of the methods used for point defects in semiconductors, see: S. T. Pantelides, Rev. mod. Phys. 50, 797 (1978).
M. Jaros, Deep Levels in Semiconductors, Adam Hilger Ltd., Bristol 1982.
M. Lannoo and J. Bourgoin, Point Defects in Semiconductors, Springer-Verlag, Berlin 1981.

[2] G. D. Watkins, in: Radiation Damage and Defects in Semiconductors, Inst. Phys., London 1973 (p. 228).

[3] For a recent review, see: G. D. WATKINS, in: Proc. 16th Internat. Conf. Phys. Semicond., Montpellier (France), Sept. 1982; Physica, **117B**, 9 (1983).
[4] G. F. KOSTER and J. C. SLATER, Phys. Rev. **95**, 1167 (1964).
[5] J. CALLAWAY, J. Math. Phys. **5**, 783 (1964).
J. CALLAWAY and A. J. HUGHES, Phys. Rev. **156**, 860 (1967).
[6] M. LANNOO and P. LENGLART, J. Phys. Chem. Solids **30**, 2409 (1969).
[7] J. BERNHOLC, N. O. LIPARI, and S. T. PANTELIDES, Phys. Rev. Letters **41**, 895 (1978); Phys. Rev. B **21**, 3545 (1980).
J. BERNHOLC and S. T. PANTELIDES, Phys. Rev. B **15**, 1780 (1977).
[8] G. A. BARAFF, E. O. KANE, and M. SCHLÜTER, Phys. Rev. Letters **43**, 956 (1979); Phys. Rev. B **21**, 3563 (1980).
[9] H. P. HJALMARSON, P. VOGL, D. J. WOLFORD, and J. D. DOW, Phys. Rev. Letters **44**, 810 (1980).
[10] M. JAROS, C. O. RODRIGUES, and S. BRAND, Phys. Rev. B **19**, 3137 (1979).
[11] E. KAUFFER, P. PEUCHER, and M. GERL, Phys. Rev. B **15**, 4107 (1977).
[12] S. T. PANTELIDES, I. IVANOV, M. SCHEFFLER, and J. P. VIGNERON, Conf. Defects in Semiconductors, Amsterdam 1982, to be published.
[13] S. DAS SARMA and A. MADHUKAR, Phys. Rev. B **24**, 2051 (1981).
[14] J. B. KRIEGER and P. M. LAUFER, Phys. Rev. B **23**, 4063 (1981).
[15] K. C. PANDEY and J. C. PHILLIPS, Phys. Rev. Letters **32**, 1433 (1974); Phys. Rev. B **13**, 750 (1976).
[16] D. A. PAPACONSTANTOPOULOS and E. N. ECONOMOU, Phys. Rev. B **22**, 2903 (1980).
[17] E. O. KANE, Phys. Rev. B **13**, 3478 (1976).
[18] D. J. CHADI, Phys. Rev. B **16**, 3572 (1977).
[19] C. TEJEDOR and J. A. VERGES, Phys. Rev. B **19**, 2283 (1979).
[20] S. G. LOUIE, Phys. Rev. B **22**, 1933 (1980).

(Received August 6, 1984)

phys. stat. sol. (b) **126**, 687 (1984)

Subject classification: 10.2 and 19; 11; 20.3; 22.5.3

Department of Physics, Memphis State University[1]) *(a) and*
Los Alamos National Laboratory[2]) *(b)*

Thermal Evolution of Low-Temperature Manganese Centers in X-Irradiated CaF_2:Mn

By

M. S. JAHAN (a) and D. W. COOKE (b)

Thermal evolution of the radiation-induced defects in CaF_2:Mn (0.1% by weight) are investigated by ESR techniques. The Mn^+ resonance is found to decay at 200 K with an activation energy of 0.44 eV and frequency factor $2.8 \times 10^8\ s^{-1}$ which agree with previous luminescence and optical absorption measurements. This activation energy and decay temperature also agree with that of the previously described H center. It is concluded that the luminescence in CaF_2:Mn results from thermal release of holes at H centers with ultimate recombination at Mn^+ ions producing excited state $Mn^{2+}(*)$. Relaxation yields the observed 500 nm emission characteristic of Mn^{2+}. A second defect which decays at 150 K is observed but not identified.

Die thermische Entwicklung der strahlungsinduzierten Defekte im CaF_2:Mn (0,1 Gew%) werden mittels ESR-Technik untersucht. Es wird gefunden, daß die Mn^+-Resonanz bei 200 K mit einer Aktivierungsenergie von 0,44 eV und einem Frequenzfaktor von $2,8 \times 10^8\ s^{-1}$ abklingt, was mit früheren Messungen der Lumineszenz und optischen Absorption übereinstimmt. Diese Aktivierungsenergie und Abklingtemperatur stimmen auch mit den schon früher für das H-Zentrum beschriebenen Werten überein. Es wird angenommen, daß die Lumineszenz in CaF_2:Mn von der thermischen Befreiung von Löchern an den H-Zentren herrührt, mit endgültiger Rekombination an Mn^+-Ionen, wobei ein angeregter Zustand mit $Mn^{2+}(*)$ erzeugt wird. Relaxation liefert die beobachtete 500 nm-Emission, die für Mn^{2+} charakteristisch ist. Ein zweiter Defekt, der bei 150 K abklingt, wird beobachtet, jedoch nicht identifiziert.

1. Introduction

Numerous studies of transition-metal ions in crystals with the fluorite structure have been conducted and the results well documented [1, 2]. One system that has received considerable attention is manganese-doped calcium fluoride, the emphasis arising primarily from its use as a thermoluminescent radiation dosimeter [3 to 6]. Depending upon the impurity concentration one observes intense luminescent glow near 200 K whose spectral content is characterized by a nearly Gaussian-shaped band with maximum near 500 nm [6]. It has been well established [7, 8] that this emission results from the deexcitation of an excited Mn^{2+} ion, i.e., a transition from the first excited level ${}^4T_{1g}({}^4G)$ to the ground state ${}^6A_{1g}({}^6S)$.

What has not been unequivocally established however is the physical mechanism associated with the Mn^{2+} excitation process. X-irradiation at 80 K produces upon heating an intense glow peak near 200 K [7] characterized by band emission near 500 nm. In addition, optical absorption studies [9] in the same temperature interval have identified a band at 450 nm attributable to Mn^+ which thermally decay at 200 K. Thus optical and thermal bleaching experiments combined with emission results suggest that the 200 K glow peak in X-irradiated CaF_2:Mn is associated with the conversion of Mn^+ to Mn^{2+} and subsequent deexcitation. Complementary charge trapping

[1]) Memphis, Tennessee 38152, USA.
[2]) Los Alamos, New Mexico 87545, USA.

and thermal stability is unknown, i.e., what is the fate of the hole removed from Mn^{2+} during irradiation?

In an attempt to shed light on the luminescence process occurring near 200 K we have performed ESR measurements on CaF_2:Mn X-irradiated at 77 K, and have monitored the thermal evolution of the radiation-induced signals. Correlation of the results with existing optical absorption and luminescence data strongly suggests that the 200 K glow peak is indeed associated with the charge conversion of Mn^+ to Mn^{2+} and its subsequent relaxation. Various possibilities for this conversion process are discussed.

2. Experimental Procedure

Single crystals of CaF_2:Mn were obtained from Optovac, Inc., with a stated manganese concentration of 0.1 wt%. Neutron activation analysis was performed on one of these crystals and the results indicated a Mn concentration of 0.11% [10]. X-irradiation was provided by an X-ray source whose effective energy was approximately 10 keV as determined by half-value layer measurements, with an exposure rate at the sample site of 3.7×10^3 R/min. Crystals were irradiated at 77 K and then inserted into a nitrogen flow cryostat without further warming. Temperature was controlled by an Oxford Instruments DTC-2 temperature controller and regulated to within ± 1 K.

ESR measurements were made with a spectrometer operating at X-band frequencies (≈ 9 GHz). Relative spin concentrations of paramagnetic species at a given temperature and time were ascertained by measuring peak-to-peak heights from first-derivative spectra, all measurements being taken on spectral lines for which the shape and width did not change during decay. Care was also taken to avoid power saturation of the samples.

3. Experimental Results

Shown in Fig. 1 are the typical patterns of the first derivative ESR spectra of CaF_2:Mn X-irradiated (2.2×10^5 R) at 77 K and measured at 90, 150, 200, and 300 K (data were recorded at 5 K intervals between 90 and 300 K). Following the RT measurement the

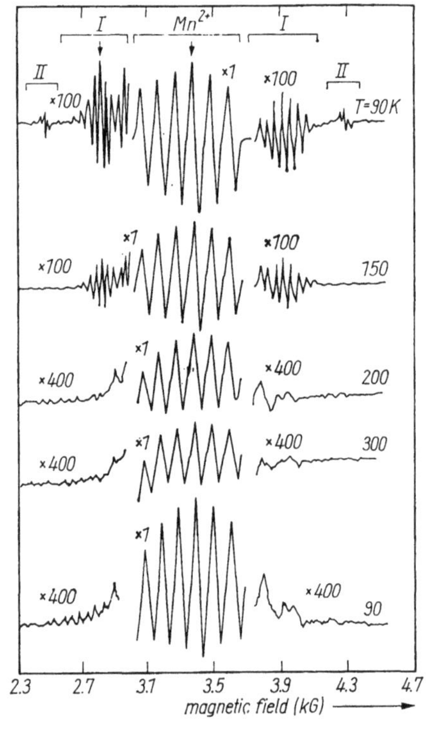

Fig. 1. Typical ESR spectra of CaF_2:Mn X-irradiated at 77 K and subsequently heated to 90, 150, 200, and 300 K. Data were taken at 5 K intervals with representative spectra shown in the figure. The bottom spectrum was recorded at 90 K following a warmup to 300 K but without further irradiation. Lines marked with arrows were used for intensity measurements. Note the scale change for centers I and II with respect to the Mn^{2+} absorption

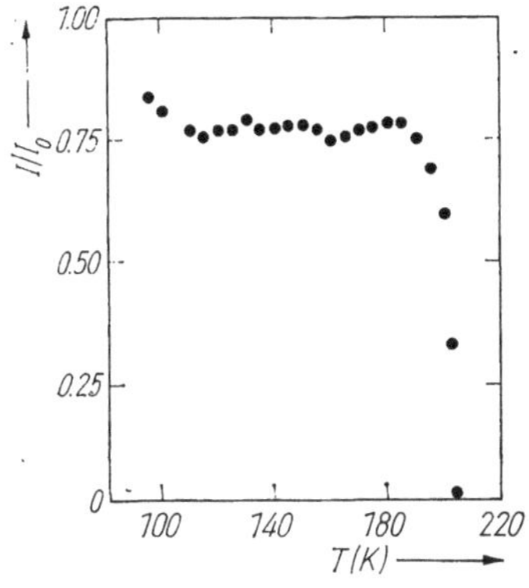

Fig. 2. Thermal decay of the Mn^+ centers. The sample was irradiated at 77 K and the intensity of absorption measured at 5 K intervals. Each intensity was corrected for the Boltzmann factor; I_0 corresponds to the intensity at 90 K

crystal was cooled to 90 K without further irradiation and the spectrum again recorded. Note the appropriate scale changes. The central six-line spectrum is well known [1] and can be attributed to Mn^{2+} ions in a substitutional site. Two additional centers which we designate as I and II are also resolved. Center I which is overlapped by the strong Mn^{2+} resonance has been studied by Baranov [11] and assigned to Mn^+ ions which exhibit paramagnetic properties due to a half-filled 4s shell. Center II is of unknown origin.

Fig. 2 illustrates an intensity–temperature plot of center I (Mn^+) resonance and clearly demonstrates that this paramagnetic species decays near 200 K, which is the temperature of intense thermally stimulated luminescence (TSL) and also the thermal decay temperature of the Mn^+ optical absorption band. Due to the low intensity of center II resonance lines it was difficult to make an accurate plot of intensity versus temperature; however, it was noted that the resonance signal completely decayed near 150 K. Those resonance lines marked by arrows in Fig. 1 were used for intensity measurements of Mn^+ signal.

The decrease in ESR intensity of a paramagnetic center as the temperature increases can be caused by an actual decay of the species and also by the Boltzmann factor. In order to correct the raw data for the Boltzmann factor, the intensities of the Mn^{2+} center (which does not undergo any measurable transformation) were measured at 5 K intervals from 90 to 300 K. An exponential fit (not shown) of the intensity–temperature data provided a slope which was used to obtain a corrected intensity–temperature plot of the decay of Mn^+ as shown in Fig. 2. As indicated there is no appreciable decay until 180 to 200 K at which point complete (and irreversible) decay occurs. Furthermore, measurements were made in this temperature interval (180 to 200 K) to determine center I rate of decay. This information allows one to deduce the parameters describing the kinetics of thermal decay. At any particular temperature the concentration of remaining paramagnetic ions in an irradiated sample at time t displays kinetics which are best described by a first-order decay equation [12]

$$\ln \frac{n}{n_0} = -k_r t \,, \qquad (1)$$

where n is the paramagnetic ion concentration at time t, n_0 the concentration at time $t = 0$, and k_r the temperature-dependent rate constant. Fig. 3 depicts the intensity versus time plots for four temperatures. Since the intensity of an ESR line is proportional to concentration of remaining ions, the measurements are given in terms of intensity.

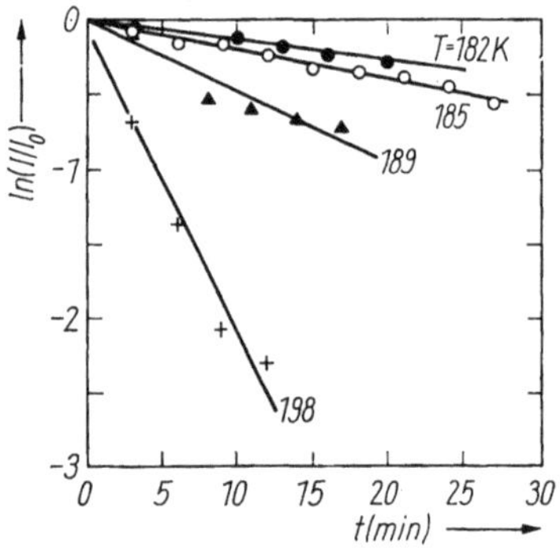

Fig. 3

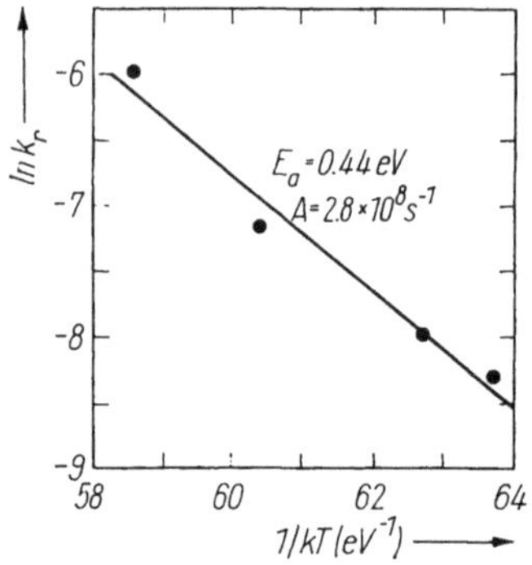

Fig. 4

Fig. 3. Intensity vs. time plots for the decay of Mn^+ centers produced in CaF_2:Mn at 77 K. Center concentrations are normalized to the concentration at $t = 0$. Solid lines represent the least-squares computer fit to equation (1)

Fig. 4. Arrhenius plot for determination of activation energy and frequency factor. Rate constants were obtained from plots such as in Fig. 3. Solid line represents the least-squares fit to equation (2)

The activation energy describing the decay process was determined from the Arrhenius equation

$$\ln k_r = -\frac{E_a}{kT} + \ln A , \tag{2}$$

where E_a is the activation energy, k the Boltzmann constant, T the absolute temperature, and A the frequency factor. A least-squares fit of the Mn^+ data is shown in Fig. 4 where the extracted parameters are: $E_a = 0.44$ eV and $A = 2.8 \times 10^8$ s^{-1}.

4. Discussion

Results presented in this work in conjunction with previous data strongly suggest that the 200 K TSL peak in irradiated CaF_2:Mn is due to the conversion of Mn^+ ions to excited state $Mn^{2+}(*)$ ions as a result of hole capture. For example, optical absorption studies of Alcalá et al. [9] identify the 450 nm absorption as due to Mn^+ which thermally decays at 200 K with an activation energy of 0.51 eV. Furthermore, TSL studies [7] reveal that the 200 K glow peak is characterized by an activation energy of 0.50 eV with spectral emission corresponding to Mn^{2+} deexcitation. Activation energies obtained from these studies, viz. 0.51 and 0.50 eV, are consistent with our value of 0.44 eV from ESR experiments although there is some discrepancy regarding the order of kinetics.[3]) Since the ESR signal also decays at 200 K and has been previously analyzed as due to Mn^+, the plethora of evidence seems to confirm the suggestion that Mn^+ conversion is responsible for the TSL glow. The mechanism incorporates thermal release of a hole and subsequent capture by Mn^+ thus producing $Mn^{2+}(*)$ in an excited state, the relaxation thereby yielding 500 nm emission as observed in TSL experiments.

---

[3]) References [7] and [9] ascribe second-order kinetics to the optical absorption and TSL decay whereas our ESR results indicate first order. It should be noted that TSL work on a 3% CaF_2:Mn sample revealed an intense peak at 175 K which is quite likely due to the same physical mechanism as associated with the 200 K peak in 0.1% CaF_2:Mn. This peak was carefully analyzed and found to obey 1.4-order kinetics (see [5]). Thus some disagreement exists concerning the kinetics order even among the TSL experiments.

The origin of the thermally released trapped hole cannot be deduced from the present work; however, existing experimental evidence suggests that the H center [13, 14][4]) is a likely candidate. This radiation-induced defect is associated with an interstitial F^- ion which captures a hole thus forming an F_2^- molecular ion whose crystallographic orientation is along the ⟨111⟩ direction. It is similar in structure to the V_K center in alkali halides [15, 16]. Although it has not been identified in CaF_2:Mn, Hall et al. [13] observed it in pure CaF_2 irradiated at 77 K and found that it annealed at 200 K with an activation energy of 0.55 eV, which agrees with the annealing temperature and activation energy extracted from luminescence, optical absorption, and the present ESR measurements. It is a bit surprising that we do not see its signature in the ESR spectrum; however, Beaumont et al. [14] point out that 90% of the holes created at 77 K in CaF_2:Tm exist at RT in non-paramagnetic form. So, it may be that they exist in CaF_2:Mn in non-paramagnetic form, or, alternatively, their resonance may be overlapped by the Mn^{2+} or Mn^+ absorption which precludes detection.

Another hole defect that is produced in CaF_2:Mn is the V_K center. We do not believe it is responsible for the Mn^+ conversion at 200 K because its optical absorption at 15 K has been identified (315 nm) and found to thermally decay near 80 K [9]. Furthermore, it could not be produced by 6 h of irradiation at 80 K.

We conclude that a plausible physical mechanism incorporating the existing data and describing the luminescence at 200 K is the following: X-irradiation at 80 K converts Mn^{2+} to Mn^+ by removing a hole which becomes trapped at an interstitial fluoride ion forming an H center. Upon heating the crystal to 200 K the hole becomes unstable and diffuses through the lattice to recombine with an Mn^+ ion forming Mn^{2+} (*) in an excited state. Relaxation of the ion yields 500 nm emission which is observed as luminescence. The role of center II in the luminescence process is unclear at this time.

In this work we have determined the thermal evolution of manganese centers responsible for luminescence and optical absorption in CaF_2:Mn and have presented a tentative model for interpreting the results. This is the first study to clearly demonstrate a correlation between the luminescence and ESR absorption in these crystals.

Acknowledgements

Work at Memphis State University was supported by Naval Surface Weapons Center. Work at Los Alamos National Laboratory was supported by U.S.D.O.E. One of us (M.S.J.) gratefully acknowledges the assistance of J. Daffron in constructing the low-temperature apparatus.

References

[1] See, for example, the excellent review article by W. Gehlhoff and W. Ulrici, phys. stat. sol. (b) **102**, 11 (1980), and references therein.
[2] W. Hayes, Crystals with the Fluorite Structure, Clarendon Press, Oxford 1974.
[3] R. J. Ginther and R. D. Kirk, J. Electrochem. Soc. **104**, 365 (1957).
[4] J. H. Schulman, F. H. Attix, E. J. West, and R. J. Ginther, Rev. sci. Instrum. **31**, 1263 (1960).
[5] D. W. Cooke, E. P. Gavathas, and M. D. Brown, J. appl. Phys. **54**, 1165 (1983).
[6] R. J. Abbundi, D. W. Cooke, V. K. Mathur, G. A. Royce, and M. D. Brown, Radiat. Protect. Dosimetry **6**, 329 (1984).
[7] P. J. Alonso and R. Alcalá, J. Lum. **21**, 147 (1980).
[8] P. J. Alonso and R. Alcalá, J. Lum. **22**, 321 (1981).

[4]) The H center was originally called a type I center [13], then a V_H center [14], and now a H center [2].

[9] R. ALCALÁ, P. J. ALONSO, G. LALINDE, and A. CARRETERO, phys. stat. sol. (b) **98**, 315 (1980).
[10] J. F. RHODES and M. D. BROWN, private communication.
[11] P. G. BARANOV, Soviet Phys. – Solid State **22**, 133 (1980).
[12] P. K. HORAN, W. D. TAYLOR, G. K. STROTHER, and W. SNIPES, Biophys. J. **8**, 164 (1968).
[13] T. P. HALL, A. LEGGEAT, and J. TWIDELL, J. Phys. C **2**, 1590 (1969).
[14] J. H. BEAUMONT, W. HAYES, D. L. KIRK, and G. P. SUMMERS, Proc. Roy. Soc. **A315**, 69 (1970).
[15] T. G. CASTNER and W. KANZIG, J. Phys. Chem. Solids **3**, 178 (1957).
[16] T. O. WOODRUFF and W. KANZIG, J. Phys. Chem. Solids **5**, 268 (1958).

(Received September 10, 1984)

phys. stat. sol. (b) **126**, 693 (1984)

Subject classification: 10.2 and 18.1; 22.5.2

Institut für Festkörperphysik der Universität Wien[1])

Magnetic Field Dependence of the Spin Polarization of the F Center

By

K. IWAHANA

A simple model for the magnetic field dependence of the spin polarization of the F center is developed, in which the spin mixing due to the hyperfine interaction both in the ground state and in the relaxed excited state is taken into account. It is shown that the spin mixing due to the hyperfine interaction plays an essential role in the interpretation of the spin polarization at low fields less than 0.5 T. Quantitative fitting of the experimental data of the magnetic field dependence of the spin polarization is performed for KBr and RbBr.

Es wird ein einfaches Modell für die Magnetfeldabhängigkeit der Spin-Polarisation in F-Zentren entwickelt, in welchem das durch die Hyperfein-Wechselwirkung verursachte Spin-Mixing sowohl im Grundzustand als auch im relaxierten angeregten Zustand berücksichtigt wird. Es wird gezeigt, daß das durch Hyperfein-Wechselwirkung verursachte Spin-Mixing eine entscheidende Rolle in der Interpretation des Vorzeichenwechsels der Spin-Polarisation bei schwachen Feldern spielt. Quantitative Anpassungen mit den experimentellen Daten der Spin-Polarisation werden für KBr und RbBr durchgeführt.

1. Introduction

The spin polarizations of the ground state (GS) and of the relaxed excited state (RES) of the F center under strong pumping have extensively been studied through the magnetic circular dichroism (MCD) of the absorption band [1 to 3] and the magnetic circular polarization (MCP) of the emission band [4 to 10], respectively. It is now generally recognized that since the spin systems of interest are brought out of thermal equilibrium by pumping intense light at sufficiently low temperatures, the spin polarizations of the GS and of the RES, P_{GS} and P_{RES}, must be described by the solution of the rate equations of the optical pumping cycle. In fact, some intriguing features of P_{GS} and P_{RES} have been successfully interpreted on the basis of Winnacker et al.'s model [3] proposed for the optical cycle of the F center [6 to 8]. The physical key quantities in their model are the different spin flipping probabilities between the effective up- and down-spin states which are due to the spin–orbit interaction in the unrelaxed excited state (URES), and which play an essential role in the interpretation of the dependence of the spin polarization upon the polarization of excitation light. However, their model has serious shortcomings on the magnetic field dependences of P_{GS} and P_{RES}, that is, it predicts only the field-independent behavior for P_{GS} and P_{RES} in strong contrast with the experimental data at low fields.

This discrepancy was partially removed by Mauser et al. [11] by taking into account the spin mixing due to the isotropic hyperfine interaction in the RES. They also successfully explained a change of sign in P_{GS} which was observed at low fields only for pumping on the low-energy side of the absorption band. Recently, on the other hand, a similar sign change at low fields was observed in P_{RES} only when excited on the

[1]) Strudlhofgasse 4, A-1090 Wien, Austria.

high-energy side of the absorption band [12, 13]. It turns out that Mauser et al.'s model cannot account for such a change of sign in P_{RES}.

In this paper we present a consistent explanation for the above-mentioned field dependences of P_{GS} and P_{RES} by taking into account the isotropic hyperfine interactions both in the GS and in the RES. In Section 2.1 the rate equations of the optical cycle in our simple model are given, and in Section 2.2 the analytical forms for the field dependences of P_{RES} and P_{GS} are derived by defining the spin polarization of the effective doublet in the URES. From the results obtained in Section 2.2 an effect of spin mixing due to the hyperfine interaction on the field dependences of P_{RES} and P_{GS} is discussed in Section 2.3. The limitations of our model are also pointed out here. Section 3 deals with numerical calculations of P_{RES} and P_{GS}, and comparison with the experimental data Section 4 gives a summary.

2. Magnetic Field Dependence of the Spin Polarization

2.1 Rate equations of the optical cycle

The F electron in the presence of a magnetic field undergoes spin mixing caused by various interactions in each stage of the optical cycle. In principle one must take into account all the possible spin mixing processes in the optical cycle. Under the particular conditions, however, it is allowed to pick up a few dominant spin mixing processes.

Since our interest lies in a spin mixing effect of the hyperfine interaction on the spin polarization, we confine ourselves to the low-field region less than 0.5 T and to the low-temperature region below 5 K. Then, we can safely neglect the spin mixing due to thermal relaxation both in the GS and in the RES [1, 9]. Furthermore, from the fact that the spin–orbit coupling constant for the 2p state in the relaxed configuration is quite small [9, 10], the spin mixing by spin–orbit coupling can be neglected in the RES. We also assume that there is no spin mixing in the unrelaxed ground state (URGS). Thus, in our model the spin mixing relevant in each stage of the optical cycle may be characterized in terms of the hyperfine interaction in the GS, the spin–orbit interaction in the URES, and the hyperfine interaction in the RES.

For the purpose of setting up the rate equations, we make the following assumptions:

(i) The URES and the URGS are described in terms of the effective spin doublet.
(ii) The distribution from the effective spin doublet to the hyperfine sublevels is governed, as was assumed in (11), by the mixing coefficients $\sqrt{\varepsilon^{G,R}(m)}$ of the up- and down-spin states in the eigenfunctions $\psi_\alpha^{G,R}(m)$ $(\alpha = \pm, |m| \leqq I - 1/2)$ which are

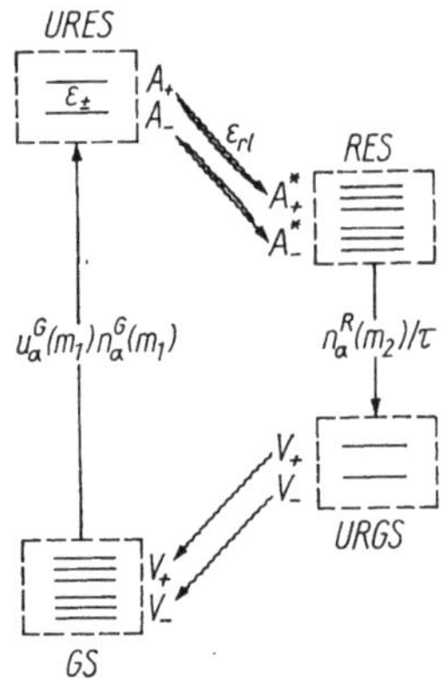

Fig. 1. Energy levels and transition rates for the optical pumping cycle of the F center including the spin mixing due to the isotropic hyperfine interactions

given by the well-known Breit-Rabi formula. Then,

$$\frac{dn_\alpha^G(m_1)}{dt} = -u_\alpha^G(m_1)\, n_\alpha^G(m_1) + [1 - \varepsilon^G(m_1)]\, \frac{V_\alpha}{2I_1 + 1} + \varepsilon^G(m_1)\, \frac{V_{-\alpha}}{2I_1 + 1}, \tag{1}$$

$$\frac{dn_\alpha^R(m_2)}{dt} = [1 - \varepsilon^R(m_2)]\, \frac{A_\alpha^*}{2I_2 + 1} + \varepsilon^R(m_2)\, \frac{A_{-\alpha}^*}{2I_2 + 1} - \frac{n_\alpha^R(m_2)}{\tau}, \tag{2}$$

with $A_\alpha^* = (1 - \varepsilon_{rl})\, A_\alpha + \varepsilon_{rl} A_{-\alpha}$ and $\varepsilon^{G,R}(\alpha(I + 1/2)) = 0$. Here $n_\alpha^{G,R}(m)$ are the occupation numbers of the states $\psi_\alpha^{G,R}(m)$, A_α and V_α the occupation numbers per unit time of the effective spin doublets in the URES and in the URGS, respectively, ε_{rl} is the spin-mixing probability during lattice relaxation from the URES to the RES [6], and τ the radiative lifetime of the RES. A schematic illustration of the optical cycle is shown in Fig. 1.

Explicit forms for $\varepsilon(m)$ are given by

$$\varepsilon(m) = \frac{1}{2}\left[1 - \left(\frac{g\beta H}{A} + m\right)\Big/\sqrt{\left(\frac{g\beta H}{A} + m\right)^2 + \left(I + \frac{1}{2}\right)^2 - m^2}\,\right], \tag{3}$$

where g and A are the g-factor and the isotropic hyperfine coupling constant, respectively, of the relevant spin system, β is the Bohr magneton, and H the magnitude of a magnetic field.

The transition rates $u_\alpha^G(m_1)$ from $\psi_\alpha^G(m_1)$ to the URES can be easily calculated by adopting the adiabatic wave function for the URES $\psi_i^\alpha(q)$ ($i = 1, 2, 3$ and $\alpha = \pm$) with the lattice coordinates of the Γ_1^+ and Γ_3^+ modes, q: Following (11), one finds that for excitation with η-polarized light with the energy E,

$$u_\alpha^G(m_1) = \frac{\sum\limits_{i=1}^{3} \sum\limits_{\alpha'=\pm} \int u_{\alpha,m_1}^{i,\alpha'}(q;\eta)\, P(q)\, \delta\big(E_i(q) - E\big)\, dq}{\sum\limits_{i=1}^{3} \int P(q)\, \delta\big(E_i(q) - E\big)\, dq}, \tag{4}$$

where $E_i(q)$ is the energy of $\psi_i^\alpha(q)$, $P(q)$ are the Gaussian probability distributions of the zero-point vibration of the Γ_1^+ and Γ_3^+ modes in the GS, $\delta(E)$ is the delta function, and

$$u_{\alpha,m_1}^{i,\alpha'}(q;\eta) = [1 - \varepsilon^G(m_1)]\, u_\alpha^i(q;\eta)\, \delta_{\alpha',\alpha} + \varepsilon^G(m_1)\, u_{-\alpha}^i(q;\eta)\, \delta_{\alpha',-\alpha}, \tag{5}$$

with the Kronecker's delta symbol $\delta_{\alpha,\beta}$. Inserting (5) into (4), one obtains

$$u_\alpha^G(m_1) = [1 - \varepsilon^G(m_1)]\, u_\alpha(E;\eta) + \varepsilon^G(m_1)\, u_{-\alpha}(E;\eta)\,. \tag{6}$$

Here $u_\alpha^i(q;\eta)$ and $u_\alpha(E;\eta)$ are given in [11].

2.2 Spin polarization

Let us now calculate the spin polarization of the RES under the steady-state condition, that is, $dn_\alpha^G(m_1)/dt = 0$ and $dn_\alpha^R(m_2)/dt = 0$. First of all, the spin polarization of the RES can be defined by

$$P_{RES}(I_1, I_2; H) = \frac{\sum\limits_\alpha \sum\limits_{m_2} \langle\psi_\alpha^R(m_2)|\, \sigma_z\, |\psi_\alpha^R(m_2)\rangle\, n_\alpha^R(m_2)}{\sum\limits_\alpha \sum\limits_{m_2} n_\alpha^R(m_2)}, \tag{7}$$

with the z component of the Pauli matrices, σ_z. With the use of the steady-state solutions of the rate equations and $\langle\psi_\alpha^R(m_2)|\, \sigma_z\, |\psi_\alpha^R(m_2)\rangle = [1 - 2\varepsilon^R(m_2)]$ one easily finds that

$$P_{\mathrm{RES}}(I_1, I_2: H) = (1 - 2\varepsilon_{\mathrm{rl}})[1 - 2\varepsilon^{\mathrm{R}}_{I_2}(H)]\,\frac{A_+ - A_-}{A_+ + A_-}\,, \tag{8}$$

where

$$\varepsilon^{\mathrm{R}}_{I_2}(H) = \left[\frac{2}{2I_2 + 1}\right] \sum_{m_2} \varepsilon^{\mathrm{R}}(m_2)\,[1 - \varepsilon^{\mathrm{R}}(m_2)]\,. \tag{9}$$

Thus, the calculation of $P_{\mathrm{RES}}(I, I_2: H)$ is reduced to that of the spin polarization of the effective doublet in the URES.

In a way similar to $P_{\mathrm{RES}}(I_1, I_2: H)$ we here define the spin polarization of the effective doublet in the URES by introducing an expectation value of σ_z in the URES, $\langle\sigma_z\rangle^{\mathrm{URES}}_{\alpha,m_1}$, as follows:

$$\frac{A_+ - A_-}{A_+ + A_-} = \frac{\sum\limits_{\alpha}\sum\limits_{m_1} \langle\sigma_z\rangle^{\mathrm{URES}}_{\alpha,m_1}\, u^{\mathrm{G}}_{\alpha}(m_1;\eta)\, n^{\mathrm{G}}_{\alpha}(m_1)}{\sum\limits_{\alpha}\sum\limits_{m_1} u^{\mathrm{G}}_{\alpha}(m_1;\eta)\, n^{\mathrm{G}}_{\alpha}(m_1)]}\,, \tag{10}$$

where η stands for the polarization of excitation light, and

$$\langle\sigma_z\rangle^{\mathrm{URES}}_{\alpha,m_1} = \sum_{\mu'}\sum_{i} \int \langle\psi^{\alpha'}_i(q)|\,\sigma_z\,|\psi^{\alpha'}_i(q)\rangle\, u^{i,\alpha'}_{\alpha,m_1}(q;\eta)\, P(q)\,\delta\big(E_i(q) - E\big)\mathrm{d}q / u^{\mathrm{G}}_{\alpha}(m_1;\eta)\,. \tag{11}$$

Note that the expectation value, $\langle\psi^{\alpha}_i(q)|\,\sigma_z\,|\psi^{\alpha}_i(q)\rangle = \alpha[1 - 2\varepsilon_i(q)]$, is weighted by the transition rate $u^{i,\alpha'}_{\alpha,m_1}(q;\eta)$ due to the fact that $u^{i,+}_{\alpha,m_1}(q;\eta) \neq u^{i,-}_{\alpha,m_1}(q;\eta)$, which naturally leads to the general definition for the spin mixing parameters in the URES.

After straightforward calculations of (10), neglecting the small terms of $P_{\mathrm{s}}(\eta)^2$ ($\ll 1$), $P_{\mathrm{RES}}(I_1, I_2: H)$ is finally obtained in a simple form

$$P^{\eta}_{\mathrm{RES}}(I_1, I_2: H) =$$
$$= (1 - 2\varepsilon_{\mathrm{rl}})\,[1 - 2\varepsilon^{\mathrm{R}}_{I_2}(H)]\,[\delta\varepsilon(\eta) - P_{\mathrm{s}}(\eta)\,(1 - 2\varepsilon_0)\,2\varepsilon^{\mathrm{G}}_{I_1}(H)]/2\varepsilon_{\mathrm{t}}(I_1, I_2: H)\,, \tag{12}$$

where

$$\delta\varepsilon(\eta) = \varepsilon_-(E;\eta) - \varepsilon_+(E;\eta)\,, \tag{13}$$

$$\varepsilon_0 = \frac{\varepsilon_+(E;\eta) + \varepsilon_-(E;\eta)}{2}\,, \tag{14}$$

$$P_{\mathrm{s}}(\eta) = \frac{u_-(E;\eta) - u_+(E;\eta)}{u_-(E;\eta) + u_+(E;\eta)}\,, \tag{15}$$

and the total spin mixing probability for one optical cycle, $\varepsilon_{\mathrm{t}}(I_1, I_2: H)$, is given by

$$\varepsilon_{\mathrm{t}}(I_1, I_2: H) = (\tfrac{1}{2})\,\{1 - [1 - 2\varepsilon^{\mathrm{G}}_{I_1}(H)]\,(1 - 2\varepsilon_0)\,(1 - 2\varepsilon_{\mathrm{rl}})\,[1 - 2\varepsilon^{\mathrm{R}}_{I_2}(H)]\}\,. \tag{16}$$

Here in deriving (12) we have used the relation $(V_+ - V_-)/(V_+ + V_-) = P_{\mathrm{RES}}(I_1, I_2: H)$ which can be obtained by defining $(V_+ - V_-)/(V_+ + V_-)$ in a similar fashion as done in $(A_+ - A_-)/(A_+ + A_-)$, and the spin mixing parameters $\varepsilon_{\alpha}(E;\eta)$ are the same as were given in [11], i.e.,

$$\varepsilon_{\alpha}(E;\eta) = \frac{\sum\limits_{i} \int \varepsilon_i(q)\, u^{i}_{\alpha}(q;\eta)\, P(q)\,\delta\big(E_i(q) - E\big)\,\mathrm{d}q}{u_{\alpha}(E;\eta)}\,. \tag{17}$$

It is obvious from (12) that at high fields, i.e., for $\varepsilon^{\mathrm{G}}_{I_1}(H) = 0$ and $\varepsilon^{\mathrm{R}}_{I_2}(H) = 0$, $P^{\eta}_{\mathrm{RES}}(I_1, I_2: H)$ reduces to $P^{*\eta}_{\mathrm{RES}} = (1 - 2\varepsilon_{\mathrm{rl}})\,\delta\varepsilon(\eta)/[1 - (1 - 2\varepsilon_0)\,(1 - 2\varepsilon_{\mathrm{rl}})]$ [6, 8].

Next, let us calculate the spin polarization of the GS, $P_{GS}(I_1, I_2: H)$. Mauser et al. [11] claimed that if one takes into account the hyperfine interaction in the GS, then it is not possible to derive an analytical form for $P_{GS}(I_1, I_2: H)$, because there is no way to project the individual occupation numbers of the hyperfine sublevels onto the occupation numbers of the effective spin doublet in the GS. However, it is straightforward to calculate $P_{GS}(I_1, I_2: H)$ in our model in analogy with the calculation of $P_{RES}(I_1, I_2: H)$. The final form for $P_{GS}(I_1, I_2: H)$ is given by

$$P^{\eta}_{GS}(I_1, I_2: H) = \frac{[1 - 2\varepsilon^{G}_{I_1}(H)]\,[P_s(\eta) + P^{\eta}_{RES}(I_1, I_2: H)]}{\{1 + P_s(\eta)\,[1 - 2\varepsilon^{G}_{I_1}(H)]\,P^{\eta}_{RES}(I_1, I_2: H)\}}. \tag{18}$$

Here again we neglected the small terms of $P_s(\eta)^2$. At high fields (18) reduces to $P^{*\eta}_{GS} = [P_s(\eta) + P^{*\eta}_{RES}]/[1 + P_s(\eta)\,P^{*\eta}_{RES}]$.

Finally, $P^{\eta}_{RES}(I_1, I_2: H)$ and $P^{\eta}_{GS}(I_1, I_2: H)$ given by (12) and (18) are summed over all the possible values of I_1 and I_2. By introducing the statistical weight for the total nuclear spin I, $p(I)$ [11], one obtains

$$P^{\eta}_{RES}(H) = [\sum_{I_1}\sum_{I_2} p(I_1)\,p(I_2)\,P^{\eta}_{RES}(I_1, I_2: H)]/[\sum_{I_1}\sum_{I_2} p(I_1)\,p(I_2)]\,, \tag{19}$$

and

$$P^{\eta}_{GS}(H) = [\sum_{I_1}\sum_{I_2} p(I_1)\,p(I_2)\,P^{\eta}_{GS}(I_1, I_2: H)]/[\sum_{I_1}\sum_{I_2} p(I_1)\,p(I_2)]\,. \tag{20}$$

2.3 Effect of hyperfine coupling on the spin polarization

From the analytical forms of $P^{\eta}_{RES}(I_1, I_2: H)$ and $P^{\eta}_{GS}(I_1, I_2: H)$ derived in Section 2.2 one may easily obtain some important features of the magnetic field dependence of the spin polarization. In the first place, due to the relations; $\delta\varepsilon(\sigma_+) = -\delta\varepsilon(\sigma_-)$, $\delta\varepsilon(\pi) = 0$, $P_s(\sigma_+) = -P_s(\sigma_-)$, and $P_s(\pi) = 0$ [11], both $P^{\eta}_{RES}(I_1, I_2: H)$ and $P^{\eta}_{GS}(I_1, I_2: H)$ have the following symmetry relations:

$$P^{\sigma_+}_{RES}(I_1, I_2: H) = -P^{\sigma_-}_{RES}(I_1, I_2: H)\,, \qquad P^{\pi}_{RES}(I_1, I_2: H) = 0\,, \tag{21}$$

$$P^{\sigma_+}_{GS}(I_1, I_2: H) = -P^{\sigma_-}_{GS}(I_1, I_2: H)\,, \qquad P^{\pi}_{GS}(I_1, I_2: H) = 0\,, \tag{22}$$

in the whole field region of interest.

Next, roughly speaking, a net effect of the spin mixing due to the hyperfine interaction is a decrease of the magnitude of the spin polarization because of the factor $[1 - 2\varepsilon_I(H)]$. A more interesting effect of such spin mixing, however, appears in the sign of the spin polarization in the low-field region. As a matter of fact, $P^{\eta}_{RES}(I_1, I_2: H)$ can undergo a change of sign at low fields, if $\delta\varepsilon(\eta)$ and $P_s(\eta)$ have the same sign and if $|\delta\varepsilon(\eta)| < |P_s(\eta)|\,(1 - 2\varepsilon_0)\,2\varepsilon^{G}_{I_1}(H)$. Note that the sign of $P^{\eta}_{RES}(I_1, I_2: H)$ at high fields is equivalent to that of $\delta\varepsilon(\eta)$. Furthermore a change of sign in $P^{\eta}_{RES}(I_1, I_2: H)$ is expected to occur only for pumping on the high-energy side of the absorption band, due to the fact that $P_s(\sigma_+) < 0$ for excitation on the low-energy side of the band and $P_s(\sigma_+) > 0$ for excitation on the high-energy side, whereas $\delta\varepsilon(\sigma_+) > 0$ over the whole absorption band [11]. Thus, it is apparent that the hyperfine coupling in the GS plays a critical role in a change of sign in $P^{\eta}_{RES}(I_1, I_2: H)$ at low fields.

As for $P^{\eta}_{GS}(I_1, I_2: H)$, on the other hand, it is easily found that a change of sign may appear at low fields if the sign of $P_s(\eta)$ is opposite to that of $P^{\eta}_{RES}(I_1, I_2: H)$ and if $|P_s(\eta)| < |P^{*\eta}_{RES}|$. These conditions may be fulfilled only for pumping on the low-energy side of the absorption band in contrast to the case of $P^{\eta}_{RES}(I_1, I_2: H)$. It should be mentioned that a change of sign in $P^{\eta}_{GS}(I_1, I_2: H)$ at low fields is caused by the spin mixing due to the hyperfine interaction in the RES rather than in the GS, as shown in [11].

The above-mentioned characteristics of the magnetic field dependence of the spin polarization are in good agreement with the experimental data [3, 12, 13].

However, there is a strong discrepancy between theoretical and experimental results near zero field. For example, $P^{\eta}_{\mathrm{RES}}(I_1, I_2\colon H)$ in (12) has a non-zero value, $(1/3) \times \times (1 - 2\varepsilon_{\mathrm{rl}}) [\delta\varepsilon(\eta) - 2P_{\mathrm{s}}(\eta) (1 - 2\varepsilon_0)/3]/[1 - (1 - 2\varepsilon_0) (1 - 2\varepsilon_{\mathrm{rl}})/9]$ at $H = 0$ except for the case when $\delta\varepsilon(\eta) = 2P_{\mathrm{s}}(\eta) (1 - 2\varepsilon_0)/3$, whereas the experimental value of the polarization is almost zero within experimental error. This discrepancy may arise from the fact that the up- and down-spin states do not completely mix up through the hyperfine interaction near zero field, since $\varepsilon_I(H = 0) \approx 1/3$ for $I \gg 1$. In this field region the spin mixing due to the other interactions such as the anisotropic hyperfine interaction, the nuclear dipole interaction, etc. can become very important, so that the states $\psi_\alpha(m)$ are no longer the eigenstates of the relevant spin system. It is then easily expected that those interactions mix the states $\psi_\alpha(m)$ resulting in the stronger spin mixing of the up- and down-spin states near zero field. However, explicit consideration of such interactions in the model is beyond the scope of the present work.

3. Numerical Calculations and Comparison with Experiment

In order to see the trend of the magnetic field dependence of the spin polarization, first of all, let us show the numerical results of $P^{\sigma_+}_{\mathrm{RES}}(I_1, I_2\colon H)$ and $P^{\sigma_+}_{\mathrm{GS}}(I_1, I_2\colon H)$ with $I_1 = I_2 = 5$ for excitation with right-circularly (σ_+) polarized light. Fig. 2a shows the calculated results of $P^{\sigma_+}_{\mathrm{RES}}(I_1, I_2\colon H)$ for the different values of $\delta\varepsilon(\sigma_+)$ as a function of $g_{\mathrm{RES}}\beta H/A_{\mathrm{RES}}$. Here R is defined by $g_{\mathrm{GS}}A_{\mathrm{RES}}/g_{\mathrm{RES}}A_{\mathrm{GS}}$. It is obvious from the figure that a change of sign occurs at low fields when the value of $\delta\varepsilon(\sigma_+)$ becomes smaller compared to that of $P_{\mathrm{s}}(\sigma_+)$. In Fig. 2b $P^{\sigma_+}_{\mathrm{RES}}(I_1, I_2\colon H)$ are plotted for the cases when $R = 0.5, 1.0, 1.5$, and 2.5. It is easily found that the degree of negative polarization at low fields is more pronounced and the critical magnetic field at which $P^{\sigma_+}_{\mathrm{RES}}(I_1, I_2\colon H) = 0$ shifts toward the higher field, when the hyperfine coupling strength of the GS becomes stronger. In Fig. 2c $P^{\sigma_+}_{\mathrm{GS}}(I_1, I_2\colon H)$ are shown for the different values of R. It turns out that $P^{\sigma_+}_{\mathrm{GS}}(I_1, I_2\colon H)$ show the opposite tendency in contrast to $P^{\sigma_+}_{\mathrm{RES}}(I_1, I_2\colon H)$ shown in Fig. 2b, namely, the smaller R, the larger negative polarization. This behavior is easily understood, as mentioned in Section 2.3, from the fact that the change of sign in $P^{\sigma_+}_{\mathrm{RES}}(I_1, I_2\colon H)$ is caused by the spin mixing due to the hyperfine interaction in the GS, while in the case of $P^{\sigma_+}_{\mathrm{GS}}(I_1, I_2\colon H)$ the spin mixing due to the hyperfine interaction in the RES plays an essential role in the change of sign. The influence of $\varepsilon_{\mathrm{rl}}$ on $P^{\sigma_+}_{\mathrm{RES}}(I_1, I_2\colon H)$ and $P^{\sigma_+}_{\mathrm{GS}}(I_1, I_2\colon H)$ are illustrated in Fig. 2d. It is found that the effect of $\varepsilon_{\mathrm{rl}}$ is the quick convergence of $P^{\sigma_+}_{\mathrm{RES}}(I_1, I_2\colon H)$ to $P^{*\sigma_+}_{\mathrm{RES}}$ after reaching a minimum value at low fields (compare to Fig. 2a). Although it was mentioned in Section 2.3 that the spin mixing due to the hyperfine interaction causes the reduction of the magnitude of the spin polarization or the change of its sign at low fields, the enhancement of the polarization can also occur in the case of $P^{\sigma_+}_{\mathrm{RES}}(I_1, I_2\colon H)$, if $|P_{\mathrm{s}}(\sigma_+)| \gg \delta\varepsilon(\sigma_+)$, $P_{\mathrm{s}}(\sigma_+) < 0$, and $\delta\varepsilon(\sigma_+) > 0$. Such cases are shown in Fig. 2e.

In order to fit the experimental data of the field dependence of the spin polarization, we have to take into account some particular shells in which the hyperfine interactions are responsible for the observed ESR linewidth. However, it is time-consuming to consider more than one shell in the numerical calculations of $P^{\eta}_{\mathrm{RES}}(H)$ and $P^{\eta}_{\mathrm{GS}}(H)$. For simplicity of computation we make the following approximations: (i) Only the single shell is taken into account in which the hyperfine coupling strength is the largest.

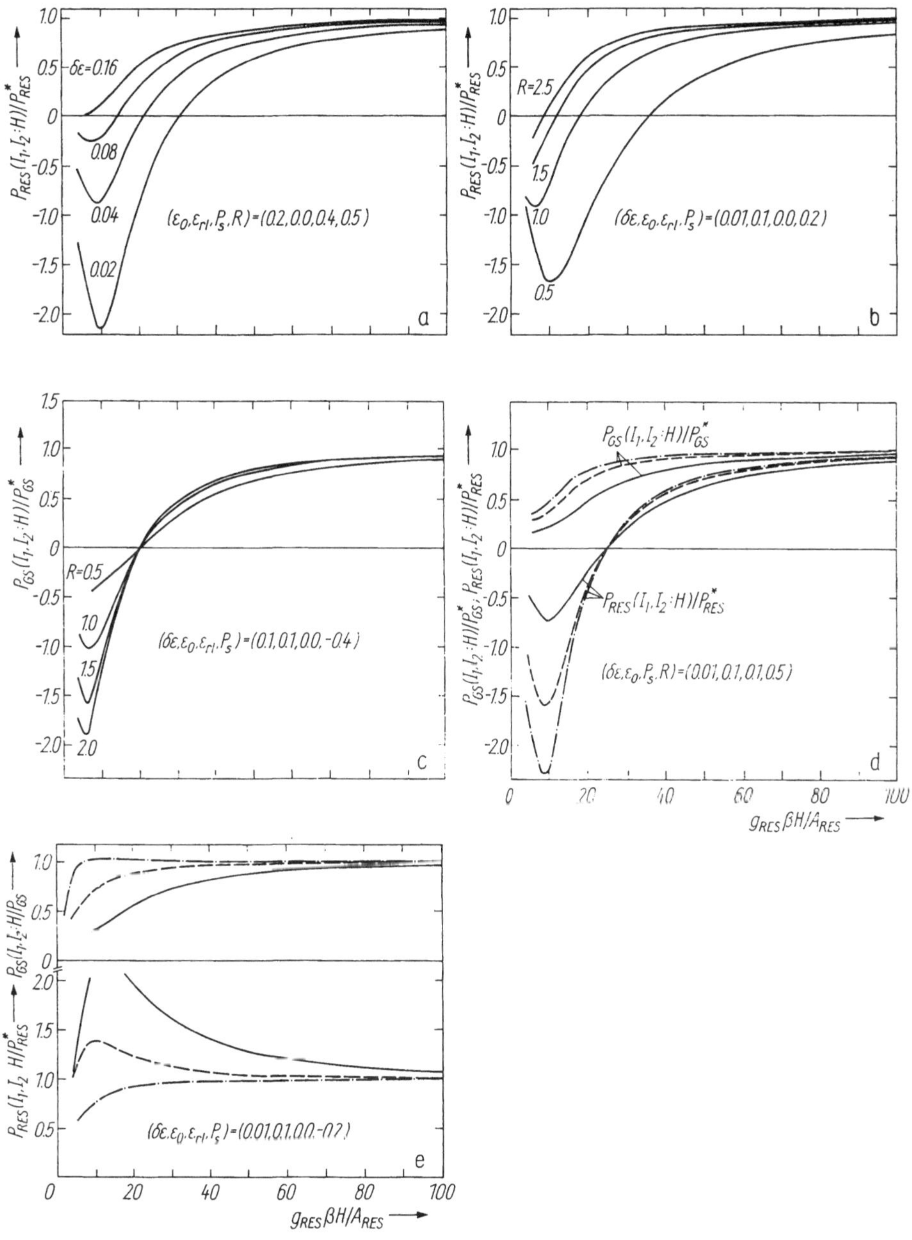

Fig. 2. Calculated results of $P^{\sigma_+}_{RES}(I_1, I_2{:}\, H)$ and/or $P^{\sigma_+}_{GS}(I_1, I_2{:}\, \mathrm{H})$ with $I_1 = I_2 = 5$ as a function of $g_{RES}\beta H/A_{RES}$ for the different values of a) $\delta\varepsilon$ when $(\varepsilon_0, \varepsilon_{rl}, P_s, R) = (0.2, 0.0, 0.4, 0.5)$; b) R when $(\delta\varepsilon, \varepsilon_0, \varepsilon_{rl}, P_s) = (0.01, 0.1, 0.0, 0.2)$; c) R when $(\delta\varepsilon, \varepsilon_0, \varepsilon_{rl}, P_s) = (0.1, 0.1, 0.0, -0.4)$; d) ε_{rl} (——— 0, — — — 0.2, — · — 0.4) when $(\delta\varepsilon, \varepsilon_0, P_s, R) = (0.01, 0.1, 0.1, 0.5)$; e) R (——— 0.5, — — — 1.0, — · — 2.0) when $(\delta\varepsilon, \varepsilon_0, \varepsilon_{rl}, P_s) = (0.01, 0.1, 0.0, -0.2)$

(ii) In the single shell model the ESR linewidth may be described by

$$\Delta H \approx \left(\frac{A}{g\beta}\right) \sqrt{\frac{\sum_I p(I)\, I(I+1)}{\sum_I p(I)}}\,, \tag{23}$$

regarding the hyperfine coupling constant as the effective one. (iii) Only the larger natural abundant isotope is considered.

Thus, in our model we have five parameters in fitting the data, i.e., $\delta\varepsilon(\eta)$, ε_0, $\varepsilon_{\mathrm{rl}}$, $P_s(\eta)$, and R. One can estimate the value of R by using the following relation:

$$R = \frac{\dfrac{\Delta H_{\mathrm{RES}}}{\Delta H_{\mathrm{GS}}} \sqrt{\sum_{I_1} p(I_1)\, I_1(I_1+1) / \sum_{I_1} p(I_1)}}{\sqrt{\sum_{I_2} p(I_2)\, I_2(I_2+1) / \sum_{I_2} p(I_2)}}\,. \tag{24}$$

The values of $P_s(\eta)$ are determined from the MCD measurements [3]. The remaining three parameters $\delta\varepsilon(\eta)$, ε_0, and $\varepsilon_{\mathrm{rl}}$ may be evaluated from the MCP measurements [10]. We will here give two examples to demonstrate the validity of our simple model.

1. KBr: In this crystal there are precise data available for the five parameters. From [3, 10, 14] one finds that $P_s(\sigma_+) = -0.08$ to -0.1, $\delta\varepsilon(\sigma_+) = 0.05$ to 0.08, $\varepsilon_0 = 0.08$ to 0.1 at 623.8 nm and $R = 1.76$ or 1.83. The value of $\varepsilon_{\mathrm{rl}}$ was estimated to be less than 0.08 at 632.8 nm, so that it was neglected in fitting the data. It was found that the above values of the parameters can reproduce the experimental data of $P^{\sigma_+}_{\mathrm{RES}}(H)$ shown in Fig. 3 of [8]. One set of the parameter values which can give a best fit to the experimental data is given in Fig. 3. In fitting the data of $P^{\sigma_+}_{\mathrm{RES}}(H)$ we also took into account the experimental fact that $P^{\sigma_+}_{\mathrm{GS}}(H) = 0$ at $H = 350$ to 380 (10^{-4} T) [11]. Then, the values of the effective coupling strengths A_{RES} and A_{GS} were estimated to be 108 and 64 MHz which correspond to $\Delta H_{\mathrm{RES}} = 265$ (10^{-4} T) and $\Delta H_{\mathrm{GS}} = 147$ (10^{-4} T), respectively. These values should be compared with the experimental ones, $\Delta H_{\mathrm{RES}} = 256$, 270 (10^{-4} T), and $\Delta H_{\mathrm{GS}} = 146$ (10^{-4} T) [14]. Note that

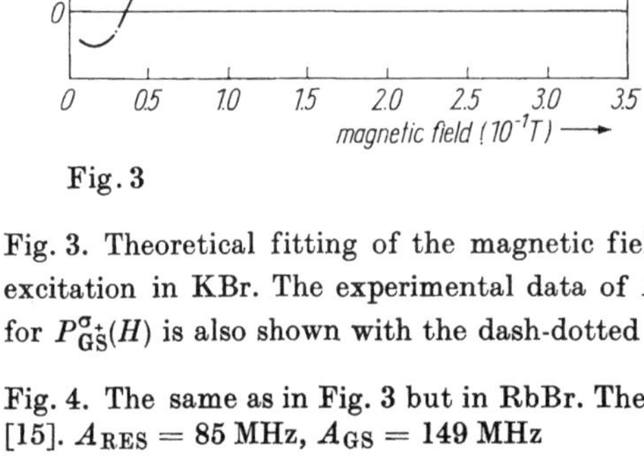

Fig. 3

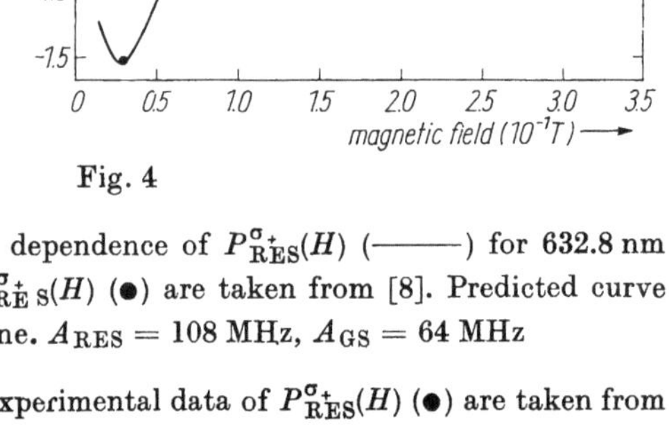

Fig. 4

Fig. 3. Theoretical fitting of the magnetic field dependence of $P^{\sigma_+}_{\mathrm{RES}}(H)$ (———) for 632.8 nm excitation in KBr. The experimental data of $P^{\sigma_+}_{\mathrm{RES}}(H)$ (●) are taken from [8]. Predicted curve for $P^{\sigma_+}_{\mathrm{GS}}(H)$ is also shown with the dash-dotted line. $A_{\mathrm{RES}} = 108$ MHz, $A_{\mathrm{GS}} = 64$ MHz

Fig. 4. The same as in Fig. 3 but in RbBr. The experimental data of $P^{\sigma_+}_{\mathrm{RES}}(H)$ (●) are taken from [15]. $A_{\mathrm{RES}} = 85$ MHz, $A_{\mathrm{GS}} = 149$ MHz

632.8 nm excitation corresponds to the pumping on the low-energy side of the absorption band in KBr, and therefore $P^{\sigma_+}_{GS}$ shows a change of sign at low fields.

2. RbBr: Though there are no data enough to fix the values of the parameters, we tried to fit the experimental data of $P^{\sigma_+}_{RES}(H)$ for excitation at 632.8 nm, which showed a change of sign at low fields [12, 13]. We put the restrictions that $P^{*\sigma_+}_{RES} = 0.03$ to 0.04 [15] and $R = 0.61$ [14]. The possible maximum value ($= 0.2$) for ε_{rl} was favorably adopted to obtain the quick convergence of $P^{\sigma_+}_{RES}(H)$ to $P^{*\sigma_+}_{RES}$ after reaching a minimum value, with increasing the magnetic field. It was found that in order to account for $P^{\sigma_+}_{RES}(H)$ in the region below 0.1 T the value of $P_s(\sigma_+)$ must be larger than 0.2 when $\delta\varepsilon(\sigma_+) = 0.02$ to 0.03. Fig. 4 shows the best fit to the experimental data of $P^{\sigma_+}_{RES}(H)$ for excitation at 632.8 nm. The calculated curve for $P^{\sigma_+}_{GS}(H)$ is also shown in the figure. From the fitting of the data the values of A_{RES} and A_{GS} were estimated to be 85 and 149 MHz, respectively, leading to $\Delta H_{RES} = 211$ (10^{-4} T) and $\Delta H_{GS} = 374$ (10^{-4}) for the ESR linewidth. These calculated ESR linewidths are about 20% smaller than the experimental ones. This discrepancy may partly come from the neglect of the effect of Rb^{87} isotope on $P^{\eta}_{RES}(H)$. The values of the parameters determined here should be checked by the experiment.

4. Summary

We have developed the simple model for the magnetic field dependences of the spin polarizations $P_{RES}(H)$ and $P_{GS}(H)$ of the F center in which the spin mixing due to the isotropic hyperfine interactions both in the GS and in the RES is taken into account. It has been shown that the model can give a consistent explanation for the correlation between the exciting energy and the sign change of $P_{RES}(H)$ and $P_{GS}(H)$ at low fields. It was clarified that the spin mixing by the hyperfine interaction in the GS plays an essential role in the sign change of $P_{RES}(H)$.

From numerical calculations of $P_{RES}(H)$ and $P_{GS}(H)$ we have also shown that the model can reproduce the experimental data of $P_{RES}(H)$ for excitation at 632.8 nm in KBr by using reasonable values for the parameters contained in the model, except the data of $P_{RES}(H)$ at very low fields less than 300 (10^{-4} T). Tentative fitting of $P_{RES}(H)$ for excitation at 632.8 nm in RbBr was performed to demonstrate the importance of the spin mixing due to the GS hyperfine interaction.

The improvement of the model near zero field is left as a future problem.

Acknowledgements

The author would like to thank M. Hirata for helpful discussions. He is also grateful for financial support from Japan Society for the Promotion of Science.

References

[1] L. F. Mollenauer and S. Pan, Phys. Rev. B **6**, 772 (1972).
[2] A. Winnacker and L. F. Mollenauer, Phys. Rev. B **6**, 787 (1972).
[3] A. Winnacker, K. E. Mauser, and B. Niesert, Z. Phys. B **26**, 97 (1977).
[4] M. P. Fontana and D. B. Fitchen, Phys. Rev. Letters **23**, 1497 (1969).
[5] G. Baldacchini, U. M. Grassano, and A. Tanga, Phys. Rev. B **16**, 5570 (1977).
[6] K. Imanaka, T. Wada, T. Iida, and H. Ohkura, Solid State Commun. **27**, 1009 (1978).
[7] G. Baldacchini, A. Tanga, and U. M. Grassano, Phys. Rev. B **20**, 4357 (1979).
[8] H. Ohkura, K. Iwahana, M. Tanaka, K. Tara, and K. Imanaka, J. Phys. Soc. Japan **49**, 2296 (1980).

[9] K. Iwahana, K. Tara, M. Hirata, and H. Ohkura, J. Phys. Soc. Japan. **51**, 1349 (1982).
[10] H. Ohkura, K. Tara, N. Akiyama, K. Iwahana, and Y. Mori, J. Phys. Soc. Japan **51**, 3615 (1982).
[11] K. E. Mauser, B. Niesert, and A. Winnacker, Z. Phys. B **26**, 107 (1977).
[12] T. Wada, Master Thesis, Osaka City University, 1979, unpublished.
[13] M. Tanaka, Master Thesis, Osaka City University, 1980, unpublished.
[14] P. G. Baranov, Yu. P. Veshchunov, and N. G. Romanov, Soviet Phys. — Solid State **20**, 1516 (1978).
[15] M. Hirata, Master Thesis, Osaka City University, 1983, unpublished.

(Received August 29, 1984)

phys. stat. sol. (b) 126, 703 (1984)

Subject classification: 14.2

Institute of Nuclear Research, Academy of Sciences of the Ukrainian SSR, Kiev[1])

Superconductivity of Metals with a Superlattice of Voids

By

V. N. RUDKO and V. I. SUGAKOV

Superconductivity of metals with a superlattice of vacancy or gas voids is investigated in the presence of magnetic fields around the second critical field H_{c2}. At $H \gtrsim H_{c2}$ the porous superconductor appears to be a system of an immense number of superconducting regions localized near the voids. The paper represents following results: (i) the dependence of the third critical field H_{c3} on the void radius; (ii) the diamagnetic properties of the system; (iii) the Bloch states of the superconducting condensate. It is shown that in type-II superconductors with a lattice of voids Josephson effects due to Cooper pair tunneling between the voids occur.

Исследована сверхпроводимость металлов, содержащих сверхрешетку вакансионных или газовых пор, в магнитных полях близких ко второму критическому полю H_{c2}. При $H \geqq H_{c2}$ пористый сверхпроводник представляет собой систему огромного числа сверхпроводящих областей, локализованных вблизи пор. В работе: (1) найдена зависимость третьего критического поля H_{c3} от радиуса пор; (2) определены диамагнитные свойства системы; (3) изучены блоховские состояния сверхпроводящего конденсата и показано, что в сверхпроводниках II рода, содержащих сверхрешетку пор, должны наблюдаться эффекты Джозефсона, связанные с туннелированием куперовских пар между порами.

1. Introduction

In the last few years physical properties of metals with voids are intensively investigated. Voids may appear either at the irradiation of a metal by neutrons and ions [1 to 3] or as the result of technological processes of metal film preparation [4]. The size of the voids depends on several conditions, mainly on material and temperature of the sample, dose and intensity of irradiation, subsequent thermal treatment. The void shape is usually almost spherical. The void radius may reach the value of tens and hundreds of nanometers. At the same time the concentration of voids may be rather high, so that the relative volume of voids can be as large as 10% and more.

In definite cases voids form a superlattice. The vacancy void superlattice was observed first by Evans in 1971 in molybdenum irradiated by nitrogen ions [1]. Later vacancy and gas void superlattices were observed in several metals (see, for example, reviews [5, 6]). The symmetry of the void lattice coincides with the symmetry of the host lattice, its lattice spacing usually equals 2 to 7 void diameters. Voids forming a superlattice have nearly equal size. Numerous attempts to describe theoretically the formation of superlattices failed (see, e.g. [7 to 9]) until 1984 [10] when it was shown that the formation of a superlattice is caused by elastic long-range interaction among defects created during irradiation. This phenomenon has a characteristic threshold: a superlattice appears when the radiation intensity becomes larger than the value at which the homogeneous distribution of defects is unstable.

The occurrence of voids changes the mechanical properties of a metal and causes some new optical [11, 12], superconducting properties [13 to 16] and some new

---

[1]) prospekt Nauki 119, 252028 Kiev, USSR.

features of neutron scattering [17 to 19]. The pinning force in porous type-II superconductors was theoretically and experimentally investigated in [13 to 15]. The dependence of the pinning force on magnetic field was studied in the region of fields less than the second critical field H_{c2}.

The present paper deals with the investigation of the properties of type-II superconductors with a superlattice of vacancy or gas voids in magnetic fields larger than the second critical field. It was noted in [16] that a porous superconductor in the presence of a strong magnetic field $H \geqq H_{c2}$ appears to be a system of numerous regions with surface superconductivity which exist near voids. Contrary to the cases of superconductivity near plane [20, 21] and cylindrical [22, 23] surfaces the superconducting regions existing near voids are localized with regard to three dimensions. In the present paper the third critical field of porous metals is obtained, the diamagnetic properties of the system are analyzed. It is shown that a superconductor with a void superlattice has to exhibit Josephson effects caused by Cooper pairs tunneling between neighbouring voids. Estimations according to the obtained expressions were undertaken for niobium in which void superlattices of good quality were repeatedly observed [2, 24, 25].

2. Third Critical Field of an Individual Void

In this section we shall presume the case of rather low void concentration, when one may assume that the order parameters corresponding to individual voids do not overlap. Let us obtain the third critical field H_{c3} for a spherical void of radius R, at which the localized superconductivity vanishes. We shall consider the temperature region in the vicinity of the critical temperature where the Ginzburg-Landau equations are valid [26, 27]. We shall also suppose that the Ginzburg-Landau parameter $\varkappa$ is rather large $\varkappa \gg 1$, so that the interaction of superconducting currents with the intrinsic magnetic field can be neglected. The qualitative results will be the same as in the case $\varkappa \approx 1$ but the calculations will be much simpler. To find the third critical field one has to solve the linearized Ginzburg-Landau equation which in spherical coordinates (r, θ, φ) with origin at the centre of the void reads as

$$\left(-\frac{1}{r^2}\frac{\partial}{\partial r}r^2\frac{\partial}{\partial r} - \frac{1}{r^2 \sin\theta}\frac{\partial}{\partial\theta}\sin\theta\frac{\partial}{\partial\theta} - \frac{1}{r^2\sin^2\theta}\frac{\partial^2}{\partial\varphi^2} + \right.$$

$$\left. + i\mathcal{H}\frac{\partial}{\partial\varphi} + \frac{1}{4}\mathcal{H}^2 r^2 \sin^2\theta\right)\psi = \psi\,, \tag{1}$$

where the symmetrical gauge of the vector potential is used; ψ is the order parameter corresponding to an individual void, r is measured in units of the coherence length ξ, $\mathcal{H} = H/H_{c2}$ is the reduced magnetic field; θ is the angle measured from the direction of the magnetic field $\boldsymbol{H}$.

The order parameter must be calculated taking into account the boundary condition which is

$$\frac{\partial\psi}{\partial r} = 0 \quad \text{at} \quad r = \frac{R}{\xi} \tag{2}$$

when the vector potential gauge is as noted above.

Equation (1) is rather complicated because the variables r and θ are inseparable. Only one solution of this equation finite at $r = 0$ has been found until now [28]. In the case under consideration the order parameter ψ is a linear combination of the function obtained in [28] and a function which is analogous (for the variables r and θ)

to the logarithmic solution for the confluent hypergeometric function which diverges at $r = 0$. The latter function being unknown we used a variational method to find ψ. Choosing the function ψ two facts were taken into account: (i) in the case of surface superconductivity the order parameter reaches the maximum in the regions where a magnetic field is parallel to the surface; (ii) using the Green function one obtains the asymptotic solutions of (1), $\ln \psi \sim -r^2 \sin^2 \theta$ at $r \sin \theta \to \infty$ and $\ln \psi \sim -(1 - H_{c2}/H)^{1/2} |r \cos \theta|$ at $|r \cos \theta| \to \infty$. Therefore, the trial function was chosen in a rather general form

$$\psi = e^{in\varphi} \sin^s \theta \frac{\exp\left[-(1 + \varepsilon \cos^2 \theta) \dfrac{(r - (R/\xi))^2}{2\gamma^2}\right]}{\mathrm{ch}^\alpha \left[\beta \cos \theta (r - (R/\xi))\right]}, \tag{3}$$

where n, s, ε, γ, α, β are the variational parameters and n is integer.

The thorough numerical checking has shown that the values ε and β do not depend on the other parameters and the system reaches its free energy minimum at $\varepsilon = -1$ and $\beta = 0$. Therefore, we can use three variational parameters instead of six. Then ψ may be rewritten as

$$\psi = e^{in\varphi} \sin^s \theta \exp\left[-\sin^2 \theta \frac{(r - (R/\xi))^2}{2\gamma^2}\right]. \tag{4}$$

Note that we can at once point out the order of magnitude of all parameters: $\gamma \approx 1$, $n \approx (R/\xi)^2$, $s \approx 2\sqrt{n}$ at $R > 2\xi$.

Using (4) one obtains the following expression for $\mathcal{H}_{c3}$:

$$\mathcal{H}_{c3} = \left(\frac{\sqrt{F_2 F_3}}{F_1} - n\right)^{-1}, \tag{5}$$

where

$$\begin{aligned} F_1 &= \frac{s}{2s - 1} + \frac{2st}{\sqrt{\pi}} \varrho + \varrho^2, \\ F_2 &= s^2 + \left(n^2 - s^2 + s + \frac{1}{4}\right) \frac{s}{s - \frac{1}{2}} + \frac{2s^2 t}{\sqrt{\pi}\,(s + \frac{1}{2})} \varrho + \frac{s + \frac{1}{2}}{2(s + 1)} \varrho^2, \\ F_3 &= \frac{3s}{2(2s - 1)} + \frac{4st}{\sqrt{\pi}} \varrho + 3\varrho^2 + \frac{4s^2 t}{\sqrt{\pi}\,(s + \frac{1}{2})} \varrho^3 + \frac{s + \frac{1}{2}}{s + 1} \varrho^4, \end{aligned} \tag{6}$$

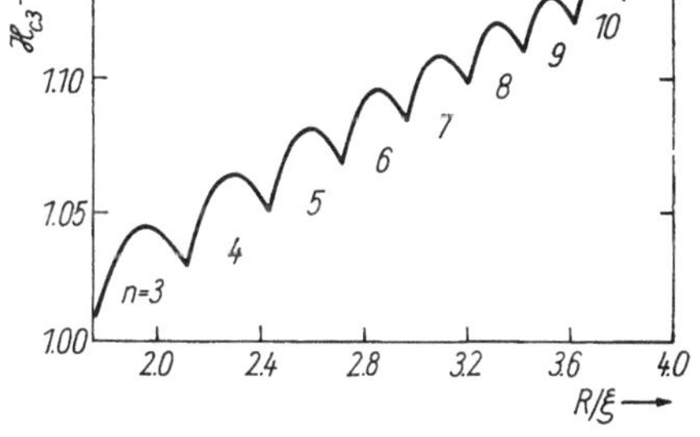

Fig. 1

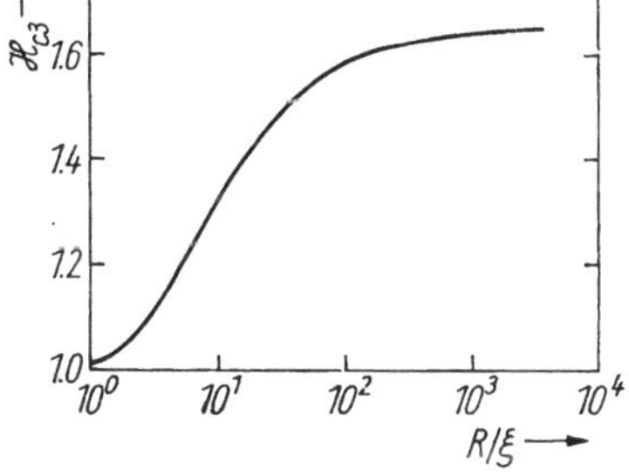

Fig. 2

Fig. 1. Dependence of the third critical field on the void radius. The values of the quantum number n, which defines the number of flux quanta trapped by the void, are written under the curve

Fig. 2. Envelope of the detailed dependence of $\mathcal{H}_{c3}$ on R/ξ

here $t = [\Gamma(s)/\Gamma(s + \frac{1}{2})]^2$, $\Gamma(s)$ is the gamma function, and the value $\varrho = R/\gamma\xi$ may be obtained from the equation

$$\varrho \sqrt{2\left(\frac{\sqrt{F_2 F_3}}{F_1} - n\right)} \sqrt[4]{\frac{F_2}{F_3}} = \frac{R}{\xi}. \tag{7}$$

The values of the parameters n and s are chosen so that one obtains the maximum value of $\mathcal{H}_{c3}$. Using (5) to (7) the dependence of the third critical field $\mathcal{H}_{c3}$ on void radius was obtained. As can be seen in Fig. 1 this dependence is nonmonotonic. The latter property is caused by the discreteness of the quantum number n which defines the number of the magnetic flux quanta trapped by the void. With the increase of R the oscillation amplitude decreases and the value $\mathcal{H}_{c3}(R)$ tends to the value of the third critical field for a plane surface in a parallel magnetic field. Fig. 2 exhibits the envelope of the detailed dependence $\mathcal{H}_{c3}(R)$.

Note that the function (4) is asymmetrical with regard to the magnetic field. Along the magnetic field the function (4) decreases much slower than in the transverse direction. For small voids ($R \approx \xi$) this asymmetry is more pronounced than for the large voids ($R \approx 10\xi$) because with increase of the R/ξ ratio the oscillating value s increases.

Above we considered the case of sufficiently low void concentration. At larger concentrations the order parameters corresponding to individual voids would overlap causing the increase of the third critical field of a superconductor. In Section 4 we shall investigate this question in detail and carry out the estimation of $\mathcal{H}_{c3}$ increase caused by the void superlattice.

3. Diamagnetism of a Porous Superconductor

The superconducting currents circulating around the voids in the magnetic field interval $H_{c2} < H < H_{c3}(R)$ give a diamagnetic contribution to magnetic susceptibility. At sufficiently high void concentration this diamagnetic contribution may be several orders of magnitude larger than the intrinsic paramagnetic metal susceptibility. Let us make corresponding estimates. For this purpose one has to consider the nonlinear Ginzburg-Landau equation

$$\left(-\Delta + i\mathcal{H}\frac{\partial}{\partial\varphi} + \frac{1}{4}\mathcal{H}^2 r^2 \sin^2\theta + \frac{8\pi e^2\xi^2\varkappa^2}{mc^2}|\psi|^2\right)\psi = \psi\,, \tag{8}$$

where Δ is the Laplace operator.

Taking into account the fact that (4) fitted the variational procedure rather well, one may expect that in the magnetic field interval under consideration reasonable estimations may be obtained using the function

$$\psi = B \sin^s\theta \exp\left[in\varphi - \frac{(r - (R/\xi))^2 \sin^2\theta}{2\gamma^2}\right], \tag{9}$$

where $B \sim (1 - H/H_{c3}(R))^{1/2}$ is the new variational parameter which characterizes the order parameter amplitude.

The magnetic moment, created by the superconducting current circulating around the void, equals

$$\boldsymbol{M} = \frac{1}{2c\xi^4}\int [\boldsymbol{r} \times \boldsymbol{J}(\boldsymbol{r})]\, d\boldsymbol{r}\,. \tag{10}$$

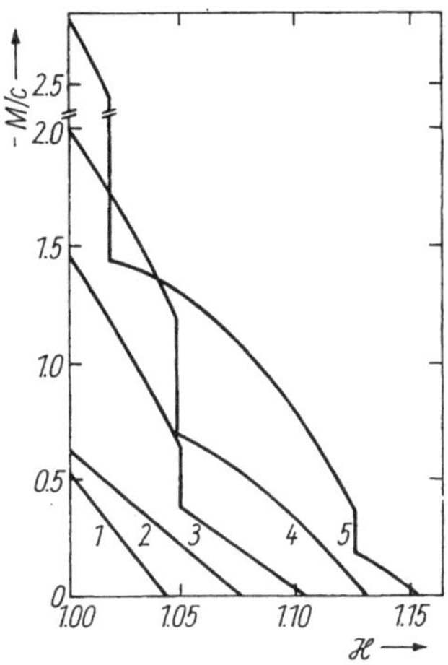

Fig. 3. Dependence of the magnetic moment M, created by the superconducting current circulating around the void, on the magnetic field $\mathcal{H}$ at several values of R/ξ: (1) 2.0, (2) 2.5, (3) 3.0, (4) 3.5, (5) 4.0; $C = \pi\hbar\xi c/8\sqrt{2}e\varkappa^2$

Inserting the usual quantum-mechanical expression for the current density $\boldsymbol{J}$ into (10) one obtains

$$\boldsymbol{M} = \frac{e\hbar\boldsymbol{z}}{2mc\xi^4} \int |\psi|^2 \left(n - \frac{1}{2} \mathcal{H}r^2 \sin^2 \theta\right) \mathrm{d}\boldsymbol{r}\,, \tag{11}$$

where $\boldsymbol{z} = \boldsymbol{H}/H$. The integration of (11) gives

$$\boldsymbol{M} = \frac{\pi^2\gamma^3}{2s\sqrt{t}} \frac{e\hbar}{mc\xi^4} B^2\boldsymbol{z} \left(nF_1 - \frac{1}{2} \mathcal{H}\gamma^2F_3\right). \tag{12}$$

One can easily show that at $H_{c3}(R) - H \ll H_{c3}(R)$

$$M \sim 1 - \frac{H}{H_{c3}(R)}. \tag{13}$$

As the calculations show the dependence $M(H)$ is linear practically in the whole interval from H_{c2} to $H_{c3}(R)$ for the case of small voids ($R < 3\xi$). For the case of larger voids ($R > 3\xi$) the dependence $M(\mathcal{H})$ becomes nonmonotonic (Fig. 3). This non-monotony is caused by the step decrease of the equilibrium number of magnetic flux quanta trapped by the void with the decrease of magnetic field. The amplitude and the frequency of oscillations of the $M(H)$ dependence increase with increasing void radius.

Estimations for niobium show that for not very small voids ($R > \xi$) and void concentrations $N_v \approx 10^{16}$ cm^{-3} the diamagnetic contribution to magnetic susceptibility, caused by localized superconductivity, in the fields $H_{c2} < H < H_{c3}(R)$ (excluding the nearest neighbourhood of $H_{c3}(R)$) is by one to two orders of magnitude larger than the intrinsic paramagnetic susceptibility of the metal. Therefore paramagnetic metals and alloys become diamagnetic in the above-mentioned range of magnetic fields.

4. Third Critical Field of a Superconductor with Void Superlattice

In Section 2 the order parameter and the third critical field $\mathcal{H}_{c3}(R)$ of an individual void were obtained. In the present section the third critical field of a superconductor with a void superlattice will be calculated.

Let us define the order parameter corresponding to an individual void by ψ_{0l} where $\boldsymbol{l}$ numbers the voids. The overlapping of order parameters of the nearest voids causes an increase of the third critical field of the superconductor. This addition to the third critical field of an individual void shall be defined by $\delta\mathcal{H}_{c3}$. To find $\delta\mathcal{H}_{c3}$ one may use the perturbation theory because overlapping of order parameters is small. In much the same way as in Section 2 the third critical field is to be obtained from the linearized Ginzburg-Landau equation

$$(-i\nabla - \tfrac{1}{2}[\vec{\mathcal{H}} \times \boldsymbol{r}])^2 \Psi = \Psi . \tag{14}$$

The order parameter of the whole system Ψ has to satisfy the boundary condition on the surface of each void

$$(-i\nabla - \tfrac{1}{2}[\vec{\mathcal{H}} \times \boldsymbol{r}])\, \boldsymbol{n}_l \Psi = 0 , \tag{15}$$

where $\boldsymbol{n}_l$ is the unit vector normal to the surface of the l-th void.

We shall try the order parameter in the form of superposition of individual void order parameters

$$\Psi = \sum_{l} C_l\, \mathrm{e}^{\frac{i}{2}[\vec{\mathcal{H}} \times \boldsymbol{l}]\boldsymbol{r}} (\psi_{0l} + f_l) , \tag{16}$$

where $\exp\left(i[\vec{\mathcal{H}} \times \boldsymbol{l}]\, \boldsymbol{r}/2\right)$ is the Peierls magnetic factor [29]; f_l is the correction to the undisturbed order parameter ψ_{0l} of the $\boldsymbol{l}$-th void, caused by the influence of neighbouring voids. Let us assume that $\delta\mathcal{H}_{c3} \ll 1$ and therefore $f_l \ll \psi_{0l}$. Inserting (16) into (15) one obtains the boundary condition for the function f_l,

$$\boldsymbol{n}_l \nabla f_l = -\frac{1}{C_l} \sum_{l' \neq l} C_{l'}\, \mathrm{e}^{\frac{i}{2}[\vec{\mathcal{H}} \times (\boldsymbol{l}' - \boldsymbol{l})]\boldsymbol{r}}\, \boldsymbol{n}_{l'} \nabla \psi_{0l'} . \tag{17}$$

Inserting (16) into (14) one can obtain the equation for f_l which in the spherical coordinate system with origin at the centre of the $\boldsymbol{l}$-th void reads as

$$\left[-\Delta + i\mathcal{H}_{c3}(R) \frac{\partial}{\partial \varphi} + \frac{1}{4} \mathcal{H}_{c3}^2(R)\, r^2 \sin^2 \theta - 1\right] f_l =$$

$$= \delta\mathcal{H} \left[n - \frac{1}{2} \mathcal{H}_{c3}(R)\, r^2 \sin^2 \theta\right] \psi_{0l} , \tag{18}$$

where $\delta\mathcal{H} = \mathcal{H} - \mathcal{H}_{c3}(R)$.

The calculation of the coefficients C_l at an arbitrary magnetic field orientation with regard to the void superlattice axes is a complicated problem. It may be easily solved in the case when a magnetic field is parallel to one of the crystallographic axes of the superlattice. This case is the most interesting one because it provides the largest value of $\delta\mathcal{H}_{c3}$. This fact may be verified taking into account the strong anisotropy of ψ_{0l} functions with regard to the magnetic field. Therefore the largest effect would be

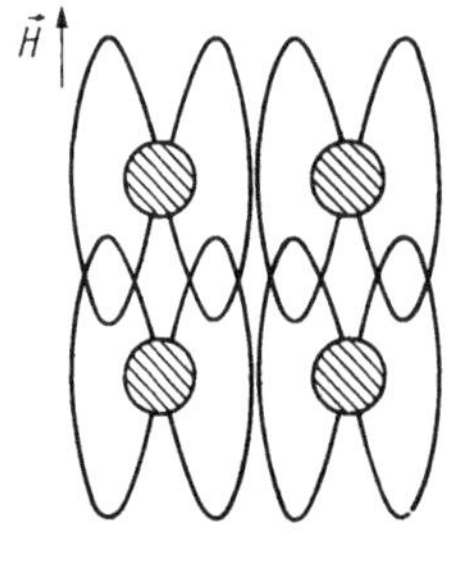

Fig. 4. Fragment of the void superlattice in the presence of a magnetic field $\boldsymbol{H}$. Voids are hatched for clarity. The schematic lines of equal order parameter modulus corresponding to different voids are represented

achieved when the magnetic field is parallel to one of the main crystallographic axes of the superlattice (see Fig. 4). In the case under consideration our system essentially becomes a quasi-one-dimensional one breaking up into numerous weakly coupled void chains oriented parallel to $\boldsymbol{H}$. Then in calculating $\delta\mathcal{H}_{c3}$ one may with sufficient accuracy neglect the overlapping of the order parameters of neighbouring chains. The spatial periodicity of voids in the chains allows to choose the coefficients C_l in the form $C_\nu\, e^{ik_\nu qD}$, where the index ν numbers the chains and the value q numbers the voids in the chain, $\boldsymbol{l} \equiv (\nu, q)$, D is the superlattice spacing (we consider the cubic lattice for simplification), k_ν is the parameter determining the increase in the phase of the order parameter of an individual void at the transfer from one void to another in the chain.

Using the Green formula [30] one obtains

$$\int \psi_{0l}^* \Delta f_l \, dV - \int f_l \, \Delta\psi_{0l}^* \, dV = \\ = \oint f_l \nabla\psi_{0l}^* \boldsymbol{n}_l \, dS - \oint \psi_{0l}^* \nabla f_l \boldsymbol{n}_l \, dS \,, \tag{19}$$

where the surface integrals are calculated over the $\boldsymbol{l}$-th void surface.

Using (1), (2), (17), (18) and the nearest-neighbour approximation one can easily obtain from (19) the expression for $\delta\mathcal{H}$,

$$\delta\mathcal{H} = 2 \cos k_\nu D \frac{\oint \psi_{0\nu,q}^* \nabla \psi_{0\nu,q+1} \boldsymbol{n}_{\nu,q} \, dS}{\int |\psi_{0\nu,q}|^2 \left[\frac{1}{2}\mathcal{H}_{c3}(R)\, r^2 \sin^2\theta - n\right] dV}. \tag{20}$$

It is obvious that the numerator of (20) is positive. The denominator is positive, too. One may check up the latter fact comparing the denominator with (11). Therefore the value $\delta\mathcal{H}_{c3}$ will be obtained provided $k_\nu D$ is aliquot to 2π. Expression (20) may be also rewritten in the equivalent form

$$\delta\mathcal{H}_{c3} = 2 \frac{\int \psi_{0\nu,q}^* \left[1 + n\mathcal{H}_{c3}(R) - \frac{1}{4}\mathcal{H}_{c3}^2(R)\, r^2 \sin^2\theta\right] \psi_{0\nu,q+1} \, dV}{\int |\psi_{0\nu,q}|^2 \left[\frac{1}{2}\mathcal{H}_{c3}(R)\, r^2 \sin^2\theta - n\right] dV}. \tag{21}$$

It was noted in Section 2 that the increase of the ratio R/ξ causes the increase of the parameter s characterizing the rate of decrease of function (4) along the magnetic field. Since $\delta\mathcal{H}_{c3} \sim (R/D)^{s+2}$ it is obvious that for the superlattice of small voids the value $\delta\mathcal{H}_{c3}$ will be larger than for the case of the superlattice of large voids. Moreover $\delta\mathcal{H}_{c3}$ quickly decreases with increasing value D/R. Let us evaluate the value $\delta\mathcal{H}_{c3}$. At $R = 1.9\xi$ one has $\mathcal{H}_{c3} = 1.042$, $n = 3$, $s = 1.88$, $\gamma = 1.29$, then at $D = 5R$ one obtains from (21) $\delta\mathcal{H}_{c3} \approx 0.010$ and at $D = 6R$ one obtains $\delta\mathcal{H}_{c3} \approx 0.006$. For $R = 2.4\xi$ we have $\mathcal{H}_{c3} = 1.056$, $n = 4$, $s = 2.19$, $\gamma = 1.15$ and at $D = 5R$ we obtain $\delta\mathcal{H}_{c3} \approx 0.006$ and at $D = 6R$ $\delta\mathcal{H}_{c3} \approx 0.003$.

At the increase of the angle θ_0 between the superlattice axis and the magnetic field the value $\delta\mathcal{H}_{c3}$ drastically decreases. For the case $D \gg \xi$ a rough estimation of the half-width $\Delta\theta_0$ of the dependence of $\delta\mathcal{H}_{c3}$ on θ_0 may be carried out assuming the difference of Peierls magnetic factors for neighbouring voids (see (17)) to be equal to $\pi/2$. Then one obtains $\Delta\theta_0 \approx \pi\xi^2/2\mathcal{H}_{c3}(R)\, D^2 \ll 1$. The dependence of $\delta\mathcal{H}_{c3}$ on θ_0 is qualitatively similar to the angular dependence of the second critical field H_{c2} of a metal containing a superlattice of superconducting inclusions [31].

It can be easily shown that the diamagnetic moment $N_v M$ created by the superconducting currents circulating around the voids has a component $N_v\, \delta M$ which depends on magnetic field orientation with regard to the void superlattice axes. Since the value M depends on $\mathcal{H}_{c3}(R)$ the value $\delta M/M$ has the same order of magnitude as $\delta\mathcal{H}_{c3}(R)$.

5. Josephson Effects in a Porous Superconductor

In the magnetic field range $H_{c2} < H < H_{c3}(R)$ the presence of the superlattice causes the existence of coherent states of superconducting condensate due to Cooper pairs tunneling between the voids. The problem of the coherent states of the superconducting condensate in the presence of the superlattice is to some extent similar to the problem of the Bloch electron states in a magnetic field [32 to 34]. It was shown by means of group theory that the electron quasi-momentum projection on the magnetic field direction is a good quantum number in the case when the magnetic field is directed along one of the crystallographic axes of the lattice [33]. Therefore we shall investigate this particular case. Hence, as was noted in Section 4, our system may be assumed to be quasi-one-dimensional. The spatial periodicity along $\boldsymbol{H}$ permits to try the order parameter of the ν-th chain of voids in the form of a Bloch superposition of the order parameters of individual voids

$$\Psi_\nu = c_\nu \sum_q e^{ik_\nu qD} \psi_{\nu,q}\,, \tag{22}$$

where the coefficients c_ν are to be obtained from (8).

The function (22) describes the state with the density of current flowing along $\boldsymbol{H}$

$$J_{\nu_z} = \frac{2e\hbar}{m} \psi^*_{\nu,q} \frac{\partial}{\partial z} \psi_{\nu,q+1} \sin k_\nu D\,. \tag{23}$$

The maximum value of the current density is reached when the magnetic field is directed along one of the main lattice axes and $k_\nu D = (2m_\nu + \frac{1}{2})\,\pi$, $m_\nu = 0, \pm 1, \pm 2 \dots$ In this case the current density averaged over the cross-section, corresponding to a chain of voids, is

$$J_{\max} = A(\mathscr{H}, s)\left(1 - \frac{H}{H_{c3}(R)}\right) \frac{\hbar c^2}{e\varkappa^2 D^2 \xi} \left(\frac{2\gamma\xi}{D}\right)^{2s+1}, \tag{24}$$

where $A(\mathscr{H}, s)$ is a smooth function (at $R < 3\xi$), $A \approx 1$.

Both $\mathscr{H}_{c3}(R)$ and s increase with increasing ratio R/ξ. The influence of s on $J_{\max}$ is more essential. The calculations show that the most advantageous experimental situation must be at $R/\xi = 1.5$ to 3. Let us evaluate $J_{\max}$ for the b.c.c. void superlattice in Nb with parameters $D = 6R = 75$ nm, which was observed in [2]. At $R = 2\xi$ one obtains $\mathscr{H}_{c3} = 1.043$, $n = 3$, $s = 1.84$, $\gamma = 1.21$. Assuming $\mathscr{H} = 1.02$ and $\varkappa = 10$ one can obtain from (24) $J_{\max} \approx 4$ A/mm^2. This value is obviously the upper estimate of $J_{\max}$, the real value should be less due to superlattice defects.

Similarly to $\delta\mathscr{H}_{3c}$ the critical current density value abruptly decreases with the increase of the angle θ_0 between the magnetic field and one of the crystallographic axes. It is rather complicated to estimate the half-width of this dependence because it depends on several parameters: ξ/D, D/L, $\delta D/D$, where L is the size of the superlattice, δD is the average deviation from the superlattice periodicity.

Under a constant electrical field the nonstationary Josephson effect will occur in the system. The characteristic frequency has the usual form of $\omega = 2eU/\hbar$ where U is the bias potential between two neighbouring voids. For the nonstationary Josephson effect the influence of superlattice defects essentially decreases due to the voltage locking effect [35 to 37].

Thus the type-II superconductor with a superlattice of voids appears to be a new Josephson medium in a strong magnetic field ($H \gtrsim H_{c2}$), which operates according to the scheme LS–N–LS, where regions of localized surface superconductivity and of normal metal alternate.

6. Conclusion

We have investigated some properties of type II superconductors containing a superlattice of voids. We should like to emphasize the following results.

1. The dependence of the third critical field H_{c3} on the void radius R was obtained. This dependence is nonmonotonic because of the discrete quantum number n which determines the increase in the phase of the order parameter on circumscribing the axis, crossing the void and being parallel to the magnetic field, along a closed contour. With the increase of the R/ξ ratio the oscillation amplitude in the $H_{c3}(R)$ dependence decreases and $H_{c3}(R)$ tends to the value of the third critical field for a plane surface of a bulk superconductor in a parallel magnetic field. The void lattice causes the appearance of a component in the $H_{c3}(R)$ dependence, which depends on magnetic field orientation with regard to crystallographic axes of the superlattice.

2. At $H \geqq H_{c2}$ the magnetic moment of the unit volume, caused by the superconducting currents flowing near the voids, was calculated. The diamagnetic moment may be by several orders of magnitude larger than the intrinsic paramagnetic moment of the metal. For a metal with a lattice of voids the diamagnetic susceptibility has a component which depends on the magnetic field orientation. Beginning with the lattices with $R/\xi \geqq 3$ the magnetization curve is nonmonotonic due to the discreteness of the quantum number n.

3. The periodicity of the system and the overlapping of the order parameters corresponding to neighbouring voids caused the existence of Bloch states of the superconducting condensate. It means that a type-II superconductor with void lattice in strong magnetic fields appears to be an array of a large number of Josephson tunnelling junctions. All collective phenomena in such a system strongly depend on the magnetic field orientation with regard to crystallographic axes. The critical current density value according to the estimations may be rather large, about 10^2 to 10^3 mA/mm^2, and it may be obtained when the magnetic field is oriented along the main crystallographic axes of the void lattice.

We have undertaken the estimation sin the case when R/ξ equals 1 to 4 because: firstly, in type-II superconductors such experimental situation is most probable [13], secondly, this case is very interesting from the theoretical point of view because for very small voids ($R \lll \xi$) the third critical field $H_{c3} = [1 + 2(R/2\xi)^6]\, H_{c2}$ [38] practically does not differ from the second critical field and for very large voids ($R \ggg \xi$) the collective phenomena in the system are rather weak.

Until now we investigated the case of magnetic fields which exceed the second critical field when there are regions of localized superconductivity in a metal. But in fields less than H_{c2} the surface superconductivity does not disappear immediately. Therefore even at $H_{c2} - H \lll H_{c2}$ Josephson effects, similar to the effects investigated in Section 5, may appear. It is also interesting to investigate the interaction of the void superlattice with the Abrikosov vortex lattice. In the fields about H_{c2} such an interaction has to cause a symmetry change of the vortex lattice. But we shall not discuss here these problems in detail.

References

[1] J. H. Evans, Nature (London) **222**, 403 (1971).
[2] J. L. Brimhall and G. L. Kulcinski, Radiat. Eff. **20**, 25 (1973).
[3] B. P. Johnson and O. J. Mazey, J. nuclear Mater. **93/94**, 721 (1980).
[4] L. S. Palatnik, P. G. Cheremskoi, and M. Ya. Fuks, Voids in Films, Energoizdat, Moscow 1982 (in Russian).

[5] V. F. ZELENSKII, I. M. NEKLYUDOV, L. S. OZHIGOV, E. A. REZNICHENKO, V. V. ROZHKOV, and T. P. CHERNYAEVA, Some Problems of Radiation Damage Physics of Metals, Naukova Dumka, Kiev 1979 (in Russian).
[6] K. KRISHAN, Radiat. Eff. **66**, 121 (1982).
[7] L. A. MAKSIMOV and A. I. RYAZANOV, Zh. eksper. teor. Phys. **79**, 2311 (1980).
[8] K. KRISHAN, Phil. Mag. A **45**, 401 (1982).
[9] J. H. EVANS, J. nuclear Mater. **119**, 180 (1983).
[10] V. I. SUGAKOV, Kiev Institute of Theoretical Physics, preprint No. 84–70R, Kiev 1984; Zh. eksper. teor. Fiz., to be published.
[11] V. I. SUGAKOV, Questions of Atomic Science and Technology, Ser. Radiation Damage Physics, No. 2 (16), Izd. Kharkov Physico-Technical Institute, 1981 (p. 71) (in Russian).
[12] YU. V. KRYUCHENKO and V. I. SUGAKOV, phys. stat. sol. (b) **111**, 177 (1982).
[13] M. P. VOLKOV, YU. N. SOKURSKII, S. I. TSIPKIN, V. I. CHUEV, and V. N. SHISHOV, Fiz. nizk. Temp. **4**, 32 (1978).
[14] S. BORKA, I. N. GONCHAROV, and I. S. KHUKHAREVA, Fiz. nizk. Temp. **4**, 962 (1978).
[15] L. N. SHEHATA, phys. stat. sol. (b) **105**, 77 (1981).
[16] V. N. RUDKO and V. I. SUGAKOV, Fiz. nizk. Temp. **9**, 440 (1983).
[17] A. YA. DZYUBLIK and V. I. SUGAKOV, phys. stat. sol. (a) **70**, K129 (1982).
[18] A. YA. DZYUBLIK and N. I. GRIGORCHUK, Zh. tekh. Fiz. **52**, 1167 (1983).
[19] J. E. EPPERSON, B. A. LOOMIS, G. H. LANDER, S. B. GERBER, R. W. HENDRICKS, J. S. LIN, and M. B. LEWIS, J. nuclear Mater. **119**, 251 (1983).
[20] D. SAINT-JAMES and P. G. DEGENNES, Phys. Letters (Netherlands) **7**, 306 (1963).
[21] I. O. KULIK, Zh. eksper. teor. Fiz. **55**, 889 (1966).
[22] D. SAINT-JAMES, Phys. Letters (Netherlands) **15**, 13 (1965).
[23] J. R. HAUSER, J. S.-Y. WANG, and C. KITTEL, Phys. Letters A **47**, 34 (1974).
[24] F. W. WIFFEN, in: Radiation-Induced Voids in Metals, Proc. Inernat. Conf., Albany (N.Y.), Oak Ridge USAEC Techn. Inform. Centre, 1972 (p. 386).
[25] B. A. LOOMIS and S. B. GERBER, J. nuclear Mater. **102**, 154 (1981).
[26] V. L. GINZBURG and L. D. LANDAU, Zh. eksper. teor. Fiz. **20**, 1064 (1950).
[27] N. R. WERTHAMER, Phys. Rev. **132**, 663 (1963).
[28] V. N. RUDKO, Phys. Letters A **101**, 41 (1984).
[29] R. PEIERLS, Z. Phys. **80**, 763 (1933).
[30] P. M. MORSE and H. FESHBACH, Methods of Theoretical Physics, McGraw-Hill, New York 1953.
[31] V. M. NABUTOVSKII, Solid State Commun. **43**, 405 (1982).
[32] J. ZAK, Phys. Rev. **136**, 776 (1964).
[33] A. RAUH, G. H. WANNIER, and G. OBERMAIR, phys. stat. sol. (b) **63**, 215 (1974).
[34] D. R. HOFSTADTER, Phys. Rev. B **14**, 2239 (1976).
[35] D. W. JILLIE, J. E. LUKENS, Y. H. KAO, and G. J. DOLAN, Phys. Letters A **55**, 381 (1976).
[36] V. A. KHLUS, A. S. ROZHAVSKII, and I. O. KULIK, Fiz. nizk. Temp. **5**, 357 (1979).
[37] R. F. VOSS and R. A. WEBB, Phys. Rev. B **25**, 3446 (1982).
[38] YU. N. OVCHINNIKOV, Zh. eksper. teor. Fiz. **79**, 1496 (1980).

(Received July 26. 1984)

phys. stat. sol. (b) **126**, 713 (1984)

Subject classification: 13.4 and 20.1; 2; 16; 22.1.2

Departments of Physics and Electrical Engineering, National Tsing Hua University, Hsin-Chu[1])

Time-Resolved Reflectivity Measurements on RF Sputtered a-Si:H Alloy[2])

By

JUH TZENG LUE

Photoinduced carrier lifetimes of RF sputtered a-Si:H alloy are measured by time-resolved reflectivity and photoconductivity decay methods. Both experiments indicate a lifetime of about 5 μs. A calculation of the relaxation time of photogenerated carriers from extended mobility edge to midgap localized states satisfactorily agrees with this value.

Photoinduzierte Trägerlebensdauern in HF-aufgestäubten a-Si:H-Legierungen werden mit zeitaufgelösten Reflexions- und Photoleitungsabklingmethoden gemessen. Beide Experimente zeigen eine Lebensdauer von etwa 5 μs. Eine Berechnung der Relaxationszeit der photoangeregten Träger aus den lokalisierten Zuständen von der ausgehenden Beweglichkeitskante bis zur Bandmitte stimmt mit diesem Wert überein.

1. Introduction

Picosecond reflectivity measurements on glow discharge a-Si:H diagnosed by a mode locked Nd:YAG laser had been reported recently [1]. The temporal behaviour of the non-equilibrium charge carriers was specified by a two-step relaxation process in which the photoexcited carriers relax from the conduction band via an intermediate state in the band gap and then to the ground state. The characteristic time to these two relaxation processes were determined to be 30 to 85 ps and 0.75 to 2.5 ns, respectively. Geminate recombination [2] is the possible mechanism for explaining such a short carrier lifetime.

In this paper a nitrogen UV laser with a pulse FWHM of 360 ps had been used to excite free carriers of a RF sputtered a:Si:H, while a He–Ne laser of wavelength 6328 Å was employed to probe the reflectivity change. The characteristic relaxation time measured by the time-resolved reflectivity is 5 μs which is closely matched to the value 4.7 μs as measured by the photoconductivity decay method. This confirms that volume recombination due to photogenerated carriers decaying from the extended mobility edge to midgap localized states is the dominant process after a rapid thermalization.

2. Theoretical Background

2.1 *Generation and recombination processes in a-Si*

A vast amount of dangling bonds will be introduced during the deposition of amorphous silicon. The lack of long-range order and structure defects are the prominent reason for forming a high density of localized states adjacent to the extended states, whereas compensated impurities bring in states near the gap centre to pin the Fermi

[1]) 855 Kuang Fu Road, Hsin-Chu, Taiwan.

[2]) This work was supported by National Science Council.

level. These localized states reduce carrier lifetime by trapping and recombination processes. Hydrogen in a-Si should saturate dangling bonds, thereby greatly reducing the density of localized states. Usually, the density of localized states of a-Si grown from RF sputtering because of ion bombardment is higher than that from CVD (chemical vapour deposition) method.

For undoped amorphous semiconductors, the thermally generated carriers are few in number resulting in low dark conductivity. Excess carriers can be generated by illuminating with light. The generation and recombination processes of photocarriers are illustrated in Fig. 1. For incident photon energy greater than the mobility gap of the amorphous semiconductors, the excess carriers are excited due to electron–hole pair production from extended valence band to extended conduction band as indicated in process (a). The electrons and holes excited well above the mobility edge will thermalize with phonons and temporarily reside in the mobility edge if they can escape from geminate recombination as shown in process (b). Thermalization with the extended states is expected to occur within 10^{-12} s [3].

The excess electrons (or holes) near the mobility edge E_c (or E_v) can transfer to the unoccupied conduction band-tail states (or occupied valence band tail) via volume recombination τ_v. The carrier within the band-tail localized states usually form small polarons [4] before recombination. Polaron formation occurs when charge carriers induce a lattice change of the equilibrium atoms. These displacements create a potential well for the carrier to occupy a bound state causing the carriers to be self-trapped. Since the self-trapped carriers are confined in the polaron potential well, the carrier can move from site to site only by phonon-assisted hopping.

The electrons in the conduction band tail state will also recombine with holes in the valence band tails by tunneling transition if these two band tails overlap with

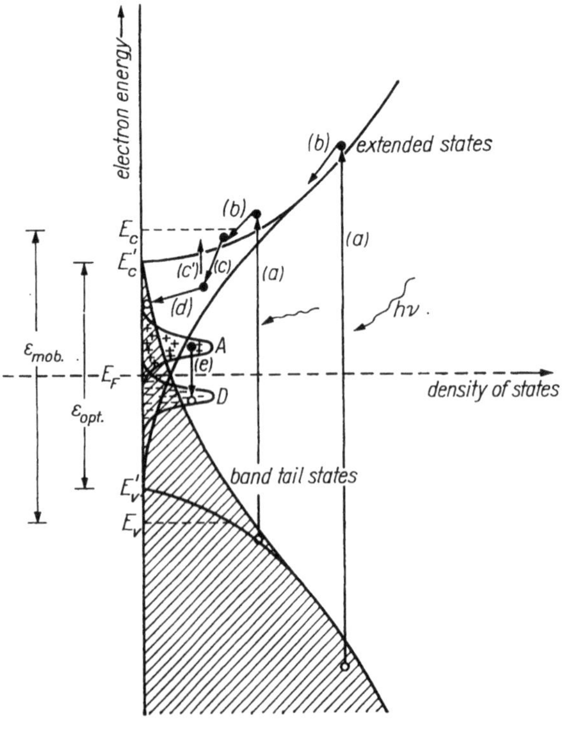

Fig. 1. Generation and recombination processes in a-Si. (a) Excitation, (b) recombination within extended states by thermalization, (c) extended to band tail recombination, (c′) thermally induced re-excitation, (d) tunneling recombination of filled and unfilled localized states, (e) recombination between acceptor and donor localized states near Fermi level. Symbols indicate: A acceptor states produced by double occupation and D deep donor states due to neutral dangling bonds

each other as shown in process (d). The radiative tunneling lifetime is given by the following equation for a-Si:H [5]:

$$\tau_d = \tau_0 \exp\left(\frac{2R}{R_0}\right), \tag{1}$$

where $\tau_0 \approx 10^{-8}$ s, R_0 is the radius of the band-tail electrons (≈ 1.0 nm), and R is the electron–hole separation. For optically excited a-Si, R is about 5.0 to 100 nm [5] after rapid thermalization consequently τ_d is estimated to be 2.2×10^{-4} s.

2.2 Volume recombination lifetime

In this experiment the observed carrier lifetime is assumed to be due to volume recombination. The steady state recombination mechanism derived by Shockley and Read [6] is not applicable to the pulsed excitation. The transient solution of the photocarrier decay time can be found from the rate equations for the excess carriers δn and δP [7],

$$\frac{d}{dt}\delta n = g - \alpha\,\delta n + \beta\,\delta p\,, \qquad \frac{d}{dt}\delta p = g - \eta\,\delta p + \gamma\,\delta n\,, \tag{2}$$

where g is the generation rate by photoexcitation, and

$$\begin{aligned}
\alpha &= \sum_i \alpha_n^i\,(N_i^c + n_0 + n_1 + \delta n) + \alpha_d(p_0 + \delta p)\,,\\
\beta &= \sum_i \alpha_n^i(n_0 + n_1 + \delta n) - \alpha_d(n_0)\,,\\
\eta &= \sum_i \alpha_p^i\,(N_i^v + p_0 + p_1 + \delta p) + \alpha_d(n_0 + \delta n)\,,\\
\gamma &= \sum_i \alpha_p^i(p_0 + p_1 + \delta p) - \alpha_d(p_0)\,.
\end{aligned} \tag{3}$$

The summation $\sum_i$ includes multiple recombination centres which may exist in the band gap. The other symbols are defined as α_n^i being the product of electron capture cross-section (σ_n) and thermal velocity of electron (V_{th}^e) for the i-th trapping centre; α_d being the radiative recombination rate; N_i^c being the density of empty recombination centres in a-Si which can be considered as localized states of conduction band tail; n_0, p_0 being the thermal equilibrium carrier concentration; and n_1, p_1 being the carrier concentration with the Fermi level at the energy level of the trap.

For undoped a-Si:H, since the intrinsic carrier densities n_0, p_0 are very small, the excess carrier density can overwhelm the intrinsic carriers at intermediately high illuminating intensity, i.e.

$$\delta n \gg n_0 + n_1\,, \quad \text{and} \quad \delta p \gg p_0 + p_1\,. \tag{4}$$

By assuming further that the radiative recombination rate is much smaller than that of the nonradiative process, i.e. $\alpha_{n,p} \gg \alpha_d$, (2) can be simplified as

$$\frac{d}{dt}\delta n = g - \sum_i \alpha_n^i N_i^c\,\delta n \quad \text{and} \quad \frac{d}{dt}\delta p = g - \sum_i \alpha_p^i N_i^v\,\delta p\,. \tag{5}$$

The solutions of (5) are trivial and are readily shown to be

$$\delta n = A + B \exp\left(\frac{-t}{\tau_n}\right), \qquad \delta p = C + D \exp\left(\frac{-t}{\tau_p}\right), \tag{6}$$

where

$$\tau_n = \frac{1}{\sum_i \alpha_n^i N_i^c} = \frac{1}{V_{th}^e \int \delta n\,(E_t)\, N_D^c \exp\left[-(E_c - E_t)/E_0\right] dE_t}$$

$$= \frac{1}{V_{th}^e N_t^c \sigma_n}$$

and

$$\tau_p = \frac{1}{V_{th}^p N_t^v\, \sigma_p}. \tag{7}$$

Here we have assumed that the density of localized states drops exponentially from the mobility edge to the midgap [8]. Actually, the electrons trapped by localized states within $k_B T$ below the conduction band edge will thermally be reexcited to the extended states, therefore, the electron lifetime τ_n will be shorter than the decay time indicated by the experimental decay curve.

2.3 Relationship of reflectivity and carrier density

Optical reflectivity is relatively easier to interpret via classical dipole oscillations [9]. The electromagnetic field of the incident light excites the free carriers into vibration, and the dipole oscillators in turn emit EM field with the same frequency as the incident light. The polarization of the dipole moment introduces a complex refractive index $n + ik = \varepsilon^{1/2}$, where ε is the dielectric constant which relates to the conductivity σ by

$$\sigma = -\frac{in}{4\pi} (\varepsilon - 1). \tag{8}$$

For normal incidence on an infinitely thick plate, the reflectivity is

$$R_\infty = \frac{(n-1)^2 + k^2}{(n+1)^2 + k^2}. \tag{9}$$

For cubic crystalline material with only one set of equivalent parabolic conduction band valleys, $\varepsilon(\omega)$ is simply written as [9]

$$\varepsilon = \varepsilon_0 - \frac{4\pi N_c e^2}{m_c^* \omega^2}, \tag{10}$$

where ε_0 is the dielectric constant in the absence of free carriers, N_c the free-carrier concentration, and m_c^* the effective mass of carriers. In the visible spectral range, the frequency ω is very high, we can therefore write

$$\varepsilon^{1/2} \approx \varepsilon_0^{1/2} \left(1 - \frac{4\pi N_c e^2}{m_c^* \omega^2 \varepsilon_0}\right)^{1/2} = \left[n_0 - \frac{2\pi N_c e^2 (n_0^2 - k_0^2)}{m_c^* \omega^2 (n_0^2 + k_0^2)}\right] +$$

$$+ i \left[k_0 - \frac{4\pi N_c e^2 n_0 k_0}{m_c^* \omega^2 (n_0^2 + k_0^2)}\right] = n + ik. \tag{11}$$

The normal reflectance of an infinitely thick plate can be calculated by exploiting the fact that $(2\pi N_c e^2)/(m_c^* \omega^2)$ is much smaller than n_0 and k_0. After some tedious algebra, we obtain

$$R_\infty = \frac{(n-1)^2 + k^2}{(n+1)^2 + k^2} \approx R_0 \left\{1 - \frac{4\pi N_c e^2}{m_c^* \omega^2} X\right\}, \tag{12}$$

where

$$X = \frac{(n_0 - 1)(n_0^2 + k_0^2) + 2n_0 k_0^2}{[(n_0-1)^2 + k_0^2](n_0^2 - k_0^2)} + \frac{(n_0 + 1)(n_0^2 + k_0^2) + 2n_0 k_0^2}{[(n_0+1)^2 + k_0^2](n_0^2 + k_0^2)}.$$

If N_c decays according to (6), then

$$R \approx R_0 \left\{1 - \frac{4\pi N_c}{m_c^* \omega^2} X \, e^{-t/\tau}\right\}. \tag{13}$$

The reflectance has an exponential time dependence with a characteristic time τ.

3. Sample Preparation and Experimental Arrangement

A-Si:H samples with thickness of 10 μm were prepared by RF sputtering in a mixture of argon and hydrogen gases at partial pressure ratio of 6:3 on quartz substrates which were held at room temperature at a deposition rate of 0.3 nm/s. Since a-Si can absorb more than 90% of visible light within a thickness of 1 μm, these samples can be considered as optically opaque. The infrared absorption dip occurring at 2090 cm^{-1} is used to estimate the hydrogen content of a-Si:H utilizing the method of Freeman and Paul [11] which turns out as 15 at% of hydrogen.

The experimental set-up for reflectivity and photoconductive decay measurement is shown in Fig. 2. A 360 ps (FWHM) N_2 laser operating at atmospheric pressure with photon energy of 3.68 eV is used to generate free carriers. The advantages of the N_2 laser are its high radiation intensity (several MW), short pulse duration, $\hbar\omega > E_g$ where E_g is the mobility band gap, and easy construction. The probing beam is a 0.5 mW cw He–Ne laser which is highly filtered so that it will not interpose the excitation. The probe beam is taken in normal incidence so that there is no distinction of reflectance between the normal and parallel components of the unpolarized light. The photodetector is an EG & G type DT-25 which has a rise time of 5 ns. The signal is directly recorded by a Tektronix 7613 storage oscilloscope.

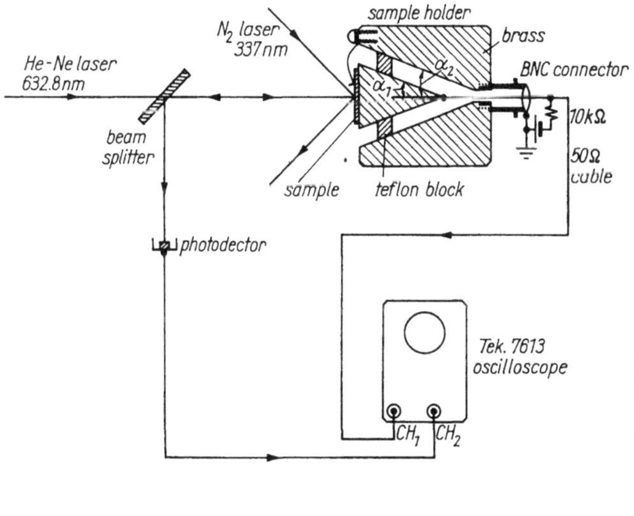

Fig. 2. Experimental arrangement for photoconductive and reflectivity measurements

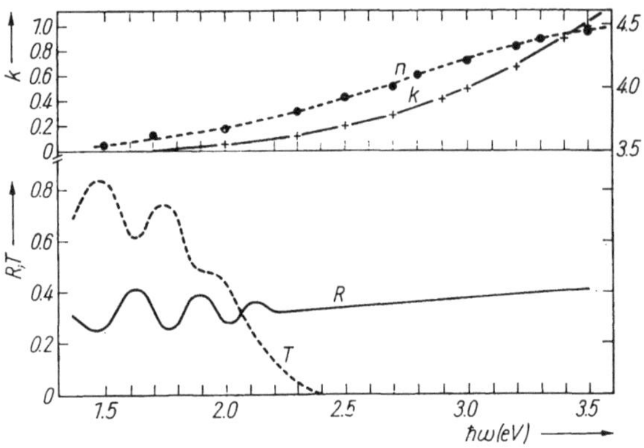

Fig. 3. Absolute transmittance (- - - -) and reflectance (———) of a-Si:H deposited by RF sputtering

4. Experimental Results and Theoretical Calculation

The refraction indices n_0 and k_0 at weak probing beam appearing in (13) were first measured by a modified double-beam single-detector wavelength modulating spectrometer [12]. As depicted in Fig. 3, the periodic oscillation of the reflectivity at photon energies below the absorption edge is due to the interference at interfaces of air–film and film–substrate. Since the fringe order of the interference maximum was unknown in this wavelength range, a procedure applied by Tubbs [13] has been used to calculate the index of refraction. First, the separation $\Delta\omega$ of each neighbouring maximum and minimum of the interference pattern after reflection was used to calculate the quantity $n^* = n + [\omega\, \Delta n/\Delta\omega]$, where $n^*(\omega)$ was fitted to a smooth curve by the dispersion relation [14]. The frequency-dependent refraction index $n(\omega)$ was then computed through the integral

$$n(\omega) = \frac{1}{\omega} \int\limits_{\omega_0}^{\omega} n^*(\omega')\, \mathrm{d}\omega' + \frac{\omega_0}{\omega}\, n(\omega_0)\,. \tag{14}$$

The constant $n(\omega_0)$ can be obtained from the extrapolated flat portion of the reflectivity near the absorption edge (e.g. 2.2 eV for a-Si) with a manipulation of (9). For photon energies greater than 2.2 eV, strong absorption due to band-to-band transition induces a fairly weak transmission, and the Kramers-Kronig analysis has been applied to calculate the refraction index. The optical constant over the entire 1.4 to 3.6 eV photon energy range is shown in the top part of Fig. 3, where $\hat{n} = n + ik$ is the complex

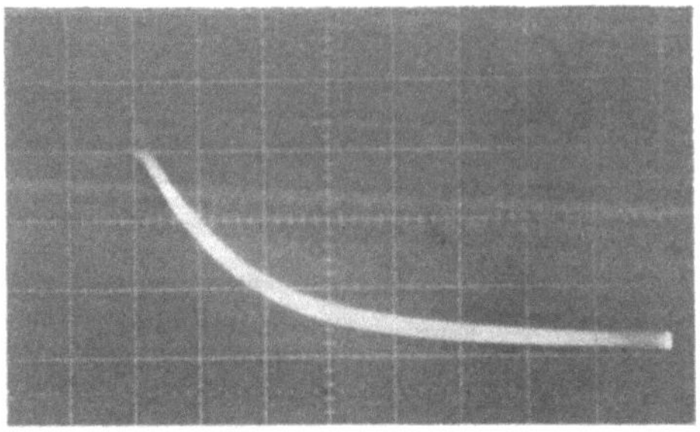

Fig. 4. Oscillogram of the time-resolved reflectivity, vertical scale: 5 mV/div, time base: 5 μs/div

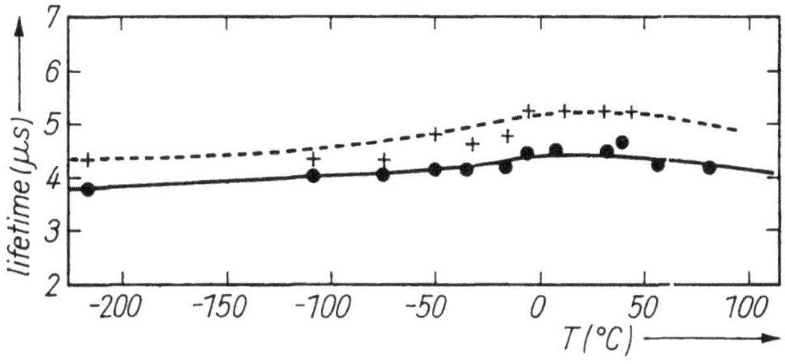

Fig. 5. Temperature dependence of the relaxation time τ_v measured by methods of photoconductivity (+) and time-resolved reflectivity (●), respectively

refraction index. With this data the R_0 and X appearing in (12) can be evaluated to be 0.327 and 0.59, respectively, at $\hbar\omega = 1.96$ eV.

A typical time-resolved reflectivity of a-Si:H measured in this experiment is shown in Fig. 4. The characteristic relaxation time is measured to be 4.16 μs which is close to the value 5 μs measured by photoconductive decay method. To check whether the observed carrier decay time is due to the volume recombination mechanism, we can estimate τ_v from (7) with the parameters $N_t^c \approx 2 \times 10^{17}$ cm^{-3}, $V_{th} \approx 10^7$ cm/s, and $\sigma_n = 10^{-19}$ cm^2 [15]. The calculated $\tau_v \approx 5$ μs satisfactorily agrees with the experimental value. The temperature dependence of the carrier recombination time measured by time-resolved reflectivity and by photoconductivity was depicted in Fig. 5. The experimental results indicate that τ_v is insensitive to temperature variation. This can be explained by (7), where the thermal velocity V_{th} increases as temperature increases, subsequently shortening the lifetime. But thermally reexcited free carriers from localized states near the mobility edge which dominate at high temperatures will compensate the effect of V_{th}.

5. Discussions

The carrier recombination lifetime of a-Si:H observed in this experiment is vastly different from that observed by Bergner et al. [1]. Different methods of sample preparation may not account for the deviation. Various excitation and recombination processes as shown in Fig. 1 may be observed by applying alternative diagnostic techniques. The relaxation time of reflectivity due to free-carrier decay measured in this experiment congruently agrees with the volume recombination lifetime. Previous works using luminescence decay method [3] found the nonradiative recombination lifetime to be less than 10 μs, and the photoconductive decay method [15] showed the free-carrier lifetime to be between 10^{-6} and 10^{-3} s, both confirming the present result.

Acknowledgement

The author is indebted to Dr. S. Y. Shaw for measuring the refraction index.

References

[1] H. Bergner, V. Brückner, D. Dietrich, F. Kerstan, and W. Nowick, phys. stat. sol. (b) 117, 197 (1983).
[2] J. Mort, A. Troup, M. Morgan, and S. Grammatica, Appl. Phys. Letters 38, 277 (1981).
[3] C. Tsang and R. A. Street, Phys. Rev. B 19, 3027 (1979).
[4] D. Emin, Polycrystalline and Amorphous Thin Films and Devices, Chap. 1, Ed. L. L. Kazmerski, Academic Press, Inc., 1980.
[5] R. A. Street and D. K. Biegelsen, Solid State Commun. 44, 501 (1982).

[6] W. SHOCKLEY and W. T. READ, JR., Phys. Rev. 87, 835 (1952).
[7] G. K. WERTHEIM, Phys. Rev. 15, 1086 (1958).
[8] J. T. LUE, Solid State Electronics 25, 869 (1982).
[9] K. SEEGER, Semiconductor Physics, Springer-Verlag, New York/Wien 1973 (p. 489).
[10] M. CARDONA, Modulation Spectroscopy, Academic Press, Inc., New York 1969 (p. 12).
[11] W. E. SPEAR, Adv. Phys. 26, 811 (1977).
[12] S. Y. SHAW and J. T. LUE, J. Phys. E 14, 1135 (1981).
[13] M. R. TUBBS, J. Phys. Chem. Solids 30, 2323 (1969).
[14] L. PAJASOVA, A. ABRAHAM, I. GREGORA, and M. ZAVETOVA, Solar Energy Mater. 4, 1 (1980).
[15] C. R. WRONSKI and R. E. DANIEL, Phys. Rev. B 23, 794 (1981).

(Received July 26, 1984)

phys. stat. sol. (b) 126, 721 (1984)
Subject classification: 14.1; 15; 21.7

Physics Laboratory I, H. C. Ørsted Institute, University of Copenhagen[1] *(a) and Department of Applied Mathematics, Faculty of Science, Ain Shams University, Cairo (b)*

Galvanomagnetic Effects of Bismuth in Non-Quantising Magnetic Fields

Pseudo-Parabolic Model

By

O. P. HANSEN (a) and I. F. I. MIKHAIL (b)

The non-parabolicity of the L-point carriers and the temperature dependence of the band structure are incorporated in a model for the transport properties of bismuth. Results are given for elements of the resistivity tensor versus magnetic field at temperatures around 80 and 300 K. A comment is made to the papers of Gitsu et al.

La non-parabolicité des porteur au point L et la dépendance de température de la structure des bandes sont incorporées dans un modèle pour les phénomènes de transport dans le bismuth. Des resultats sont présentés pour des composants du tenseur de résistivité en fonction de champ magnetique aux températures en environ 80 et 300 K. Un commentaire est fait sur les articles de Gitsu et al.

1. Introduction

Although much work on the transport properties of bismuth has been done, still some basic problems remain unsolved. One of these concerns the influence of the L-point energy gap on the scattering rate of the charge carriers. The scattering rate is known to depend on the energy gap through its proportionality to the density of states. However, it was found [1] that if this is the only way in which the energy gap enters, then the usual (Boltzmann) transport theory gives results at variance with the experimental data for the thermomagnetic properties. It was further found in [1] that the experimental data were best accounted for if the energy gap was completely neglected, corresponding to a parabolic dispersion relation.

This latter fact prompted Heremans and Hansen [2] to ascribe to the scattering matrix element an energy dependence which cancelled other influences of the energy gap. This phenomenological way of solving the problem was denoted pseudo-parabolic model. The model was subsequently used to analyse the thermopower of pure as well as tin-doped bismuth [3, 4]. In both cases it was found to give results in reasonable agreement with the experimental data. However, two weaknesses of the model have to be mentioned: First, it lacks theoretical justification, and second, it was combined with a temperature dependence of the band structure which has features still open to experimental testing.

The incomplete knowledge of the temperature dependence of the band structure led Heremans and Hansen [2] to adjust the Fermi energy within the pseudo-parabolic model in such a way that the experimental thermopower in zero magnetic field was reproduced. The resulting Fermi energy increases with temperature, as is the case for

[1]) Universitetsparken 5, 2100 Copenhagen, Denmark.

the Fermi energy obtained with parabolic dispersion. It is denoted pseudo-parabolic Fermi energy.

Mikhail et al. [3] evaluated the dependence on magnetic field of a number of elements of the thermoelectric tensor. In order to obtain these elements, also elements of the resistivity tensor were calculated; but these latter elements were not reported, as they are less informative as regards the energy dependence of the scattering rate. Recently, however, Gitsu et al. [5] have published a series of papers reporting very comprehensive theoretical work and numerical results on the thermomagnetic as well as the galvanomagnetic properties. The method used by Gitsu et al., and also the type of input parameters, differs in several essential ways from [3]. For this reason we want in the present communication to report on the galvanomagnetic properties as found from the pseudo-parabolic model and the pseudo-parabolic Fermi energy. Further we want to comment on the papers by Gitsu et al. [5].

The temperature dependence of the band structure starts around 80 K, and for this reason we shall present the results at this temperature and at room temperature only.

2. Calculation of Transport Properties from the Pseudo-Parabolic Model and the Pseudo-Parabolic Fermi Energy

For the details we shall refer to [3], and here we just give a summary of the basic features.

The elements of the transport tensors are calculated using the Boltzmann transport equation with a dispersion relation (Lax et al. [6]),

$$\gamma = \varepsilon\left(1 + \frac{\varepsilon}{\varepsilon_G}\right) = \frac{1}{2}\,\boldsymbol{p}\mathrm{m}^{-1}\bar{\boldsymbol{p}}\,, \tag{1}$$

and a relaxation time tensor,

$$\boldsymbol{\tau} = \boldsymbol{\tau}_0 \left(\frac{\gamma}{kT}\right)^{-1/2} \gamma'\,, \tag{2}$$

Table 1

Parameters used in the calculations

T	ε_F^e	ε_F^h	ε_G	N	μ_1	μ_2	μ_3	μ_4	ν_1
(K)	(meV)			(10^{23} m^{-3})	(m^2/Vs)				
77	35	12.8	15.2	5.24	59.1	1.12	32.7	−3.88	9.38
100	38.5	13.6	16.3	6.60	37.8	0.746	19.7	−2.58	6.11
120	43.1	15.7	17.5	8.31	25.2	0.516	13.9	−1.74	4.17
140	47.7	17.4	18.8	9.85	18.0	0.378	10.1	−1.34	3.03
180	58.5	20.9	22.1	13.8	9.96	0.242	5.85	−0.749	1.81
220	71.1	25.8	26.2	17.7	6.51	0.153	3.57	−0.472	1.18
260	85.6	28.5	31.0	22.3	4.33	0.0983	2.17	−0.321	0.767
300	102	31.5	36.7	30.2	2.44	0.0876	1.48	−0.223	0.529

ε_F^e pseudo-parabolic L-point electron Fermi energy [2]; ε_F^h T-point hole Fermi energy [1]; ε_G L-point energy gap [8]; N, μ's, and ν_1 electron density (equal to hole density), and drift mobilities of electrons and holes [3]. The drift mobilities are the elements of M (see (3)), for the electrons they pertain to that one of the three ellipsoids which has one of its main axes parallel to a binary direction. The element ν_3 is found from ν_1 and the hole mass ratio.

where ε_G denotes the energy gap, $\boldsymbol{p}$ the momentum, and $\mathbf{m}$ the band-edge mass tensor, and where prime on γ means differentiation with respect to energy.

No attempt to calculate $\boldsymbol{\tau}_0$ is made. Instead the drift mobility tensor $\mathbf{M}$ is introduced (upper sign for electrons, lower sign for holes),

$$\mathbf{M} = \mp \frac{\langle \mathbf{U} \rangle}{\langle 1 \rangle}, \tag{3}$$

where

$$\mathbf{U} = \frac{q}{\gamma'} \mathbf{m}^{-1} \boldsymbol{\tau}, \tag{4}$$

q denotes the charge with its sign, and $\langle \; \rangle$ means an energy average. The weak-field galvanomagnetic coefficients are expressed in terms of $\mathbf{M}$, of the carrier density, of certain energy integrals that account for the non-degenerate statistics, and of the magnetic field. From these expressions the elements of $\mathbf{M}$, along with the carrier density, are calculated using a least-squares minimisation by which the experimental weak-field galvanomagnetic coefficients, as measured by Michenaud and Issi [7], are reproduced. The results are given in Table 1 along with the Fermi energies and the L-point energy gap.

The transport tensors for arbitrary field values are expressed in terms of the elements of $\mathbf{M}$, the carrier density, certain energy integrals, and the magnetic field, and are calculated numerically at temperatures from 77 to 300 K and for fields up to 6 T.

3. Comparison of Calculated Results and Experimental Data. Discussion

Below we present a side by side comparison of calculated results and experimental data for elements of the resistivity tensor $\boldsymbol{\rho}(\boldsymbol{B})$ with $\boldsymbol{B}$ directed successively along each of the crystallographic directions: binary (1), Fig. 1 to 3; bisectrix (2), Fig. 4 and 6; and trigonal (3), Fig. 7.

Although the calculations were performed up to 6 T we can expect the results to be adequate only as long as the Landau level spacing is less than the thermal energy. The field strength at which this happens is denoted by B_L,

$$B_L = \frac{kT\gamma' m_b}{\hbar c}, \tag{5}$$

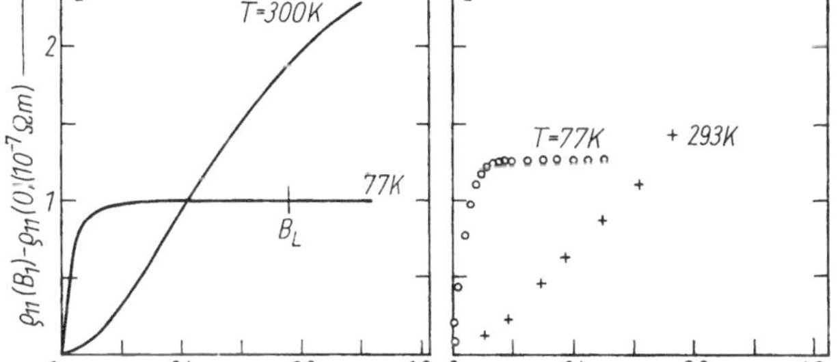

Fig. 1. Experimental data at 77 K reproduced from Saunders and Sümengen [10], Fig. 2; at 293 K from Stierstadt [11], Fig. 2. The field value B_L is calculated from equation (5). Here and in the other figures a) for theory, b) for experiment

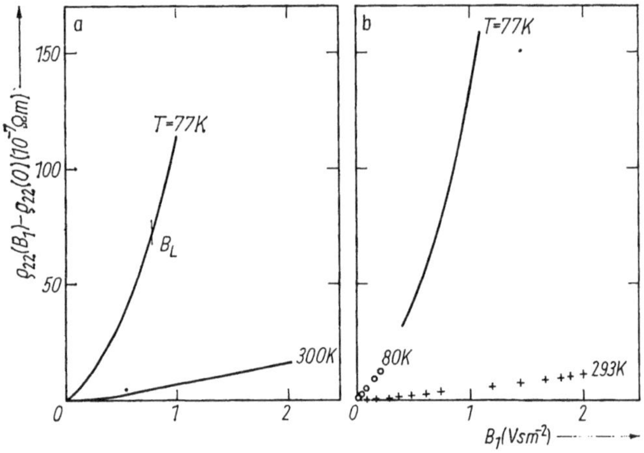

Fig. 2. Experimental data at 77 K (continuous curve) reproduced from Bogod et al. [12], Fig. 3, curve 2; at 80 K from Abeles and Meiboom [13], Fig. 3; at 293 K from Stierstadt [11], Fig. 8 and [14], Fig. 8

where

$$m_b = \left(\frac{\det \mathbf{m}}{\boldsymbol{b}\mathbf{m}\boldsymbol{b}}\right)^{1/2}; \qquad \boldsymbol{b} = \frac{\boldsymbol{B}}{|\boldsymbol{B}|}. \tag{6}$$

Values of B_L corresponding to the lightest electron mass for each field direction are marked by vertical bars on the curves showing the calculated results at 77 K. At 300 K they fall beyond the displayed field ranges.

In fields below B_L we can still obtain the classical strong-field behavior characterised by the mobility–field product being much larger than one. The powers of B for the limiting behavior of $\boldsymbol{\rho}(\boldsymbol{B})$ are for the three directions:

$$\begin{matrix} \boldsymbol{B} = (B_1, 0, 0) & \boldsymbol{B} = (0, B_2, 0) & \boldsymbol{B} = (0, 0, B_3) \\ \begin{pmatrix} 0 & & \\ & 2 & 2 \\ & 2 & 2 \end{pmatrix} & \begin{pmatrix} 2 & 1 & 1 \\ 1 & 0 & 0 \\ 1 & 0 & 2 \end{pmatrix} & \begin{pmatrix} 2 & 1 & \\ 1 & 2 & \\ & & 0 \end{pmatrix}. \end{matrix} \tag{7}$$

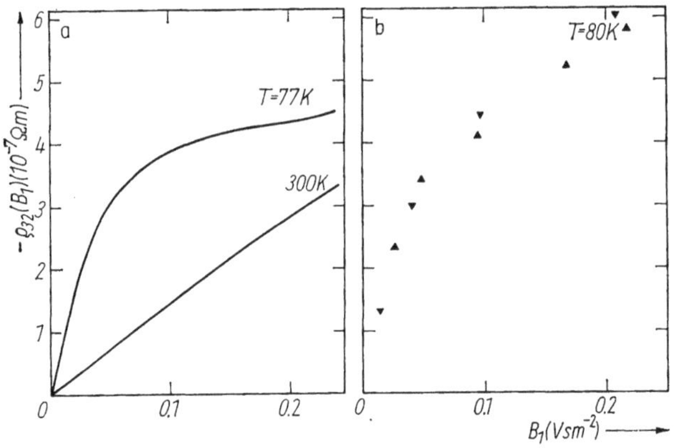

Fig. 3. Experimental data reproduced from Abeles and Meiboom [13], Fig. 1. ▲ $-\varrho_{32}(+B_1)$, ▼ $-\varrho_{23}(-B_1)$

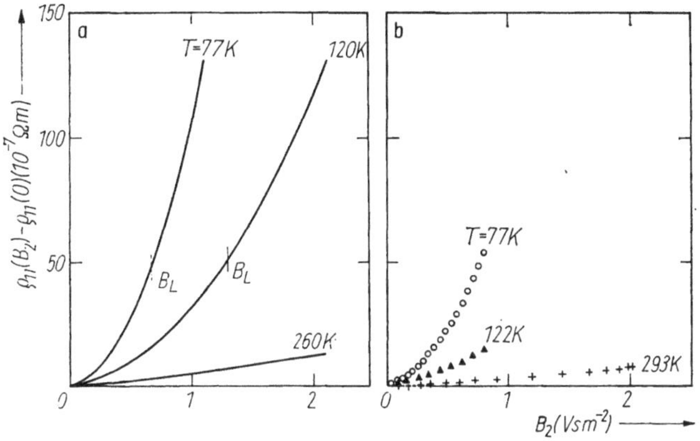

Fig. 4. Experimental data at 77 and 122 K reproduced from Cheruvier [15], Fig. 4.7 a and b; at 293 K from Stierstadt [11], Fig. 8 and [14], Fig. 11

These powers of B are characteristic of the geometry of the Fermi surface, not of the pseudo-parabolic model. The deviations from the corresponding powers in an isotropic, compensated material are due to the tilt of the electron ellipsoids, i.e. to the existence of μ_4. This pertains to the elements $\varrho_{23}(B_1)$, $\varrho_{12}(B_2)$, $\varrho_{23}(B_2)$, and their transposed ones. These elements are all proportional to μ_4 in the strong-field limit considered.

The element $\varrho_{32}(B_1)$ is displayed in Fig. 3. As for the calculated values at 77 K we want to add that outside the field range displayed, around 1.7 Vs/m², there is a change in sign, after which the element is positive, and the B^2-behaviour shown in (7) starts. Our results for this element agree with the remark made by Akgöz and Saunders [17] concerning its even and odd parts at 0.5 Vs/m².

Gitsu et al. [5] report their results for $\varrho_{ii}(B_k)$ at 77 K in the field range up to 0.6 T. We have made a comparison (not displayed) between these results and those we find for the same temperature and field range: There is a good agreement, except that for B_1 we get $\varrho_{33} > \varrho_{22}$ in the range above 0.15 T. Analysing this, we note that in the strong-field limit the ratio $\varrho_{33}(B_1)/\varrho_{22}(B_1)$ is strongly dependent on the smaller one of the two

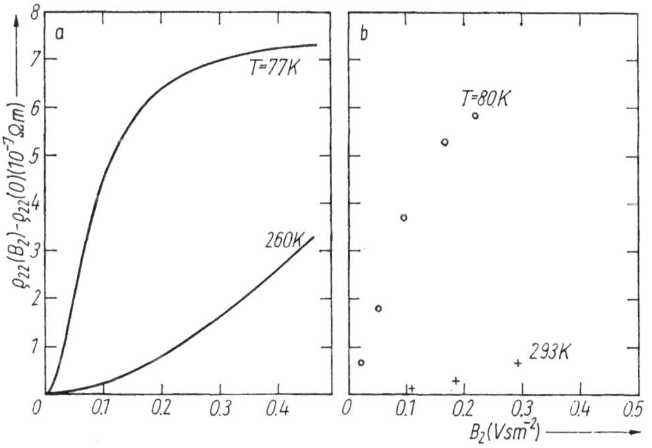

Fig. 5. Experimental data at 80 K reproduced from Abeles and Meiboom [13], Fig. 3; at 293 K from Stierstadt [11], Fig. 2

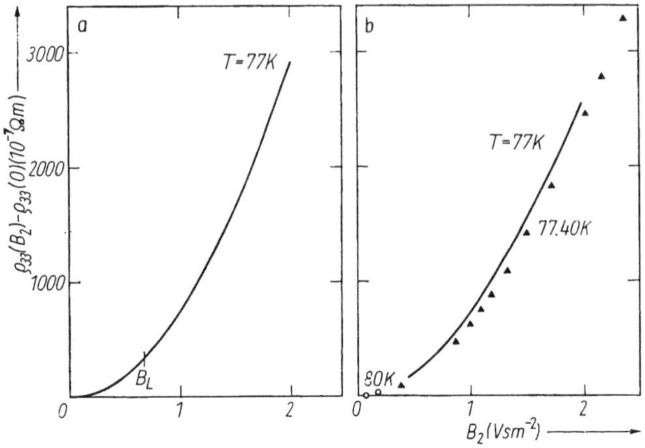

Fig. 6. Experimental data at 77.40 K (▲) reproduced from Shubnikov and de Haas [16], Table 1; at 77 K (continuous curve) from Bogod et al. [12], Fig. 3, curve 1; at 80 K (○) from Abeles and Meiboom [13], Fig. 2

hole mobilities, ν_3. This mobility could not be determined by the minimisation mentioned in Section 2, but was found from ν_1 and the mass ratio of the holes.

For the elements $\varrho_{ii}(B_i)$ the observation of an "anomalous longitudinal magnetoresistance", which behaves as a decreasing function of magnetic field, has been reported. This phenomenon was analysed experimentally as well as theoretically by Yoshida [9]. Through this analysis it was convincingly shown that the "anomalous" behavior originates from the use of current electrodes with small (point-like) contact areas, giving rise to an inhomogeneity of currents flowing through the samples. Obviously, such a behavior is not accounted for in the present calculations.

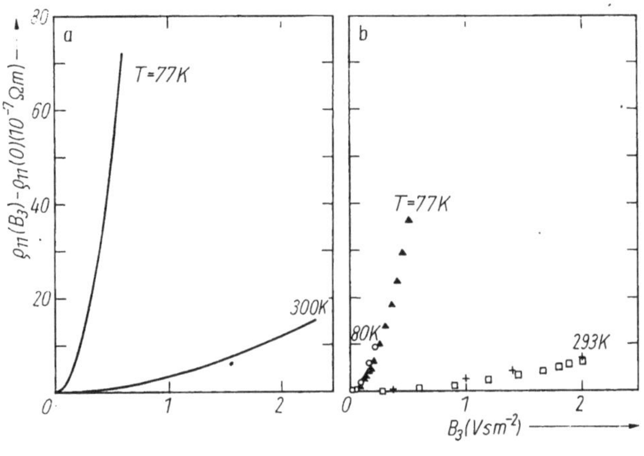

Fig. 7. Experimental data at 77 K reproduced from Saunders and Sümengen [10], Fig. 1; at 80 K from Abeles and Meiboom [13], Fig. 3 (the original figure displays $\varrho_{22}(B_3)$, but according to the $\bar{3}$m symmetry this is equal to $\varrho_{11}(B_3)$); at 293 K from Stierstadt [11], Fig. 6 (+ $\varrho_{11}(B_3)$) and [14], Fig. 8 (□ $\varrho_{22}(B_3)$)

4. Comments on the Papers by Gitsu et al. [5]

Mikhail et al. [3] adjusted the drift mobilities along with the carrier density, to reproduce the weak-field galvanomagnetic data. Contrary to this, Gitsu et al. [5] calculated the electron–phonon scattering probability from deformation potentials and elastic constants. However, as these authors write, they had to allow for an unknown temperature dependence of the deformation potentials, and they did so by adjusting them to reproduce the zero-field resistivities.

In [5] a variational calculation is used to find the transport tensors, i.e. the relaxation time approximation, as expressed by (2), is avoided. In connection with their variational calculation, Gitsu et al. report a problem: In the case of an ellipsoidal, non-parabolic model the convergence of the power series for the trial functions was deteriorated. We think that a possible reason for this deterioration may be traced back to the expression for the scattering probability (see (4.1) in [18]), which seems to be independent of the electron energy. This feature is just the essential one for the pseudo-parabolic model. Apparently it was the problem of deteriorated convergence that led Gitsu et al. to use parabolic dispersion in their final, numerical calculations.

Along with the parabolic dispersion, Gitsu et al. use mass elements which are independent of temperature. This brings up the general problem of temperature dependence of the band structure, as the magnetoreflection measurements of Vecchi and Dresselhaus [8] revealed some of the mass elements to be dependent on temperature. This latter feature is incorporated in the pseudo-parabolic model.

5. Conclusion

For the range of classical magnetic fields and the temperatures considered the transport properties are reasonably well accounted for, both in the pseudo-parabolic model and in the work by Gitsu et al. [5]. However, some basic problems remain. Thus, the pseudo-parabolic model lacks theoretical justification, and the pseudo-parabolic Fermy energy has not had its temperature dependence experimentally verified. As regards the work of Gitsu et al., the L-point energy gap is not taken into account, nor is the temperature dependence of the electron mass elements. Further, the temperature dependence of the deformation potentials is used as an adjustable parameter.

Acknowledgement

This work was in part supported by the cultural agreements between Egypt and Denmark.

References

[1] O. P. Hansen, E. Cheruvier, J.-P. Michenaud, and J.-P. Issi, J. Phys. C 11, 1825 (1978).
[2] J. Heremans and O. P. Hansen, J. Phys. C 12, 3483 (1979).
[3] I. F. I. Mikhail, O. P. Hansen, and H. Nielsen, J. Phys. C 13, 1697 (1980).
[4] J. Heremans and O. P. Hansen, J. Phys. C 16, 4623 (1983).
[5] D. V. Gitsu, I. M. Golban, and V. G. Kantser, phys. stat. sol. (b) 112, 473 (1982); 113, 59, 497 (1982).
[6] B. Lax, J. G. Mavroides, H. J. Zeiger, and R. J. Keyes, Phys. Rev. Letters 5, 241 (1960).
[7] J.-P. Michenaud and J.-P. Issi, J. Phys. C 5, 3061 (1972).
[8] M. P. Vecchi and M. S. Dresselhaus, Phys. Rev. B 10, 771 (1974).
[9] K. Yoshida, J. appl. Phys. 51, 4226 (1980).
[10] G. A. Saunders and Z. Sümengen, J. Phys. F 2, 972 (1972).
[11] O. Stierstadt, Z. Phys. 95, 355 (1935).

[12] YU. A. BOGOD, V. V. EREMENKO, and L. K. CHUBOVA, Zh. eksper. teor. Fiz. **56**, 32 (1969); Soviet Phys. — J. exper. theor. Phys. **29**, 17 (1969).
[13] B. ABELES and S. MEIBOOM, Phys. Rev. **101**, 544 (1956).
[14] O. STIERSTADT, Z. Phys. **85**, 310 (1933).
[15] E. CHERUVIER, Thesis, Laboratoire PCES, Université Catholique de Louvain (Belgium), 1975.
[16] L. SCHUBNIKOW and W. J. DE HAAS, Commun. Phys. Labor. Univ. Leiden **210b**, 22 (1930).
[17] Y. C. AKGÖZ and G. A. SAUNDERS, J. Phys. C **8**, 1387 (1975).
[18] D. V. GITSU, I. M. GOLBAN, and V. G. KANTSER, Preprint, Kishinev 1982 (in Russian).

(Received July 4, 1984)

phys. stat. sol. (b) **126**, 729 (1984)

Subject classification: 2 and 14.1; 21.1; 21.6

International Centre for Theoretical Physics, Trieste[1])

Electrical Resistivity of Liquid Noble Metal Alloys

By

M. A. ALAM[2]) and M. TOMAK[3])

Calculations of the dependence of the electrical resistivity in liquid Ag–Au, Cu–Ag, Cu–Au binary alloys on composition are reported. The structure of the binary alloy is described as a hard-sphere system. A one-parameter local pseudopotential, which incorporates s–d hybridization effects phenomenologically, is employed in the resistivity calculation. A reasonable agreement with experimental trends is observed in cases where experimental information is available.

Berechnungen der Abhängigkeit des elektrischen Widerstandes von der Zusammensetzung werden für flüssige binäre Ag–Au-, Cu–Au-, und Cu–Au-Legierungen durchgeführt. Die Struktur der binären Legierung wird als System harter Kugeln behandelt. Ein einparametriges lokales Pseudopotential, das s–d-Hybridisierung phänomenologisch einschließt, wird zur Widerstandsberechnung benutzt. Es wird eine vernünftige Übereinstimmung mit experimentellen Trends in den Fällen, wo experimentelle Informationen vorhanden sind, beobachtet.

1. Introduction

The pseudopotential theory continues to provide a powerful tool for the study of electronic properties of simple liquid metals. Difficulties arise when the study involves noble, transition, and rare-earth metals. In noble metals Cu, Ag, and Au, narrow d-bands are filled and lie below the Fermi energy. Since the d-bands are not seriously affected by melting, as suggested by experimental studies, pseudopotentials for noble metals have to be energy dependent and non-local if they are to explain fully all the structural and electronic properties. For simple transport properties such as electrical resistivity, however, local, energy independent pseudopotentials seem to give results in reasonable agreement with experiments. It is in this spirit that the present study has been undertaken to calculate the resistivity of noble metal alloys using a single parameter, energy independent, local pseudopotential recently proposed by Idrees et al. [1].

Relatively few studies for liquid noble metal alloys have been undertaken, principally because of the non-availability of experimentally determined partial structure factors. These have therefore to be calculated using a theoretical model. In the present study a hard-sphere model has been used.

The plan of the present paper is as follows. In Section 2, a brief discussion of the model pseudopotential, partial structure factors and the Faber-Ziman theory of electrical resistivity of liquid binary alloys is given. In the following section we present results on the composition dependence of electrical resistivity for Ag–Au, Cu–Ag, and Cu–Au systems, with a brief summary of the conclusions.

---

[1]) P.O.B. 586, 43100 Trieste, Italy.

[2]) Permanent address: Physics Department, Punjab University, Lahore, Pakistan.

[3]) Permanent address: Physics Department, Middle East Technical University, Inonu Bulvari, Ankara, Turkey.

2. The Electrical Resistivity of Noble Metal Binary Alloys

The Faber-Ziman [2] well-known formula for the electrical resistivity of a liquid binary alloy modified by Ashcroft and Langreth [3] can be written as

$$\varrho = \frac{4\pi^3\hbar Z^*}{e^2k_F}\int_0^1 dy\, y^3\{xV_2^2(y)\, S_{22}(y) + 2[x(1-x)]^{1/2}\, V_1(y)\, V_2(y)\, S_{12}(y) + + (1-x)\, V_1^2(y)\, S_{11}(y)\}\,, \tag{1}$$

where $y = q/2k_F$; x concentration of species 2; Z^* effective valence. It requires a knowledge of the electron–ion pseudopotential V_i and the partial structure factor S_{ij}. In the following a brief discussion of each of these factors is given.

2.1 Model pseudopotentials

In their study of electrical resistivity of liquid noble metal binary alloys Ashcroft and Langreth [3] used the Ashcroft empty-core potential

$$V_i(y) = -\frac{\lambda^2 \cos(s_i y)}{\lambda^2 + \lambda^2 f(y)}\,, \tag{2}$$

where $s_i = 2k_F R^i_{\text{core}}$, $\lambda^2 = (\pi a_0 k_F)^{-1}$, and $f(y)$ is the Lindhard dielectric function. In their calculation of the electrical resistivity, they have chosen R_{core} to fit the resistivity of the molten metal at both ends of the concentration. They claim reasonable agreement with the experimental results.

In the present study, a local, energy independent, single parameter pseudopotential recently introduced by Idrees et al. [1] is employed. The potential has the form,

$$V^b(r) = \begin{cases} \dfrac{Z\,e^2(R_m^2 - R_w r)}{R_w^2(R_w - R_m)}\,; & 0 < r < R_m\,, \\[2ex] -2\dfrac{Z\,e^2}{R_w}\,; & R_m < r < R_w\,, \\[2ex] -\dfrac{Ze^2}{r}\,; & R_w < r\,, \end{cases} \tag{3}$$

and is repulsive in the region from $r = 0$ to $r = R_m$ (core radius), attractive and constant in the region $r = R_m$ to $r = R_w = (3\Omega/4\pi)^{1/3}$, the Wigner-Seitz radius and purely Coulombic beyond. The potential is claimed to give an improved account of the s–d mixing which has been introduced phenomenologically. The Fourier transform of (3) is given by

$$V^b(q) = \frac{4\pi}{\Omega q^3}\Big[(AR_m^2 + B - AR_wR_m)(\sin qR_m - qR_m \cos qR_m) - -B(\sin qR_w - qR_w\cos qR_w) - -\frac{AR_w}{q}\big(qR_m \sin qR_m + 2(\cos qR_m - 1)\big) - Z\,e^2\,q\cos qR_w\Big]\,, \tag{4}$$

where

$$A = \frac{Z\,e^2}{R_w^2(R_w - R_m)}\,,$$

$$B = \frac{2Z\,e^2}{R}\,.$$

The parameter R_m is fixed by fitting the theoretical factor (4) to the experimental one or to a better calculation. The electron–ion pseudopotential to be used in (1) is the screened one. The screening has been done with the Vashista and Singwi [4] dielectric function.

2.2 Partial structure factors

For the partial structure factor of the binary alloys, the hard-sphere solution of the Percus-Yevick equation has been used. The analytic expression for these has been given by Ashcroft and Langreth [5].

3. Numerical Results and Conclusions

The input parameters and the model pseudopotentials appropriate for Cu, Ag, and Au are taken from [1], since calculations based on these have been done for electrical resistivities, thermoelectric power, and effective ion–ion interactions for liquid Cu, Ag, and Au. The results seem to be somewhat better than those of other investigations. Thus, the present study may be considered as another test of the proposed pseudopotentials.

The concentration dependence of the calculated electrical restistivity for liquid Cu–Ag, Cu–Au, and Ag–Au systems is shown in Fig. 1a to c.

In the case of Ag–Au and Cu–Au systems, the experimental trend is well produced. At intermediate concentrations the model pseudopotential and the hard-sphere structure factors used seem to be poor as far as the absolute values of the resistivity are concerned. In an earlier study of the Ag–Au system [9], the hard-sphere diameter of the reference system was varied so as to reproduce as closely as possible the experimental structure factors for Ag and Au. The resulting values for the resistivity in the intermediate concentration range were overestimated, the experimental trend

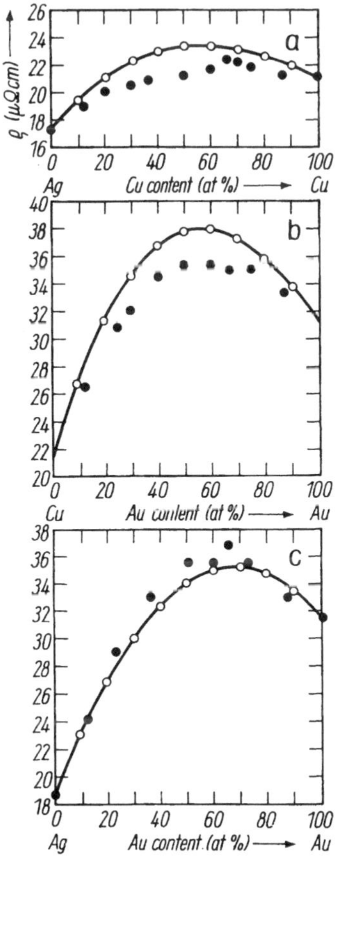

Fig. 1. Concentration dependence of the calculated electrical resistivity for liquid a) Cu–Ag, b) Cu–Au, and Ag–Au systems. ○ Present results, ● Roll and Motz [6]

was, however, well generated. In the present study, the hard-sphere diameter ratio was varied so as to reproduce the experimental values of the resistivity at the two ends of the concentration range. This resulted in the underestimation of the resistivity of the Ag–Au system. To get better numerical agreement with experimental resistivity values [6 to 8] one is forced to use hard-sphere diameters which tend to result in structure factors that are in disagreement with the experimental one [9].

In the present calculation the atomic volume of the alloy is linearly interpolated between the values of the pure components. The Fermi wave number is calculated with this atomic volume. The packing function is taken to be 0.45 for all the three systems. The hard-sphere diameter ratio for Cu–Ag, Cu–Au, and Ag–Au has been taken to be 0.86, 0.86, and 0.99, respectively.

The results of the present study seem to confirm the generalization that the resistivity of binary alloys is a smooth function of concentration at constant temperature. The use of a local, energy independent pseudopotential even in the case of noble metals seems to be reasonable.

Acknowledgements

The authors would like to thank Prof. Abdus Salam, the International Atomic Energy Agency, and UNESCO for hospitality at the International Centre for Theoretical Physics, Trieste.

References

[1] M. Idrees, F. A. Khwaja, and M. S. K. Razmi, Solid State Commun. **41**, 469 (1982).
[2] T. E. Faber and J. M. Ziman, Phil. Mag. **11**, 153 (1965).
[3] N. W. Ashcroft and D. C. Langreth, Phys. Rev. **159**, 500 (1967).
[4] P. Vashista and K. S. Singwi, Phys. Rev. B **6**, 875 (1972).
[5] N. W. Ashcroft and D. C. Langreth, Phys. Rev. **156**, 685 (1967); Erratum: **166**, 934 (1968).
[6] A. Roll and H. Motz, Z. Metallk. **48**, 495 (1957).
[7] R. A. Howe and J. E. Enderby, Phil. Mag. **16**, 467 (1967).
[8] O. Drierach, R. Evans, H. J. Güntherodt, and H. Kunzi, J. Phys. F **2**, 709 (1972).
[9] M. A. Alam and M. Tomak, Solid State Commun. **49**, 891 (1984).

(Received September 3, 1984)

phys. stat. sol. (b) 126, 733 (1984)

Subject classification: 6 and 15; 14.3; 22.1.1

Institut für Festkörperphysik der Universität Hannover[1])

The Thermoelectric Power of p-Ge at Low Temperatures

By

E. KADEN and H.-L. GÜNTER

The thermoelectric power of lightly doped p-Ge is measured in the temperature range of 5 to 300 K for the investigation of phonon scattering. The application of a simple theoretical model requires some knowledge of the dominant carrier scattering processes obtained by measurements of electrical conductivity and Hall effect. In terms of that model, the temperature dependence of the phonon drag part of the thermoelectric power can be attributed to only two phonon scattering processes: phonon–phonon scattering at higher temperatures and phonon–boundary scattering at lower temperatures. The constant phonon relaxation time of boundary scattering and its dependence on sample diameter has been observed at temperatures below 10 K.

Zur Untersuchung von Phononen-Streuprozessen wurde die Thermokraft von niedrig dotiertem p-Ge im Temperaturbereich 5 bis 300 K untersucht. Die Anwendung eines einfachen theoretischen Modells erforderte Informationen über die dominierenden Ladungsträgerstreuprozessse, die aus Messungen von elektrischer Leitfähigkeit und Hall-Effekt gewonnen wurden. Nach diesem Modell kann die Temperaturabhängigkeit des Phonon-drag-Anteils der Thermokraft zwei Phononen-Streuprozessen zugeordnet werden: Phonon-–Phonon-Streuuung bei hohen Temperaturen und Phonon–Grenzflächenstreuung im Bereich tiefer Temperaturen. Die konstante Relaxationszeit der Phononen bei Grenzflächenstreuung und ihre Abhängigkeit von den Probenabmessungen wurden im Temperaturbereich unter 10 K beobachtet.

1. Introduction

In measurements of the thermoelectric power of semiconductors at low temperatures several workers [1, 2] have seen that the values of the thermoelectric power are enormously increased, which cannot be explained by the usual theory of the thermoelectric power. Frederikse [3] and Herring [4, 5] found the explanation of these high values of the thermoelectric power in the so-called phonon-drag effect. In the presence of a temperature gradient the distribution of phonon motions is not isotropic. Therefore the scattering of charge carriers by phonons is not isotropic: the charge carrier are more often pushed by phonons from hot to cold than reverse. The effect depends on the strength of the electron–phonon coupling and on the relaxation time for randomization of the motions of the long-wavelength phonons, with which the carriers interact. Therefore measurements of the phonon-drag part of the thermoelectric power enable one to learn something about electron–phonon interaction and to get information on interaction of phonons with e.g. phonons or boundaries.

The observed thermoelectric power α can be expressed as

$$\alpha = \alpha_e + \alpha_\varphi \,, \tag{1}$$

where α_e is the electronic contribution and α_φ that contribution caused from the phonon-drag effect. In the low temperature range the phonon-drag part α_φ predominates; this enables one to analyse the temperature dependence of scattering processes of phonons.

[1]) Appelstr. 2, D-3000 Hannover 1, FRG.

In the last years investigations of the influence of the phonon-drag effect to the charge transport [6, 7] and of the phonon-drag thermoelectric power of highly doped p-germanium [8] have been reported. At low temperatures scattering at boundaries can dominate all scattering processes of phonons. Up to now the constant relaxation time characteristic of boundary scattering has been observed only in measurements of the thermal conductivity [9].

In this work therefore special attention is given to the influence of boundary scattering of phonons to the thermoelectric power at low temperatures, as this is the predominant scattering process.

2. Phonon-Drag Part of the Thermoelectric Power

The phonon-drag part of the thermoelectric power α_φ depends on the electron–phonon interaction and the scattering process of the phonons. In a non-degenerate semiconductor with simple band structure and with only one type of charge carriers, according to Anselm [10], α_φ is given by

$$\alpha_\varphi = \pm \frac{s^2 g}{T \mu_\varphi} \langle \tilde{\tau}_\varphi \rangle , \tag{2}$$

where s is the velocity of sound of longitudinal acoustic phonons, μ_φ the mobility of carrier exclusively scattering by phonons, $\tilde{\tau}_\varphi$ the relaxion time of phonons averaged over the wave number of phonons,

$$\tilde{\tau}_\varphi(k) = \frac{1}{4k^4} \int\limits_0^{2k} \tau_\varphi(q)\, q^3 \, \mathrm{d}q , \tag{3}$$

$\langle \tau \rangle$ the relaxation time average over the energy ε of the charge carriers,

$$\langle \tau \rangle = \frac{4}{3\pi} \int\limits_0^\infty \tau(x)^{3/2}\, \mathrm{e}^{-x}\, \mathrm{d}x ; \qquad x = \frac{\varepsilon}{k_0 T},$$

g the dimensionless number in the order of 1 defined as

$$g = \frac{\langle \tilde{\tau}_\varphi \tau_\mathrm{e} / \tau_\mathrm{e}^\varphi \rangle}{\langle \tilde{\tau}_\varphi \rangle \, \langle \tau_\mathrm{e} \rangle / \langle \tau_\mathrm{e}^\varphi \rangle} .$$

If the dependence on the wave number and on the energy of the relaxation times of the predominant scattering process is known, g can be calculated. This makes it possible to evaluate the measurements quantitatively. Such a treatment, however, requires data on Hall effect and conductivity additional to those on thermoelectric power.

3. Experimental

For the determination of the dependence of the differential thermoelectric power on temperature measurements of the absolute temperature, the temperature difference and the thermovoltage of the sample are necessary. These measurements have been carried out in the temperature range of 5 to 300 K. The sample has been mounted in an evacuated chamber located in a continuous flow cryostat as shown in Fig. 1.

With this arrangement the gradient of temperature was adjusted, so that the temperature difference did not exceed 1% of the measured temperature. The absolute temperature has been measured by a calibrated Ge resistor and the temperature

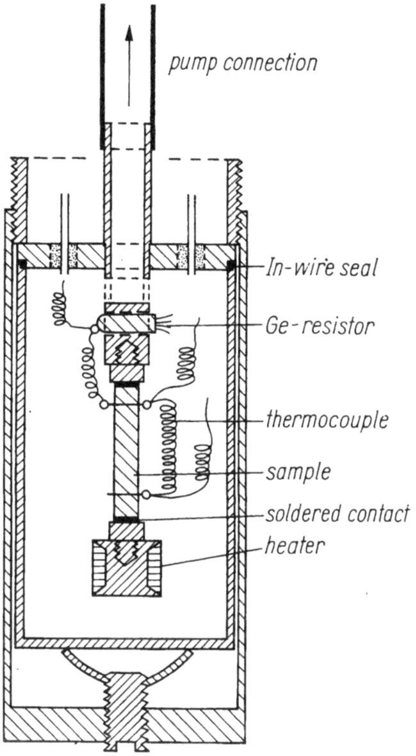

Fig. 1. Sample holder

gradient by an Au–Fe chromel thermocouple with a thermal coefficient between 13 and 20 μV/K in the investigated temperature range. The soldered joints at the end surfaces are used as measurement contacts for the thermoelectric power and as thermal contacts between the sample and the cooled chamber and the heater, respectively.

A special problem has been to achieve an extremely high electrical insulation and at the same time a sufficiently high thermal conductivity between the thermocouple and the sample. The following technique was found to be suitable: Across the sample two 1 mm deep and 0.3 mm wide grooves have been cut. Into these grooves 0.25 mm thick sheets of silver were glued, using an epoxy with good thermal conductivity but high electrical resistivity. The thermocouple then was soldered to the silver sheets.

The bar-shaped samples were cut from dislocation-free single crystal material with the heat flow perpendicular to the (111) face.

4. Experimental Results

At room temperature the specific resistance of the germanium crystal was 0.2 Ωm. From measurements of conductivity and Hall effect a doping concentration of

$$N_A = 2 \times 10^{20}\ \mathrm{m}^{-3},$$

$$N_D = 6 \times 10^{18}\ \mathrm{m}^{-3}$$

was derived.

Despite the limited strength of the applied magnetic field of 0.6 T it was assumed that the mobility of the carriers is given from the measured data as the ratio of Hall constant and the conductivity. The plot (Fig. 2) shows the typical trend: With decreasing temperature the mobility increases due to lattice scattering, and decreases in the low-temperature range ($T < 10$ K) due to ionized impurity scattering.
Measurements of the differential thermoelectric power have been taken at two samples with different cross-sectional area A and length L, sample 1: $A = 3.8 \times 4.4\ \mathrm{mm}^2$; $L = 20.0$ mm, 2: $A = 1.6 \times 4.0\ \mathrm{mm}^2$; $L = 24.1$ mm.

The results of the measurements (Fig. 3) are almost identical for both samples in the temperature range above 40 K, while in the range below 40 K a difference due to the cross-secting is clearly seen.

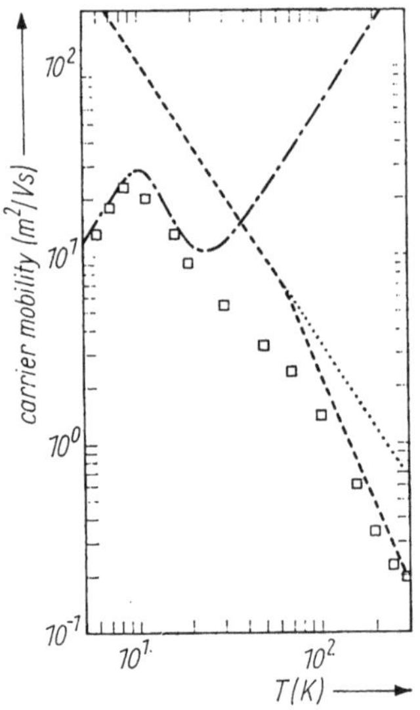

Fig. 2

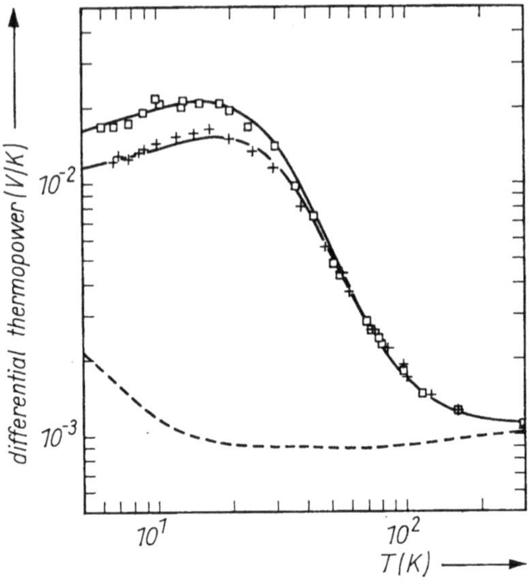

Fig. 3

Fig. 2. Measurement of charge carrier mobility (squares) with scattering asymptotes for scattering by acoustic and optical phonons (— — —), scattering only by acoustic phonons (· · · · · ·) and impurity scattering (—·—)

Fig. 3. Measurement of differential thermopower of a large sample (squares) and a small sample (crosses) combined with calculated curves (———), and the charge carrier part of differential thermopower (— — —)

5. Discussion

According to (1) the phonon drag part $\alpha_\varphi(T)$ can be calculated from the experimental data $\alpha(T)$ by subtracting the charge carrier part α_e. The charge carrier part is given by [5, 10]

$$\alpha_e = \frac{k}{e}\,(a + \ln N_V R_H\, e)\,, \tag{4}$$

where N_V is the effective density of states of the valence band, R_H the Hall coefficient, a a dimensionless number, nearly independent of temperature ($1.7 < a < 4$ for $250\ \text{K} > T > 5\ \text{K}$).

The temperature dependence of α_e is shown in Fig. 3. The difference between experimental data α and the calculated carrier part α_e is the phonon drag part α_φ shown in Fig. 4.

The slope of α_φ can be described by $\alpha_\varphi \sim T^{0.5}$ in the low-temperature range and by $\alpha_\varphi \sim T^{-2.5}$ for high temperatures. In order to describe these results, the temperature dependence of the lattice mobility of charge carriers μ_φ has to be known according to (2).

For the scattering of carriers by acoustic phonons from theory follows:

$$\mu_\varphi \sim T^{-3/2}\,. \tag{5}$$

The temperature dependence of the lattice mobility, however, cannot be measured directly. For temperatures below 50 K the scattering of charge carriers by ionized impurities dominates the mobility. For higher temperatures an exponent of —2.33 is found. This behaviour is attributed to scattering by optical phonons. Since optical

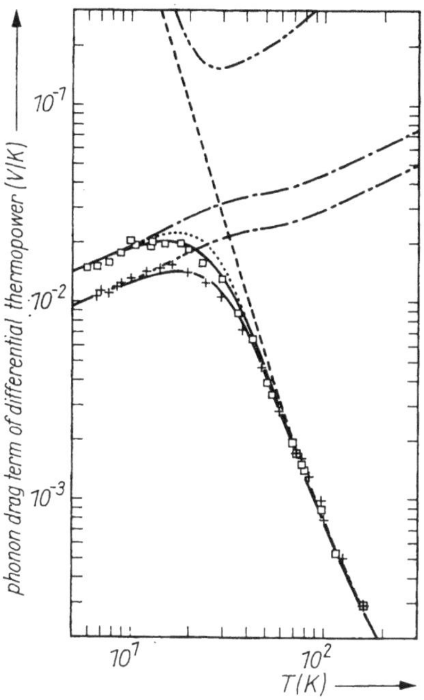

Fig. 4. Measurement of phonon-drag part of differential thermopower of the large sample (squares) and the small sample (crosses) with scattering asymptotes for charge carrier scattering (—··—), boundary scattering (—·—), phonon–phonon scattering (— — —), and their reciprocal sum with (———), and without (·····) carrier scattering

phonons are not excited at low temperatures, the lattice mobility shows the temperature dependence as given by (5). In accordance with [11] an estimation of

$$\mu_\varphi = 3.4 \times 10^3 \times T^{-1.5}\ \mathrm{m^2/Vs} \tag{6}$$

is assumed. The temperature dependence including acoustic and optical phonons is shown in Fig. 2 (dashed line, μ_φ). Because of the negligibly small sound velocity there are no optical phonons involved in the phonon drag. One therefore has to use also in the high-temperature range the dependence of the mobility as given in (5).

This leads to the final description for the temperature dependence of the phonon-drag part as

$$\alpha_\varphi \approx T^{1/2}\,\langle \tilde{\tau}_\varphi \rangle\,. \tag{7}$$

The temperature dependence of the relaxation time τ_φ within the phonon–phonon scattering region can be described by the Herring formula [12]

$$\langle \tilde{\tau}_\varphi^\varphi \rangle \sim T^{-4+\gamma}\,, \tag{8}$$

γ varies from $\gamma = 0$ for $T \ll \theta$ to $\gamma = 2$ for $T = \theta$ (θ is the Debye temperature, $\theta_{\mathrm{Ge}} \approx 360$ K).

From our measurements (Fig. 4) a temperature exponent -2.5 ($\gamma = 1$) was found in agreement with results of Mochan et al. [13] and Herring et al. [14].

For scattering of phonons by boundaries the relaxation time depends only on the dimension of the crystal. Therefore, the temperature behaviour can be described by

$$\alpha_\varphi^{\mathrm{b}} \sim T^{1/2} \tag{9}$$

in almost quantitative agreement with our measured data shown in Fig. 4.

The effect of scattering of phonons by carriers on the phonon-drag term is quite small because of the very low doping of the samples and might therefore be neglected. Calculations [15] show that there is only a noticeable influence between 20 and 40 K as can be seen in Fig. 4.

A quantitative evaluation of the corresponding equations shows that other scattering processes are negligible [15]. Impurity scattering occurs only in very highly doped material. Phonons may also be scattered by isotopes, however, this process has no influence on the thermoelectric power despite the fact that it has noticeable effect on the thermal conductivity. The conservation of energy and momentum requires, that a phonon with a wave vector $\boldsymbol{q}$ can interact with electrons having a wave vector $\boldsymbol{k}$ only if $q < 2k$. These are long-wavelength phonons. Scattering by isotopes, however, is a more important process for short-wavelength phonons. Therefore, scattering by isotopes affects the thermal conductivity [9], which depends on scattering processes of all phonons, but it does not influence the phonon drag effect.

The temperature dependence of the thermoelectric power in Fig. 4 thus is completely described by phonon–phonon scattering and phonon–boundary scattering. In order to carry out a quantitative evaluation of the thermoelectric power, the factor g in (2) has still to be calculated. In the range $T > 100$ K τ_e is approximately equal to τ_e^φ (only phonon scattering), thus it follows from (3) that $g = 1$. For $T < 10$ K $\tau_e \approx \tau_e^i$ (scattering by ionized impurities) and τ_φ^b (boundary scattering) is valid. Calculations [15] yield: $g = 1.46$. Thus for T between 10 and 100 K g ranges between these values.

The two main scattering processes of the phonons can be separated by asymptotic extrapolations for the corresponding branches of the measured curve. The asymptotes can be described by

$$\alpha_\varphi^b = \frac{s^2 g}{T\mu^\varphi} \langle\tilde{\tau}_\varphi^b\rangle \sim T^{1/2} \quad (\text{for } g = \text{const}) \tag{10}$$

and

$$\alpha_\varphi^\varphi = \frac{s^2 g}{T\mu^\varphi} \langle\tilde{\tau}_\varphi^\varphi\rangle \,. \tag{11}$$

A superposition for low and high temperatures, respectively, of these two scattering processes is approximately given by

$$\frac{1}{\alpha_\varphi} = \frac{1}{\alpha_\varphi^\varphi} + \frac{1}{\alpha_\varphi^b}. \tag{12}$$

In Fig. 4 the scattering asymptotes are calculated premissing the following discussion. In (11) and (8) only the temperature dependence is given but no absolute value. In this case an adapting is made in a way that the reciprocal sum of all three scattering asymptotes (solid line) fits the experimental data.

A quantitative comparison between theory and experiment can be done by considering the mean free path of the phonons l_φ in the range of boundary scattering

$$l_\varphi^b = s \langle\tilde{\tau}_\varphi^b\rangle \,. \tag{13}$$

In an infinitely long sample with a square-shaped cross-section A the mean lifetime of phonons is given by [16]

$$\langle\tau_\varphi^b\rangle \approx 1.1 \frac{\sqrt{A}}{s}. \tag{14}$$

In a sample with any other cross-section and a finite length L $\langle\tau_\varphi^b\rangle$ can be approximately described by

$$\langle\tau_\varphi^b\rangle = \frac{d'}{s}\frac{2}{\pi}\left(\frac{\pi}{2} - \arctan\frac{2d'}{L} - \frac{L}{2d'} \ln \operatorname{cosarctan} \frac{2d'}{L}\right), \tag{15}$$

$$d' = 1.1 \sqrt{A} \,.$$

From the measurements of the thermoelectric power we find a mean free path for the two samples of $l_{\varphi,1} = 2.7$ mm, $l_{\varphi,2} = 1.9$ mm.

This agrees very well with the theoretical values $\bar{l}_{\varphi,1} = 3.9$ mm, $\bar{l}_{\varphi,2} = 2.6$ mm calculated by using (15) and support the validity of the model describing the boundary scattering. Besides other possible explanations for the small difference between $\bar{l}_\varphi$

and l_φ the uncertainty of the constant factor in (6) seems to be the most probable one. To fit our values a constant factor of 4.7×10^{-3} is necessary. Measurements of this factor [11] have only asymptotic character, because the true value is always veiled by impurity scattering. Any higher value than the factor in (6) fits the measurements of [11]. The influence on carrier mobility in Fig. 2 is negligible on account of the dominance of impurity scattering and optical phonons. The calculated curves in Fig. 4 based on the equation

$$\mu_\varphi = 4.7 \times 10^{-3} \times T^{-3/2}$$

are given to have a comparison with experimental data.

6. Conclusions

The thermoelectric power of lightly doped p-Ge has been investigated in the temperature range $5\text{ K} < T < 300\text{ K}$ with the following results: At high temperatures the measurements are in accordance with the results of other groups [1, 2, 13]. In the low-temperature range, for the first time the temperature dependence of the phonon-drag term $\alpha_\varphi \sim T^{1/2}$ due to boundary scattering of phonons is observed. These results can be described by a relative simple theoretical model. Using additional data from measurements of Hall effect and conductivity, a quantitative comparison between theory and experimental data is possible.

In particular, it is shown that the small influence of scattering by charge carriers can be omitted and the observed temperature dependence can be explained with the influence of only two scattering processes: phonon scattering by phonons and by boundaries.

As these two processes are independent of doping, the study of other scattering processes (e.g. scattering by dislocations or charge carriers) might only be possible after establishing some knowledge of these two scattering processes.

Furtheron measurements of the thermoelectric power make it possible to determine the influence of scattering by acoustic phonons on the mobility of charge carriers. This might be important, as direct measurements of the mobility allow only an asymptotic determination. It should also be mentioned that measurements on n-Ge at temperatures of nearly 100 K exhibit nearly the same value of this mobility as for p-Ge [6, 10, 11]. Unfortunately this has not yet been verified experimentally in the temperature range around 10 K, where boundary scattering dominates and a quantitative comparison between n-Ge and p-Ge is much easier.

Acknowledgement

The authors wish to thank Prof. Dr. M. Henzler for his interest in the present work and many helpful discussions.

References

[1] H. P. R. Frederikse, Phys. Rev. **92**, 248 (1953).
[2] H. Geballe and G. W. Hull, Phys. Rev. **94**, 1134 (1954).
[3] H. P. R. Frederikse, Phys. Rev. **91**, 491 (1953).
[4] C. Herring, Phys. Rev. **92**, 257 (1953).
[5] C. Herring, Phys. Rev. **96**, 1163 (1954).
[6] N. Perrin and H. Budd, Phys. Rev. B **9**, 3454 (1974).
[7] P. Kocevar, in: Phonon Scattering in Solids, L. J. Challis, V. W. Rampton, and A. F. G. Wyatt, Plenum Press, New York 1976 (p. 299).
[8] K. S. Dubey, phys. stat. sol. (b) **100**, 671 (1980).
[9] T. H. Geballe and G. W. Hull, Phys. Rev. **110**, 773 (1958).

[10] A. I. ANSELM, Einführung in die Halbleitertheorie, Akademie-Verlag, Berlin 1964.
[11] D. M. BROWN and R. BRAY, Phys. Rev. 127, 1593 (1962).
[12] C. HERRING, in: Halbleiter und Phosphore, Ed. M. SCHÖN and H. WELKER, Vieweg, Braunschweig 1958.
[13] I. V. MOCHAN, Y. N. OBRAZTSOV, and T. V. SMIRNOVA, Soviet Phys. – Solid State 1, 1239 (1960).
[14] C. HERRING, T. H. GEBALLE, and J. E. KUNZLER, Phys. Rev. 111, 36 (1958).
[15] E. KADEN, Dissertation, Universität Hannover, 1979.
[16] R. BERMAN, F. E. SIMON, F. R. S. ZIMAN and J. M. ZIMAN, Proc. Roy. Soc. A 220, 171 (1953).

(Received August 27, 1984)

phys. stat. sol. (b) 126, 741 (1984)

Subject classification: 14.1; 13; 21; 21.1

Sektion Mathematik/Physik der Pädagogischen Hochschule „N. K. Krupskaja", Halle[1])

Effective Medium Approach for the Residual Resistivity of Alloys Containing Guinier-Preston Zones

By

U. H. GLÄSER and H. LÖFFLER

Different models for describing the residual resistivity of alloys with precipitates are briefly reviewed. The microscopic theories are able to reproduce the initial increase in the resistivity appearing during the ageing in some alloys forming Guinier-Preston zones but fail in the large-zone limit, whereas just the opposite situation is known for the macroscopic effective medium theories. The classical Maxwell-Garnett theory for inhomogeneous materials is extended to particle sizes comparable or smaller than the electron mean free path. Regarding only the inner part of a GP zone as a macroscopic inclusion and treating the outer part microscopically as a part of the matrix one obtains a model for the resistivity containing only two essential parameters: The extra resistivity due to coherent scattering and the mean free path of the electrons inside the zones. The values of the parameters are derived from the electronic structure of the alloy. The model is found to be in a good agreement with experimental data for an Al–5 at% Zn alloy.

Es werden verschiedene Modelle zur Beschreibung des Restwiderstandes ausscheidungsbildender Legierungen kurz diskutiert. Die mikroskopischen Theorien können zwar den zu Beginn der Alterung bei einigen Legierungen mit Guinier-Preston-Zonen auftretenden Widerstandsanstieg erklären, sie versagen aber bei großen Ausscheidungen. Die umgekehrte Situation liegt dagegen bei den makroskopischen Effektiv-Medium-Theorien vor. Die klassische Maxwell-Garnett-Theorie für heterogene Systeme wird für den Fall erweitert, daß die Teilchenabmessungen vergleichbar sind mit der mittleren freien Weglänge der Elektronen. Dabei wird nur das Innere der GPZ als makroskopischer Einschluß betrachtet, während der Rand mikroskopisch und als Bestandteil der Matrix behandelt wird. Man erhält ein Modell, in dem nur zwei wesentliche Parameter auftreten: Der Zusatzwiderstand hervorgerufen durch kohärente Streuung und die mittlere freie Weglänge der Elektronen innerhalb der Zonen. Die Werte beider Parameter werden aus der Elektronenstruktur der Legierung abgeleitet. Am Beispiel einer Al–5 At% Zn-Legierung wird gezeigt, daß sich das Modell in guter Übereinstimmung mit dem Experiment befindet.

1. Introduction

It is well known from experiments that clustering and decomposition processes in supersaturated alloys are accompanied by characteristic changes in the physical properties. This is widely used in material sciences and technology in order to control the hardening of aluminium alloys. In the following we refer to AlZn alloys with zinc contents between about 3 and 20 at%. After homogenization and subsequent quenching of the samples to a temperature T_a situated below the upper temperature limit of rapid homogeneous nucleation the decomposition of the supersaturated alloy proceeds by the formation of Guinier-Preston zones (GP zones).

In the case of AlZn the GP zones are nearly spherical particles with a large Zn concentration of about 70 at%. They are fully coherent with the Al f.c.c. lattice. Both the volume fraction of the precipitates and their mean diameter D increase during the ageing process.

---

[1]) Kröllwitzer Str. 44, DDR-4020 Halle, GDR.

This microstructure of the alloy and its kinetics can be studied by means of small-angle X-ray scattering, transmission electron microscopy, or microcalorimetry (see, e.g., [1, 2]). A supplementary and comparably simple experimental tool is the measurement of the electrical resistivity. But this technique does not yield direct information about the microstructure, thus a detailed knowledge about the transport mechanisms in these alloys is necessary. The situation is by far not clear today. Not exceeding $T_a \approx 400$ K in AlZn for example the resistivity passes a maximum during the isothermal ageing. This "anomalous" behaviour can be independently explained by a couple of competing models which are briefly reviewed and discussed in Section 2. As a consequence of this analysis a simple macroscopic model with only two essential parameters is proposed and checked in Section 3. Section 4 links the parameters to the electronic structure of the alloy and thus proves that the model is more than an interpolation formula. Some conclusions are drawn in Section 5, especially the applicability to other Al alloys is discussed.

2. Existing Theoretical Descriptions

We want to discuss in this section the established methods for describing the contribution of clustering to the residual resistivity of alloys. Methods yielding only general relations between ordering and resistivity but no quantitative data for GP zones in Al alloys are not considered here. But it is important to know that the appearance of a resistivity maximum is by far not "anomalous" [3].

The nature of the subject, a rather smooth resistivity curve versus the ageing time with at best only one broad maximum and no other structures, makes it hard to decide experimentally between different theoretical models. So it is not surprising that we find in the literature a wide variety of different and often mutually excluding models. This has caused some confusion in the question for the dominating mechanisms.

With the only exception of that in Section 2.5 all models discussed here are microscopic ones basing on the relaxation time solution of the linearized Boltzmann equation

$$\varrho = \frac{\Omega_0 m}{Z\,\mathrm{e}^2}\,\frac{1}{\langle\tau\rangle} \tag{1}$$

with

$$\frac{1}{\tau(\boldsymbol{k})} = \frac{N\Omega_0}{4\pi^2\hbar} \oiint_{S_\mathrm{F}} \frac{\mathrm{d}S'}{|\mathrm{grad}\,\varepsilon_{\boldsymbol{k}'}|}\,|T_{\boldsymbol{k},\boldsymbol{k}'}|^2\,(1 - \cos\vartheta)\,, \tag{2}$$

where ϑ is the scattering angle between $\boldsymbol{k}$ and $\boldsymbol{k}'$ and $\langle\tau\rangle$ indicates averaging of the relaxation time on the Fermi surface. All other symbols have their usual meaning. Equations (1) and (2) origin in the well-known Ziman-Faber theory for the resistivity. The approximations involved and further details can be found for example in [4]. The t-matrix T of the system is written in single-site approximation as the product of the structure factor S and the t-matrix t of a single impurity

$$T_{\boldsymbol{k},\boldsymbol{k}'} = S(\boldsymbol{k} - \boldsymbol{k}')\,t(\boldsymbol{k} - \boldsymbol{k}')\,. \tag{3}$$

The models to be discussed in the following Sections 2.1 to 2.4 only differ with respect to the approximations used for evaluating (1) to (3).

2.1 Isotropic interference

For randomly distributed impurities, i.e. for the solid solution, the scattering of the conduction electrons caused by the impurities will be diffusely spread over the Fermi surface. The same holds in the presence of small clusters when the Bragg-like maxima of the structure factor in (3) are broad. Then the relaxation time $\tau(\boldsymbol{k})$ will be nearly isotropic in $\boldsymbol{k}$ space and consequently the averaging in (1) can be replaced by

$$\frac{1}{\langle\tau\rangle} \approx \left\langle\frac{1}{\tau}\right\rangle. \tag{4}$$

Assuming now a free electron host material, (1) to (4) turn into the Ziman equation for the electrical resistivity

$$\varrho = Ng \int_0^2 \mathrm{d}x\, x^3 \langle|S(\boldsymbol{q})|^2\rangle\, |t(q)|^2 \tag{5}$$

with $k_{\mathrm{F}}x = q = |\boldsymbol{k} - \boldsymbol{k}'|$ and $g = \Omega_0^2 m^2 k_{\mathrm{F}}/Z\, e^2\, 4\pi^2\hbar^3$.

The structure factor

$$|S(\boldsymbol{q})|^2 = (1/N^2) \sum_m \sum_n \exp\left(i\boldsymbol{q}\cdot(\boldsymbol{R}_m - \boldsymbol{R}_n)\right) \tag{6}$$

has now to be averaged over all directions of $\boldsymbol{q}$ or over all cluster orientations. The summations in (6) run over all lattice sites being occupied by solute atoms. The averaging yields

$$\langle|S(\boldsymbol{q})|^2\rangle = \frac{c_{\mathrm{imp}}}{N} \sum_{d\geqq 0} n(d)\, \mathrm{j}_0(qd)\,, \tag{7}$$

where $n(d)\, c_{\mathrm{imp}}/N$ is the number of impurity pairs with a distance d in the alloy.

Evaluating (5) and (7) for a single GP zone with a diameter D we obtain a relative extra resistivity (compared with the residual resistivity of the corresponding solid solution $\varrho_0 = c_{\mathrm{imp}}\varrho_{\mathrm{imp}} = c_{\mathrm{imp}}gI(0)$) of

$$\varkappa(D) = \frac{\varrho(D) - \varrho_0}{\varrho_0} = \sum_{d>0} n^D(d)\, I(d)/I(0)\,. \tag{8}$$

This extra resistivity is a result of coherent scattering due to clustering.

The interference integrals

$$I(d) = g \int_0^2 \mathrm{d}x\, x^3 \mathrm{j}_0(k_{\mathrm{F}}xd)\, |t(k_{\mathrm{F}}x)|^2 \tag{9}$$

were first given by Beal and Friedel [5] for impurity pairs and in first-order Born approximation instead of the t-matrix in (9). Rossiter and Wells [6] used the same analysis in their calculation of the resistivity increase for impurity clusters. However, contrary to their view the averaging procedure (7) yields always and independent of the shape of the clusters the spherical Bessel function $\mathrm{j}_0(qd) = \sin(qd)/(qd)$ provided that the zones are randomly oriented. The shape of the zones is entering the resistivity only via the pair distribution function $n(d)$. The interference of coherently scattered electrons is also the basis for the initial increase of the resistivity in the Bragg-scattering model of Hillel [8], although the analysis described above is not carried out explicitly.

It can be seen from (8) and (9) that the initial change in the resistivity during ageing depends on the values of the interference integrals for the first few co-ordination shells. Because of the oscillating spherical Bessel function, however, the interference integrals are sensitively influenced by the t-matrix $t(q)$, by the Fermi momentum k_{F}, and by the lattice constant within the clusters as it was pointed out in [7]. It is therefore perilous to apply first-order Born approximation in replacing the t-matrix in (9) by the matrix element of the pseudopotential difference.

Due to the rapidly decaying oscillations of the interference integrals (9) for larger distances d the relative extra resistivity tends already for comparatively small clusters to an asymptotic value. Thus the isotropic interference is to be completed by the models described in the following in order to yield a resistivity maximum.

Finally it should be emphasized that clustering can affect either an increase or a decrease in resistivity depending on the given alloy. Of course the first case is of greater interest.

2.2 Bragg scattering

With increasing zone size additional mechanisms become more and more important. One of them is the increasing anisotropy of the scattering keeping approximation (4) for the averaged relaxation time no longer valid. The scattered electrons are concentrated within Bragg-like peaks but only those intersecting the Fermi surface are still contributing to the resistivity. The intersections become smaller with increasing zone size leading to a decreasing resistivity.

This mechanism was intensively studied by Hillel, Edwards, and Wilkes [8 to 11] and will therefore be referred to as HEW mechanism. The width of a Bragg peak in $\boldsymbol{k}$-space is of the order R^{-1} of the cluster radius R. Being Δk the mean separation of the Fermi surface from the Brillouin zone boundaries, the HEW mechanism begins to operate when the cluster radii exceed values in the order of

$$R_{\text{HEW}} \approx (\Delta k)^{-1} \tag{10}$$

[12], producing a resistivity maximum if the initial change had been positive. Parametrizing the Bragg scattering on the Fermi surface leads to very handy formulas [9 to 11] which are often used for evaluating experimental data.

There are also detailed calculations including the band structure of the host and the full $\boldsymbol{k}$-dependent relaxation time [13]. With regard to the scattering potential used there and in view of the fact that other mechanisms are important too, however, the reliability of such calculations seems to be doubtful despite the considerable effort.

2.3 Attenuation of the incident wave

The second important effect is multiple scattering. It was mainly investigated by Rossiter and Wells [6, 14, 15] and will therefore be referred to as RW mechanism. Multiple scattering diminishes the number of incident electrons that reach the inner parts of the clusters. For large zones the scattering reduces to surface scattering only resulting in a decrease in the resistivity. The RW mechanism begins to operate when the cluster radii exceed values in the order of

$$R_{\text{RW}} \approx \lambda_{\text{int}} \,. \tag{11}$$

The mean free path λ_{int} within a cluster is much shorter than that of the overall alloy. There are various reasonable aspects for calculating λ_{int}. We will return to this point once more in Section 4.

The simplest way to take into account multiple scattering is to introduce a damping factor into the structure factor (6). Besides attenuation the damping also prevents coherent scattering from impurities with a separation larger than λ_{int} so that the HEW mechanism is something depressed. Hillel and Rossiter [12] pointed out that both the HEW and the RW mechanisms act in the same direction and moreover are of the same order of magnitude.

2.4 Scattering by compact zones

An alternative model is to consider the GP zones not as accumulations of individual impurity atoms but as compact structures with a mean electron density and crystal potential being different from that of the matrix. In an early work Gerold and Merz [16] regarded the GP zones as spherically shaped, square-well potentials. They obtained indeed a resistivity maximum, but unphysically large potentials were necessary to reproduce the experimental values, or in other words: The scattering at the interface between zones and matrix does not give a significant contribution to the resistivity. Furthermore only the resistivity of the GP zones was calculated, but taking into account the resistivity of the depleting matrix one finds always a resulting monotonous decrease of the overall resistivity. The latter difficulty appears also if the sum in the structure factor (6) is replaced by the corresponding integral [17].

The reason for this failure of the compact zone models is the following: The scattering potential of one impurity incorporated into a cluster can be approximately smeared out over the whole Wigner-Seitz cell if the directionally averaged structure factor is of the order of $+1$,

$$\langle \exp (iqr_{\mathrm{WS}}) \rangle = \frac{\sin (k_{\mathrm{F}} r_{\mathrm{WS}} x)}{k_{\mathrm{F}} r_{\mathrm{WS}} x} \approx 1 . \tag{12}$$

Inserting the parameters for Al, (12) is fulfilled up to about $x = q/k_{\mathrm{F}} = 0.8$. But in the case of the resistivity the main contribution comes from large-angle scattering ($x \lesseqgtr 2$) so that the replacement of the individual scatterers by a compact zone leads to large errors.

However, despite this failure the difference in k_{F} between zones and matrix has a large influence on the resistivity via the interference integrals (9) [7].

2.5 Effective medium theories

At the final stages of ageing, i.e. for large zone sizes the alloy can be regarded as a two-component heterogeneous system. The resulting resistivity ϱ of a material with a resistivity ϱ_{m} containing macroscopic spherical particles with a resistivity ϱ_{G} is given by Maxwell-Garnett [18],

$$\varrho = \varrho_{\mathrm{m}} \frac{1 - va}{1 + 2va} \quad \text{with} \quad a = \frac{\varrho_{\mathrm{m}} - \varrho_{\mathrm{G}}}{\varrho_{\mathrm{m}} + 2\varrho_{\mathrm{G}}} . \tag{13}$$

The resistivity only depends on the volume fraction v of the enclosements and not on their size. This is caused by the absence of a characteristic length in the classical theory. Equation (13) only applies for large GP zones and is not able to describe a resistivity maximum [19, 20].

Improvements of the theory concerning large volume fractions and nonspherical particle shapes were particularly developed with respect to the optical properties of fine particles [21] but did not change the situation in our field remarkably.

Looking at the generally accepted microscopic models discussed in Sections 2.1 to 2.3, we find that they do not ensure the right asymptotic behaviour (13) for large zones. Furthermore, following classical electrodynamics the current density becomes strongly inhomogeneous in the environment of a GP zone. As the result the current flows around the zones using low-resistivity paths in the matrix. This self-shielding effect of the zones from the current, which diminishes the influence of the other mechanisms discussed, is also not included in any of them. We therefore were encouraged to start from the Maxwell-Garnett theory and modify it in order to describe the resistivity behaviour in the whole GP zone size range.

3. Effective Medium Model

It becomes transparent from Section 2 that the Maxwell-Garnett formula (13) is valid in the case of large inclusions, whereas the microscopic theories have to be applied for small clusters. The question where to draw the boundary between the macroscopic and the microscopic descriptions, has not been answered.

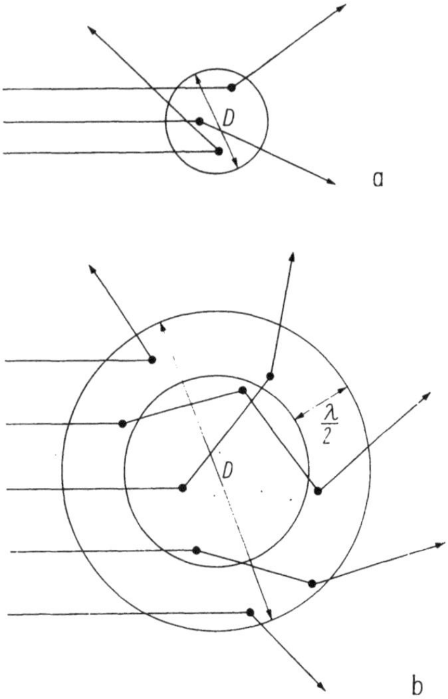

Fig. 1. Scattering processes at GP zones of different sizes. a) For small zones ($D < \lambda$) the electrons coming from the matrix are scattered directly back into the matrix. The GP zone can therefore be treated as a distortion of the matrix. b) With increasing zone size ($D > \lambda$) an increasing core is formed which can be reached only via multiple scattering. This is taken as the criterion for the deployment of an own macroscopic resistivity of the GP zone core

In the case of small clusters the matrix electrons are scattered directly back into the matrix via a single scattering process (Fig. 1a). Thus at this stage the clusters can be treated like single impurities in the Al matrix. Using Matthiessen's rule and (8) we obtain for the resistivity of the alloy containing single solute atoms (c_s) and small clusters ($c_c = c_0 - c_s$),

$$\varrho = c_s \varrho_{Zn} + c_c(1 + \varkappa)\, \varrho_{Zn}\,, \tag{14}$$

ϱ_{Zn} being the residual resistivity of Zn impurities in an Al host. The dependence of $\varkappa$ on the zone diameter D has been neglected for reasons that will become clear in the next section. The situation changes when the cluster diameter D exceeds the mean free path λ of an electron within the material the GP zones are composed of. Then double and later also multiple scattering processes will occur within a GP zone. Obviously this is just the criterion for deploying an own macroscopic resistivity of the clusters. However, the outer part of a GP zone still has the chance to scatter the matrix electrons directly back into the matrix, consequently it is regarded as a part of the matrix (Fig. 1b). Only the core of a GP zone is an inclusion in the spirit of the Maxwell-Garnett formula (13). Following Fig. 1 its volume fraction

$$v = \gamma f, \qquad \gamma = \begin{cases} 0\,; & D < \lambda\,, \\ \left(1 - \dfrac{\lambda}{D}\right)^3; & D > \lambda \end{cases} \tag{15}$$

is different from the volume fraction f of the whole GP zone and in addition is size dependent.

An effective medium approach taking into account the particle size has been developed already to describe the optical properties of dielectrics containing small metallic spheres [22]. Contrary to our situation in that model the electrons cannot escape from the metallic spheres into the surrounding insulator resulting in a change in the plasma frequency of a small particle compared with the bulk material. Thus the "free path effect" is clearly distinguished from our mechanism (15).

According to the considerations above the spherical cores are imbedded in an "effective matrix" consisting of the Al host with the single solute Zn atoms (true matrix) and of the outer parts of the zones. In analogy to (14) we obtain for the residdual resistivity of the effective matrix

$$\varrho_m = \left(c_m(1 - f) + c_G(1 + \varkappa)(f - v)\right) \frac{\varrho_{Zn}}{1 - v}, \tag{16}$$

c_m and c_G being the Zn concentration in the true matrix and in the zones, respectively. Writing for the residual resistivity of the core of a GP zone $\varrho_G = \beta\varrho_{Zn}$ we obtain from (13) with (15) and (16) the following expressions for the relative extra resistivity:

$$\Delta\varrho = \frac{\varrho - \varrho_0}{\varrho_0} = \frac{c_m'}{c_0} \frac{1 - f\gamma a}{1 + 2f\gamma a} - 1\,,$$
$$c_m' = c_0 \frac{1 + (\varkappa - \gamma - \varkappa\gamma) fc_G/c_0}{1 - f\gamma}, \tag{17}$$
$$a = \frac{c_m' - \beta}{c_m' + 2\beta}.$$

In order to check the usefulness of (17), as an example, an Al–5 at% Zn alloy is considered which distinctly shows maxima of resistivity in its early stages of ageing when isothermally aged in the room temperature region (see, e.g., [1]). In order to simplify matters a monodisperse system of spherical zones is considered. (The influences of size distributions as well as deviations from the spherical shape of the GP zones will be treated in the near future.) At room temperature the zinc concentration inside the zones is $c_G \approx 0.7$ (see e.g. [23, 24]) and that of the matrix $c_m \approx 0.02$ [23] in metastable equilibrium with the existing zones. Taking these values and applying the lever rule for the saturation value of the volume fraction f_0 it follows $f_0 \approx 0.9c_0$. The experimentally determined increase of the volume fraction with time follows in many cases the Johnson-Mehl-Avrami formula [25 to 27],

$$f = f_0 \left\{1 - \exp\left(-\left(\frac{t}{\tau}\right)^n\right)\right\}, \tag{18a}$$

with values of n not far from unity [28] and for the growth of the GP zones

$$D = D_0\left(\frac{t}{\tau}\right)^m \tag{18b}$$

frequently is obtained, where m is in the order of about 0.4 [29].

Furthermore, as will be shown in Section 4, $\varkappa = 0.2$ is a reasonable value for Al–Zn alloys. Using these parameters, the course of the resistivity–time curves sensitively depends on the chosen values of D_0 in relation to the mean free path λ of the electrons inside the zones. That can be easily seen from the results presented in Fig. 2b where the course of $\Delta\varrho$ is plotted for four different values of D_0. The case $D(t) \gg \lambda$, $\Delta\varrho$ is monotonously decreasing (see e.g. [17]), reflects the pure macroscopic behaviour observed in Al–Zn alloys when the critical size of the initially nucleated zones exceeds λ (see the dash-dotted line in Fig. 2b). According to Becker's theory of nucleation [30] this is true close to, but below the critical temperature for homogeneous nucleation of GP zones. Notice that a decrease of $\Delta\varrho$ from beginning on can also happen for alloys with $\varkappa < 0$, i.e. when the impurities are less effective scatterers inside the zones than in the matrix.

The other limiting case is that $D(t)$ does not exceed λ (dotted line in Fig. 2b). Independent of the special course of $D = D(t)$ the extra resistivity is proportional to the course of the volume fraction $f = f(t)$. Also this course of the resistivity is often experimentally observed, for instance when the growth of the zones is fairly early

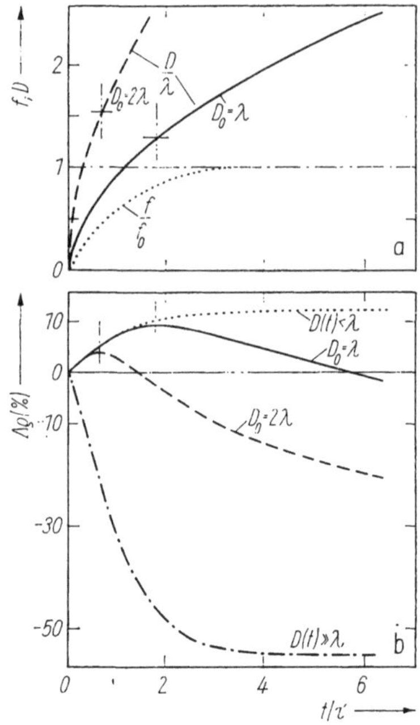

Fig. 2. Results of (18) and (17) for Al–5 at% Zn. a) Supposed ageing kinetics: Time evolution of the volume fraction f and two examples for the growth of the zone diameter D. b) Time evolution of the relative extra resistivity. The curves at the bottom (at the top) hold for an extremely fast (slow) growth of the zones. The two central curves correspond to the kinetics given in Fig. 2a

stopped by lack of movable vacancies. This happens frequently in Al–Zn alloys containing highe radditions of Mg (see, e.g., [28]). In the region between these two limiting cases (see solid and dashed lines in Fig. 2b) the extra resistivity passes a maximum before decreasing if $D(t)$ in the course of ageing sufficiently exceeds λ. As transparent from Fig. 2b the critical value D_{crit} where the maximum of the resistivity appears, depends on D_0 (D_{crit} increases with D_0 as indicated by the vertical marks in the figure). Also in these cases at the beginning of the ageing $\Delta\varrho$ exactly increases proportional to f. After passing the maximum of resistivity $\Delta\varrho$ is decreasing even below its starting value if the zone growth is not stopped before. Also this feature is frequently observed (see e.g. [1]).

Finally we should remark that (17) can be simplified for small Zn concentrations as in our example. Expanding (17) into a power series for c_0 and terminating with the linear term we arrive at

$$\Delta\varrho \approx \left\{\gamma(1 - 3a) + \frac{(\varkappa - \gamma - \varkappa\gamma)\, c_{\text{G}}}{c_0}\right\} f\,. \tag{19}$$

The parameter a depends only weakly on the volume fraction f so that in (19) the f- and the D-dependence (via (15)) are factorized.

4. Microscopic Verification of the Parameters

The model which has been established in the previous section contains the two basic parameters $\varkappa$ and λ. It will be demonstrated now that they can be calculated from the electronic structure of the alloy.

The parameter $\varkappa$ has been defined as the relative extra resistivity per impurity atom of a precipitate. It can be obtained from (8) and (9). Within a partial wave expansion the t-matrix of a single scatterer is expressed in terms of the difference η_l in the phase shifts between impurities and host atoms

$$t_{\boldsymbol{k},\boldsymbol{k}'} = -\frac{2\pi\hbar^2}{\Omega_0 m k_{\text{F}}} \sum_l (2l+1)\sin\eta_l\, \mathrm{e}^{i\eta_l}\, \mathrm{P}_l(z)\,, \tag{20}$$

$\mathrm{P}_l(z) = \mathrm{P}_l(\cos\vartheta)$ being the Legendre polynominal for the angular momentum l and the scattering angle ϑ. Using the relation ([31] p. 540, $x = \sqrt{2 - 2z}$)

$$\mathrm{j}_0(k_{\text{F}} x d) = \sum_l (2l+1)\, \mathrm{j}_l^2(k_{\text{F}} d)\, \mathrm{P}_l(z)\,, \tag{21}$$

equation (9) yields

$$I(d) = \sum_l f_l j_l^2(k_F d) \,,$$

$$f_l = \frac{h(2l+1)}{e^2 Z k_F} \sum_{l'} \sum_{l''} (2l'+1)(2l''+1) \sin \eta_{l'} \sin \eta_{l''} \cos(\eta_{l'} - \eta_{l''}) \, c_{l,l',l''} \,, \tag{22}$$

$$c_{l,l',l''} = \int_{-1}^{1} dz (P_0 - P_1) P_l P_{l'} P_{l''} \,.$$

Inserting into (22) the phase shifts for zinc impurities in aluminium [32], taking into account the difference between the Fermi momenta of zones and matrix and finally evaluating (8) for different GP zone diameters D one arrives at the results given in Fig. 3. We see that for AlZn the extra resistivity approaches its asymptotic value of abou t20% per Zn atom already for very small clusters of about ten atoms. Thus the assumption of a parameter $\varkappa = 0.2$ independently of the zone size is justified, i.e. a Zn atom within a Zn environment gives a contribution to the residual resistivity being 20% larger than it would be isolated in the matrix. Notice that this result has been obtained from first principles without fitting any data. Details of the calculation can be found in [7].

The other problem is to estimate the residual resistivity ϱ_G of the material the GP zones are composed of. ϱ_G determines both the parameter β which is of minor influence on the results of our model and the parameter λ which is responsible for the position of the resistivity maximum. We first evaluate a reasonable value for β. Unfortunately there are no experimental values available for the residual resistivity of a f.c.c. Zn–30 at% Al alloy because it does not exist outside GP zones. The only information is 0.26 μΩcm/at% [13] for the Al side of the concentration range and 0.4 μΩcm/at% [33] for the Zn side. The latter value is of course for Al in h.c.p. Zn. Drawing the corresponding straight lines in a concentration diagram, a rough extrapolation yields about $\varrho_G = 10$ μΩcm and consequently $\beta = 0.4$. This value is sufficiently higher than the phonon contribution even at room temperature so that the temperature at which $\Delta\varrho$ of the decomposing alloy is measured, has no significant influence on the position of the resistivity maximum. Usually the mean free path is calculated from the resistivity via the relation

$$\lambda = \frac{3\pi^2 h}{e^2 k_F^2 \varrho} \,. \tag{23}$$

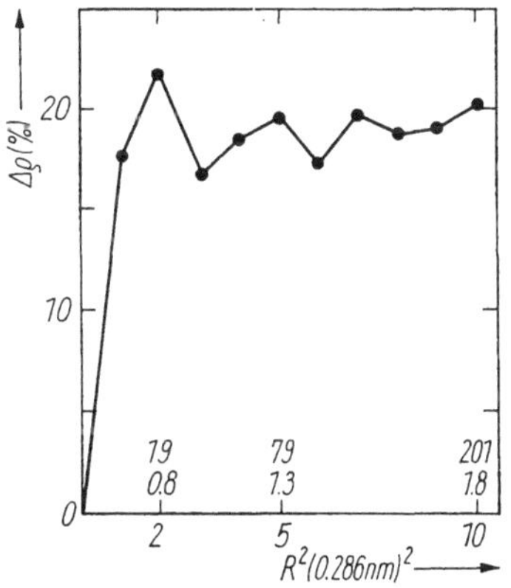

Fig. 3. Relative extra resistivity per Zn atom of a cluster in an Al matrix. The cluster size is given by its diameter in nm (second figure from top), by the number of atoms, (top figure), or by the square of the radius R in units of the Al nearest neighbour distance

In our case we obtain a mean free path inside a GP zone of about 5 nm. Following the arguments of the previous section the resistivity maximum should appear for zone diameters of about 6 nm which is much too large in comparison with the experimental results (1.5 to 2.0 nm). To our opinion the reason for this discrepancy is the definition of the mean free path via (23): The resistivity is dominated by scattering processes of large angles ϑ (see (2)). However, in our case we are interested in the mean free path between the scattering events independently of whether being small- or large-angle scattering.

The omission of the weighting factor $(1 - \cos \vartheta)$ in (2) leads to a higher "resistivity" and via (23) to a lower mean free path. In the low-concentration range this omission enlarges the "resistivity" by a factor of about 2.5. Assuming that this ratio does not change at higher zinc concentrations we arrive at a mean free path $\lambda \approx 2$ nm being close to the experimental estimation of the resistivity maximum.

5. Conclusions

The classical Maxwell-Garnett theory for inhomogeneous materials was extended to small particle sizes. The order of magnitude of the two main parameters $\varkappa$ and λ appearing in our model was estimated from the electronic structure in the case of AlZn alloys. A higher theoretical effort for calculating the parameters more accurately seems to us not very promising because of the many simplifications that had to be made compared with the real technical alloys. For example we assumed a monodispersed system of spherical GP zones the concentration of which is independent of their sizes, zero measuring temperature, no imperfections other than GP zones and single Zn impurities, no anisotropy in the scattering, and free electrons. On the one hand, it should be possible to describe some of the mentioned effects within an improved model, on the other hand, it is easier to take them into account by a suitable adjustment of the two parameters and to use the model as an interpolation scheme.

It is known from experiment (see, e.g., [34]) that the zones are surrounded by coherency stress fields. Of course, these stress fields also affect the resistivity to a certain extent. Furthermore, in cases the precipitates are incoherent with respect to the ambient matrix an additional contribution to the resistivity is given by the particle–matrix interface. Both effects make the GP zones appear larger than in reality. They can be regarded by suitably diminishing the value of λ, thus calculating λ more accurately than in Section 4 does not seem very useful to us.

With respect to its results our model is quite similar to that of Rossiter and Wells discussed in Section 2.3. However, they differ in their basic ideas and in the details. We approach the problem from the classical point of view and therefore obtain automatically the right large zone limit. The model covers the whole concentration range and all ratios between the resistivities of zones and matrix. The extension to ellipsoidal GP zones, which appear for AlZn at later stages of ageing, is straightforward.

Some problems arise when the ellipsoidals are extremely flattened forming layer GP zones as they do in AlCu. Here one would expect that for such thin zones all effects related to the mean free path should be not important [35]. However, the residual resistivity of copper within an aluminium host is about three times larger than that of zinc, consequently the mean free path is of the order of about 0.6 nm only. This value is already very close to the layer thickness. Reminding that the Al matrix around such a Cu layer is highly distorted so that the actual thickness of a GP zone seen by the conduction electrons is larger, we conclude that our model should describe the situation also in AlCu. In the later stages of ageing multilayers are formed, they can be treated as macroscopic particles anyway.

By a collection of the alloys with a resistivity maximum as was given in [20] we are confirmed that a maximum only appears if (i) the parameter $\varkappa$ describing the effect of isotropic interference is positive and (ii) the mean free path λ is comparably large preventing the deployment of an own macroscopic resistivity already for the small clusters in the initial stage of ageing. AlZn (26), AlAg (120), AlCu (80), AlMg (33), and CuBe (62) fulfil this condition whereas CuFe (1490), CuCo (650), CuCr (1990), AlFe (550), AlCr (720), and AlMn (720) independently of the first condition have a mean free path too short to yield a resistivity maximum. The numbers in brackets are the residual resistivities in μΩcm.

Acknowledgements

The authors are indebted to Prof. P. Rennert and Prof. G. Paasch for fruitful discussions.

References

[1] H. Löffler and H.-G. Fabian, phys. stat. sol. (a) **33**, 345 (1976).
[2] H.-G. Fabian, R. Kroggel, H. Löffler, and O. Simmich, phys. stat. sol. (a) **56**, 143 (1979).
[3] J. Richter and G. Schubert, phys. stat. sol. (b) **116**, 597 (1983).
[4] J. M. Ziman, Principles of the Theory of Solids, Cambridge University Press, 1972.
[5] M. T. Beal and J. Friedel, Phys. Rev. **135**, A466 (1964).
[6] P. L. Rossiter and P. Wells, Phil. Mag. **24**, 425 (1971).
[7] U. H. Gläser, phys. stat. sol. (b) **124**, 423 (1984).
[8] A. J. Hillel, Acta metall. **18**, 253 (1970).
[9] A. J. Hillel, J. T. Edwards, and P. Wilkes, Phil. Mag. **32**, 1231 (1975).
[10] J. T. Edwards and A. J. Hillel, Phil. Mag. **35**, 1221 (1977).
[11] A. J. Hillel and J. T. Edwards, Phil. Mag. **35**, 1231 (1977).
[12] A. J. Hillel and P. L. Rossiter, Phil. Mag. B **44**, 383 (1981).
[13] N. Luiggi, J. P. Simon, and P. Guyot, J. Phys. F **10**, 865 (1980).
[14] P. L. Rossiter and P. Wells, J. Phys. C **4**, 354 (1971).
[15] P. L. Rossiter, Phil. Mag. **33**, 1015 (1976).
[16] V. Gerold and W. Merz, Z. Metallk. **53**, 341 (1962).
[17] C. Radomsky and H. Löffler, phys. stat. sol. (a) **54**, 615 (1979).
[18] J. C. Maxwell-Garnett, Phil. Trans. Roy. Soc. (London) **203**, 385 (1904); **205**, 237 (1906).
[19] J. Lašek, Czech. J. Phys. **B27**, 1059 (1977).
[20] J. Merlin and G. Virgier, phys. stat. sol. (a) **58**, 571 (1980).
[21] C. G. Granquist and O. Hunderi, Phys. Rev. B **16**, 3513 (1977).
[22] U. Kreibig and C. V. Fragstein, Z. Phys. **224**, 307 (1969).
[23] V. Gerold and W. Schweizer, Z. Metallk. **52**, 76 (1961).
[24] G. Dlubek, O. Kabisch, O. Brümmer, and H. Löffler, phys. stat. sol. (a) **55**, 509 (1979).
[25] W. Johnson and R. F. Mehl, Trans. AIME **135**, 416 (1939).
[26] M. Avrami, J. chem. Phys. **7**, 1103 (1939).
[27] M. Avrami, J. chem. Phys. **8**, 212 (1940).
[28] W. Oettel, C. Radomsky, and H. Löffler, Kristall und Technik **15**, 713 (1980).
[29] H.-G. Fabian, H. Löffler, R. Kroggel, and B. Gueffroy, Kristall und Technik **15**, 755 (1980).
[30] R. Becker, Z. Metallk. **29**, 245 (1937).
[31] P. Ziesche und G. Lehmann, Ergebnisse in der Elektronentheorie der Metalle, Akademie-Verlag, Berlin 1983.
[32] J. Deutz, P. H. Dederichs, and R. Zeller, J. Phys. F **11**, 1787 (1981).
[33] D. E. Farell, J. H. Tripp, and T. J. Harding, Phys. Rev. B **1**, 4533 (1970).
[34] R. Ramlau and H. Löffler, Crystal Res. Technol. **19**, 1279 (1984).
[35] A. J. Hillel, Phil. Mag. B **48**, 237 (1983).

(Received August 16, 1984)

Erratum

To the paper by U. Walzer

Elastic and Bonding Properties of Transition Metals

phys. stat. sol. (b) **125**, 55 (1984)

the author requests the following rectification:
On page 58, equation (17) read

$$p = k_1 \tfrac{2}{3} \Omega^{-5/3} - k_2 \tfrac{1}{3} \Omega^{-4/3} + \dots$$

instead of

$$p = k_1 \tfrac{2}{3} \Omega^{-5/3} - k_2 \Omega^{-4/3} + \dots .$$

phys. stat. sol. (b) 126, K109 (1984)

Subject classification: 6; 8; 12.1; 21.2

Department of Applied Physics and Mathematics, Mining College, Akita University[1)]

Mode Grüneisen Parameter for Local and Band Modes of Rb-K, Cs-K, and Cs-Rb Alloy Systems

By

T. SOMA and H.-MATSUO KAGAYA

Nearly, we /1/ have presented a simplified treatment to study quantitatively the lattice dynamics of alkali solid solutions. Considering the volume effect on the lattice dynamics of the pure constituent, we have obtained the phonon dispersion relations of the local and band modes for $Rb_{0.71}K_{0.29}$, $Rb_{0.3}K_{0.7}$, $Cs_{0.7}K_{0.3}$, $Cs_{0.3}K_{0.7}$, $Cs_{0.7}Rb_{0.3}$, $Cs_{0.3}Rb_{0.7}$ systems, and the concentration x-dependence of the local and band mode-frequencies in the $Rb_{1-x}K_x$, $Cs_{1-x}K_x$, and $Cs_{1-x}Rb_x$ solid solutions. In the present note, we report the concentration dependence of the local and band mode Grüneisen parameters of these alloy systems.

The mode Grüneisen parameter $\gamma_i(q)$ is the measure for the volume Ω-dependence of the i-th mode phonon frequency $\nu_i(q)$ and defined by

$$\gamma_i(q) = -\frac{d\left[\ln \nu_i(q)\right]}{d[\ln \Omega]} \quad . \tag{1}$$

When, for instance, a $Rb_{1-x}K_x$ solid solution is formed, K atoms in the solid solution are in a state of volume expansion compared with those in pure K, and Rb atoms in a state of volume compression compared with those in pure Rb. Considering the volume dependence of phonon frequencies in our treatment /2, 3/ from first principle based on the perturbational treatment and the local Heine-Abarenkov model potential /4/, we obtained the corresponding mode Grüneisen parameter $\gamma_i^{(1 \text{ or } 2)}(q)$ for the local and band modes. In Fig. 1, we show the obtained results at the representative points H, P, and N in the first

1) Akita 010, Japan.

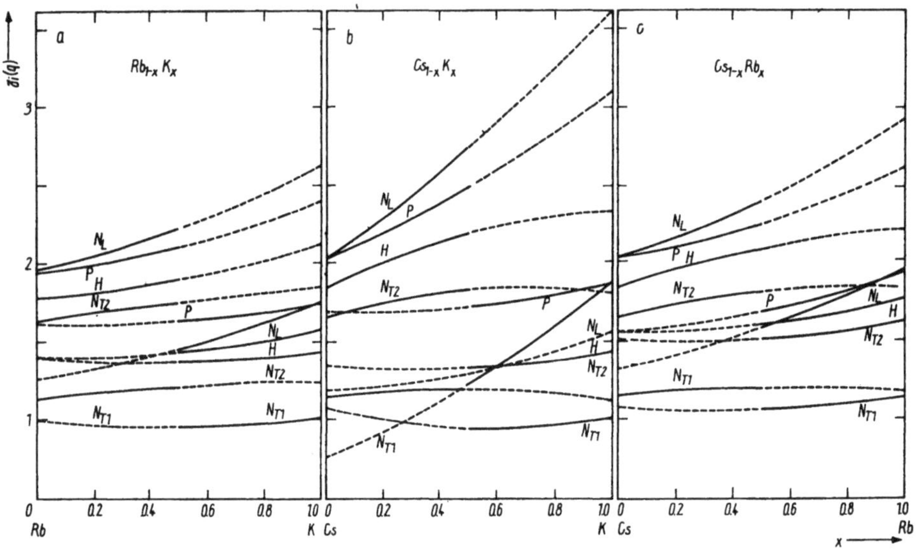

Fig. 1. The concentration x-dependence of the local (dashed) and band (solid) mode Grüneisen parameters $\gamma_i^{(1 \text{ or } 2)}(q)$ at the representative points at H, P, and N for a) $Rb_{1-x}K_x$, b) $Cs_{1-x}K_x$, and c) $Cs_{1-x}Rb_x$ systems

Brillouin zone for the $Rb_{1-x}K_x$, $Cs_{1-x}K_x$, and $Cs_{1-x}Rb_x$ alloys. In Fig. 1, the results with Vashishta-Singwi screening function /4/ are given, and the maximum variation width of $\gamma_i^{(1 \text{ or } 2)}(q)$ using other screening functions /4/ amounts to 0.2 to 0.3.

Over the long-wavelength region $q \to 0$, we cannot calculate the corresponding mode Grüneisen parameter $\gamma_i^{(1 \text{ or } 2)}(q \to 0)$ because of the very small shift of the phonon frequency in (1). In the long-wavelength limit, we obtain the corresponding mode Grüneisen parameters $\gamma_i(q \to 0)$ from the pressure dependence of the elastic stiffness constants dC_j/dP in the elastic theory of a continuum. In the symmetrical directions such as [100], [111], and [110], we obtain

$$\gamma_i(q \to 0) = -\frac{1}{6} + B_T \sum_j b_j \frac{dC_j}{dP} \Big/ \left(2 \sum_j b_j C_j\right) , \tag{2}$$

where b_j, C_j are the linear coefficients and elastic constants for the i-mode in

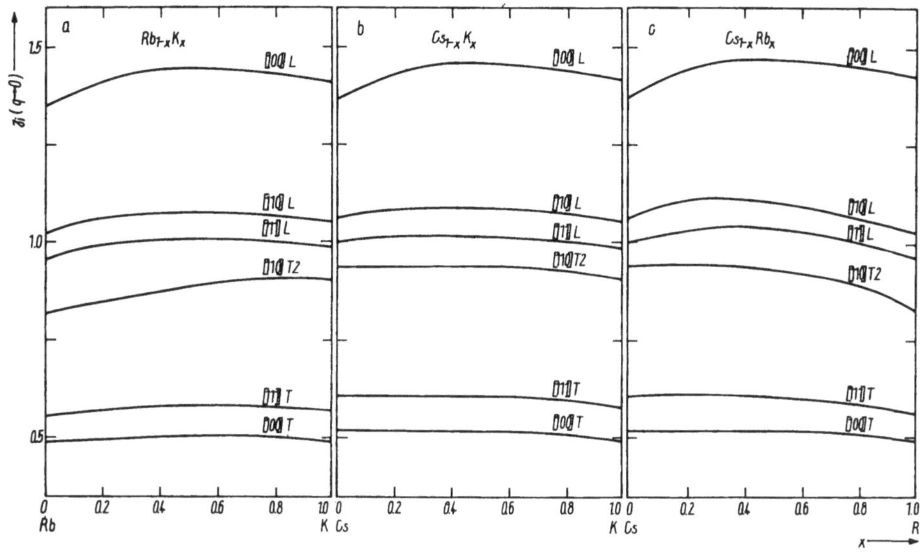

Fig. 2. The concentration x-dependence of the long-wave mode Grüneisen parameters $\gamma_i(q\to0)$ along [100], [111], and [110] directions for a) $Rb_{1-x}K_x$, b) $Cs_{1-x}K_x$, and $Cs_{1-x}Rb_x$ systems

question, and B_T the isothermal bulk modulus. Recently, using the pseudo-alloy atom model and the homogeneous deformation method, we /5/ have obtained theoretically the pressure derivatives dB/dP, dC_{44}/dP, and dC'/dP of $Rb_{1-x}K_x$, $Cs_{1-x}K_x$, and $Cs_{1-x}Rb_x$ systems as a function of the atomic fraction x. From these data, we obtain the corresponding long-wave mode Grüneisen parameter $\gamma_i(q\to0)$ along [100], [111], and [110] directions. In Fig. 2, we show the results with Vashishta-Singwi screening function, and the variation width $\gamma_i(q\to0)$ using other screening functions remains to be below 0.05. From Fig. 2, we see that the concentration dependence of $\gamma_i(q\to0)$ for these alloy systems is small compared with that of $\gamma_i(q)$ in Fig. 1. We investigated theoretically the concentration dependence of the mode Grüneisen parameters for Rb-K, Cs-K, and Cs-Rb solid solutions, and our obtained data are useful in studies of the thermal properties and the anharmonicity of these alloy systems.

References

/1/ T. SOMA, H. OHSUGI, and H.-MATSUO KAGAYA, phys. stat. sol. (b) 124, 525 (1984).

/2/ T. SOMA, phys. stat. sol. (b) 99, 195 (1980).

/3/ T. SOMA, Y. KIMURA, and H.-MATSUO KAGAYA, phys. stat. sol. (b) 112, 151 (1982).

/4/ T. SOMA, Physica (Utrecht) 97B, 76 (1979).

/5/ T. SOMA, Y. KIMURA, and H.-MATSUO KAGAYA, phys. stat. sol. (b) 111, 679 (1982).

(Received October 8, 1984)

phys. stat. sol. (b) 126, K113 (1984)

Subject classification: 6 and 20.1; 22.2.2

Laboratoire de Physique des Solides de l' Université P. et M. Curie, associé au C.N.R.S., LA 154, Paris[1)]

Raman Detection of the Nitrogen Localized Mode in GaP:N

By

M.P. DOS SANTOS[2)], C. HIRLIMANN, and M. BALKANSKI

For several decades now the study of impurity effects in semiconductors has been very active. More specifically the doping of gallium phosphide with the nitrogen isoelectronic impurity has been at the origin of multiple experiments due to the complexity of its emission spectrum /1/. Strangely enough, there has never been any report on the Raman observation of the localised vibration mode of nitrogen in GaP. This note is devoted to fix this lack.

We have performed resonant Raman scattering experiments in GaP:N at 5 K, by tuning the excitation energy through some of the nitrogen pair level energies, specially NN_3.

We used as an exciting source a coherent Radiation CR 490 model dye laser; the dye was Rhodamine 110 pumped by the 515 nm line of an Ar^+ laser and delivered a maximum output power of 20 mW. The spectral analysis was made by means of a Coderg PH_0 double monochromator, a FW-130 (S20 cathode) ITT photomultiplier, and a photon counting system. Our results were obtained on a unique (100) oriented, epitaxial GaP sample doped with nitrogen at a concentration of 9×10^{18} cm^{-3}, cooled down to 5 K by a helium vapor flow.

Fig. 1 shows the luminescence spectrum of the GaP:N sample when excited by the 515 nm (2.41 eV) line of an Ar^+ laser, therefore far from resonance. One can see the usual nitrogen pairs trapped exciton series, noted NN_i, up to the eight order /1/. On the high energy side the absence of the free exciton line is a clear indication of the high nitrogen concentration in the sample. The series of LO phonon assisted recombination lines (NN_i^*) is also clearly visible. The NN_3^* line, for instance, is shifted from the NN_3 electronic level, toward low energy, by an energy equal to one longitudinal optical phonon (W_{LO} = 405 cm^{-1},

---

1) 4 place Jussieu, 75230 Paris Cédex 05, France.

2) Present address: Laboratório de Física, Universidade do Minho Braga, Portugal. Research supported by a grant of the French Government.

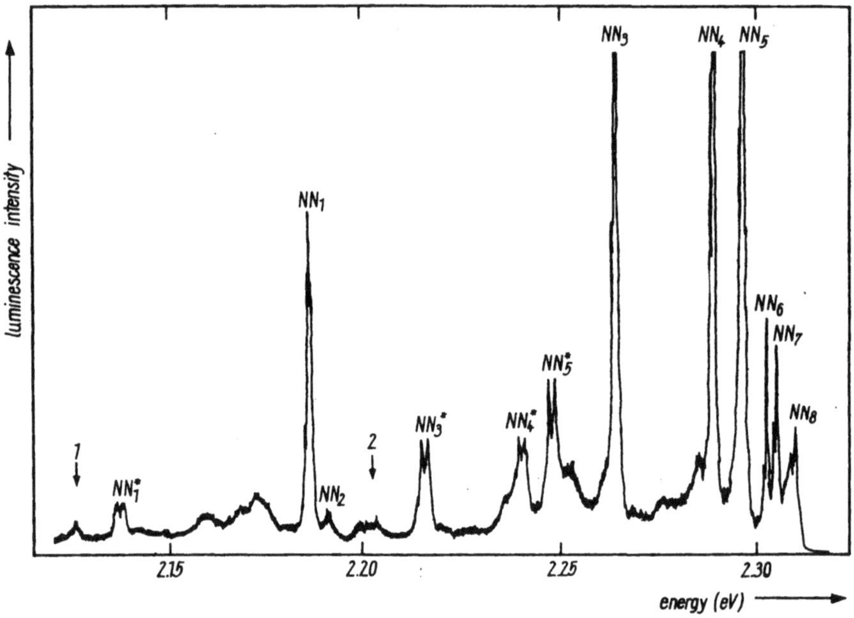

Fig. 1. Emission spectrum of GaP:N at 5 K when excited by the 514.5 nm line of an Ar^+ laser; N = $8x10^{19}$ cm^{-3}

50.2 meV). One can also observe phonon side bands involving one localised nitrogen mode: NN_3^{loc}, NN_1^{loc}. From these side bands one can already deduce an energy of 491 cm^{-1}, 60.9 meV for the localised mode /1/.

In order to see the Raman line associated to the localised vibration of nitrogen it is necessary to work under resonant condition to compensate for the low concentration of the impurity atoms. Raman scattering by localised modes needs concentrations as high as 1% at least two orders of magnitude greater than in our case - nitrogen pair levels are well suited for such a purpose; they fall in easy-to-use dyes tunability range and their wave function at the center of the Brillouin zone is finite so that one can expect large resonant enhancement of the Raman scattered light.

Fig. 2 focuses attention on the surrounding of the localised mode assisted emission under quasi-resonant condition. For this experiment an exciting photon energy of 2.271 eV was chosen which falls 6.2 meV (50 cm^{-1}) above the NN_3 nitrogen trapped excitonic level. Two different peaks can be distinguished:

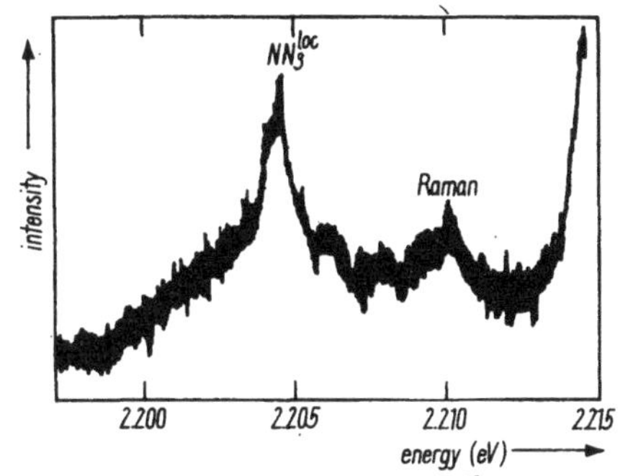

Fig. 2. Detail of the emission spectrum of GaP:N at 5 K under quasi-resonant condition. The exciting energy 2.271 eV is 6.2 meV greater than the NN_3 level energy. Beside the localised mode assisted emission peak one can see the Raman peak due to the nitrogen localised mode

the most intense one is the NN_3^{loc} phonon assisted line, the weak one is shifted toward higher energies by an amount, 6 meV (46 cm^{-1}), comparable with the excess energy of the exciting photons compared to the NN_3 line. This stands for the first direct Raman observation of the nitrogen localised mode in GaP. Further evidence of the Raman nature of the weak peak is given by its shifting energetic position when the exciting photon energy is shifted. From that observation one deduces a localised nitrogen mode shift of $(495 \pm 2) cm^{-1}$ (61.3 meV).

This result is in good agreement with both theoretical and experimental values previously reported in the literature, as can be seen from Table 1.

Table 1

N localised mode (cm^{-1})	determination
498	theory /2/
492	theory /3/
496	IR absorption /4/
495	IR reflectivity /5/
491	luminescence side band
495	Raman

References

/1/ D.G. THOMAS and J.J. HOPFIELD, Phys. Rev. 150, 680 (1966).

/2/ G. LUCOVSKY, M.H. BRODSKY, and E. BURSTEIN, Phys. Rev. B 2, 3295 (1970).

/3/ D.G. THOMAS, J.J. HOPFIELD, and C.J. FROSCH, Phys. Rev. Letters 15, 853 (1965).

/4/ A.H. KACHARE, J.M. CHERLOW, T.T. YANG, W.G. SPITZER, and F.K. EULER, J. appl. Phys. 47, 161 (1976).

/5/ F. THOMPSON and R.G. NEWMAN, J. Phys. C 4, 3249 (1971).

(Received November 1, 1984)

phys. stat. sol. (b) 126, K117 (1984)
Subject classification: 10.2; 22.5.2

Department of Physics, University of Kalyani[1)]

Solution Enthalpies of Trivalent Defects in Alkali Chlorides

By
S. BANDYOPADHYAY and S.K. DEB

Introduction In our previous paper /1/ methods were suggested to find the Born-Mayer repulsive parameters (RPs) of a divalent defect cation in alkali halide crystal and these parameters were successfully utilised to calculate the solution enthalpies of divalent defects in alkali halides. Here, we have extended that method to find the B-M RPs and the solution enthalpies of trivalent defects in alkali halides. A trivalent impurity ion occupies a substitutional position as a triply charged ion in an alkali halide crystal. Here we report calculations of the heat of formation of dilute solid solutions of $AlCl_3$, $ScCl_3$, $GaCl_3$, YCl_3, and $InCl_3$ in each of the three alkali chlorides NaCl, KCl, and RbCl.

Method The principle of the method for finding the defect ion-host ion RPs from the correlation between the B-M RPs and the charge density overlap parameters of the interacting ions is as given earlier /1 to 3/ and will not be repeated here. However, for nearest-neighbour interaction, the correlations are given by

$$\ln(b\beta_{ij}) - -5.2818 + 0.0227 \ln (h^2 r_0^5) \quad (1)$$

and

$$\varrho = 0.7452 - 0.0579\ s\ , \quad (2)$$

where all the quantities are expressed in atomic units.

The free ionic charge density of the triply charged defect ion (having similar closed shell electronic configuration as the singly charged monovalent host ion which it replaces) and that of the neighbouring host anion are plotted at a distance of r_0 of the host lattice, as in Fig. 1 in /1/. An overlap region is obtained and the overlap parameters h and s are thus obtained. Putting these parameters in the already obtained correlations (1) and (2), a set of $(b\beta_{ij})$ and ϱ values is obtained. Though this ϱ value represents the trivalent ion-host

---

1) Kalyani 741235, West Bengal, India.

ion range parameter, $(b\beta_{ij})$ obtained from (1) should be modified, as in (3), to represent the trivalent ion-host ion strength parameter, the reason being clearly discussed in /1/,

$$\ln\left(\frac{b\beta_{ij}}{Z}\right) = -5.2818 + 0.0227 \ln (h^2 r_o^5) , \tag{3}$$

where Z, defined earlier as "lattice stability factor", is a number equal to the ratio of the charge of the defect ion to the charge of the host ion which is being replaced.

When a trivalent ion occupies a substitutional position as a triply charged ion in an alkali chloride crystal, two cation vacancies are also created to keep the host lattice neutral. The solution enthalpies that we will calculate correspond to the "free" trivalent cation impurities /1, 4/ which occur at sufficiently high temperature. Hence, in this calculation, we have not considered any interaction of the vacancies with the trivalent defect.

Following the definition in our previous papers /1, 5/, the heat of formation of a solid solution of say, YCl_3 in RbCl (designated as $RbCl:Y^{3+}$) is given by

$$\Delta H_f = \Delta E_o + \Delta E_{min} + 2 E_v , \tag{4}$$

where

$$\Delta E_o = 3 W_{LA} - W_{LB}$$

$= 3$ (energy of a Rb^+ ion at Rb^+ ion site in RbCl) - (energy of a Y^{3+} ion in YCl_3),

ΔE_{min} energy required to replace a Rb^+ ion by a trivalent Y^{3+} ion in RbCl lattice, and E_v vacancy formation energy of a cation in RbCl.

The detailed expressions for obtaining the different energy terms are given in /1, 5/. The relaxation σ is confined to the nearest neighbours of the defect ion and is radial in nature. When the Y^{3+} ion occupies a substitutional position in a RbCl lattice, we also require RPs for Y^{3+}-Rb^+ interaction which are obtained from the charge density distributions of the Y^{3+} and Rb^+ ions and the corresponding correlations /1, 3/.

Calculation and results The equilibrium interionic separation r_o and the RPs of the alkali chlorides are taken from the work of Boswarva and Simpson /6/. To utilise the data for the ionic radii for monovalent and trivalent ions

Table 1

Calculated values of the repulsive parameters and ΔH_f

crystal	RPs from charge density overlap method				vacancy formation energy	$\Delta E_0 = 3W_{LA} - W_{LB}$	ΔE_{min}	ΔH_f
	defect cation-host anion		defect cation-host cation					
	$(b\beta_{ij})$ $(10^5$ J)	ϱ(nm)	$(b\beta_{ij})$ $(10^5$ J)	ϱ(nm)	E_{v+} (eV)	(eV)	(eV)	(eV)
$NaCl:Al^{3+}$	0.9646	0.03614	1.2793	0.03597	4.884	31.836	-40.737	0.867
$NaCl:Sc^{3+}$	0.9948	0.03817	1.2423	0.03583	4.884	26.629	-36.024	0.373
$NaCl:Ga^{3+}$	0.9848	0.03567	1.2396	0.03563	4.884	30.121	-38.343	1.546
$NaCl:Y^{3+}$	1.0097	0.03366	1.2216	0.03572	4.884	22.199	-31.879	0.088
$NaCl:In^{3+}$	0.9794	0.03604	1.2644	0.03599	4.884	25.197	-34.129	0.836
$KCl:Al^{3+}$	0.9503	0.03611	1.2575	0.03587	4.560	33.954	-41.645	1.429
$KCl:Sc^{3+}$	0.9836	0.03816	1.2257	0.03573	4.560	28.747	-37.200	0.667
$KCl:Ga^{3+}$	0.9738	0.03565	1.2248	0.03547	4.560	32.305	-39.384	2.041
$KCl:Y^{3+}$	0.9983	0.03559	1.2101	0.03541	4.560	24.317	-33.344	0.093
$KCl:In^{3+}$	0.9663	0.03603	1.2453	0.03583	4.560	27.315	-35.456	0.979
$RbCl:Al^{3+}$	0.9456	0.03629	1.2501	0.03581	4.402	34.608	-42.310	1.102
$RbCl:Sc^{3+}$	0.9776	0.03798	1.2199	0.03555	4.402	29.401	-37.897	0.308
$RbCl:Ga^{3+}$	0.9706	0.03583	1.2195	0.03530	4.402	32.959	-40.060	1.703
$RbCl:Y^{3+}$	0.9939	0.03567	1.2057	0.03529	4.402	24.971	-34.109	-0.334
$RbCl:In^{3+}$	0.9598	0.03599	1.2515	0.03565	4.402	27.969	-36.176	0.597

from the same source, we have taken these values from /7a/. The electronic charge densities of the interacting ions are calculated from the wave functions given by Clementi and Roetti /8/. For triply charged defect ions, the neighbouring ions feel a stronger attraction and move sufficiently inward to make the overlap between the ions large enough so as to overwhelm the positive repulsive energy by the polarization energy, in the polarizable point-ion model. This breakdown in the physical model is known as "polarisation catastrophe" and this problem is solved as in /1/. The vacancy formation energy E_{v+} has been calculated following the method of /9/. The lattice energies of the monovalent and trivalent crystals are taken from the theoretical calculations of Jenkins from /7b/. The values of ΔH_f calculated with the RPs obtained from the charge density overlap method are shown in Table 1.

For monovalent defects, enough calculated and experimental values of solution enthalpies are available in the literature /10, 11/. For divalent defects, some theoretical calculations and experimental values also are available /4/. But we have not been able to find any theoretical calculation on the solution enthalpies of trivalent defects and we find only one datum relating to the experimental value of the heat of formation of YCl_3 in NaCl /12/. Thus, we cannot compare our values with those of other workers except for one system NaCl:Y^{3+}. The calculated value is 0.09 eV as against the experimental value of the free enthalpy of Y^{3+} in NaCl being (2.91 - 13 kT)eV or nearly 2.6 eV at moderately high temperature. Any error in the radii of the triply charged ions greatly effects the value of ΔH_f. In /7/ there are two sources for obtaining the lattice energy of the trivalent crystals - one from thermochemical cycle and the other by theoretical methods and there is wide variation between these values. We have taken in this work lattice energy values obtained by theoretical methods. With all these uncertainties in the input data, our values seem to be quite physical. Some corrections which will increase the accuracy of the theoretical calculations are in progress and will be reported in time.

References

/1/ S. BANDYOPADHYAY and S.K. DEB, phys. stat. sol. (b) 124, 545 (1984).

/2/ S.K. DEB and A.K. GHOSH, Indian J. Phys. 59, 528 (1975).

/3/ S. BANDYOPADHYAY and S.K. DEB, Indian J. pure appl. Phys. 22, 273 (1984).

/4/ R.C. BOWMAN, Jr., J. chem. Phys. 59, 2215 (1973).
/5/ C. NEOGY and S.K. DEB, J. Phys. Chem. Solids 41, 1381 (1980).
/6/ I.M. BOSWARVA and J.H. SIMPSON, Canad. J. Phys. 51, 1923 (1973).
/7a/ R.A. WEAST (Ed.), Handbook of Chemistry and Physics, Vol. 60, CRC Press, Cleveland (Ohio) 1979-1980 (p. F-214).
/7b/ ibid. (p. D-89).
/8/ E. CLEMENTI and C. ROETTI, Atomic Data nuclear Data Tables 14, 177 (1974).
/9/ S. BANDYOPADHYAY and S.K. DEB, Indian J. Phys. 57A, 422 (1983).
/10/ M.P. TOSI and M. DOYAMA, Phys. Rev. 151, 642 (1966).
/11/ B.G. DICK and T.P. DAS, Phys. Rev. 127, 1053 (1962).
/12/ F. BENIERE and R. ROKBANI, J. Phys. Chem. Solids 36, 1151 (1975).

(Received September 28, 1984)

phys. stat. sol. (b) 126, K123 (1984)
Subject classification: 10.2; 22.5.3

Department of Physics, University of Athens[1]

Connection of the Defect Volumes to the Bulk Properties in CaF_2

By
M. LAZARIDOU

A macroscopic model has been proposed for the calculation of the formation (v^f), the activation (v^{act}), and the migration (v^m) volume of defects in solids /1, 2/. One of the main predictions of this model /2, 3/ was that the volumes for the migration process of various defects in the same host crystal have to be proportional to the corresponding migration enthalpies h^m. The ratio $v^{m,i}/h^{m,i}$ depends on the bulk properties according to the relation

$$v^{m,i} = h^{m,i} \frac{1}{B_0^{SL}}\left(\frac{dB}{dP} - 1\right) \exp \int_0^T \beta dT \quad , \tag{1}$$

where B is the isothermal bulk modulus, B_0^{SL} is the intercept of the linear part of the graph B = f(T) with the vertical axis, and β the thermal volume expansion coefficient; the superscript (i) denotes the corresponding migration mechanism.

A comparison of the experimental results in SrF_2 obtained by the dielectric loss technique /4/ and by ionic conductivity under various pressures /5/ confirmed this prediction /6/. It is the object of the present paper to show that the experimental results in CaF_2 by Fontanella et al. /7/ in conjunction with earlier measurements by Lallemand /8/ also give a constant ratio v^m/h^m for the various conduction mechanisms.

The temperature and pressure variation of the dielectric loss peak of CaF_2 doped with sodium /7/ indicates that the nn bound fluorine vacancy has a migration enthalpy $h^{m,b} = 0.502$ eV and a migration volume (at ambient pressure) $v^{m,b} = 1.72\ cm^3/mol$. Conductivity measurements by Lallemand /8/ in CaF_2 under various pressures gave for the migration enthalpy of the free anion vacancy motion the value $h^{m,f} = 0.84$ eV whereas the corresponding migration volume is $v^{m,f} = 3.2\ cm^3/mol$.[2] The experimental results, for

1) Knossou Str. 36, Ano Glyfada, 165 61 Athens, Greece.

2) If one considers the recent analysis of the conductivity plot of fluorides in /5/, these values of Lallemand might not correspond to the free anion vacancy; however, whatever the mechanism, it does not influence the conclusions of the present paper.

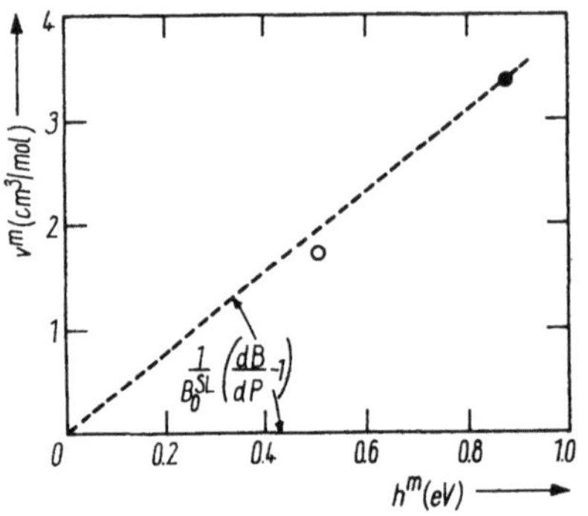

Fig. 1. Experimental migration volumes v^m for a free (anion) vacancy (○) and a nn bound fluorine vacancy (●) in CaF_2. The line represents the prediction of (1)

both conduction mechanisms, are plotted in Fig. 1; considering the experimental errors they practically lie on a line which passes through the origin of the axis.

The question remains to examine if the slope of this line is compatible with the theoretical expectation of (1). The elastic data of Wong and Schuele /9/ indicate that $B_{T_0}^{SL} = 0.93 \times 10^{12}$ dyn/cm^2 and $\frac{dB}{dP} \approx 4.9$. Taking also into account that $\exp \int_0 \beta \, dT \approx 1.04$ (corresponding to the "mean" temperature of the experiments mentioned above) (1) suggests that the plot of $v^{m,i}$ versus $h^{m,i}$ must be a straight line which passes through the origin of the axis and has a slope 4.4×10^{-12} cm^2/dyn. This line has been plotted in Fig. 1 and if one takes into account the experimental errors of the quantities used, it is in very good agreement with the experimental data.

References

/1/ P. VAROTSOS, W. LUDWIG, and K. ALEXOPOULOS, Phys. Rev. B 18, 2683 (1978).

/2/ P. VAROTSOS and K. ALEXOPOULOS, Phys. Rev. B 24, 904 (1981).

/3/ P. VAROTSOS and K. ALEXOPOULOS, phys. stat. sol. (b) 110, 9 (1982).

/4/ C. ANDEEN, L.M. HAYDEN, and J. FONTANELLA, Phys. Rev. B 21, 794 (1980).

/5/ J.M. OBERSCHMIDT and D. LAZARUS, Internat. Conf. Fast Ion Transport in Solids, Lake Geneva (Wisconsin), May 1979.

/6/ P. VAROTSOS and K. ALEXOPOULOS, J. Phys. Chem. Solids 42, 409 (1981).

/7/ J. FONTANELLA, M. WINTERSGILL, and C. ANDEEN, phys. stat. sol. (b) 97, 303 (1980).

/8/ M. LALLEMAND, Thesis, Université de Paris VI, 1972.

/9/ C. WONG and D. SCHUELE, J. Phys. Chem. Solids 29, 1309 (1968).

(Received September 13, 1984)

phys. stat. sol. (b) 126, K125 (1984)
Subject classification: 18.2; 18.4

Department of Physics, Suzhou University[1)]

Magnetic Properties of a Random Mixture of a One-Dimensional System with Anisotropic Heisenberg Exchange

By
ZHEN-YA LI and CHUAN-ZHANG YANG

For a random mixture of magnetic atoms with Ising spins /1 to 4/ and with anisotropic Heisenberg exchange (AHE) /5, 6/ the magnetic properties have been investigated by some authors. In this note, we use the technique of partition function for the AHE spin system to calculate the specific heat and magnetic susceptibility of a random binary mixture (RBM) of a one-dimensional AHE spin system. Here we only consider the site random problem with nearest-neighbour anisotropic interaction.

A linear chain composed of two kinds of magnetic atom A and B with AHE is considered. The atoms are numbered as i = 1, 2, ..., N, and A and B atoms are denoted by $s_i = 1$ and $s_i = -1$. The magnetic moments of A and B atoms are denoted by m_A and m_B, and the exchange integrals between them by J_{AA}, J_{BB}, and J_{AB}. For a definite configuration s_i of the RBM system (site problem), the Hamiltonian is represented by

$$\frac{\mathcal{H}\{s_i\}}{k_B T} - \sum_{i=1}^{N-1} K_{i,i+1}\left\{\sigma_i^z\sigma_{i+1}^z + r(\sigma_i^x\sigma_{i+1}^x + \sigma_i^y\sigma_{i+1}^y)\right\} - \sum_{i=1}^{N} L_i\sigma_i^z , \qquad (1)$$

where $0 \leqq r \leqq 1$, σ_i is the Pauli spin operator for the i-th atom and /4/

$$K_{i,i+1} = \begin{bmatrix}\frac{1+s_i}{2} & \frac{1-s_{i+1}}{2}\end{bmatrix}\begin{bmatrix}K_{AA} & K_{AB}\\ K_{BA} & K_{BB}\end{bmatrix}\begin{bmatrix}\frac{1+s_i}{2}\\ \frac{1-s_i}{2}\end{bmatrix} , \qquad (2)$$

$$L_i = \begin{bmatrix}\frac{1+s_i}{2} & \frac{1-s_i}{2}\end{bmatrix}\begin{bmatrix}L_A\\ L_B\end{bmatrix} = \begin{bmatrix}L_A & L_B\end{bmatrix}\begin{bmatrix}\frac{1+s_i}{2}\\ \frac{1-s_i}{2}\end{bmatrix} , \qquad (3)$$

1) Suzhou, Jiangsu, People's Republic of China.

$$K_{AA} = J_{AA}/2k_BT\,, \qquad L_A = m_AL\,, \qquad L_B = m_BL\;; \tag{4}$$

$$L = H/k_BT\,, \qquad m_i = g_i\mu_B\,, \qquad i = A,\ B\,, \tag{5}$$

where H is an external magnetic field along the z-axis, g and μ_B are the Landé factor and Bohr magneton, respectively.

With the operator P_{ij} defined by

$$P_{ij} = \sigma_i^z\sigma_j^z + r(\sigma_i^x\sigma_j^x + \sigma_i^y\sigma_j^y)\,, \tag{6}$$

$\exp\left[K_{i,i+1},\ P_{i,i+1}\right]$ can be given by

$$\exp\left[K_{i,i+1},\ P_{i,i+1}\right] = a_o(K_{i,i+1}) + a_1(K_{i,i+1})\sigma_i^z\sigma_{i+1}^z + a_2(K_{i,i+1})(\sigma_i^x\sigma_{i+1}^x + \sigma_i^y\sigma_{i+1}^y) \tag{7}$$

with

$$a_o(K_{i,i+1}) = \frac{1}{2}\left[\exp(K_{i,i+1}) + \exp(-K_{i,i+1})\cosh(2\,rK_{i,i+1})\right],$$
$$a_1(K_{i,i+1}) = \frac{1}{2}\left[\exp(K_{i,i+1}) - \exp(-K_{i,i+1})\cosh(2\,rK_{i,i+1})\right],$$
$$a_2(K_{i,i+1}) = \frac{1}{2}\left[\exp(-K_{i,i+1})\sinh(2\,rK_{i,i+1})\right]. \tag{8}$$

We use the method of the pair approximation proposed in /5/, the specific heat and the magnetic susceptibility at zero external field can be obtained by

$$\frac{C}{N} = k_B\sum_{\alpha,\beta} K_{\alpha\beta}^2 p_\alpha p_\beta\Big\{\left[\exp(K_{\alpha\beta}) + \exp(-K_{\alpha\beta})\cosh(2\,rK_{\alpha\beta})\right]^{-1} \times \left[\exp(K_{\alpha\beta}) + \exp(-K_{\alpha\beta})\cosh(2\,rK_{\alpha\beta}) - 4\,r\exp(K_{\alpha\beta})\sinh(2\,rK_{\alpha\beta}) + 4\,r^2\exp(-K_{\alpha\beta})\cosh(2\,rK_{\alpha\beta})\right] - \left[\exp(K_{\alpha\beta}) + \exp(-K_{\alpha\beta})\cosh(2\,rK_{\alpha\beta})\right]^{-1} \times \left[\exp(K_{\alpha\beta}) - \exp(-K_{\alpha\beta})\cosh(2\,rK_{\alpha\beta}) + 2\,r\exp(-K_{\alpha\beta})\sinh(2\,rK_{\alpha\beta})\right]^2\Big\} \tag{9}$$

and

$$\frac{k_BT}{N}\chi = D^{-1}\Big\{(p_Am_A^2 + p_Bm_B^2) + p_A(p_Am_A^2 - p_Bm_B^2)\tanh R_{AA}^{(1)} + p_B(-p_Am_A^2 + p_Bm_B^2)\tanh R_{BB}^{(1)} + 4p_Ap_Bm_Am_B\tanh R_{AB}^{(1)} - p_Ap_B(p_Am_A^2 + p_Bm_B^2)(\tanh R_{AA}^{(1)}\tanh R_{BB}^{(1)} - \tanh^2 R_{AB}^{(1)})\Big\}, \tag{10}$$

where p_A and p_B are concentrations of A and B atoms, respectively, and

$$D = 1 - p_A \tanh R^{(1)}_{AA} - p_B \tanh R^{(1)}_{BB} + p_A p_B (\tanh R^{(1)}_{AA} \tanh R^{(1)}_{BB} - \tanh^2 R^{(1)}_{AB}) \qquad (11)$$

with

$$\tanh R^{(1)}_{\alpha\beta} = \frac{a_1(K_{\alpha\beta})}{a_o(K_{\alpha\beta})} . \qquad (12)$$

Now we extend the (q + 1) approximation in /5/ to the linear RBM chain with AHE. In the (q + 1) approximation, all spins are grouped into clusters of the three i-th, (i + 1)-th, and (i + 2)-th spins. Therefore, we must distinguish the types of exchange interaction between nearest-neighbour magnetic atoms. We neglect the uncommutability that $\left[P_{i,i+1}, P_{j,j+1}\right]$ ($j = i \pm 1$) is non-zero for $r \neq 0$, and put as follows:

$$\exp\left\{K_{i,i+1} P_{i,i+1} + K_{i+1,i+2} P_{i+1,i+2}\right\} \approx a_o a'_o + a_o a'_1 \sigma^z_{i+1} \sigma^z_{i+2} +$$
$$+ a_1 a'_o \sigma^z_i \sigma^z_{i+1} + a'_o a'_2 (\sigma^x_i \sigma^x_{i+1} + \sigma^y_i \sigma^y_{i+1}) + a'_2 a_o (\sigma^x_{i+1} \sigma^x_{i+2} + \sigma^y_{i+1} \sigma^y_{i+2}) +$$
$$+ a_1 a'_1 \sigma^z_i \sigma^z_{i+2} + a_2 a'_2 (\sigma^x_i \sigma^x_{i+2} + \sigma^y_i \sigma^y_{i+2}) , \qquad (13)$$

$$a_o = a_o(K_{i,i+1}) , \quad a_1 = a_1(K_{i,i+1}), \quad \dots ,$$
$$a'_o = a'_o(K_{i+1,i+2}) , \quad a'_1 = a'_1(K_{i+1,i+2}), \quad \dots . \qquad (14)$$

For a defined configuration of the system, the partition function of the RBM system can be calculated and then the specific heat in vanishing magnetic field is given by

$$\frac{C}{N} = \frac{1}{k_B T^2} \frac{\partial^2}{\partial \beta^2} \sum_{\alpha,\beta} p_\alpha p_\beta \ln a_o(K_{\alpha\beta}), \qquad \alpha,\beta = A, B . \qquad (15)$$

In fact, this result is in agreement with that of the pair approximation. When we calculate the magnetic susceptibility of the RBM system, we must consider the different kind of the (i + 1)-th atom and using corresponding matrix representation, then we get the susceptibility, which is given by

$$
\begin{aligned}
\frac{k_B T}{N}\chi = \frac{1}{2}\Bigg\{ & p_A\Bigg[p_A^2\,\frac{a_o^2(K_{AA})-a_1^2(K_{AA})}{a_o^2(K_{AA})+a_1^2(K_{AA})}\,m_A^2+p_B^2\,m_B^2\ + \\
&+p_Ap_B\,\frac{a_o^2(K_{AA})-a_1^2(K_{AA})}{a_o^2(K_{AA})}\,m_A^2\Bigg]+p_B\Bigg[p_Bm_B^2\,\frac{a_o^2(K_{BB})-a_1^2(K_{BB})}{a_o^2(K_{BB})+a_1^2(K_{BB})}\ + \\
&+p_A^2m_A^2+p_Ap_B\,\frac{a_o^2(K_{BB})-a_1^2(K_{BB})}{a_o^2(K_{BB})}\,m_B^2\Bigg]\Bigg\}+\frac{1}{2}\Big\{p_A^2(p_AF^{+2}_{AA}m_A^2+p_BF^{-2}_{AA}m_B^2)\ + \\
&+p_B^2(p_AF^{+2}_{BB}m_A^2+p_BF^{-2}_{BB}m_B^2)+p_Ap_B(p_AF^{+2}_{AB}m_A^2+p_BF^{-2}_{AB}m_B^2+p_AF^{+2}_{AA}m_A^2\ + \\
&+p_BF^{-2}_{BA}m_B^2)+p_Am_A^2+p_Bm_B^2+2p_A\Big[p_A(p_AF^{+}_{AA}m_A+p_BF^{-}_{AA}m_B)\ + \\
&+p_B(p_AF^{+}_{BA}m_A+p_BF^{-}_{BA}m_B)\Big]m_A+2p_B\Big[p_A(p_AF^{+}_{AB}m_A+p_Bm_BF^{-}_{AB})\ + \\
&+p_B(p_AF^{+}_{BB}m_A+p_BF^{-}_{BB}m_A)\Big]m_B\Big\}\ + \\
&+\begin{bmatrix}1-p_AE_{AA} & -p_AE_{AB}\\ -p_BE_{BA} & 1-p_BE_{BB}\end{bmatrix}^{-1}\Bigg\{\begin{bmatrix}m_A & m_B\end{bmatrix}\begin{bmatrix}p_A & \\ & p_B\end{bmatrix}\Bigg(\begin{bmatrix}C_{AA} & C_{AB}\\ C_{BA} & C_{BB}\end{bmatrix}\ + \\
&+\begin{bmatrix}E_{AA} & E_{AB}\\ E_{BA} & E_{BB}\end{bmatrix}\Bigg)\begin{bmatrix}p_A & \\ & p_B\end{bmatrix}\begin{bmatrix}m_A\\ m_B\end{bmatrix}+\begin{bmatrix}p_A & p_B\end{bmatrix}\begin{bmatrix}G_{AA} & G_{AB}\\ G_{BA} & G_{BB}\end{bmatrix}\begin{bmatrix}p_A & \\ & p_B\end{bmatrix}\ \times \\
&\times\Bigg(\begin{bmatrix}C_{AA} & C_{AB}\\ C_{BA} & C_{BB}\end{bmatrix}\begin{bmatrix}p_A\\ p_B\end{bmatrix}+\begin{bmatrix}E_{AA} & E_{AB}\\ E_{BA} & E_{BB}\end{bmatrix}\begin{bmatrix}p_A & \\ & p_B\end{bmatrix}\begin{bmatrix}m_A\\ m_B\end{bmatrix}\Bigg)\Bigg\}\,,
\end{aligned}
\tag{16}
$$

where

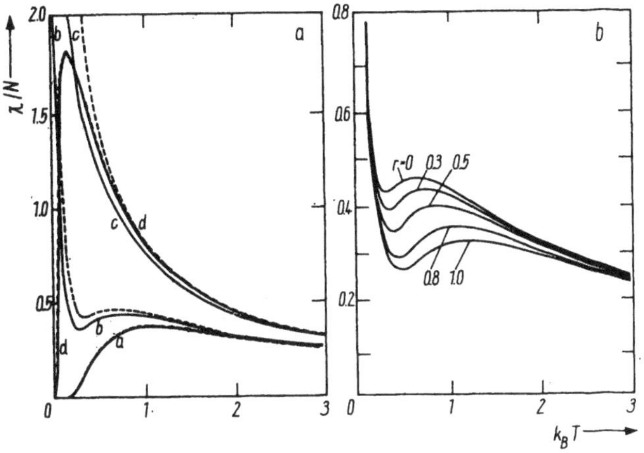

Fig. 1. a) The susceptibilities of the one-dimensional RBM of an antiferromagnetic system with a) Ising spins; $J_{AA} = -1$, $J_{BB} = -0.2$, $J_{AB} = 0$, $m_A = m_B = 1$, $p_B = 1 - p_A$. (a) $p_A = 1$, (b) $p_A = 0.95$, (c) $p_A = 0.5$, (d) $p_A = 0$. The full lines and dashed lines show the results of calculation by the pair approximation and by the (q + 1) approximation, respectively. b) AHE by the (q + 1) approximation; $J_{AA} = -1$, $J_{BB} = -0.2$, $m_A = m_B = 1$, $p_A = 0.95$

$$C_{\alpha\beta} = p_A F^+_{\alpha\beta} m_A \tanh R^+_{\alpha\beta} + p_B F^-_{\alpha\beta} m_B \tanh R^-_{\alpha\beta} , \tag{17}$$

$$E_{\alpha\beta} = p_A \tanh R^+_{\alpha\beta} + p_B \tanh R^-_{\alpha\beta} , \qquad G_{\alpha\beta} = p_A F^+_{\alpha\beta} m_A + p_B F^-_{\alpha\beta} m_B , \tag{18}$$

$$F^{\pm}_{\alpha\beta} = \frac{b_1^{\pm}(K_{\alpha\beta}) + b_1'^{\pm}(K_{\alpha\beta})}{b_0^{\pm}(K_{\alpha\beta}) + b_3^{\pm}(K_{\alpha\beta})} , \quad \tanh R^{\pm}_{\alpha\beta} = \frac{b_3^{\pm}(K_{\alpha\beta})}{b_0^{\pm}(K_{\alpha\beta})} , \tag{19}$$

$$b_0 = a_0 a_0' , \quad b_1 = a_1' a_0 , \quad b_1' = a_0 a_1' , \quad b_3 = a_1 a_1' .$$

Here "+" and "-" represent the situation in which $s_i = 1$ and -1, respectively. The magnetic susceptibilities at various values of r from r = 0 to r = 1 can be obtained from (15). The magnetic susceptibility of the linear RBM system with Ising spin (r = 0) is shown in Fig. 1a which is in excellent agreement with what is obtained by means of the pair approximation and /4/, respectively, especially as $k_BT > 3$, they almost overlap. For a certain concentration of magnetic atoms, the magnetic susceptibilities with various values of r are shown in Fig. 1b.

In conclusion, we have shown that the techniques proposed by /5/ can be extended to the one-dimensional RBM with AHE, and to calculations of the specific heat and magnetic susceptibility at various values of r. These results can be reduced to the linear chain composed of one kind of magnetic atoms with pure Ising spin, when they are reduced, they apparently agree with those obtained by other authors. For a one-dimensional spin system with AHE, the (q + 1) approximation does not give rise to any improvement for specific heat calculations, but for the calculation of the magnetic susceptibility at various values of r, the (q + 1) approximation has expressed some distinct behaviours. It is obvious that the technique can be extended to random mixture systems of many kinds of magnetic atoms.

References

/1/ F. MATSUBARA, K. YOSHIMURA, and S. KATSURA, Canad. J. Phys. 51, 1053 (1973).
/2/ D.C. RAPAPORT, J. Phys. C 5, 1830 (1972).
/3/ M. THOMSEN and M.F. THORPE, J. Phys. C 16, 4191 (1983).
/4/ S. KATSURA and F. MATSUBARA, Canad. J. Phys. 52, 120 (1974).
/5/ Y. TANAKA and NRYU, Physica 105 A, 493 (1981).
/6/ J.C. BONNER and M.E. FISHER, Phys. Rev. 135, A640 (1964).

(Received October 24, 1984)

phys. stat. sol. (b) 126, K131 (1984)

Subject classification: 13.5.3 and 18.4

Section of Low Temperature Physics, Department of Physics, Moscow State University[1)]

Generalized Dispersion Relations for Surface Polaritons in Uniaxial Antiferromagnetic Crystals

By

N.B. IVANOV[2)]

In the last few years a great number of papers has been devoted to the problems of surface polaritons on nonmagnetic crystals (see /1, 2/). On the other hand, the properties of these modes on magnets, and, especially, on magnetic metals, are scarcely examined so far. Recently, however, far-infrared laser transmission and reflection experiments on several antiferromagnets have been published in /3 to 5/, where the resonance spectra are described in terms of magnetic polaritons. It is natural to expect that in the near future the surface magnetic polaritons /6/ can be observed in the far-infrared laser experiments.

The presence of conduction electrons provides a great variety of properties of the electromagnetic waves, spreading in a metal /7/, and, especially, on its surface /8/. Notice that because of the special form of the spin-wave dispersion law, the role of the spatial dispersion in the magnetic susceptibility tensor $\hat{\mu}$ for metals is, as a rule, very important /9/. Namely, the dependence on wave vector is determined by the strong exchange interactions, while the activation energy is connected with the weak relativistic interactions. Therefore, in considering surface polaritons in magnetic metals, it is necessary, in general, to take into account the spatial dispersion both in the dielectric permittivity $\hat{\varepsilon}$ and the magnetic susceptibility $\hat{\mu}$ tensors. In this note dispersion relations for surface polaritons are obtained, supposing time and spatial dispersions are important both in $\hat{\varepsilon}$ and $\hat{\mu}$.

Let us have a uniaxial antiferromagnet, occupying the half-space $y > 0$ ($y < 0$ vacuum), and the easy axis be z-directed. In this case the tensor $\hat{\mu}$ is diagonal $\mu_{xx} = \mu_{yy} \equiv \mu$, $\mu_{zz} \approx 1$ (in zero external magnetic field), and $\hat{\varepsilon}$ is as-

---

1) 117234 Moscow, USSR.

2) Present address: Institute of Solid State Physics, Bulgarian Academy of Sciences, 72 blvd. Lenin, 1184 Sofia, Bulgaria.

sumed to be diagonal, too ($\hat{\varepsilon} = \varepsilon\delta_{ij}$). To begin with, consider a TM-surface wave, which is characterized by the field vectors $\vec{h} = (h_x(\vec{r}, t), 0, 0)$, $\vec{e} = (0, e_y(\vec{r}, t), e_z(\vec{r}, t))$ and the wave vector $\vec{k}_\parallel = (0, 0, k_\parallel)$. Because of the fact that the surface impedance slightly depends on the character of the electron reflection /7/, we can consider, without loss of generality, h_x and e_y as being even and e_z odd functions of y. Then, supposing $\vec{h}, \vec{e} \sim \exp(i k_\parallel z - i\omega t)$, after Fourier transformation, we obtain from the Maxwell equations

$$e_z(\omega, k_\parallel, k) = \frac{2ic}{\omega} h_x(\omega, k_\parallel, y = 0_+) \times$$
$$\times \frac{k_\parallel^2 - \varepsilon(\omega, k_\parallel, k)\mu(\omega, k_\parallel, k)\omega^2/c^2}{\varepsilon(\omega, k_\parallel, k)\left[k_\parallel^2 + k^2 - \varepsilon(\omega, k_\parallel, k)\mu(\omega, k_\parallel, k)\omega^2/c^2\right]}, \tag{1}$$

$$\frac{e_z(\omega, k_\parallel, y = 0_-)}{h_x(\omega, k_\parallel, y = 0_-)} = \frac{c\varkappa_0}{i\omega}, \quad \varkappa_0 = (k_\parallel^2 - \omega^2/c^2)^{1/2}, \tag{2}$$

where c is the light velocity. Using ordinary boundary conditions at y = 0, as well as (1, 2), we derive the following dispersion equation for the TM-surface polariton:

$$\varkappa_0(\omega_s, k_\parallel) + \frac{2}{\pi}\int_0^\infty \frac{k_\parallel^2 - \varepsilon(\omega_s, k_\parallel, k)\mu(\omega_s, k_\parallel, k)\omega_s^2/c^2}{\varepsilon(\omega_s, k_\parallel, k)\left[k_\parallel^2 + k^2 - \varepsilon(\omega_s, k_\parallel, k)\mu(\omega_s, k_\parallel, k)\omega_s^2/c^2\right]} dk = 0 \tag{3}$$

(the boundary condition for magnetization $\vec{m}|_{y=0} = 0$ is assumed).

Analogically, if h_x and e_y are assumed to be even and e_z odd functions of y, we obtain

$$\frac{1}{\varkappa_0(\omega_s, k_\parallel)} + \frac{2}{\pi}\int_0^\infty \frac{\varepsilon(\omega_s, k_\parallel, k)\,dk}{k_\parallel^2 + k^2 - \varepsilon(\omega_s, k_\parallel, k)\mu(\omega_s, k_\parallel, k)\omega_s^2/c^2} = 0, \quad \left.\frac{d\vec{m}}{dy}\right|_{y=0} = 0. \tag{4}$$

In such way one can derive relations for the TE-surface polariton, characterized by the vectors $\vec{h} = (h_x(\vec{r}, t), h_y(\vec{r}, t), 0)$, $\vec{e} = (0, 0, e_z(\vec{r}, t))$, and $\vec{k}_\parallel = (k_\parallel, 0, 0)$:

$$\varkappa_0(\omega_s, k_\parallel) + \frac{2}{\pi}\int_0^\infty \frac{k_\parallel^2 - \varepsilon(\omega_s, k_\parallel, k)\mu(\omega_s, k_\parallel, k)\omega_s^2/c^2}{\mu(\omega_s, k_\parallel, k)\left[k_\parallel^2 + k^2 - \varepsilon(\omega_s, k_\parallel, k)\mu(\omega_s, k_\parallel, k)\omega_s^2/c^2\right]} dk = 0, \tag{5}$$

$$\left.\frac{dm_x}{dy}\right|_{y=0}, \quad m_y\big|_{y=0} = 0 \quad ;$$

$$\frac{1}{\varkappa_0(\omega_s, k_\parallel)} + \frac{2}{\pi}\int_0^\infty \frac{\mu(\omega_s, k_\parallel, k)\,dk}{k_\parallel^2 + k^2 - \varepsilon(\omega_s, k_\parallel, k)\mu(\omega_s, k_\parallel, k)\omega_s^2/c^2} = 0\,, \qquad (6)$$

$$\left.\frac{dm_y}{dy}\right|_{y=0} = 0, \qquad m_x\Big|_{y=0} = 0\,.$$

Equations (3) to (6) determine the dispersion function $\omega_s(k_\parallel)$ for the surface polaritons in the general case. Consider the case of static electromagnetic fields. Substituting formally $c \to \infty$ in (3), we obtain the well-known dispersion relation for surface plasmons, derived by Richie and Marusak in /10/:

$$1 + \frac{2}{\pi}k_\parallel \int_0^\infty \frac{dk}{\varepsilon(\omega_s, k_\parallel, k)(k_\parallel^2 + k^2)} = 0\,, \quad \vec{m}\Big|_{y=0} = 0\,.$$

Equation (5) yields a similar relation for the surface magnetostatic waves in antiferromagnets:

$$1 + \frac{2}{\pi}k_\parallel \int_0^\infty \frac{dk}{\mu(\omega_s, k_\parallel, k)(k_\parallel^2 + k^2)} = 0\,, \quad \left.\frac{dm_x}{dy}\right|_{y=0} = 0\,, \quad m_y\Big|_{y=0} = 0\,.$$

The author is grateful to Prof. M.I. Kaganov for stimulating discussions.

References

/1/ V.M. AGRANOVICH, Uspekhi fiz. Nauk 115, 199 (1975).

/2/ V.M. AGRANOVICH and D.L. MILLS (Ed.), Surface Polaritons, North-Holland Publ. Co., Amsterdam/New York/Oxford 1982.
A. BOARDMAN (Ed.), Electromagnetic Surface Modes, Wiley & Sons, New York 1982.

/3/ R.W. SANDERS, V. JACCARINO, and S.M. REZENDE, Solid State Commun. 28, 907 (1978).

/4/ K.M. HÄUSSLER, A. LEHMEYER, and L. MERTEN, phys. stat. sol. (b) 111, 513 (1982).

/5/ V.V. EREMENKO, V.M. NAUMENKO, V.G. PASHKEVICH, and V.V. PISHKO, Zh. eksper. teor. Fiz., Pisma 38, 97 (1983).

/6/ A. HARSTEIN, E. BURSTEIN, A.A. MARADUDIN, R. BREWER, and R.F. WALLIS, J. Phys. C 6, 1266 (1973).

/7/ E.A. KANER and V.G. SKOBOV, Uspekhi fiz. Nauk 89, 367 (1966).

/8/ N.B. IVANOV, Zh. Fiz. nizkikh Temp. 10, 608 (1984).

/9/ A.YA. BLANK and M.I. KAGANOV, Uspekhi fiz. Nauk 92, 583 (1967).
/10/ R.H. RICHIE and L. MARUSAK, Surface Sci. 4, 243 (1966).

(Received July 5, 1984)

phys. stat. sol. (b) 126, K135 (1984)

Subject classification: 13.5.1 and 20.1; 22.4.1; 22.4.2

Department of Semiconductor Physics,
V. Kapsukas State University, Vilnius[1] (a)
and Sektion Physik der Humboldt-Universität zu Berlin,
Bereich Halbleiteroptik[2] (b)

Sample-Dependent Luminescence Spectra in Highly Photoexcited CdS and CdSe

By

R. BALTRAMIEJŪNAS (a), I. RÜCKMANN (b), G. TAMULAITIS (a), and A. ŽUKAUSKAS (a)

Highly excited direct gap semiconductors are still an object of intense investigations in many laboratories to study the properties of dense non-equilibrium quasi-particle populations and to find new collective quantum phenomena occurring under particular conditions. In spite of progress in this field, the problem of comparability of the results obtained in different laboratories employing various samples is still actual, however /1/.

Below we present the results of luminescence spectra investigation of typical samples of CdS and CdSe. The heating of the photoexcited exciton gas /2/ and of the free carrier gas /3/ was used as probe effect for comparison. Two groups of samples were employed. The first group (I) contains freshly cleaved block single crystals grown from the melt specially selected for the dominance of free exciton emission under moderate excitation. The second group (II) consists of thin (3 to 10 μm) and pure ($N_D \leqq 10^{16}$ cm^{-3} /4/) platelet samples obtained from the vapour phase. The experimental conditions were similar in both cases: harmonics of YAG-laser radiation ($\tau_L \approx 10$ ns, repetition 12 Hz, spot diameter ≈ 250 μm) were used for an interband excitation of CdSe ($\hbar\omega = 2.34$ eV), and CdS ($\hbar\omega = 3.5$ eV) samples immersed in liquid nitrogen. The luminescence was measured from the excited surface in the geometry $\vec{E} \perp \vec{c}$, $\sphericalangle(\vec{k},\vec{c})$, and spectra were recorded by means of a boxcar averager and a logarithmic amplifier.

Fig. 1 presents the obtained luminescence spectra of the two groups of CdS samples. The same behaviour was found in the case of CdSe. The mean

---

1) Knyvos g. 4a, 232006 Vilnius, USSR.

2) Invalidenstr. 110, DDR-104 Berlin, GDR.

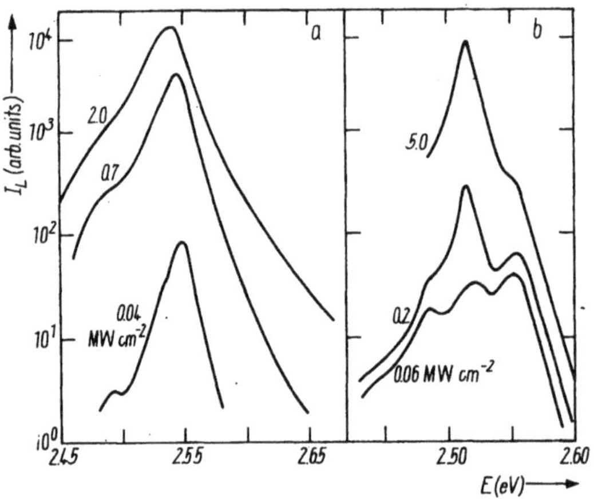

Fig. 1. Luminescence spectra of CdS a) block crystals and b) thin platelet samples at different excitation intensities (77 K, $\vec{E} \perp \vec{c}, \measuredangle(\vec{k}, \vec{c})$)

feature of group (I) luminescence spectra is the presence of a single broad band, whose peak position corresponds to the A-exciton ground state energy E_{1s}^{A}. The emission-intensity dependence (ID) on excitation intensity (Fig. 2, curve a) is square up to I_{ex} = 0.1 MWcm^{-2}, and becomes linear at higher excitation intensities. The peak position, halfwidth, and ID show that this band is due to elastic exciton-exciton collisions of two 1s-excitons scattered into the same band (P_1-line) /1, 2, 5/. At highest pump intensities the band shifts slightly towards lower energies due to the onset of further exciton-exciton collision processes (P_2 to P_∞) /1, 5/. The exciton effective temperature determined from the near high energy slope of the band /2/ exceeds the lattice temperature (Fig. 2, curve b). The second exponential slope in the far high energy region $\hbar\omega > E_g$ indicates a significant heating of the excited electron gas at the surface of the sample /3/.

Another situation is observed in the group (II) samples as shown in Fig. 1b. The luminescence intensity of the A-exciton line is saturated quickly (Fig. 2, curve c). At very low excitation a linear dependence of the ID of the A-exciton line can be observed in agreement with /6/. In the spectra of group (II) samples the A-LO band located on the low energy side of the exciton line already starts to dominate at relatively moderate pump intensity. The strong increase of this band points out its stimulated origin (Fig. 2, curve d) /7/. The

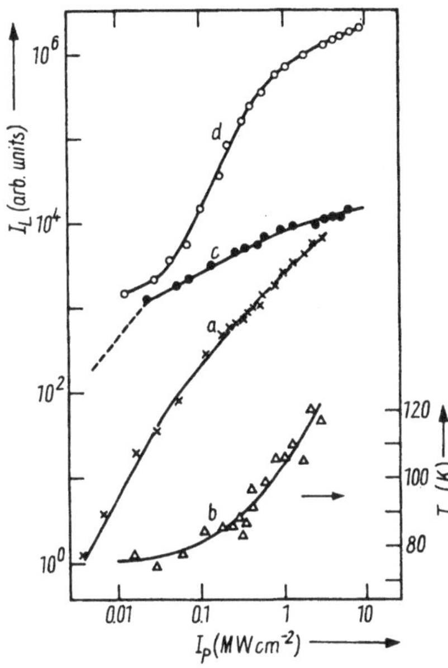

Fig. 2. Dependence of emission intensity on excitation (a) of the A-exciton line of CdSe block crystals, (c) of the A-exciton line of CdSe platelet samples, and (d) of the A-LO phonon line of CdSe platelet samples; (b) dependence of the exciton effective temperature T_{exc} on the excitation intensity (CdSe block crystal)

striking feature of group (II) spectra is the absence of a significant deviation of the high energy exponential slope from the value given by the lattice temperature even at highest pump intensities.

Investigating the luminescence lineshape the influence of the strong increase of the exciton damping Γ_{exc} with exciton density up to $\Gamma_{exc} \gtrsim k_B T_{exc}$ due to many-exciton processes as found in /8/ for group (II) samples must be discussed. To do this, we have calculated a convolution lineshape considering the thermal distribution of excitons /5/ as well as the exciton damping Γ_{exc}. The calculation shows (i) that the consideration of damping Γ_{exc} results in a symmetrical luminescence lineshape, (ii) which is broadened with increasing Γ_{exc} at given T_{exc}, and (iii) whose slope on both sides is influenced only by T_{exc} in the damping region under consideration.

To explain the obtained spectral differences between group (I) and (II) samples the different properties as quality, geometry, lifetime shortening due to stimulation, diffusion, and, hence, different non-equilibrium quasi-particle densities must be taken into account. In block crystals the macroscopic inhomogeneity and the geometry prevent stimulated recombination. The quasi-particle density increases here as a square root of the excitation intensity and can be estimated from the observed heating to reach $\approx 10^{18}$ cm^{-3} at highest intensities /3/. The homogeneous and macroscopically pure platelet samples have low thresholds of stimulated recombination and, hence, the increase of the quasi-particle density is a weak function of pumping /9/. Under peak excitation intensity densities of $\approx 10^{17}$ cm^{-3} are available in these samples /8/.

The tested electron or exciton thermalisation with their own temperature likely occurs at long lifetimes of excited species and at densities of 10^{17} to 10^{18} cm^{-3}. The quasi-particle photoheating effects are difficult to observe in group (II) crystals under the used experimental conditions.

In conclusions it should be noted that the selection of the crystal type with local or bulk purity has to depend on the aim of investigation: Group (I) crystals are preferable for generation of extremely high non-equilibrium quasi-particle densities, while group (II) crystals are preferable for the high efficiency of stimulated emission and intrinsic bulk transport properties.

References

/1/ R. BALTRAMIEJŪNAS and E. KUOKŠTIS, phys. stat. sol. (b) 111, 187 (1982).
/2/ R. BALTRAMIEJŪNAS, E. KUOKŠTIS, and A. ŽUKAUSKAS, Solid State Commun. 44, 955 (1982); phys. stat. sol. (b) 119, 453 (1983).
/3/ R. BALTRAMIEJŪNAS, A. ŽUKAUSKAS, and E. KUOKŠTIS, Zh. eksper. teor. Fiz. 83, 1215 (1982).
/4/ J. PULS and J. VOIGT, phys. stat. sol. (b) 94, 199 (1979).
/5/ T. MORIYA and T. KUSHIDA, J. Phys. Soc. Japan 40, 1668 (1976).
/6/ J. VOIGT and F. SPIEGELBERG, phys. stat. sol. (b) 30, 659 (1968).
/7/ J. VOIGT and F. MIR, phys. stat. sol. (b) 68, K195 (1975).
/8/ I. RÜCKMANN, V. MAY, and J. VOIGT, phys. stat. sol. (b) 102, 97 (1980).
/9/ V.D. EGOROV, phys. stat. sol. (b) 102, 317 (1980).

(Received October 19, 1984)

phys. stat. sol. (b) 126, K139 (1984)

Subject classification: 13.1 and 20.1; 12.2; 22; 22.8

Institute of Physics, Academy of Sciences of the Azerbaidzhan SSR, Baku[1])

The Influence of Hydrostatic Pressure on the Fundamental Absorption Edge of Crystals with TlSe-Type Structure

By

K.R. ALLAKHVERDIEV, S.S. BABAEV, N.A. BAKHYSHOV, T.G. MAMEDOV, E.YU. SALAEV, and I.K. EFENDIEVA

$A^{III}B^{VI}$ group monocrystals with chain-type crystal structure (TlSe, $TlInSe_2$, etc.) /1/ occupy a special place in the row of compounds with strongly anisotropic crystal structure. They are perspective materials for optoelectronics /2/.

The influence of hydrostatic pressure on the energy gaps of TlSe and $TlInSe_2$ was investigated in /3 to 7/. The values of different authors for the pressure coefficients of indirect (dE_g^i/dP) and direct (dE_g^d/dP) energy gaps for TlSe and $TlInSe_2$ are presented in Table 1. It is seen that these data are contradictory for TlSe.

The aim of the present note is the investigation of the pressure behaviour of energy band gaps of crystals with TlSe-type crystal structure: TlSe, $TlInSe_2$, and mixed crystals $TlInSe_{2(1-x)}S_{2x}$ ($0 \leq x \leq 0.25$) on their basis.

The crystals for the investigations were grown by the Bridgman and the horizontal zone level methods in evacuated quartz tubes. The samples with thickness 35 to 150 μm for measurements were cleaved from block single crystals and then gently polished. The method of registration of the transmission spectra at hydrostatic pressure did not differ from that described earlier /4, 5/. The spectra measured upon increasing pressure (from atmospheric to 5.5×10^8 Pa) coincide with the spectra measured when decreasing the pressure (from 5.5×10^8 Pa to atmospheric). The absorption coefficient was calculated from transmission spectra using the value of the refractive index /8, 9/. The variation of the refractive index and of the isothermal compressibility with pressure was neglected. With the aim of obtaining information about the pressure behaviour of energy band gaps the absorption spectra were analyzed with

1) prospekt Narimanova 33, 370143 Baku, USSR.

Table 1

Values of pressure coefficients of crystals with TlSe-type structure (the accuracy of pressure coefficient determination was not worse than $\pm 4\%$)

compound	dE_g^i/dP (10^{-10} eV/Pa)	dE_g^d/dP (10^{-10} eV/Pa)	ref.
TlSe	-1.0	-1.6	/3/
	-1.4	-2.1	/6/
	-1.85	-	/7/
	$-(2.0\pm 0.2)$	-	/4/
	-2	-1.25	present
$TlInSe_2$	$-(1.48\pm 0.08)$	-1.10	/5/
	-		present
$TlInSe_{1.9}S_{0.1}$	-1.45	-	present
$TlInSe_{1.8}S_{0.2}$	-1.43	-	present
$TlInSe_{1.7}S_{0.3}$	-1.37	-	present
$TlInSe_{1.5}S_{0.5}$	-1.3	-	present

the formula

$$(\alpha h\nu)^m \sim h\nu - E_g ,$$

where m = 1/2 and 2/3 for allowed indirect and forbidden direct transitions, respectively. Fig. 1a shows a plot of $(\alpha h\nu)^{2/3} \sim f(h\nu)$ of $TlInSe_2$ in the ($600 < \alpha < 1600$ cm^{-1}) absorption region at some pressures. The value of $dE_g^d/dP = -(1.10 \pm 0.03) \times 10^{-10}$ eV/Pa was determined from the pressure dependence of the direct band gap of $TlInSe_2$ (Fig. 1a, inset). The value of the minimum direct band gap of $TlInSe_2$ at P = 0 is $E_g^d = 1.358$ eV. This result is in accordance with $E_g^d = 1.36$ eV obtained in /10/ from theoretical band structure calculations.

The values of dE_g^i/dP and dE_g^d/dP for TlSe determined by us from the pressure dependence of indirect and direct band gap energies (Fig. 1b) and the value of dE_g^i/dP for mixed crystals $TlInSe_{2(1-x)}S_{2x}$ ($0 \leq x \leq 0.25$) are presented in

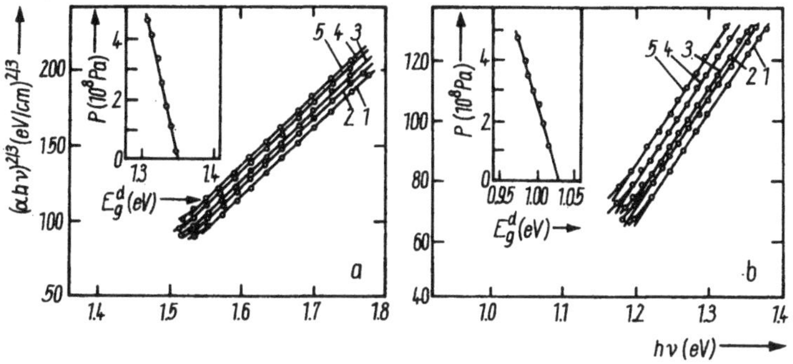

Fig. 1. a) Dependences of $(\alpha h\nu)^{2/3} = f(h\nu)$ at different pressures for a) $TlInSe_2$ crystals ((1) 1.2×10^8, (2) 1.9×10^8, (3) 2.5×10^8, (4) 3.5×10^8, (5) 4.7×10^8 Pa); the inset shows the energy band gap E_g^d of $TlInSe_2$ crystals at different pressures and b) TlSe crystals ((1) 0.3×10^8, (2) 1.8×10^8, (3) 2.6×10^8, (4) 4.2×10^8, (5) 4.7×10^8 Pa); the inset shows the energy band gap E_g^d of TlSe crystals at different pressures

Table 1. Both in thallium selenide and in $TlInSe_{2(1-x)}S_{2x}$ $(0 \leqq x \leqq 0.25)$ the indirect and direct band gaps linearly decrease with pressure. As seen from the table, the pressure coefficient dE_g^i/dP of mixed crystals $TlInSe_{2(1-x)}S_{2x}$ $(0 \leqq x \leqq 0.25)$ decreases with increasing sulphur concentration.

It is necessary to note that the relation $dE_g^i/dP > dE_g^d/dP$ of pressure coefficients obtained by us for TlSe and $TlInSe_2$ differs from the corresponding one predicted by theory for TlSe /6/. At the same time the value $dE_g^i/dP = -(2 \pm 0.05)\times10^{-10}$ eV/Pa obtained by us in the present work for TlSe is in accordance with $-(2.3 \pm 0.2)\times10^{-10}$ eV/Pa /4/ and -1.85×10^{-10} eV/Pa /7/ measured by us in previous optical and electrical investigations.

Using our value of the pressure coefficient of the indirect band gap for $TlInSe_2$ and the temperature coefficient reported by us previously /11/, we determine the contributions due to lattice expansion $(dE/dT)_p$ and electron-phonon interaction $(dE/dT)_{e-ph}$ to the temperature coefficient dE/dT of the fundamental absorption edge.

It is known that

$$dE_g/dT = (dE_g/dT)_p + (dE_g/dT)_{e-ph} ,$$

where

$$(dE_g/dT)_p = -(\beta/\varkappa)(dE_g/dP)$$

and β is the volume thermal expansion coefficient, $\varkappa$ the isothermal compressibility.

For polycrystalline $TlInSe_2$ $\beta = 3x9.65x10^{-6}\ K^{-1}$, $\varkappa = 1.763x10^{-11}\ m^2/N$ determined by us in /12/ at T = 290 K. The contribution due to lattice expansion to the temperature coefficient of the indirect band gap of $TlInSe_2$ is

$$(dE_g^i/dT)_p = 2.43x10^{-4}\ eV/K.$$

The contribution due to electron-phonon interaction is

$$(dE_g^i/dT)_{e-ph} = -1.18x10^{-4}\ eV/K.$$

These contributions to the temperature coefficient of indirect and direct band gaps for TlSe are

$$(dE_g^i/dT)_p = 2.31x10^{-4}\ eV/K (3.24x10^{-4}\ eV/K)\ ,$$

$$(dE_g^i/dT)_{e-ph} = -6.26x10^{-4}\ eV/K (-7.19x10^{-4}\ eV/K)\ ,$$

$$(dE_g^d/dT)_p = 1.32x10^{-4}\ eV/K (1.94x10^{-4}\ eV/K)\ ,$$

$$(dE_g^d/dT)_{e-ph} = -6.89x10^{-4}\ eV/K (-7.52x10^{-4}\ eV/K)\ .$$

We use the values of dE_g^i/dT and dE_g^d/dT reported by us for TlSe in /13/ and the values of β, $\varkappa$ given in /14/. The quantities obtained using the values of β and $\varkappa$ from /15/ are presented in brackets.

References

/1/ J.A. KETELAAR, W.H. T'HART, M. MOEREL, and D. POLDER, Z. Krist. 101, 396 (1939).

/2/ G.D. GUSEINOV and G.B. ABDULLAEV, Dokl. Akad. Nauk SSSR 208, 1052 (1973).

/3/ G.R. VALYUKONIS, A.S. MEDEISHIS, and A.YU. SHILEIKA, Fiz. Tekh. Poluprov. 16, 1137 (1982).

/4/ K.R. ALLAKHVERDIEV, T.G. MAMEDOV, E.YU. SALAEV, and I.K. EFENDIEVA, phys. stat. sol. (b) 113, K127 (1982).

/5/ K.R. ALLAKHVERDIEV, T.G. MAMEDOV, E.YU. SALAEV, and I.K. EFENDIEVA, phys. stat. sol. (b) 117, K109 (1983).

/6/ F.M. GASHIMZADE and G.S. ORUDZHEV, phys. stat. sol. (b) 106, K67 (1981).

/7/ K.R. ALLAKHVERDIEV, SH.G. GASYMOV, T.G. MAMEDOV, M.A. NIZAMETDINOVA, and E.YU. SALAEV, Fiz. Tekh. Poluprov. 17, 203 (1983).

/8/ K.R. ALLAKHVERDIEV, M.A. NIZAMETDINOVA, R.M. SARDARLY, E.A. VINOGRADOV, and G.N. ZHIZHIN, Proc. Internat. Conf. Lattice Dynamics, Flammarion, Paris 1978 (p. 95).

/9/ E.A. VINOGRADOV, N.M. GASANLY, B.M. DZHAVADOV, and V.I. TAGIROV, Fiz. tverd. Tela 21, 2793 (1979).

/10/ F.M. GASHIMZADE and G.S. ORUDZHEV, Dokl. Akad. Nauk Azerb. SSR No. 12, 18 (1980).

/11/ K.R. ALLAKHVERDIEV, T.G. MAMEDOV, E.YU. SALAEV, and I.K. EFENDIEVA, phys. stat. sol. (b) 113, K43 (1982).

/12/ K.R. ALLAKHVERDIEV, D.D. GUSEINOV, T.G. MAMEDOV, T.S. MAMEDOV, and I.K. EFENDIEVA, Dokl. Akad. Nauk Azerb. SSR No. 2, 18 (1984).

/13/ K.R. ALLAKHVERDIEV, T.G. MAMEDOV, E.YU. SALAEV, and I.K. EFENDIEVA, Preprint No. 84, Inst. Phys. Acad. Sci. Azerb. SSR, Baku 1983.

/14/ K.P. MAMEDOV and S.B. BAGIROV, Izv. Akad. Nauk Azerb. SSR, Ser. fiz. tekh. i mat. Nauk No. 1, 73 (1967).

/15/ K. SHIN and K. HASHIMOTO, Technol. Repis Kynshi Univ. 45, 820 (1972).

(Received June 5, 1984)

phys. stat. sol. (b) 126, K145 (1984)

Subject classification: 13.1 and 20.1; 12.2 ; 22.8.2

Department of Electrical Engineering, Faculty of Science and Technology (a) and Faculty of Engineering (b), Science University of Tokyo

Pressure Effect on the Absorption Edges of $CdIn_2S_4$ and Related Compounds

By

H. NAKANISHI (a) and T. IRIE (b)

A study of the non-magnetic chalcogenide spinels is important since it furnishes fundamental knowledge on magnetic semiconductors. Electrical and optical studies have been made on several chalcogenide spinels. However, the effect of hydrostatic pressure on the semiconducting properties has not been reported yet.

The present work is the first investigation of the hydrostatic pressure effect on the optical absorption edges of $CdIn_2S_4$ and related semiconductors with spinel structures.

The single crystals used for the measurements are prepared by the normal freezing method, cut into sufficiently small pieces, and polished down to thicknesses of about 100 to 10 μm. The hydrostatic pressure was applied by a diamond-anvil pressure cell using a mixture of methyl-ethyl alcohol (4:1) as the pressure transmitting medium. The absorption spectra were measured by a Nihon-Bunko SS-50 monochromator equipped with a Hamamatsu photonics R636 or 7102 photomultipliers.

It is known that the substances adopted in this study are all indirect-gap semiconductors /1 to 3/. Thus, the absorption spectra were measured only in the wavelength region where the absorption coefficients vary between 5×10^2 and 5×10^3 cm^{-1}. All measurements were made at room temperature.

Fig. 1 shows the absorption spectra of $CdIn_2S_4$ at various pressures in $(\alpha\hbar\omega)^{1/2}$ versus $\hbar\omega$ coordinates. It can be seen that the absorption edge shifts to higher energy with increasing pressure.

Fig. 2 shows the absorption edge determined from the curves in Fig. 1 as a function of pressure, yielding a pressure coefficient of the indirect energy gap, $(\partial E_g/\partial P)_T$, of $+7.1 \times 10^{-11}$ eV/Pa for $CdIn_2S_4$.

In general, the observed temperature dependence of the energy gap $(\partial E_g/\partial T)_P$ can be written as the sum of two terms:

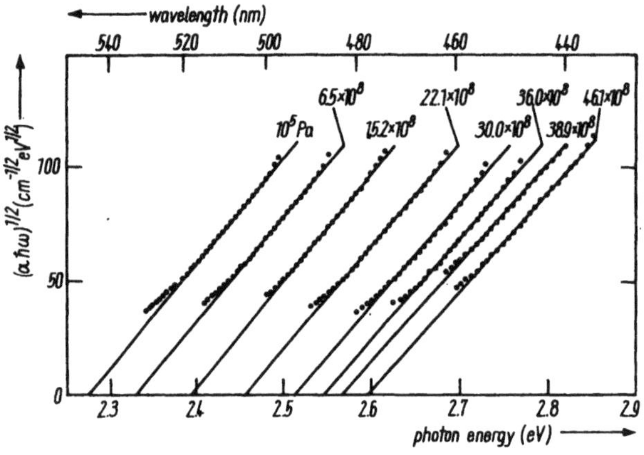

Fig. 1. Absorption spectra of $CdIn_2S_4$ at various pressures and RT

$$\left(\frac{\partial E_g}{\partial T}\right)_P = \left(\frac{\partial E_g}{\partial T}\right)_V - \frac{3\alpha_T}{K}\left(\frac{\partial E_g}{\partial P}\right)_T \quad , \tag{1}$$

where the first and the second terms are the electron-lattice and thermal expansion contributions to $(\partial E_g/\partial T)_P$, respectively, α_T is the linear thermal expansion coefficient, and K is the isothermal compressibility.

For $CdIn_2S_4$, the values of $(\partial E_g/\partial T)_P$ /1/, α_T /4/, and K /5/ at room temperature are known and are listed in Table 1 along with the measured value of $(\partial E_g/\partial P)_T$ in the present work. These values yield

$$\frac{3\alpha_T}{K}\left(\frac{\partial E_g}{\partial P}\right)_T = 1.0\times10^{-4}\ eV/K \tag{2}$$

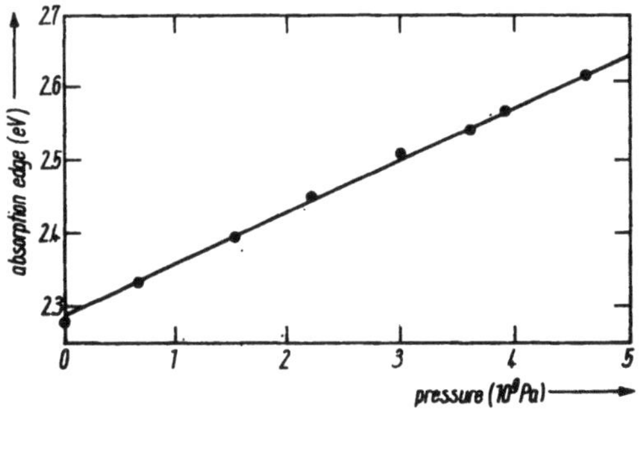

Fig. 2. Absorption edge as a function of pressure for $CdIn_2S_4$ at RT

Table 1

Temperature and pressure coefficients of energy gap, linear thermal expansion coefficient, and compressibility of $CdIn_2S_4$

$(\partial E_g/\partial T)_P$ (10^{-4} eV/K)	$(\partial E_g/\partial P)_T$ (10^{-11} eV/Pa)	α_T (10^{-6} K^{-1})	K (10^{-11} m^2/N)
-7.5	+7.1	8.4	1.76

and

$$\left(\frac{\partial E_g}{\partial T}\right)_V = -6.5\text{x}10^{-4}\ \text{eV/K}\ , \tag{3}$$

indicating that the electron-lattice interaction mainly contributes to the temperature shift of the indirect energy gap of $CdIn_2S_4$ at room temperature.

Fig. 3 shows the energy gap versus pressure of the spinel-type semiconductors In_2S_3, $CuIn_5S_8$, and $AgIn_5S_8$. For these three materials, the energy gaps at various pressures were determined from the photon energies at which the absorption coefficients had a fixed value, i.e. 500 cm^{-1} for In_2S_3 and $AgIn_5S_8$ and 300 cm^{-1} for $CuIn_5S_8$. Such a way to determine the energy gap as a function of pressure was justified by applying it to $CdIn_2S_4$, the value of which was already determined in an accurate way, i.e. the $(\alpha\hbar\omega)^{1/2}$ versus $\hbar\omega$ plots (Fig. 1).

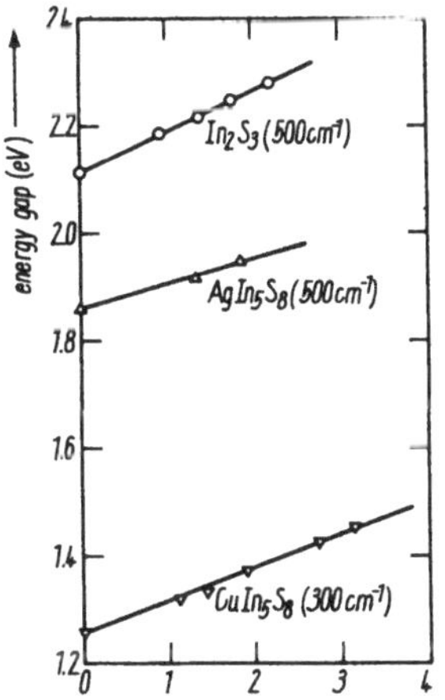

Fig. 3. Energy gap versus pressure of the spinel-type semiconductors In_2S_3 (o), $CuIn_5S_8$ (Δ), and $AgIn_5S_8$ (∇) at RT

Table 2

Pressure coefficients of energy gap of non-magnetic chalcogenide spinels

	$(\partial E_g/\partial P)_T$ (10^{-11} eV/Pa)
$CdIn_2S_4$	+7.1
In_2S_3	+7.6
$CuIn_5S_8$	+6.1
$AgIn_5S_8$	+4.7

The pressure coefficients of the energy gaps of the spinels we have studied are summarized in Table 2. It can be seen that the pressure coefficient $(\partial E_g/\partial P)_T$ is positive for all the compounds.

References

/1/ H. NAKANISHI, Japan. J. appl. Phys. 19, 103 (1980).

/2/ H. NAKANISHI, S. ENDO, and T. IRIE, Japan. J. appl. Phys., Suppl. 3, 19, 261 (1980).

/3/ A. USUJIMA, S. TAKEUCHI, S. ENDO, and T. IRIE, Japan. J. appl. Phys. 20, L505 (1981).

/4/ T. TOYODA, Y. SEKI, and M. YABE, Proc. Fourth Japan. Symp. Thermophysical Properties, Yokohama 1983 (p. 47).

/5/ M. YAMADA, T. SHIRAI, K. YAMAMOTO, and K. YABE, J. Phys. Soc. Japan 48, 874 (1980).

(Received June 5, 1984)

phys. stat. sol. (b) 126, K149 (1984)
Subject classification: 13.1 and 13.5.2; 22.8

Sektion Physik der Bergakademie Freiberg[1)]

Electron Energy Loss Measurements on $CdSnP_2$: Plasmon Data, Dielectric Function, Band Structure Data

By

H. BOUDRIOT, R. GRÜNDLER[2)], K. DEUS, T. STÖCKERT, and H. A. SCHNEIDER

The interest in the optical properties of the ternary analogy of $A^{III}B^{V}$ compounds, the $A^{II}B^{IV}C_2^{V}$ compounds, arises from theoretical as well as practical reasons. The band structure of a ternary compound can be obtained in its gross features by backfolding of the corresponding $A^{III}B^{V}$ band structure into the first Brillouin zone of its ternary analog. Then at $\vec{k} = 0$ interband transitions may occur that cannot be observed in the corresponding binary compound (e.g., in the indirect gap GaP and pseudodirect gap ternary compound $ZnGeP_2$ /1/). This led to a practical interest in this compound class as a potential material basis for optoelectronics.

The optical properties of many ternary compounds are intensively investigated in the energy region below 6 eV mainly by modulation spectroscopic methods (e.g. /2/). At higher energies reliable data are nearly exclusively missing. There is one report on reflectivity measurements up to 12 eV /3/ from which the dielectric function $\varepsilon(E) = \varepsilon_1(E) + i\,\varepsilon_2(E)$ is determined by Kramers-Kronig analysis affected with all the uncertainties that arise from the narrow-limited energy range and the error in the reflection coefficient values at higher energies.

In this short note we report on transmission electron energy loss (TEL) measurements on monocrystalline $CdSnP_2$ thin films. From the experimental energy loss spectrum $\varepsilon(E)$ was computed. The results are discussed under considerations of two aspects: comparison with results obtained from reflectivity measurements /3, 4/ and with data on InP (binary analog of $CdSnP_2$) as far as available /5, 6/.

---

1) Silbermannstr. 1, DDR-9200 Freiberg, GDR.

2) Zentralinstitut für Elektronenphysik der Akademie der Wissenschaften der DDR, Bereich Dünne Schichten, Hausvogteiplatz 5/7, DDR-1086 Berlin, GDR.

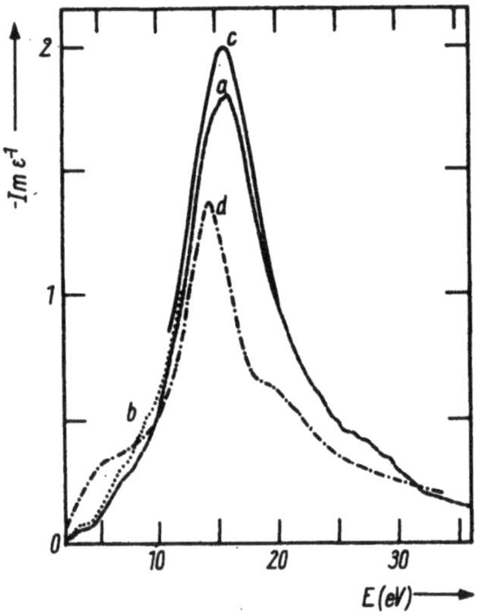

Fig. 1. Energy loss function $-Im\ \varepsilon^{-1}(E)$ of $CdSnP_2$: (a) present TEL measurements, (b) reflectivity measurements /3, 4/, (c) Drude model values. For comparison $-Im\ \varepsilon^{-1}(E)$ for InP /5/ (d) is shown

The TEL spectra of $CdSnP_2$ thin films have been measured up to 40 eV using 35 keV electrons with an energy resolution of about 0.8 eV. The details of the TEL spectrometer were described elsewhere /7/. Thin films of an average thickness of about 50 nm have been prepared by chemical etching of thin slabs of monocrystalline $CdSnP_2$ nearly down to perforation. From electron diffraction patterns we selected the monocrystalline region from which the TEL spectra were recorded. To avoid instabilities of the beam current and decrease the statistical error (<2% in the whole measured energy region) numerous single spectra were accumulated to obtain a TEL spectrum for analysis. To calculate the dielectric function and quantities derived from it, a number of correction and numerical computation procedures were carried out that have been described elsewhere /8/. The correct values of $\varepsilon(E)$ were determined by fitting $\varepsilon_1(0)$ to the value of 9.47 /9/.

Fig. 1 shows the energy loss function $-Im\ \varepsilon^{-1}(E)$ of $CdSnP_2$ calculated as described above together with that obtained from $\varepsilon_1(E)$ and $\varepsilon_2(E)$ of /3, 4/ and of InP /5/. The volume plasmon energy E_p that is given as usually by the energy of the main maximum of $-Im\ \varepsilon^{-1}(E)$ is $E_p = 15.7$ eV. The corresponding free electron model value is $E_p^{fe} = 14.6$ eV taking N = 16 valence electrons per molecule ($5s^2$ of Cd, $5p^2$ and $5s^2$ of Sn, $3p^3$ and $3s^2$ of P). It is somewhat surprising that the free electron value E_p^{fe} is essentially smaller than E_p since in most $A^{III}B^V$ compounds this difference is very small. To explain it we used the Drude energy loss and dielectric function to model the corresponding experimental functions in the plasmon energy region as described in /10/. The fitting procedure (curve c in Fig. 1, 2, and 4) yields $E_p^D = 16.2$ eV

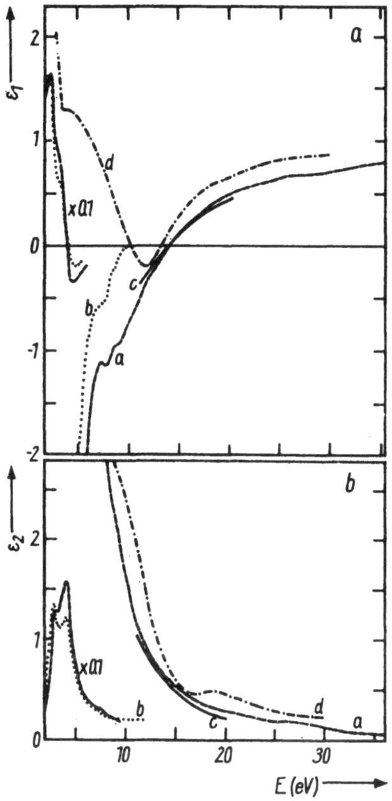

Fig. 2. Dielectric function part a) $\varepsilon_1(E)$ and b) $\varepsilon_2(E)$ of $CdSnP_2$, the same notation as in Fig. 1 is valid

and a damping constant $\Gamma^D = 8.3$ eV. From E_p^D we obtained N = 20.5. This N is the saturation value of the quantity

$$n_{eff}(E) = c \int_0^E \varepsilon_2(E')E'dE'$$

for large E. As seen from Fig. 4, in the plasmon energy region the experimental $n_{eff}(E)$ can be well reproduced by both parameters. The value of N is clearly above 16. This could mean that $4d^{10}$ electrons of Cd with a binding energy of about 9 eV located inside the valence band take part to certain extent in the volume plasmon excitation, too.

In Fig. 2 a and b $\varepsilon_1(E)$ and $\varepsilon_2(E)$ are plotted obtained as described above and after Kramers-Kronig analysis of the reflection coefficient measured between 1 and 12 eV /3, 4/. This is a too small energy range to get reliable data from a Kramers-Kronig analysis especially in the plasmon energy region where the zero crossing of ε_1 occurs. In /3, 4/ $\varepsilon_1 = 0$ at about 10 eV, we found for this energy 14.0 eV. To get more information about structures in the dielectric function parts we calculated their first energy derivatives and performed a line shape analysis according to /11/. From the shape and the magnitude of prominent structures in these first energy derivatives we identified critical points of the joint density of states, the energy and type of which are listed in Table 1. Together with them the energies of structures of the reflection coefficient are shown obtained by the present authors and in /3, 4/. For comparison corresponding data of the binary analog InP are tabulated /6/. We discuss only structures above 3 eV because the accuracy of the TEL spectrum is too poor at lower energies. Most of the structures can be correlated one to

Table 1

Energies (in eV) of structures of reflectivity measurements and of critical points of the band structure

$CdSnP_2$				InP			
critical point		R	R	critical point		R	R'/R
energy	type		/4/	energy	type /6/	/6/	/6/
3.85	M_1	3.8	3.54	4.70	M_0	4.80	
				4.71	M_1		
4.2	M_2	4.5	4.35	5.02	M_2	5.05	
				5.29	M_0	5.25	
5.5	M_0	5.75	5.5	5.5		5.6	5.5
				5.82	M_1		
				6.2		6.57	
7.4	M_1	7.75	7.3				
9.0	M_2	9.1					
10.9	M_3	11.0					
11.8	M_1						

another for both compounds according to the type of the corresponding critical point. The general trend can be confirmed that for the ternary compound the critical points have distinctly lower energies than the corresponding ones for the binary compound /4/. The behaviour of $\varepsilon_2(E)$ and of R(E) (see Fig. 3) can be divided into three parts: interband transition region mainly up to 10 eV, $4d^{10}$ Cd to conduction band transitions above about 10 eV, $4d^{10}$ Sn to conduction band transitions above about 24 eV corresponding to the $4d^{10}$ Sn binding energy. This is confirmed

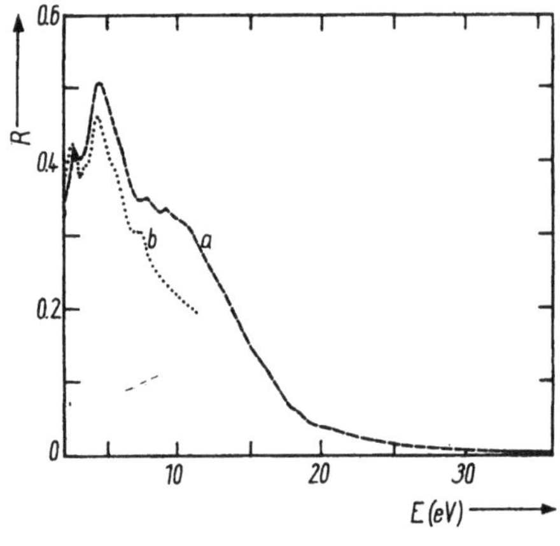

Fig. 3. Reflection coefficient R(E) of $CdSnP_2$: (a) present TEL measurements, (b) reflectivity measurements /3, 4/

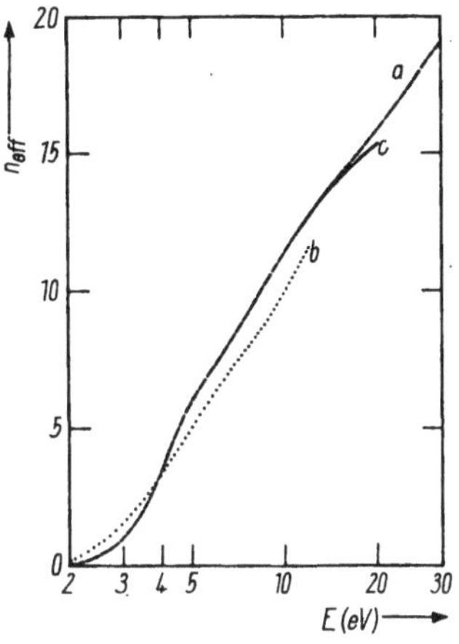

Fig. 4. n_{eff} of $CdSnP_2$: (a) present TEL measurements, (b) reflectivity measurements /3, 4/, (c) Drude model values

by the behaviour of $n_{eff}(E)$ (Fig. 4). At 8.5 and 23 eV the indicated saturation of $n_{eff}(E)$, that is typical of an oscillator strength exhaustion of a subgroup of the electron ensemble, changes to a new rise of the slope of n_{eff}. This rise of slope is indicative of the onset of transitions of a respective new subgroup. Together with our experimental $n_{eff}(E)$ the corresponding curve of /3, 4/ is plotted obtained from reflectance measurements. It has been recalculated to obtain the number of electrons per $CdSnP_2$ molecule. It is very similar to ours. The deviation amounts to about 1 electron per molecule between 4 and 12 eV.

References

/1/ C.V. DE ALVAREZ, M.L. COHEN, A.E. KOHN, Y. PETROFF, and Y.R. SHEN, Phys. Rev. B 10, 5175 (1974).

/2/ J. Physique, Suppl. 9, C3 (1975).

/3/ YU.F. KAVALYAUSKAS, G.Z. KRIVAITE, and A.YU. SHILEIKA, Liet. Fiz. Rink. 15, 605 (1975).

/4/ G. AMBRAZEVICHIUS, G. BABONAS, J. KAVALIAUSKAS, G. KRIVAITE, and A. SHILEIKA, J. Physique, Suppl. 9, C3-115 (1975).

/5/ J. FRANDON, B. BROUSSEAU, and L. DURRIEU, phys. stat. sol. (b) 72, K157 (1975).

/6/ C.V. DE ALVAREZ, J.P. WALTER, M.L. COHEN, J. STOKES, and Y.R. SHEN, Phys. Rev. B 6, 1412 (1972).

/7/ G. KAISER and B. KUBIER, Dissertation, Freiberg 1979.

/8/ W. TEWS and R. GRÜNDLER, phys. stat. sol. (b) 109, 255 (1982).

/9/ S.E. STOKOWSKI, Phys. Rev. B 6, 1294 (1972).

/10/ B. KUBIER, H. BOUDRIOT, K. DEUS, R. GRÜNDLER, H. ZSCHEILE, and E. BUHRIG, phys. stat. sol. (b) 106, K103 (1981).
H. BOUDRIOT, R. GRÜNDLER, B. KUBIER, K. DUES, and H.A. SCHNEIDER, phys. stat. sol. (b) 108, K129 (1981).

/11/ M. CARDONA, Solid State Phys., Suppl. 11 (1969).

(Received August 14, 1984)

phys. stat. sol. (b) 126, K155 (1984)
Subject classification: 13.1; 14.3; 15; 22

Faculty of Physics, University of Sofia[1)]

Density-of-States Effective Mass in Solid Solutions of the $(Ag_{x/2}Pb_{1-x}Sb_{x/2})Te$ Type

By
L. BORISOVA

The solid solutions with the general formula $(Ag_{x/2}Pb_{1-x}Sb_{x/2})Te$ are promising thermoelectric materials. Some of them show a thermoelectric efficiency higher than that of PbTe /1/.

In the present note we have studied the density-of-states effective mass m^*_{do} in the valence band extremum as a function of composition and temperature of the alloys. The calculations refer to single-phase samples of p-type conductivity obtained by the Bridgman method with carrier concentrations $p = (1 \text{ to } 9) \times 10^{18}\ cm^{-3}$ /1, 2/.

The effective masses in this solid solution system have not been studied. As they have the rocksalt structure, it may be respected that their electronic band structure is similar to that of the lead salts and the other $A^{IV}B^{VI}$ tellurides. On the basis of the results for PbTe we have used the Kane approximation /4/. Such an approach is justified on the basis of the experimental results obtained in studies on transport phenomena and optical properties of this system /1 to 3/. For all studied compositions (x = 0 to 1) an additional extremum in the valence band was found along the Σ direction with a higher effective mass and lower mobility. The energy separation between the extremum of the conduction band at the L point and the valence band along directions L and Σ has been determined, and it is found to be a monotonous function of the composition. It was observed that the maximum of the valence band on the Σ axis at 330 K is energetically higher than that at the L point.

Since at temperatures $T \leq 100$ K the valence band in Σ direction is energetically lower than that at L, we assume that at these temperatures the sample properties will be determined by the light p-carriers at L only. In such a case one might expect the dependence of m^*_{do} on the composition of solid solutions to correspond to the dependence of the energy gap width on the composition /4/.

---

1) blvd. Anton Ivanov 5, 1126 Sofia, Bulgaria.

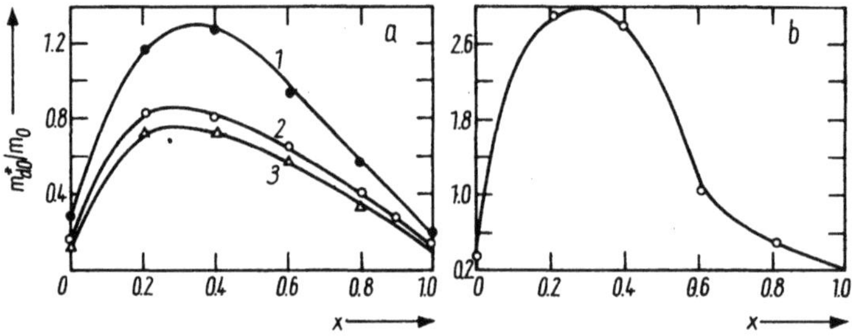

Fig. 1. The band edge density-of-states effective mass versus composition at a) 100 K and b) 350 K under the assumption of (1) parabolic band structure /4/, (2) non-parabolic band model /4/, (3) according to /5/

The band edge density-of-states effective mass has been found from the experimental values of the thermoelectric power, carrier concentration, and the band gap width E_g.

Fig. 1a and b represent a plot of the ratio m^*_{do}/m_o versus composition (m_o is the free electron mass) at temperatures 100 and 350 K, respectively. Curve 1 has been obtained under the assumption of a parabolic band structure /4/. Curve 2 represents the results obtained by accounting for the non-parabolicity of the bands by the Kane model. In such a case, the carrier concentration and the thermoelectric power are given by /4/

$$n = \frac{\left[2\, m^*_{do}\, k_o\, T\right]^{3/2}}{3\pi^2\, \hbar^3}\; {}^{o}L_{o}^{3/2}\,,$$

$$\alpha = \frac{k_o}{e}\left[\frac{{}^{1}L_{2}^{1}}{{}^{o}L_{2}^{1}} - E_F^*\right],$$

where ${}^{n}L_{k}^{m}(E_F^*, \beta)$ are the generalized Fermi integrals, $E_F^* = E_F/k_oT$, and $\beta = k_oT/E_g$ are the reduced Fermi energy and non-parabolicity factor, respectively. The results obtained by taking into account the energy dependence of the square of the transition probability matrix element /5/ are given in Fig. 1a by curve 3.

The temperature dependence of the ratio m^*_{do}/m_o is shown in Fig. 2. As seen, the m^*_{do} dependence on the solid solution compositions is characterized by a maximum at x = 0.2 to 0.4 (Fig. 1) which does not correspond to the

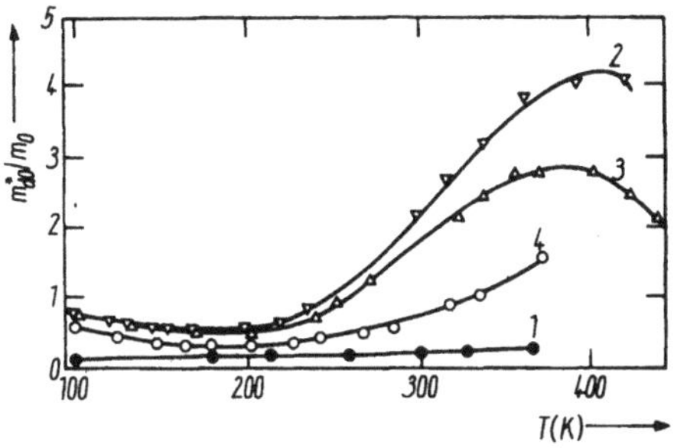

Fig. 2. Temperature dependence of the band edge density-of-states effective mass. x (1) = 0, (2) 0.2, (3) 0.4, (4) 0.8

dependence of the band gap on the composition /2, 3/. Only for PbTe samples (curve 1, Fig. 2) the results obtained are in agreement with the Kane model and $dm^*/dT \sim dE_g/dT$ /4/. For solid solutions at T ≧ 200 K, m^*_{do} increases considerably more stronger. The rate of increase depends on the carrier concentration - it is the highest for the highest concentration. Despite the specification made in the adopted model to find m^*_{do}, the nature of the dependence is not changed.

The results obtained are explained in terms of the changes in the band structure with composition. We suppose that the strong increase in m^*_{do} with concentration and temperature of the solid solutions is due to the effect of the band of heavy p-carriers in Σ direction, which with increasing temperature gets closer to that at L and at a higher rate than for PbTe /3, 6/. As a result of this its effect on the kinetic coefficients is greater. At temperatures higher than 330 K, in the transport phenomena only carriers from the valence band in Σ direction take part and the determined m^*_{do} values at T ≧ 330 K refer only to the heavy p-carriers characterized by a higher effective mass and lower mobility (Fig. 1b).

The results obtain are in qualitative agreement with the data of optical studies on the band structure of these solid solutions /7/. The substitution of Pb atoms by Ag and Sb atoms in the series $PbTe-AgSbTe_2$ is shown to change the spin-orbit interaction which affects the interaction matrix element in the expression for m^*_{do} /4/ and produces a change in the ordering of the levels in the bands. Thus, while in PbTc thc conduction band bottom is described mainly by the state L^{6-}_{3-}, in the solid solutions it is given by the state L^{6-}_{2-} i.e. the levels have interchanged their positions. It is concluded that the anisotropy coefficient in the series $PbTe-AgSbTe_2$ may decrease, expecting for $AgSbTe_2$ to be close to unity. From these results it may be concluded that the physical properties of both end compounds PbTe and $AgSbTe_2$ and the solid solutions existing between them cannot be described by application of the same model.

I am thankful to G. Ghristakudis for the computer treatment of the data by the method described in /5/ and to M. Baleva for useful discussions.

References

/1/ L. BORISOVA and S. DIMITROVA, phys. stat. sol. (a) 53, 403 (1979).

/2/ L.D. BORISOVA and ST.K. DIMITROVA, J. Phys. (Bulg.) 1, 50 (1975).

/3/ M. BALEVA and L. BORISOVA, phys. stat. sol. (b) 101, K57 (1980).

/4/ YU.I. RAVICH, B.A. EFIMOVA, and I.A. SMIRNOV, Metody issledovanya poluprovodnikov v primenenii k khalkogenidam svintsa PbTe, PbSe, PbS, Nauka, Moscow 1968.

/5/ L.V. VASSILEV and G.GH. GHRISTAKUDIS, phys. stat. sol. (b) 117, K53 (1983).

/6/ L.D. BORISOVA-LICHEVA, Ph.D. Dissertation, Sofia 1984.

/7/ M. BALEVA, Ph.D. Dissertation, Sofia 1982.

(Received July 26, 1984)

phys. stat. sol. (b) 126, K159 (1984)
Subject classification: 20.3; 22.4.1

Sektion Physik der Humboldt-Universität zu Berlin,
Bereich Halbleiteroptik[1)]

Non-Exponential ZnS:Mn Luminescence Decay Due to Energy Transfer

By
O. GOEDE, W. HEIMBRODT, and DANG DINH THONG

In some recent papers /1 to 3/ we have shown that for the ZnS:Mn luminescence at higher Mn concentrations c_{Mn} the energy transfer between the various centres is of great importance. The measured temperature and Mn-concentration dependence of the intensity and decay time of the yellow $Mn^{2+}(^4T_1 \rightarrow {}^6A_1)$ emission band could be explained assuming a radiationless resonance energy transfer from the primarily excited Mn^{2+} ions via several other Mn^{2+} ions finally to non-radiative centres. Furthermore, the additional red emission band appearing at $c_{Mn} \gtrsim 1$ mol% is found to be excited by an energy transfer from the usual Mn^{2+} centres as proved especially by the observed characteristic time delay of the red emission intensity after pulse excitation (see /3/).

In the present paper it is shown that also the observed non-exponentiality of the decay of the yellow emission band in ZnS:Mn can be explained quantitatively in dependence on the Mn concentration on the basis of the energy transfer model as also suggested recently in /4 to 6/. In the past the non-exponential decay of this emission band measured in photo- or electroluminescence was ascribed to the occurrence of n-atomic Mn clusters (n = 2, 3, 4) at higher c_{Mn} which were assumed to have decreasing radiative lifetimes with increasing n /7 to 10/.

In Fig. 1 the measured decay of the yellow emission intensity is given both for ZnS:Mn single crystals grown by I_2 transport reactions and electron-beam evaporated thin films for various Mn concentrations. The excitation was carried out by 2 ns pulses of 2.48 eV photons from a dye laser pumped by a N_2 pulse laser. As already pointed out in /3/ in contrast to the films this type of single crystals does not show the additional red emission band also at the highest Mn concentrations of 26.7 mol%. Further experimental details are described in /3/. As can be seen in Fig. 1 the decay time constant τ_∞ taken at sufficiently long times after the excitation pulse strongly decreases with increasing c_{Mn} as dis-

---

1) Neue Schönhauser Str. 20, DDR-1020 Berlin, GDR.

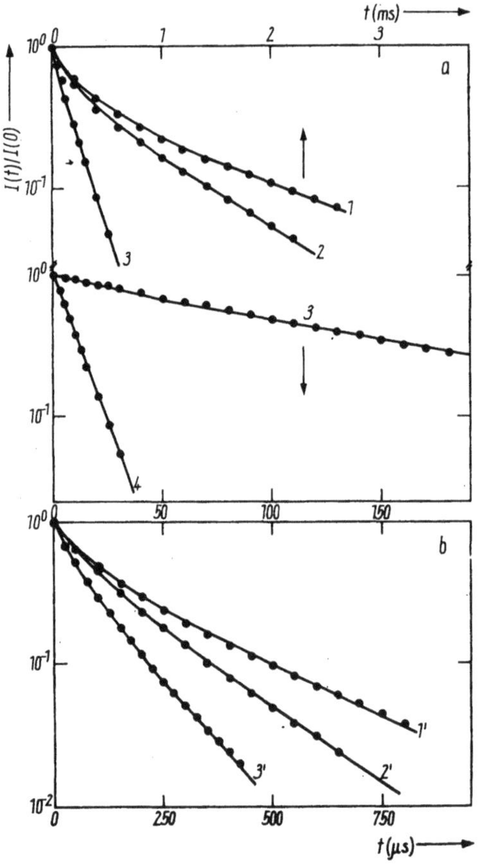

Fig. 1. Measured decay of the yellow emission band after pulse excitation at 77 K for ZnS:Mn a) single crystals and b) thin films with various Mn concentrations c_{Mn} (experimental points). a) c_{Mn} = (1) 0.02, (2) 1.0, (3) 9.5, and (4) 26.7 mol%; b) c_{Mn} = (1') 2.4, (2') 3,6, and (3') 5.5 mol%. Excitation energy 2.48 eV. Solid lines calculated after (1) with τ_y = 3.3 ms and c_n/c_o = = a) 1.2, b) 2.0 for curves 1', 2' and 2.4 for curve 3'

cussed in /1 to 3/. At shorter times a remarkable non-exponentiality of the decay is observed also for the lowest Mn concentration, which decreases with increasing c_{Mn}. For $c_{Mn} \gtrsim$ 8 mol% practically a pure exponential decay is found (see curves 3 and 4 in Fig. 1a).[2)]

Just this behaviour can be explained on the basis of the energy transfer model proposed in /1 to 3/. The radiationless energy transfer between the Mn^{2+} ions being randomly distributed on the cation sites should be described in a continuum approximation as an energy diffusion process with a diffusion constant D depending on the Mn concentration and temperature. For the following no special assumptions are necessary about the mechanism of this Mn^{2+}-Mn^{2+} energy transfer. For the energy transfer from an excited Mn^{2+} ion to a centre of radiationless transitions in a distance R a dipole-dipole interaction is supposed with a transfer rate proportional to $(R_o/R)^6$ where R_o is a range parameter. Then a theoretical expression can be applied which was obtained by Yokota and Tanimoto /12/ by solving the corresponding diffusion equation in a Padé approximation. This yields for the decay of the yellow Mn^{2+} emission band after a δ-shaped pulse excitation

2) This may correspond to the observed purely exponential decay of the Mn^{2+} luminescence band in $Cd_{0.49}Mn_{0.51}Te$ /11/.

$$I(t)/I(0) = \exp\left\{-\frac{t}{\tau_y} - \sqrt{\pi}\,\frac{c_n}{c_o}\left(\frac{t}{\tau_y}\right)^{1/2}\left[\frac{1 + 10.87\,\frac{D}{D_o}\left(\frac{t}{\tau_y}\right)^{2/3} + 15.5\left(\frac{D}{D_o}\right)^{2}\left(\frac{t}{\tau_y}\right)^{4/3}}{1 + 8.743\,\frac{D}{D_o}\left(\frac{t}{\tau_y}\right)^{2/3}}\right]^{3/4}\right\}. \qquad (1)$$

Here τ_y is the radiative lifetime of an isolated excited Mn^{2+} ion, c_n is the concentration of the nonradiative centres. $c_o = ((4\pi/3)R_o^3)^{-1}$ and $D_o = R_o^2/\tau_y$ are independent of c_{Mn}. Equation (1) yields at sufficiently long times an exponential decay with the time constant

$$\tau_\infty = \tau_y\left[1 + 1.773\sqrt{\pi}\,\frac{c_n}{c_o}\left(\frac{D}{D_o}\right)^{3/4}\right]^{-1} \qquad (2)$$

but for shorter times and sufficiently small D a strong non-exponential decay. Other theoretical treatments of the considered energy transfer problem do not lead to explicit formulas for the decay curves /13, 14/ or cannot explain the observed Mn concentration dependence /15/.

In the limiting case $c_{Mn} \to 0$, i.e. $D \to 0$, (1) simply reduces to the expression $I(t)/I(0) = \exp\left\{-t/\tau_y - \sqrt{\pi}\,\frac{c_n}{c_o}\left(\frac{t}{\tau_y}\right)^{1/2}\right\}$, which can be used to fit the measured decay curve for the lowest Mn concentration (see Fig. 1, curve 1). The obtained value for τ_y is somewhat larger than the usually accepted value between 1.7 and 2.0 ms. The decay curves measured for higher c_{Mn} can well be fitted by the general expression (1) both for the single crystals and thin films as shown in Fig. 1a and b, respectively. Especially the observed decrease of the non-exponentiality with increasing c_{Mn} is well represented. In the case of the single crystals the same c_n/c_o value could be used for all samples whereas for the thin films as may be expected a somewhat higher concentration c_n of the non-radiative centres has to be assumed. The obtained D/D_o values are plotted in Fig. 2 as a function of c_{Mn}. Because the range parameter R_o is unknown c_n and D cannot be determined absolutely. For a rough estimation a (maximum) value $R_o \lesssim 10^{-6}$ cm can be assumed leading to $c_n \gtrsim 3\times10^{17}$ cm^{-3} and $D \lesssim 2.6\times10^{-9}$ cm^2 s^{-1} for the ZnS:Mn single crystal with c_{Mn} = 9.5 mol% as an example.

As shown by the solid line in Fig. 2 for $c_{Mn} \lesssim 10$ mol% the Mn concentration dependence of the energy diffusion constant in ZnS:Mn can be described by $D \sim c_{Mn}^{4/3}$ as derived theoretically for dipole-dipole energy transfer processes /16/. In the case of a purely exponential decay according to (2) only the product $\frac{c_n}{c_o}\left(\frac{D}{D_o}\right)^{3/4}$ can be determined where $c_n/c_o \ll D/D_o$ must hold.

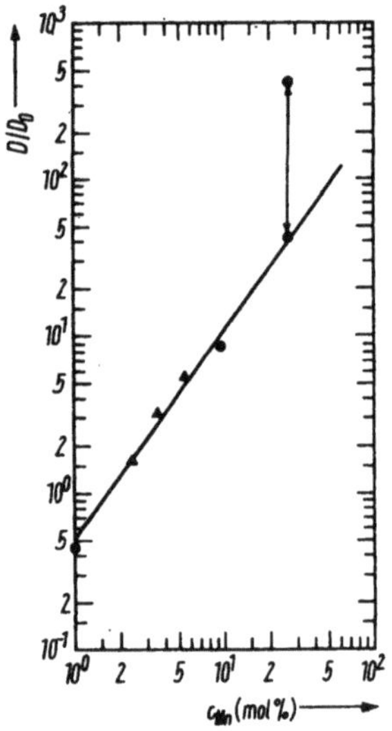

Fig. 2. Diffusion constant D of the excitation energy as a function of c_{Mn} for the ZnS:Mn single crystals (●) and thin films (▲), respectively. Solid line: $D \sim c_{Mn}^{4/3}$

For the highest investigated c_{Mn}, therefore, D/D_o can be fixed only into the range indicated in Fig. 2 by the arrow. It cannot be decided at present if a steep increase of D/D_o occurs connected with the approach to the percolation threshold of Mn in ZnS:Mn, or c_n/c_o increases by about a factor 6 at these c_{Mn} values. Further experimental investigations are in progress to answer this question which is strongly related to the nature of the nonradiative centres in ZnS:Mn.

The quantum efficiency for the yellow Mn emission band $\eta = \tau_y^{-1} \int_0^\infty I(t)/I(0)\,dt$ calculated by numerical integration of (1) is shown in Fig. 3. η strongly decreases both with increasing c_n and $D(c_{Mn})$. It should be pointed out that also for the lowest c_{Mn} a quantum efficiency $\eta < 1$ is obtained and that for the thin films a clearly smaller η results than for single crystals with the same Mn concentration.

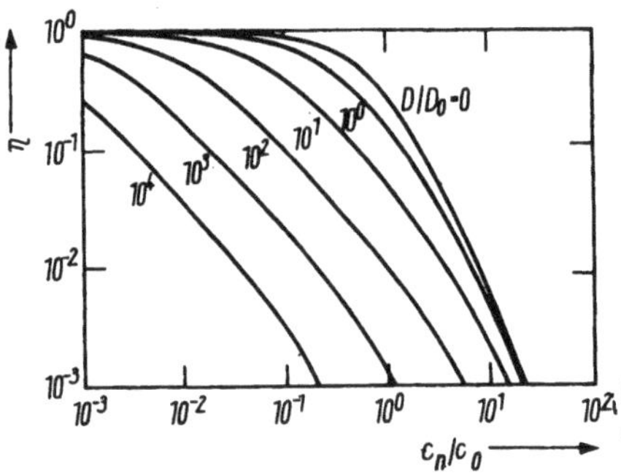

Fig. 3. Quantum efficiency η as a function of c_n/c_o for various D/D_o calculated after (1)

References

/1/ O. GOEDE and DANG DINH THONG, phys. stat. sol. (b) 124, 343 (1984).

/2/ DANG DINH THONG and O. GOEDE, phys. stat. sol. (b) 120, K145 (1983).

/3/ DANG DINH THONG, W. HEIMBRODT, D. HOMMEL, and O. GOEDE, phys. stat. sol. (a) 81, 695 (1984).

/4/ W. BUSSE, H.E. GUMLICH, M. KRAUSE, H.J. MOROS, J. SCHLIWINSKI, and D. TSCHIERSE, Preprint, TU Berlin (West), 1984.

/5/ J. BENOIT, P. BENALLOUL, A. GEOFFROY, N. BALBO, C. BARTHOU, J.P. DENIS, and B. BLANZAT, phys. stat. sol. (a) 83, 709 (1984).

/6/ P. BENALLOUL, J. BENOIT, J. DURAN, P. EVESQUE, and A. GEOFFROY, Solid State Commun. 51, 389 (1984).

/7/ N.A. VLASENKO, S.A. ZYNIO, and YU.V. KOPYTKO, phys. stat. sol. (a) 29, 671 (1975).

/8/ N.A. VLASENKO, YU.V. KOPYTKO, and V.S. PEKAR, phys. stat. sol. (a) 81, 661 (1984).

/9/ W. BUSSE, H.E. GUMLICH, B. MEISSNER, and D. THEIS, J. Lum. 12/13, 693 (1976).

/10/ H.E. GUMLICH, J. Lum. 23, 73 (1981).

/11/ E. MÜLLER, W. GEBHARDT, and V. GERHARDT, phys. stat. sol. (b) 113, 209 (1982).

/12/ M. YOKOTA and O. TANIMOTO, J. Phys. Soc. Japan 22, 779 (1967).

/13/ H. SIEBOLDT and J. HEBER, J. Lum. 22, 297 (1981).

/14/ A. BLUMEN, J. KLAFTER, and R. SILBEY, J. chem. Phys. 72, 5320 (1980).

/15/ J. HEBER, phys. stat. sol. 48, 319 (1971).

/16/ M. TRLIFAJ, Czech. J. Phys. 8, 510 (1958).

(Received November 7, 1984)

phys. stat. sol. (b) 126, K165 (1984)

Subject classification: 14.3.4 and 20.1; 23

Bereich Strahlenchemie, Hahn-Meitner-Institut für Kernforschung Berlin (West)[1]

Charge Carrier Generation in Solid Xenon by Two-Photon Absorption

By

E.-H. BÖTTCHER[2] and W.F. SCHMIDT

The study of electronic transport properties of solidified rare gases is of fundamental interest since this class of materials represents one of the simplest kinds of solid insulators. Besides, these materials are being used as detection media in ionization chambers for high energy physics.

Experimentally, excess charge carriers have to be introduced by external agents, for instance by irradiation with high energy X-ray or electrons or by illumination with UV light. The electron drift velocity and the energy of the band gap have been published /1, 2/. In this note we should like to report on the two-photon ionization of solid xenon, which was attained by illumination with light from an excimer laser at 193 and 249 nm.

Xenon was liquified in a parallel plate conductivity cell (electrode separation d = 1.75 mm) which had been placed in a cryostat with suprasil windows. Crystals were slowly grown from the melt. The experiments on the crystals were carried out at 160 K. Voltage was connected to one electrode while the other was connected to ground via the input resistor of a fast Tektronix oscilloscope. The energy of the laser pulse was monitored with a laser energy meter (Laser precision, model LP 152). The time profile of the laser pulse exhibited a full width at half maximum of 6 ns. Neutral density filters were used to reduce the energy of the laser pulse. The laser light was not focused. A typical oscilloscopic trace of the electronic component of the photoionization signal is shown in Fig. 1. The triangular decay is characteristic of homogeneous volume ionization /3/. From the knick point the drift time t_d and the drift velocity $v_d = d/t_d$ are obtained. Our data were obtained for the range of electric field strengths of 5.7 to 34.3 kV/cm where the drift velocity saturates with respect

---

1) Postbox, D-1000 Berlin 39, Berlin (West).

2) Present address: Institut für Festkörperphysik, Technische Universität Berlin, D-1000 Berlin 12, Berlin (West).

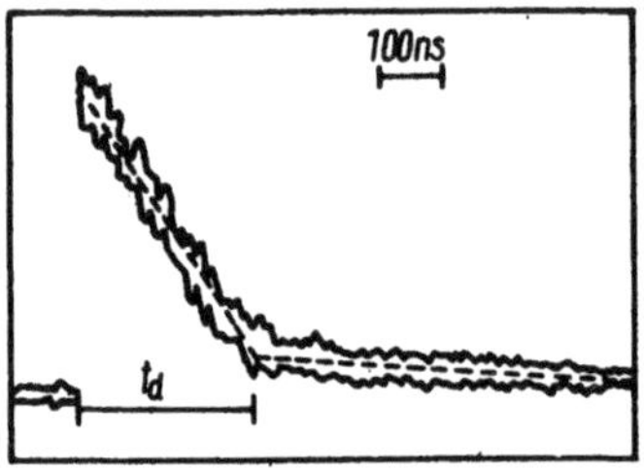

Fig. 1. Oscilloscopic trace of the electronic component of the photoionization current in solid Xe. F = 34.3 kV/cm, T = 160 K, λ = 248 nm

to the field strength. A value of $v_s = 6.3 \times 10^5$ cm/s was determined which agrees well with literature data /2, 4/.

The collected charge generated per laser pulse Q as a function of the energy E_p of the laser pulse entering the photoconductivity cell is shown in Fig. 2. E_p was varied from 0.5 to 0.015 mJ at 248 nm (5 eV) and from 0.25 to 0.02 mJ at 193 nm (6.4 eV). A quadratic dependence of Q on E_p was observed for both wavelengths.

In Fig. 3 a schematic representation of the band structure of solid xenon near the Γ-point ($\vec{k} = 0$) is shown. The positions of the n = 1 excitons originating from the first conduction band and the valence bands, which are split by spin-orbit interaction, are indicated by broken lines. The data of the band structure

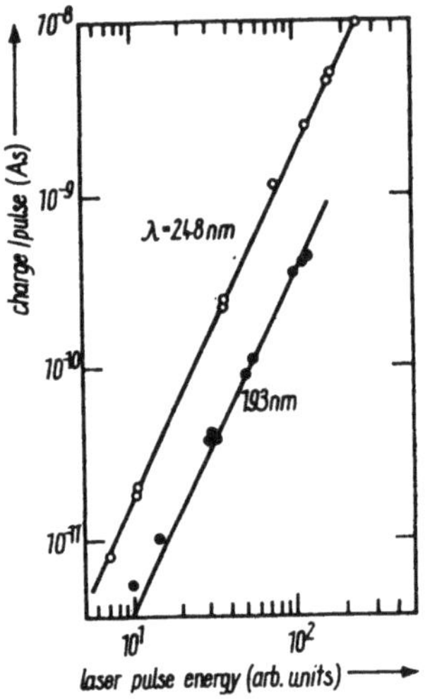

Fig.2

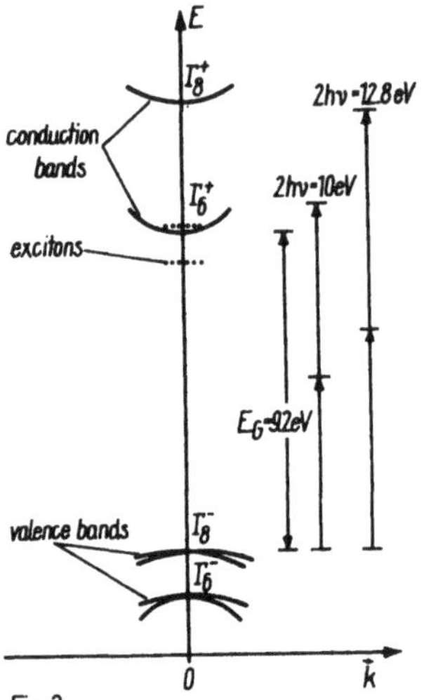

Fig.3

Fig. 2. Generated charge in solid Xe at T = 160 K as a function of the laser energy per pulse. Applied electric field strength 34.3 kV/cm

Fig. 3. Schematic band structure of solid Xe near $\vec{k} = 0$. The positions of the n = 1 excitons are indicated by broken lines are based on experimental and theoretical results which were summarized in a review article /5/. The energy of two photons at 248 nm is sufficient to excite electrons to the conduction band since the band gap is E_g = 9.28 eV /1/. The photon energy at 193 nm is smaller than the threshold energy for one photon absorption into the first exciton state at 8.4 eV /1/. Therefore, the charge carrier generation in solid Xe by irradiation with photons at 193 and 248 nm proceeds via the simultaneous absorption of two photons. The generated charge Q with respect to constant energy per laser pulse represents the charge carrier yield. The yield is shown in Fig. 4 as a function of the electric field strength F at 193 nm. In Fig. 4 also data are included which show the field dependence of the charge carrier yield of liquid xenon a 163.4 K measured with the same set-up. Data for the charge carrier yield of γ-irradiated liquid xenon are included /6/. It is interesting to note that in liquid xenon saturation of charge carrier collection is more easily approached than in solid xenon although the thermal electron mobility and the saturation drift velocity are both greater in solid xenon /2, 6/. It seems that superthermal electrons from the ionization act lose more energy per collision in solid xenon than in liquid xenon. Support for this point of view comes from measurements of the electron thermalization time τ, tor which in liquid xenon, τ = 6.5 ns and in solid xenon τ = 4.4 ns were measured /7/. The authors explain this effect by the much smaller effective phonon mass in the solid.

Since only a small fraction of the incident laser light is absorbed by the sample the generated charge per laser pulse Q is related to the two-photon absorption coefficient δ by the following relation /8/:

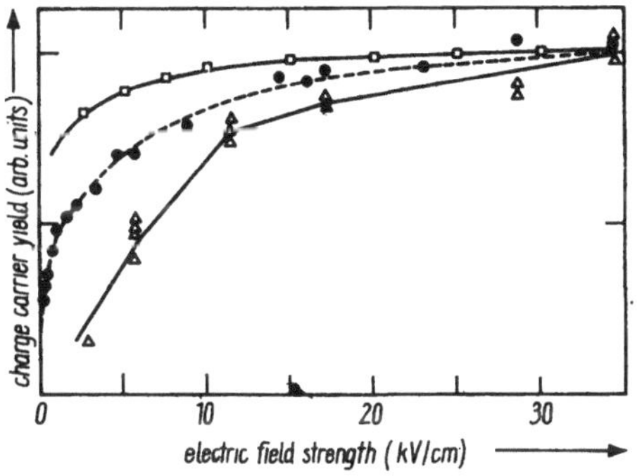

Fig. 4. Relative charge carrier yield as a function of the applied electric field strength; △ data for solid xenon; ● data for liquid xenon; □ data for γ-irradiated liquid xenon, after /6/. The data have been normalized at the highest value of the field strength

$$Q = e\Phi\delta(E/h\nu)^2 (\Delta t)^{-1} A^{-1} n l ; \qquad (1)$$

δ is given in units of $cm^4 s/(photon\ atom)$. Φ is the probability that a two-photon absorption at energy $h\nu$, with an externally applied field strength F and at a temperature T results in an electron-hole pair which escapes geminate recombination. A is the area of the incident laser beam between the cell electrodes, and n is the number of xenon atoms per cm^3. $E/h\nu$ denotes the number of photons per laser pulse, l is the electrode separation, and Δt is the FWHM of the laser pulse. The time profile of the laser pulse was approximated by a square wave pulse with the pulse width Δt.

From our data the following values for the product $\Phi\delta$ at 34.3 kV/cm are inferred: $\Phi\delta \approx 5\times10^{-50}\ cm^4 s\ photon^{-1} atom^{-1}$ at 248 nm and $\Phi\delta \approx 1.5\times10^{-50}\ cm^4 s\ photon^{-1} atom^{-1}$ at 193 nm. In solid Xe we can assume that every two-photon transition at the photon energies studied creates an electron-hole pair, which has a small recombination probability at the field strength applied. If we assume $\Phi \approx 1$ then the δ-values obtained are of the typical order of magnitude of δ-values obtained for electric dipole allowed two-photon transitions in solids and liquids /9, 10/.

A theoretical model /11/ can be used to describe the two-photon absorption (tpa) in solid Xe at 248 and 193 nm. In this model it is assumed that the tpa leads to final p-exciton states via discrete s-exciton states as intermediates. This theoretical approach has been successfully applied to the explanation of the experimental tpa spectra of the alkali halides /12/. In the case of tpa in solid Xe at 248 and 193 nm the n = 1 exciton states will mainly contribute to the virtual s-exciton state. By KrF laser radiation ($2\,h\nu = 10$ eV) continuum p-exciton states excited from the first valence band (Γ_8^-) and autoionizing p-exciton states from the second valence band (Γ_6^-) are the final states of the tpa in solid Xe. For ArF laser radiation ($2\,h\nu = 13.8$ eV) the situation is more complicated, because electrons from the first valence band (Γ_8^-) can be excited to the second conduction band (cf. Fig. 3). From the theoretically predicted band structure of solid Xe /13, 14/ one derives that the tpa at 193 nm out of the first valence band, leads only to continuum exciton states of the second conduction band (Γ_8^+), since the excited states lie above the top of the first conduction band (Γ_6^+). If the electrons are excited at 193 nm from the second valence band (Γ_6^-) they are lifted to p-exciton continuum states of the first conduction band. Solid xenon is a crystal with a face centered cubic structure. Electric dipole allowed tpa, therefore, requires the initial and final electronic state to have the same parity. Since the tpa transitions discussed here, fulfil this requirement, they are allowed /12/. This condition is also reflected by the order of magnitude of the values of δ extracted from the present experimental results.

References

/1/ U. ASAF and I.T. STEINBERGER, Phys. Letters A 41, 19 (1972).

/2/ L.S. MILLER, S. HOWE, and W.E. SPEAR, Phys. Rev. 166, 871 (1968).

/3/ W.F. SCHMIDT and G. BAKALE, Chem. Phys. Letters 17, 617 (1972).

/4/ E.M. GUSHCHIN, A.A. KRUGLOV, and I.M. OBODOVSKII, Soviet Phys.- J. exper. theor. Phys. 55, 650 (1982).

/5/ U. RÖSSLER, in: Rare Gas Solids, Vol. I, Ed. M.L. KLEIN and J.A. VENABLES, Academic Press, New York 1976 (p. 505).

/6/ S.S. HUANG and G.R. FREEMAN, Canad. J. Chem. 55, 1838 (1977).

/7/ U. SOWADA, J.M. WARMAN, and M.P. DE HAAS, Phys. Rev. B 25, 3434 (1982).

/8/ E.-H. BÖTTCHER and W.F. SCHMIDT, Proc. 7th ICDL, Berlin 1981 (p. 89).

/9/ W.M. MC CLAIN and R.A. HARRIS, in: Excited States, Vol. 3, Ed. E.C. LIM, Academic Press, New York 1977.

/10/ G. KOBBE and C. KLINGSHIRN, Z. Phys. B 37, 9 (1980).

/11/ G.H. MAHAN, Phys. Rev. 170, 825 (1968).

/12/ D. FRÖHLICH, in: Festkörperprobleme, Vol. X, Ed. O. MADELUNG, Pergamon/Vieweg, 1970.

/13/ U. RÖSSLER, phys. stat. sol. 42, 345 (1970).

/14/ M.H. REILLY, J. Phys. Chem. Solids 28, 2067 (1967).

(Received September 13, 1984)

phys. stat. sol. (b) 126, K171 (1984)

Subject classification: 14.3.2

Department of Physics, Illinois Institute of Technology, Chicago[1] (a) and Department of Physics, King Saud University, Riyadh[2] (b)

Scattering of Electrons by a Sheet of Impurities in a Quasi-Two-Dimensional Heterostructure

By

H.N. SPECTOR (a) and V.K. ARORA (b)

Presently, there is a growing interest in quantum transport studies of semiconducting layered heterostructures, thin films, and inversion layers /1/. In quantum well heterostructures which have been modulation doped so as to separate the carriers from the ionized impurities, scattering of the carriers has been greatly reduced /2/ and high values of the mobility have been observed experimentally /3 to 5/. Drummond et al. /4/ have assumed the existence of a temperature independent scattering mechanism arising from the presence of interface charge. Similarly, in semiconducting inversion layers, interface charge is found to be effective in limiting the mobility in these systems /6, 7/. Therefore, it is of interest to investigate the Coulomb scattering of the confined carriers by a sheet of impurities which models the effect of the interface charge on the scattering of the carriers.

As in /2/, we take the momentum relaxation rate due to the scattering of carriers by ionized impurities to be given by

$$\tau_k^{-1} = \frac{2\pi}{\hbar} \sum_{k'} \int_0^\infty dz_0 \, N_I(z_0) \left|(k'|V(\vec{r}, z_0)|k)\right|^2 (1-\cos\Theta)\,\delta(E_{k'} - E_k) \,, \qquad (1)$$

where $N_I(z_0)$ is the density of ionized impurities at $z = z_0$ and Θ is the angle through which the carriers are scattered in the plane of the quasi-two-dimensional system. If the sheet of impurities is outside the quantum well of thickness d, $N_I(z_0)$ is given by

$$N_I(z_0) = N_{2D}\,\delta(z_0 - L) \,, \qquad (2)$$

where N_{2D} is the number of impurities per unit area and L is the distance of the sheet of impurities from the well. If we assume a delta function dis-

1) Chicago, Illinois 60616, USA.

2) P.O.B. 2455, Riyadh, Saudi Arabia 11451.

tribution of the confined carriers in the well (DFM) as is assumed in /4/, the momentum relaxation rate is

$$\tau_k^{-1} = \frac{m^*}{2\pi \hbar^3} N_{2D} \left[\frac{e^2}{\varepsilon \varepsilon_o}\right]^2 \int_0^\pi \frac{d\Theta \sin^2\Theta \exp(-4kL\sin\Theta)}{(2k \sin\Theta + S_1)^2} . \quad (3)$$

The notation used in (3) is the same as that used in /2/. We have assumed the conditions for the validity of the size-quantum limit ($kd \ll 1$) in arriving at (3). If the sheet of impurities is separated from the quantum well by a distance which is much larger than the de Broglie wavelength of the electron ($kL \gg 1$), (3) simplifies to give us

$$\tau_k^{-1} = \frac{0.01\ m^* N_{2D}}{8\pi\hbar^3 k^3 S_1^2 L^3} \left[\frac{e^2}{\varepsilon \varepsilon_o}\right]^2 . \quad (4)$$

Equation (4) is valid in the strong screening limit where the screening length is much smaller than the width of the quantum well ($S_1 L \gg 1/2$). To evaluate the integral in (3), we have used the fact that the integrand decreases exponentially for $kL \gg 1$, limiting the values of Θ for which the integrand is non-vanishing to $\sin\Theta \lesssim (4kL)^{-1} \ll 1$. As expected, the momentum scattering rate decreases with increasing distance L of the impurities from the quantum well. Equation (3) is in agreement with a result obtained by Price /8/ apart from a numerical factor.

In the sine-function model (SFM) considered in /2/, and using the same type of approximations concerning the strength of screening and the separation of the sheet of impurities from the quantum well as used in DFM, the scattering rate is obtained is

$$\tau_k^{-1} = \frac{3.5\mathrm{x}10^{-4}\ m^* N_{2D}\, d^3 e}{4\hbar^3 S_1^2 k^3 L^5} \left[\frac{e^2}{\varepsilon \varepsilon_o}\right]^2 . \quad (5)$$

Thus, the momentum relaxation has a different dependence on the well width and the distance from the well to the sheet of impurities in SFM than in DFM, although in both cases the scattering is suppressed as the impurities are removed further from the quantum well.

The mobility for the two models is given by the expressions

$$\mu(\mathrm{DFM}) = \frac{1500\ S_1^2 L^3\ (2\pi k_B T)^{3/2}\ e}{m^{*1/2}\ N_{2D}} \left[\frac{\varepsilon \varepsilon_o}{e^2}\right]^2 \quad (6)$$

and

$$\mu(\mathrm{SFM}) = \frac{42,857\,(2\pi)^{1/2}\; e S_1^2\, L^5\, (k_B T)^{3/2}}{m^{*1/2}\; N_{2D}\; d^3} \left[\frac{\varepsilon\varepsilon_o}{e^2}\right]^2 , \tag{7}$$

when the carriers in the heterostructures are nondegenerate in their statistics and by the expressions

$$\mu = \frac{1,600\,\pi e\; S_1^2\, L^3}{N_{2D}} \left[\frac{2}{m}\right]^{1/2} \left[\frac{\varepsilon\varepsilon_o}{e^2}\right]^2 E_f^{3/2}$$

and

$$\mu = \frac{22,857\, e\; S_1^2\, L^5}{N_{2D}\; d^3} \left[\frac{2}{m}\right]^{1/2} \left[\frac{\varepsilon\varepsilon_o}{e^2}\right]^2 E_f^{3/2} , \tag{8}$$

when the carriers are degenerate in their statistics. For the case when the confined carriers are degenerate, the relation between the Fermi energy and the density of carriers per unit area n is $E_f \doteq (\pi\hbar^2\, n/m)$. For the case where the carriers are nondegenerate, the mobility due to scattering from a sheet of ionized impurities varies as $T^{3/2}$ with increasing temperature while when the carriers are degenerate, the mobility is temperature independent. The existing experimental data on the electron mobility in heterostructures indicate either no or a weak temperature dependence of the mobility at low temperatures and a steep decrease of the mobility with temperature at high temperatures /9, 10/. The temperature of the mobility at low temperatures is consistent with our calculations if the carriers are degenerate at these temperatures. At higher temperatures, the mobility will be limited by phonon scattering which will lead to a decrease of the mobility with increasing temperature which is observed /9, 10/. There are considerable differences in the mobilities predicted from these two models. In the SFM, the mobility increases more rapidly as the distance between the sheet of impurities and the quantum well increases (as L^5 in SFM versus a L^3 dependence in DFM). The temperature dependence of the mobility is the same in both models and is also the same as for ionized impurity scattering in the bulk. Since the SFM actually takes account of the variation of the carrier concentration across the width of the well, the use of this model is more realistic than the use of DFM. However, since in this model, the finite height of the potential barrièr at the boundaries of the well is neglected, the use of this model should only be valid when the sheet of impurities is well removed from the well so that the penetration of the electrons into the classically forbidden region can be neglected /11/.

In conclusion, we have shown that the scattering of confined carriers from a sheet of impurities, which approximately models the effect of interface charge, is different depending upon what model we use for the distribution of the carriers in the quantum well. Therefore, in the interpretation of experimental data, one should use the most realistic models for the distribution of the confined carriers.

References

/1/ T. ANDO, A.B. FOWLER, and F. STERN, Rev. mod. Phys. 54, 437 (1982).

/2/ J. LEE, H.N. SPECTOR, and V.K. ARORA, J. appl. Phys. 54, 6995 (1983).

/3/ R. DINGLE, H.L. STORMER, A.C. GOSSARD, and W. WIEGMANN, Appl. Phys. Letters 33, 665 (1978).

/4/ T.J. DRUMMOND, H. MORKOC, K. HESS, and A.Y. CHO, J. appl. Phys. 52, 5231 (1981).

/5/ H.L. STORMER, A. PINCZUK, A.C. GOSSARD, and W. WIEGMANN, Appl. Phys. Letters 38, 691 (1981).

/6/ F. STERN, Phys. Rev. Letters 44, 1469 (1980).

/7/ K.M. CHAM and R.G. WHEELER, Phys. Rev. Letters 44, 1472 (1980).

/8/ P.J. PRICE, J. Vacuum Sci. Technol. 19, 599 (1981).

/9/ T.J. DRUMMOND, W. KOPP, H. MORKOC, and M. KEEVER, Appl. Phys. Letters 41, 277 (1982).

/10/ N. SANO, H. KATO, and S. CHIKA, Solid State Commun. 49, 123 (1984).

/11/ S. OGALE and A. MADHUKAR, J. appl. Phys. 55, 486 (1984).

(Received October 11, 1984)

Erratum

To the Short Note by D.V. KAPOR, J.P. ŠETRAJČIĆ, and M.J. ŠKRINJAR

Temperature Dependence of the Energy Gap of an Anisotropic Heisenberg Ferromagnetic Model

phys. stat. sol. (b) 125, K45 (1984)

the authors request the following rectifications:

On page K46 formula (4b) reads

$$\gamma(\vec{k}) = 2 \sum_{r=x,y,z} \cos k_r a \quad \text{instead of} \quad \gamma(\vec{k}) = \sum_{r=x,y,z} \cos k_r a .$$

On page K48 line 6 read

$$-2\pi \varrho^{5/2} \zeta_{5/2} \quad \text{instead of} \; -\frac{\pi}{2} \varrho^{5/2} \zeta_{5/2}$$

and the twelfth line from below reads

$$\coth x \approx 1 + 2\, e^{-2x} \quad \text{instead of} \quad \coth x \approx 1 + e^{-2x}$$

The equation before (11) reads

$$Y_1 \left[\frac{(q_m a)^2}{75\, \pi^2} + \frac{8\pi}{5} \varrho^{5/2} \right] \quad \text{instead of} \quad Y_1 \left[\frac{(q_m a)^2}{75\pi} + \frac{3\pi}{5} \varrho^{5/2} \right]$$

and the correct expression (11) is

$$E(0) = \sigma^2 \left[0.256363 + \pi\, (0.337486 + \zeta_{5/2}) \varrho^{5/2} \right]$$

instead of

$$E(0) = \sigma^2 \left[0.256363 + \frac{\pi}{4}\, (0.506228 + \zeta_{5/2}) \varrho^{5/2} \right] .$$

Erratum

To the Short Note by P. ENDERS

Electronic Structure of Group-V Semimetals:
Tight-Binding Parameter Estimation from Chemical Trends

phys. stat. sol. (b) 125, K133 (1984)

the author requests the following rectifications:

On p. 133, second line below formula (1) read $\xi_1 \approx -\xi_0/4$ instead of $\xi_1 \approx \xi_0/4$.

On p. 135, in Table 2 the comparison is made with the results of /1/ and not with those of /4/.

Pre-Printed Titles

Papers to be published
in "physica status solidi (b) - basic research"
Vol. 127, No. 1

Review Article

H.W. STREITWOLF, Berlin
Physical Properties of Polyacetylene

Original Papers

Y. NAKAOKA, Osaka
Crystal Structure Dependence of Hole Stability in Ionic Crystals

G. PAASCH, Ilmenau
Alternative Quasi-Classical Transport Equations (I)

H. RAU and R. GEICK, Würzburg
Crystal-Independent Interionic Potentials and the Lattice Dynamics of the Perovskite-Type Layer Material K_2ZnF_4

G.V. KOZLOV, E.B. KRYUKOVA, B. WYNCK, F. BREHAT, and M.E. SHERIF, Moscow and Nancy
Low-Frequency Lattice Dynamics of $NaNO_2$ Crystals

H. KÖSTER, JR., O.V. KURNOSOVA, and I.N. YASSIEVICH, Rostock and Leningrad
Tunneling from Deep Levels of 1-c Type in Electric Fields

A.I. AGAFONOV, A.F. PLOTNIKOV, and V.N. SELEZNEV, Moscow
Conductivity of Amorphous A_xB_{1-x} Semiconductors with Potential Fluctuations of the Mobility Edges

M.C. BÖHM, Stuttgart
Consistent Evaluation of Long-Range and Short-Range Contributions to the Quasi-Particle Energies of Transition Metal Solids

V.D. GOROBCHENKO and F.J. LITTERST, Moscow, Munic, and Jülich
Mössbauer Relaxation Spectra Due to Electric Field Gradient Reorientation and Nuclear Hopping for a I = 5/2 to 7/2 Nuclear Transition

M. TOMAK, HANUMAN SINGH, B.K. SHARMA, and S. MANNINEN, Ankara, Jaipur, and Warwick
Electron Momentum Distribution in Niobium and the Renormalised Free Atom Model

A. AKILBEKOV, A. DAULETBEKOVA, and A. ELANGO, Tartu
Photo- and Thermochemical Reactions with Participation of Br_3^- Centres in X-Rayed KBr

A.G. MIRONOV, V.A. MOROZOVA, and V.S. VAVILOV, Moscow
Some Peculiarities of the Electrooptical Spectra of n-GaAs(Cr) at Liquid Helium Temperatures (Wannier Levels?)

J. SCHREIBER and D. SCHULZ, Dresden
Self-Consistent Dynamical Theory for Heisenberg Spin Glasses

J. ŠLECHTA, Leeds
On SCCF Theory of Falicov-Ramirez-Kimball Transitions in 1D Binary Alloys

D. KONTOS, Colchester, Essex
Spin Green Functions for Ferromagnetic Films with Asymmetric Boundary Conditions

E.B. FELDMAN and K.T. SUMMANEN, Chernogolovka
On Quasistationary States in the Multipulse Spin-Locking

N. DUTT, O.P. SHARMA, and J. SHANKER, Agra
Analysis of Photoelastic Behaviour of Fluorite-Structure Crystals

G. HEIN, Braunschweig
Analysis of Shubnikov-De Haas Oscillations in HgSe

N. KRISTOFFEL, Tartu
On the Possibility of Anomalous Bulk Photovoltaic Effect and of Induced Absorption in the Nominal Transparency Region of a Noncentrosymmetric Crystal in a Magnetic Field

W.N. BHENDE, P.S. BAKHSHI, and J. SHANKER, Agra
Thermal Energy and the Equation of State for Alkali Halide Crystals

A. BHATTACHARYA, A. GHOSAL, and D. CHATTOPADHYAY, Calcutta
Small-Signal AC Response of Two-Dimensional Hot Electrons in Quantum-Well Structures of Polar Semiconductors

V.O. ESIN and V.I. DANILYUK, Sverdlovsk
Cluster Structure of a Two-Dimensional Lattice Gas Model

P. REINEKER, J. KÖHLER, and U. SCHMID, Ulm
Triplet Excitons in Non-Resonant Molecular Pairs

S. GHOSH and S. DIXIT, Ujjain
Effect of Relativistic Mass Variation of the Electron on the Parametric Instability of Acoustic Waves in Transversely Magnetised Piezoelectric Semiconducting Plasmas

G. GERBETH, E. MROSAN, R. ZELLER, and P.H. DEDERICHS, Dresden and Jülich
Calculations of the Change in Knight Shift Due to Zn, Ga, and Ge Impurities in Copper

Ç. TUNCAY and M. TOMAK, Ankara and Trieste
Empirical Tight-Binding Band Structure of Wurtzite Semiconductors SiC, ZnSe, and ZnTe

A.A. VAKHNENKO, YU.B. GAIDIDEI, and A.A. EREMKO, Kiev
Dissociation-Accompanied Raman Scattering by Davidov Solitons

E.A. GOREMYCHKIN, E. MÜHLE, B. LIPPOLD, O.D. CHISTYAKOV, and E.M. SAVITSKII, Dubna, Leipzig, and Moscow
Crystal Field Study of $HoNi_5$ by Neutron Scattering

E. HEINER, Rossendorf
A Dynamical Theory for a Disordered Hubbard Chain far from Equilibrium (I)

S.V. NISTOR, D. SCHOEMAKER, and B.R. YANG, Bucharest
The Study of Tl_2^+<110> Centres in NaCl and RbCl

P.V. SMITH, Shortland, New South Wales
Adequacy of the Nearest-Neighbour LCAO Model Hamiltonian for the F.C.C. Transition and Noble Metals

B.I. RESER, Sverdlovsk

Mixed Basis Method with Overlapping Atomic Orbitals

P.C. MAITHANI and M.D. TIWARI, Gharwal

Lattice Specific Heat and Superconducting Transition Temperature of a V:W Alloy

N.V.K. PRABHAKAR and R.K. SINGH, Jabalpur

Anharmonic Properties of Sodium Halides

N. DASS and M. KUMARI, Roorkee

Derivation of Some Equations of State for Solids

G.CH. CHRISTAKUDIS, S.K. PLACHKOVA, E.S. AVILOV, L.E. SHELIMOVA, and N.KH. ABRIKOSOV, Sofia and Moscow

Transport Phenomena of Ferroelectric Solid Solutions $(GeTe)_{1-x}((Ag_2Te)_{1-y}(Sb_2Te_3)_y)_x$ with $y = 0.60$, $0 \leqq x \leqq 1$ at 300 K

J. DUSZEWSKI, Łódź

Influence of Perpendicular Field on Magnetic Fluctuations in Thin Films

J.A. TUSZYNSKI and J.M. DIXON, St. John's, Newfoundland, and Warwick

Tensorial Coupling and the Reduced Matrix Elements of One- and Two-Body Operators

P. KLEINERT, Berlin

Self-Consistent Field Approximation for the Vibrational Spectra of Quaternary Alloys

Papers to be published
in "physica status solidi (a) - applied research"
Vol. 86, No. 2

Original Papers

N. V. RUMAK and V. V. KHATKO, Minsk
Structure and Properties of Silicon Dioxide Thermal Films (II)

D. FASOLD, K. HEHL, and S. JETSCHKE, Jena
Refractive Index and Absorption Measurement in a Thin Layer by Two Reflection Measurements

T. OHKAWA and H. HASHIMOTO, Kanagawa and Osaka
An X-Ray Wave Theory for Heavily Distorted Crystals (II)

M. G. BLANCHIN and L. A. BURSILL, Villeurbanne and Parkville, Victoria
Extended versus Small Defect Equilibria in Non-Stoichiometric Rutile (II)

A.J. MUHAMMAD, D.V. MORGAN, and A.E. GUILE, Leeds and Arbil
Contact Overlap in MIM Structures with Distributed Traps

M.S.C. BOSE, Bombay
Effect of Saturated Magnetic Field on Fatigue Life of Carbon Steel

A. ZOUL, Prague
Influence of Deep Surface Levels on C-U Schottky Doping Profiles of n-GaAs VPE Layers

H.J. ENGELMANN, K. BERNDT, and F. STORBECK, Dresden
Quantitative Chemical Analysis of 3d Transition Metal Alloys by SXAPS

H. TERLETZKI, A. BODEN, F. WULF, and W.R. FAHRNER, Berlin (West)
Laterally Inhomogeneous Charge-Build-Up in CMOS Inverters during Ionizing Irradiation

I. THURZO and E. PINČÍK, Bratislava
Some Peculiar Features of Small-Signal Charge DLTS Response of GaAs MOS Capacitors

S. IHARA, Y. WARITA, and K. SUZUKI, Tokyo and Kanagawa
Ionic Conductivity in $AgI_{1-x}Cl_x$

A.YU. LEIDERMAN, Tashkent
On the Generation-Recombination Currents in p-n Junctions of Semiconductors with Continuous Gap-State Spectrum

R.P. GOYAL, D. RAVIENDRA, and B.R.K. GUPTA, Agra and Pantnagar
Electroless Deposition of In_2O_3 and In_2O_3:Sn (ITO)

D.M. VANDERWALKER, Stony Brook, New York
The Nucleation of Ni Silicides on Dislocations and Twins in Silicon

R. DÖHL, M.-P. MACHT, and V. NAUNDORF, Berlin
Measurement of the Diffusion Coefficient of Cobalt in Copper

A.U. SHELEG and V.V. ZARETSKII, Minsk
X-Ray Study of the Commensurate-Incommensurate Phase Transitions in α-ZnP_2

YU.S. BOYARSKAYA, D.Z. GRABKO, and D.S. PISHKOVA, Kishinev
Correlation between Yield Stress and Microhardness in Crystals such as NaCl

R.C. ECOB, Berkeley
The Measurement of Dislocation Link Length Distributions in a Tilted TEM Foil

T.S. KÊ, P. CUI, S.C. YAN, and Q. HUANG, Hefei
Internal Friction Peaks of Super-High-Purity Aluminium at Medium Temperatures

M.E. ABU-ZEID, M.M. EL-OKER, A.I. ABU-SHADY, and M.S. ADILI, Kuwait
Photoacoustic and Ellipsometric Study of Black Nickel Coatings on Copper Substrates

C.I. NISTOR, V. SPÂNU, S. MANDACHE, G. FILOTI, and V. FLORESCU, Bucharest
Hyperfine Field Distribution in Metglasses

V.M. BABICH, N.P. BARAN, A.A. BUGAI, YU.P. DOTSENKO, V.B. KOVALCHUK, and V.A. SHERSHEL, Kiev
On the Properties of Thermodonors-II in Cz-Si Crystals of High Carbon Content

19. Magnetic Resonance

20. Optical Properties
    - 20.1 Spectra. Optical Constants (X-Ray Spectra See 20)
    - 20.2 Lasers
    - 20.3 Luminescence (X-Ray Spectra See 20)

21. Metals. Alloys
    - 21.1 Transition Metals (Group Sc to Zn) and Their Alloys
        - 21.1.1 Fe and Fe Alloys
    - 21.2 Alkali Metals
    - 21.3 Alkali-Earth Metals
    - 21.4 Rare-Earth Metals
    - 21.5 Actinides
    - 21.6 Noble Metals
    - 21.7 Semi-Metals (Semiconducting Alloys 22.7)

22. Semiconductors and Ionic Crystals
    - 22.1 Elements
        - 22.1.1 Germanium
        - 22.1.2 Silicon
        - 22.1.3 Group VI Elements
    - 22.2 III–V Compounds
        - 22.2.1 Arsenides
        - 22.2.2 Phosphides
        - 22.2.3 Antimonides
        - 22.2.4 Mixed Crystals
    - 22.3 IV–IV, II–V, and III–VI Compounds
        - 22.3.1 Mixed Crystals
    - 22.4 II–VI Compounds
        - 22.4.1 Sulphides
        - 22.4.2 Selenides
        - 22.4.3 Tellurides
        - 22.4.4 Mixed Crystals
    - 22.5 Halides
        - 22.5.1 Silver Halides
        - 22.5.2 Alkali Halides
        - 22.5.3 Alkali-Earth Halides
        - 22.5.4 Mixed Crystals
    - 22.6 Simple Oxides (CdO, ZnO See 22.4)
        - 22.6.1 Mixed Crystals
    - 22.7 Semiconducting Intermetallic Compounds
        - 22.7.1 Mixed Crystals
    - 22.8 Tertiary and Higher Compounds
        - 22.8.1 Salts of Oxy-Acids (Carbonates, Phosphates, Silicates Including Mica etc.)
        - 22.8.2 Spinels, Garnets, Ferrites
    - 22.9 Organic Semiconductors

23. Solidified Gases

Attention

The preceding classification scheme is closely akin to the scheme used by Physics Abstracts to classify the various fields of solid state physics. A paper will either be classified under one or more of the subsections or, if none of these are relevant, it will just appear under the general section heading (i. e. sections and subsections are equivalent in this sense).

The *subject classification* given at the head of each paper begins with the main subject matter of the paper, and this is followed, if necessary, by figures referring to subsidiary subject matter and the type of solids investigated.

The papers published in this issue are listed in the order of their subject classification at the end of the table of contents.

32 913

Manuscripts and letters for physica status solidi (b) — basic research and physica status solidi (a) — applied research

should be sent to

DDR-1086 Berlin, Leipziger Straße 3–4, P. B. 1233

or

Internationaler Buchversand GmbH, 1000 Berlin 19, Eschenallee 21.

The object of "physica status solidi (b)" and "physica status solidi (a)" is to publish rapidly important and topical results concerning fundamental research in the field of solid state physics and aspects of application and experimental techniques of fundamental solid state research, respectively. The issues of "physica status solidi (b)" and "physica status solidi (a)" will be published on the 1st and the 16th day of each month, respectively. Either series will contain the following categories:

1. Review articles written by arrangement with the editors. They will be remunerated.
2. Original papers selected from those which have not been submitted for publication elsewhere, after being reviewed. (The latest date for acceptance of contributions is the 8th and the 23rd day of each month for "physica status solidi (b)" and "physica status solidi (a)", respectively, 50 days before the delivery of the journal.)
3. Short notes published in the off-set part of the journal. They are, if necessary after being reviewed, selected by editorial staff. (The latest date for acceptance of manuscripts for this part is the 12th and the 27th day of each month for "physica status solidi (b)" and "physica status solidi (a)", respectively, 16 days before the delivery of the journal.)
4. Pre-printed titles of papers to be published in the next issue of both series.

Editorial Information

Manuscripts should be submitted preferentially in English. Besides, papers in German, French, and Russian are admitted. The final decision on the incorporation of the papers in the series (b) or (a) is taken by the editorial staff after consultation of the referee. Authors are asked to observe the following rules:

1. Manuscripts have to be written as concisely as possible. As most, 20 and 4 pages (28 lines per page; abstracts, tables, references, figures, and captions of figures included, where 3 figures are counted as about one page) are admitted for original papers and short notes, respectively. All manuscripts should be type-written on one side only. Two copies are required. Badly arranged manuscripts are returned.
2. The author's name, initials, and institution should be given.
3. One abstract of about ten lines in English and one in German, French, or Russian are required for original papers.
4. Because of lack of space conclusions (if any) must not have the character of a summary (i. e. substantially a repetition of the contents of the abstract).
5. Formulae should not be written too small and not with pencil. Separate lines for formulae are desirable. Indistinct writing will lead to a return of the manuscript.
6. Drawings should be enclosed on separate sheets of paper, not exceeding the maximum and minimum sizes of 20×30 and $10 \times 15\ cm^2$, respectively. Photographs should be submitted, on white paper, only for review articles and original papers. Duplicates are required.
7. Captions for the figures should be submitted on separate sheets of paper, and special instructions with regard to the size of illustrations should be given on these.
8. References should be numbered in order of appearance and should be given, on a separate sheet, in the form used in physica status solidi.
9. SI-units should be used.
10. Each paper should contain a proposal for the subject classification according to the classification scheme appearing on the second and third cover of each issue.

The authors will be sent galley proofs of review articles and original papers. Proof reading of short notes and pre-printed titles will be done by the editorial staff.

For each original paper and short note a total of 75 reprints is available free of charge. The number of reprints free of charge per review article is 25. Requests for additional reprints (against payment) should be given together with the submitted manuscripts. **The deadline for orders of additional reprints is the 15th and the 30th of each month** for "physica status solidi (b)" and "physica status solidi (a)", respectively, two weeks before the delivery of the journal.

Bei Fragen zur Produktsicherheit wenden Sie sich bitte an:
If you have any questions regarding product safety,
please contact:

Walter de Gruyter GmbH
Genthiner Straße 13
10785 Berlin
productsafety@degruyterbrill.com